Syntax on the Edge

Empirical Approaches to Linguistic Theory

VOLUME 21

The titles published in this series are listed at *brill.com/ealt*

Syntax on the Edge

A Graph-Theoretic Analysis of Sentence Structure

By

Diego Gabriel Krivochen

BRILL

LEIDEN | BOSTON

Library of Congress Cataloging-in-Publication Data

Names: Krivochen, Diego Gabriel, author.
Title: Syntax on the edge : a graph-theoretic analysis of sentence structure / by Diego Gabriel Krivochen.
Description: Leiden ; Boston : Brill, 2023. | Series: Empirical approaches to linguistic theory, 2210-6243 ; volume 21 | Includes bibliographical references and index.
Identifiers: LCCN 2023029773 (print) | LCCN 2023029774 (ebook) | ISBN 9789004541429 (hardback ; acid-free paper) | ISBN 9789004542310 (e-book)
Subjects: LCSH: Grammar, Comparative and general–Syntax. | Grammar, Comparative and general–Sentences.
Classification: LCC P291 .K75 2023 (print) | LCC P291 (ebook) | DDC 415–dc23/eng/20230721
LC record available at https://lccn.loc.gov/2023029773
LC ebook record available at https://lccn.loc.gov/2023029774

Typeface for the Latin, Greek, and Cyrillic scripts: "Brill". See and download: brill.com/brill-typeface.

ISSN 2210-6243
ISBN 978-90-04-54142-9 (hardback)
ISBN 978-90-04-54231-0 (e-book)

This book is printed on acid-free paper and produced in a sustainable manner.

Contents

Editorial Foreword

For many, maybe most, of us, graphs are simply visual representations of data involving two axes and various points connected by lines. Many definitions of *graph* bear this out, as they are highly descriptive. For instance, the *Oxford English Dictionary*, on-line, gives the following definition: "a kind of symbolic diagram (used in *Chemistry*, *Mathematics*, etc.) in which a system of connections is expressed by spots or circles, some pairs of which are colligated by one or more lines", and the *New Oxford American Dictionary*, on-line, defines *graph* in these terms: "a diagram showing the relation between variable quantities, typically of two variables, each measured along one of a pair of axes at right angles". Some characterizations of *graph* that one can find border on the vague, e.g. the Wikipedia definition from discrete mathematics of "a structure amounting to a set of objects in which some pairs of the objects are in some sense 'related'". But this Wikipedia definition actually moves us in a different direction, namely one in which graphs are not just a practical visualization aid, but rather are also a highly technical entity, one that serves as the basis for a branch of mathematics known as *graph theory*, which, www.sciencedirect.com tells us, "deals with connection amongst points (vertices/nodes) by edges/lines".

It is this last sense of *graph* that is most relevant for linguists, for graph theory is also a basis for a theoretical approach to the analysis of syntactic structures in language, and it is within such a *graph theory* approach that the present volume for the Empirical Approaches to Linguistic Theory (EALT) series, *Syntax on the Edge: A Graph-Theoretic Analysis of Sentence Structure*, takes its place. In this work, author Diego Gabriel Krivochen clearly satisfies the theoretical imperative of the series by justifying the use of graph theory for linguistic purposes while at the same time satisfying the series' empirical imperative by analyzing a wide range of syntactic phenomena, mostly from English but from Spanish as well. In the pages that follow, accordingly, we see analyses of "Raising" and "Equi" (i.e., "control") structures, of binding effects, of WH-questions, of coordination, of passivization, and of dative shift, all in English, but Dr. Krivochen adds in an examination of clitic climbing in Spanish as well. Especially regarding the English structures, this collection represents a veritable who's who of constructions that are interesting in their own right but which have also played a key role in the development of syntactic theories over the years. These analyses thus shed new light on the structures in question while also demonstrating the utility of a graph-theoretic approach to syntactic analysis and in this way advance the theory. Author Krivochen is not the first linguistic scholar to adopt such a framework, but this volume represents perhaps the most ambitious application of the principles of this theory to syntactic phenomena.

As series editor, I am pleased to be in a position to allow such a empirically rich and theoretically insightful work as this one to have a place among the volumes in EALT.

Brian D. Joseph
EALT Series Managing Editor
Columbus, Ohio USA, 25 July 2023

Preface

This book originally started as an exploration of a 'what if' question pertaining to the way in which the transformational rules of generative grammar work: McCawley (1982) noted that certain transformations only change the linear order of the elements of sentence structure without disrupting syntactic relations. So—I wondered—, what if *nearly all* transformations required to provide an empirically adequate analysis of English in fact only changed *linear order*, leaving syntactic relations unaffected? (As we will see in **Chapter 13**, exceptions include, most notably, *Passivisation* and *Dative Shift*) Can we get an adequate description of the major syntactic phenomena of English while assuming that syntactic relations, once created, are not disrupted?

Linguists who study natural language syntax agree that sentence structure is hierarchical in nature and therefore that a representation of hierarchical relations between the expressions in a sentence is necessary; disagreements arise with respect to (a) how, and (b) what kinds of relations are really needed. Perhaps the most familiar way of representing hierarchical structure in theories of natural language is by means of *tree diagrams* which correspond to derivations of sentences obtained by either top-down rewriting rules (as in Phrase Structure Grammars) or bottom-up recursive discrete combinatorics (as in Minimalist syntax): in both cases we are dealing with formal systems born out of the Immediate Constituency (IC) approach to syntax. Trees represent relations of *containment* and *precedence* between syntactic objects: branching nodes in a tree dominate segments that are linearly contiguous and also correspond to aspects of semantic interpretation. These trees belong to the larger class of mathematical objects known as *graphs*, which are subject to a set of conditions that we will discuss in **Chapter 1** of this book and which favour binary-branching endocentric structures with no closed cycles or loops: from classical phrase markers and X-bar theory to Bare Phrase Structure and everything in between. This approach represents the state-of-the-art in generative syntax, whereby syntactic structure is the result of a mechanism of unordered binary set formation, and trees are to be understood strictly as *diagrams of sets*.

The IC approach sketched above finds difficulties when dealing with expressions that, although not contiguous linearly, do seem to form a constituent insofar as there are grammatical processes that take them as a unit. We can give two simple examples of this phenomenon: on the one hand, we may consider a case like *heat the soup up* (where the particle may appear after the direct object despite forming a constituent with the verb). On the other, in the transformational treatment of an interrogative sentence like *what did Emma say that*

Beth bought? the embedded verb phrase contains a variable that is bound by the *wh*-word *what.* A transformation-enhanced IC grammar needs to impose underlying contiguity to these cases: create a level of representation where we have *heat up* as a V and *buy what* as a VP, and which gets disrupted by reordering rules. Non-transformational models can take strings as they come and provide a WYSIWYG constituent structure (such is the case of trace-less versions of Lexical Functional Grammar or—for the most part—Simpler Syntax), but represent discontinuity at other, parallel levels (in Lexical Functional Grammar, *functional structure*; in Simpler Syntax, a combination between the levels of *constituent structure* and *conceptual structure*). The core of the issue is the existence of *mismatches* between underlying representations (where constituency, thematic relations, and grammatical functions would be defined) and superficial representations (which represent word order). In accounting for structural relations, superficial or not, most existing syntactic theories appeal to a number of non-overt symbols in their representations: branching nodes that do not correspond to any expression in an overt string (e.g., NP, VP, etc.), phonologically null categories (e.g., traces, unpronounced copies, empty subjects), functional material (e.g., functional heads such as Tense, Aspect, little *v*, Degree, Number, Focus, Topic, Force, etc.). This has as a consequence that the structural description assigned to a string of symbols will always have more symbols than the string it is modelling. In the transformational tradition, we also need a set of reordering operations, plus a way to re-instate an underlying order.

Is it possible to simplify syntactic theory and improve empirical adequacy by adopting different assumptions about what syntax is and how it is formalised? To propose an answer to this question, this book uses a larger graph-theoretic toolset to explore a novel framework for syntactic analysis.

Our approach to syntactic structure is based on two basic ideas:

- That structural descriptions of sentences are, unless strictly necessary, restricted to specifying relations among overt expressions, and
- That grammatical functions are primitives of the theory

From here, there are several paths we could take. A theory could, for example, provide a level of representation where overt expressions and relations are specified and another level where grammatical functions (subject, object, oblique, adjunct…) are defined; these levels can then be put in correspondence. An example of this kind of approach is Lexical Functional Grammar (LFG) (Kaplan & Bresnan, 1982; Bresnan et al., 2016; Dalrymple et al., 2019). Or, we could define a single level of representation where expressions, relations, and grammatical functions are specified but allow for operations to *change* the relations between these expressions, by ordering grammatical functions in a hierarchy and promoting or demoting expressions (making subjects into objects or objects into subjects). Examples of theories that develop this latter approach are Relational Grammar (RG) (Perlmutter, 1980; Perlmutter & Postal, 1983a, b) and Arc Pair Grammar (APG) (Johnson & Postal, 1980); antecedents of such a view include Jespersen (1985 [1937]), as noted in McCawley's introduction to the 1985 edition of *Analytic Syntax*. It is important to note that Johnson & Postal's theory is explicitly couched in graph-theoretic terms.

The alternative we adopt here follows Occam's Razor in a very specific way: do the most you can with as few elements as possible. If we are talking about graphs, these elements are nodes and edges. The formal underpinnings of theories based on phrase structure grammars require representations in which nodes are minimally connected. Whether we are dealing with Government and Binding's X-bar trees, LFG's *c(onstituent)-structures*, Tree Adjoining Grammar's elementary trees, etc., the usual state of affairs is one in which there are two kinds of symbols in a grammar's alphabet (and thus two kinds of nodes in phrase structure trees): terminals and non-terminals. In a Phrase Structure tree, non-terminal nodes dominate other nodes, whereas terminal nodes do not dominate anything. In most versions of generative theory (transformational or not) any given node, terminal or not, may be immediately dominated by only one other node. There is no self-domination (no loops) and no closed cycles: in this respect, phrase structure trees are indeed graph-theoretic trees. As we will see, this entails that in most situations we must multiply the number of nodes, adding new non-terminals or terminals as appropriate to syntactic representations, to maintain the fundamental properties of trees. However, nothing prevents us from considering that an alternative to trees as the format for structural descriptions of natural language sentences is well motivated. This book investigates the empirical and theoretical consequences of allowing a syntactic structure to grow by keeping the number of nodes to a minimum while exploiting all possible connections between those nodes: we remove the restrictions on how many nodes a given node may be directly connected to. The result is a model of grammar in which accounting for empirical aspects of natural language that have proven problematic for what we will call 'Mainstream Generative Grammar' (MGG[1]) has required supplemental *ad hoc* assumptions

1 Following Culicover and Jackendoff (2005: 3), we use the term Mainstream Generative Grammar (MGG) throughout the present work as a shorthand '*to refer to the line of research most closely associated with Noam Chomsky*', from Syntactic Structures (1957) through to the Minimalist Program (1995 et seq.), as well as those extensions by authors closely related to Chomsky's theoretical position, and whom Chomsky or collaborators have recognised as part of their enterprise. When referring to MGG, we will focus on its grammatical commitments (that is, MGG as an empirical theory of natural language syntax) rather than on aspects of the the-

such as disallowing discontinuities in constituent structure or crossing dependencies. In the alternative model we present in this book, the possibility of these hitherto problematic aspects of natural language fall out.

Generative grammar's approach to structure building and transformations since the early 1980's is based on the idea that the syntactic component of the grammar should create *unambiguous paths* in strictly binary-branching trees. However, as pointed out before, that forces us to have more nodes than expressions: if we have two expressions, generative grammar combines them by creating a binary branching tree with *three* nodes: a mother node that dominates our two expressions. So, in a sense, if we combine A and B, there is no direct relation between A and B: it is mediated by a mother node (usually called the *label* of the object that results from the combination of A and B). What if we required that expressions be connected *directly*? In other words: what if we dispensed with intermediate nodes in syntactic structures? We could still use terms like VP or NP or S, but they would have no formal value. They would be mnemonics for sequences of terms or expressions, but not part of the syntactic representation *sensu stricto* and would this not be part of the input or output of any syntactic operation (nor could they be referred to by any declarative constraint). Syntactic representations would be drastically simplified in terms of number of nodes. However, we would need to allow for less restrictive relations between the nodes that we have left: we want to connect the nodes we have as densely as possible, so as to avoid having to introduce more nodes in syntactic representations. This approach entails an economy metric that differs from current generative grammar's: having more nodes connected in very limited ways is more costly than having more connections between a smaller number of nodes.

This takes us to the issue of what the format of those representations can be. Here is where our commitments are strongest: we want to formulate a theory of grammar where all relations between expressions are defined in *graphs*. Furthermore, the theory we have in mind is fundamentally *constraint-based* (or *declarative*): the aim is to define a set of conditions that will separate well-formed graphs (which correspond to well-formed derived expressions in a language) from ill-formed graphs (which do not). Constraints, unlike derivational operations, are statements that can be true or false for a given graph: for example, we can take the statement *every node has only one mother node*—which is distinct from itself—(the so-called Single Mother Condition;

ory of the human language faculty (innateness, specificity, modularity), its phylogenetic and ontogenetic development, etc.

Sampson, 1975) and create a configuration such that it makes the statement *false*. Similarly, as pioneered in McCawley (1968), PS rules may be interpreted as well-formedness constraints over portions of trees, and as such may be true or false applied to a specific configuration. In contrast, there is no way to make the generative operation Merge as defined in Minimalist works (e.g., Chomsky, 1995, 2021) 'true' or 'false', as it defines sequences of derivational steps, procedurally. The theory of grammar, in the constraint-based view, provides a set of statements and a conjunction of those statements applies to the structure assigned to a specific derived expression: in this sense, individual expressions do or do not satisfy a set of constraints (a 'model'). In our case, sets of constraints specify what a well-formed directed graph looks like, and what kinds of relations can be established between nodes in those graphs. If we consider the Single Mother Condition, a graph where a node is immediately dominated by more than one other node will be ill-formed: it cannot correspond to a well-formed derived expression of the language. We have chosen to start with this condition because it is one of the cornerstones of syntactic representations in generative grammar, and a condition that we will dispense with in this monograph: we will explore the empirical and theoretical consequences of adopting a formal model for syntactic structures where the goal is, as we have stated, to maximise the connections between available nodes, such that we can keep the number of nodes to a minimum.

The present monograph does not aim at 'reinventing' syntax: we firmly stand on the shoulders of giants. Classical transformational Grammar, Relational Grammar, Arc Pair and Metagraph Grammars, Lexical Functional Grammar, Tree Adjoining Grammars, (pure) Categorial grammars, and Dependency grammars provide us with empirical insights and technical tools. Our approach, hopefully, can contribute to enrich the already lush forest of syntactic theory.

Acknowledgments

The first person to hear the core idea of this book was, as usual, Doug Saddy (back in 2017). Since then, he has consistently provided insightful commentary and even suggested the title *Syntax on the Edge* (which is perhaps the best part of this book). My debt to him extends far beyond this monograph, and is difficult to overstate. Also an early interlocutor for the ideas in this work, in addition to a mentor and a friend, Susan F. Schmerling has painstakingly read, commented, and edited many parts of the manuscript (sometimes more than once!). She has also been patiently guiding me through the dense forest of Montague grammar for some years now, the results of which guidance can be seen throughout the book. She is not to blame, of course, for my misunderstandings.

More recently, I have greatly benefitted from lengthy discussions with Andrea Padovan and Denis Delfitto about the details of graph theoretic approaches to syntax. Many of the later chapters were written during my stay in fair Verona, which I have Denis to thank for. The sections on Spanish syntax owe much to joint work with Luis García Fernández and Ana Bravo, from whom I learnt much about the art and science of doing grammar.

At some point or another, the following individuals have commented on different (technical, empirical) aspects of the framework explored here and/or the presentation of the argument (in alphabetical order): Víctor Acedo-Matellán, Tom Bever, Probal Dasgupta, Julie Franck, Daniel Fryer, Hans-Martin Gärtner, Elly van Gelderen, Dave Medeiros, Ebrahim Patel, Massimo Piattelli-Palmarini, Paul Postal, Haj Ross, Whit Tabor, and Juan Uriagereka. They all have my thanks. Needless to say, I alone am responsible for the contents of the book, and someone's appearance in the lists above should not be interpreted as an endorsement of or agreement with the framework explored here.

An anonymous reviewer provided comments that helped improve the content and structure of the monograph, and for that they have my sincere gratitude. The index owes much to Susan F. Schmerling and Merve Odabaşı, minus the errors it may contain. For editorial assistance, and patience with my queries, thanks are also due to Elisa Perotti at Brill. Finally, my gratitude extends also to Eline Badry for her tireless work in making my manuscript into the beautifully typed piece you, reader, hold in your hands.

Figures

Abbreviations

ACC	Accusative case
AOR	Aorist
AP	Adjective Phrase
APG	Arc Pair Grammar
Aux	Auxiliary verb
CFG	Context-free grammar
CFL	Context-free language
CG	Categorial Grammar
CL	Clitic
COND	Conditional
CNP	Complex Noun Phrase
CP	Complementiser Phrase
DAT	Dative case
DG	Dependency Grammar
DOC	Double Object Construction
DP	Determiner Phrase
EC	Extension Condition
ECP	Empty Category Principle
EST	Extended Standard Theory
FS	Finite-state (language/grammar)
FP	Functional Phrase
GB	Government and Binding theory
GEN	Genitive case
GER	Gerund
GF	Grammatical Function
GPSG	Generalised Phrase Structure Grammar
HAB	Habitual aspect
HPSG	Head-Driven Phrase Structure Grammar
IA	Item-and-Arrangement
IC	Immediate Constituency
IMPERF	Imperfective aspect
INF	Infinitive
IP	Item-and-Process
IP	Inflection Phrase
LFG	Lexical Functional Grammar
LTAG	Lexicalised Tree Adjoining Grammar
MG	Metagraph Grammar

MGG	Mainstream Generative Grammar
MP	Minimalist Program
NOM	Nominative case
NONPAST	Non-past tense form
NP	Noun Phrase
NTC	No Tampering Condition
OC	Obligatory Control
PART	Participle
PASS	Passive voice
PAST	Past Tense
PERF	Perfective aspect
PIOC	Prepositional Indirect Object Construction
PRES	Present Tense
PROG	Progressive aspect
PRT	Particle
PSG	Phrase Structure Grammar
RCT	Relation Changing Transformation
REST	Revised Extended Standard Theory
RG	Relational Grammar
RPT	Relation Preserving Transformation
ST	Standard Theory
TAG	Tree Adjoining Grammar
TP	Tense Phrase
VP	Verb Phrase
*v*P	Light verb Phrase

CHAPTER 1

Introduction: Setting the Scene

1.1 Methodological and Historical Context

Natural language sentences are, superficially, sequences of symbols. From a formal perspective, these symbols constitute what we can call the *lexicon* or *alphabet* of a given language: the cards we have to play with. In formal language theory and early generative grammar, languages are indeed defined as sets of *strings* (Hopcroft & Ullman, 1969; Chomsky, 1957), where a string is defined as a finite concatenation of symbols from an alphabet. In this context, the first and most evident relation that can be defined between symbols in a sentence is, precisely, *precedence*. The linear character of the spoken or written word has been a crucial aspect of linguistic inquiry since at least de Saussure (1983) [1916], and aspects of linear order between units in natural language is still an active area of research.

However, natural language sentences feature relations among their constituent parts that go beyond linear precedence: in a sentence like *the man who wears a black coat arrives today* there is morphological agreement between *man* and *arrives*, even though they are not linearly adjacent (cf. **The man who wears a black coat arrive today*, where the ungrammaticality is due to the fact that *arrive* does not agree with the third person singular subject). Similarly, in *what did John say that Mary wanted him to think?* The *wh*-word *what* establishes a relation with the verb *think* (not with *say* or *want*, even though those are linearly closer), and *John* and *him* may denote the same individual. The questions raised by cases like these have guided formal grammar since its inception (within generative grammar, Chomsky's, 1956, 1957 work is the *locus classicus* for arguments against strictly finite-state models of natural language grammar). How are these relations established? How can a theory of the grammar provide an adequate characterisation of the objects that make up a sentence and the relations that they establish with one another? This is perhaps the most important question in syntax, and there is no shortage of answers. In particular, let us consider the following two, which we may think of as defining kinds or families of syntactic theories:

1. An adequate characterisation of syntactic relations takes the form of a procedure that takes atomic objects and builds structure stepwise (possibly but not necessarily following a format specified *a priori*). The basic ingredients of a theory are *symbols* and *rules of combination*.

 | DOI:10.1163/9789004542310_002

II. An adequate characterisation of syntactic relations takes the form of a set of declarative constraints over relational structures: the basic ingredients of a theory are *expressions*, *relations*, and *statements* about the internal structure of expressions.

The scenario that emerges is an extremely rich landscape of syntactic theories, which differ from one another in the specific assumptions made about the nature of symbols, rules of combination, relations, etc. However, there are some basic points of agreement.

Among these points of agreement we may focus on the following. First, the syntactic analysis of objects and relations in natural language sentences cannot proceed in terms of linear predecence only; it becomes necessary to assign an abstract description to sentences where non-adjacent relations can be adequately characterised. Furthermore, symbols in a string need to be grouped in some way, and assigned to a category that allows us to formulate universally quantified generalisations over categories (as opposed to existential statements over lexical tokens): we want to say that there is a level of abstraction at which *the table*, *some smart students*, and *John* belong to a natural class. Let us call these the problems of *segmentation* and *categorisation*. In syntactic analysis, natural language sentences are assigned structural descriptions: a structural description for a string is an annotated analysis of that string, in which relations and dependencies between elements are indicated by various means. For example, *phrase structure grammars* (PSGs), perhaps the best known and most widely used of grammatical formalisms, appeal to a combination of phrasal labels (NP, VP, AP, PP, and the like) and configurational information (the configurational primitives are 2-place predicates: *dominates*, and *precedes*; see e.g. McCawley, 1968: 244). In the use of PSGs made by early generative grammar, structural descriptions are formal objects obtained *procedurally* by the ordered application of rules over the alphabet (e.g. Chomsky, 1956: 117): [Σ, F] PSGs (where Σ is a finite set of initial strings, and F is a set of production rules of the form 'rewrite X as Y', notated X → Y) are an example of theories of the Type I, in the (very broad) classification above.

For purposes of clarity, the process whereby symbols from the alphabet are operated over by rules is often represented by means of diagrams. These may adopt various forms: bracketings, boxes, etc., depending on the grammatical theory being used and the information that is deemed most relevant. Most frequently, however, these configurations are defined in *tree diagrams*. For example, (1) is a constituent structure tree diagram of the sentence *Mary will buy books*:

(1)

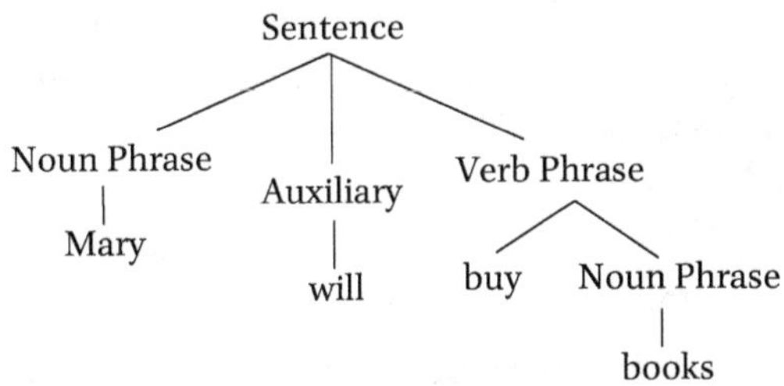

FIGURE 1.1 Phrase structure sentential tree

It is important to note that such a structural description not only contains the elements in the string *Mary will buy books*, but also symbols such as Verb Phrase and Noun Phrase: these are not part of the sentence to be analysed; rather, they are theoretical entities part of the syntactician's toolbox. Both, however, feature as nodes in the tree diagram. Some contemporary syntactic theories have a high number of theoretical entities intertwined with expressions of a language in tree diagrams, to the point where it is difficult to determine what is in the string and what is not from the structural description. Why is this important? Because if we distinguish between what is in the string and what is in the theory, we can aim at formulating an approach where syntactic representations do not contain anything other than the expressions in the input sentence. In (1), 'Verb Phrase' is a useful abbreviation for something along the lines of 'a syntactic object that contains a verb and its direct object, but excludes the subject, …'. Here, … stands for whatever distributional and/or morphological criteria are used to determine that a sequence of expressions is a VP in a particular language (whatever syntactic tests are used to establish whether a verb and its object behave as a unit: fronting, pronominalisation, etc.). This last remark is important to the extent that some languages, such as Warlpiri, do not have VPs at all in the sense that the sequence V + Direct Object does not behave like a constituent for purposes of grammatical operations (see e.g. Simpson, 1991).

We can look at (1) in terms of the formal properties of the object that we are defining: is (1) a mathematical object or a notational variant of another kind of representation (a bracketing, for example)? Generative grammar has used tree diagrams since the beginning, and they have become the standard syntactic diagram in other approaches as well (e.g., Dependency Grammars, 'pure' Categorial Grammars, Lexical Functional Grammar, to name but a few); therefore, they merit close inspection. When trees are used *as mathematical objects* to represent the structure of sentences in natural languages, we will follow Postal (2010: 7) and refer to them as *L-trees*. An L-tree is a mathematical construct, a *set of nodes and edges* where relations between nodes can be unambiguously defined and which respect certain well-formedness conditions: the exact formulation of these conditions determines what counts as an allowable structure, and one goal of syntactic theory is to define the set of conditions that yield descriptively adequate structures. These questions, in turn, raise others, pertaining to the nature of the connection between these relations and the meanings of sentences (the problem of *compositionality*). In the view that we will pursue in the present monograph, it is the goal of syntax to uncover the ways in which natural languages yield structural relations between meaningful parts.

Now is a good time to make the first crucial distinction in this book: McCawley (1998: 47–48) and Postal (2010: 7) warn that L-trees must not be confused with *diagrams* of L-trees. As we said before, L-trees are mathematical objects where formal relations between elements can be defined, but *diagrams of L-trees* are merely pictures composed of lines and symbols (thus, they are typographical rather than mathematical objects): *diagrams* of L-trees are notational variants of other kinds of representations, and we need to keep them apart. This is important because taking drawings too seriously may mislead us into thinking that typographical properties are formal properties. We will come back to this point several times throughout the book, but we can illustrate the confusions that may arise (and hopefully start clarifying them) by considering the following diagram (technically, (2) is a 'non-planar' graph):

(2) 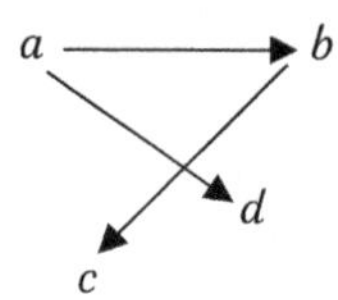

FIGURE 1.2 Directed graph with crossing edges

In (2), we will call *a, b, c*, and *d nodes* or *vertices* indistinctly: they are the expressions relations between which we aim to define. The only relation that we will define for this figure is an asymmetric irreflexive binary predicate *dominates*(*x*, *y*) for all *x*, *y*, and which we represent with arrows, or *directed edges*. An L-tree as a formal object, then, is a set of nodes and edges (we will provide formal definitions in **Section 1.4**, for now this will suffice). Note that the arrows stand for edges, which connect nodes: these are part of the formal definition of the mathematical object we are analysing. This is a crucial aspect of the theory of grammar developed in this monograph. We can now fully specify the relations established in (2) between the elements of our alphabet A = {*a, b, c, d*} as *dominates*(*a*, *d*), *dominates*(*a*, *b*), and *dominates*(*b*, *c*). Now let us examine a second diagram:

(3) 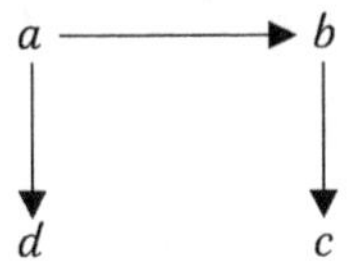

FIGURE 1.3 Directed graph without crossing edges

The question we need to ask now is: do the formal relations between elements in the alphabet vary between (2) and (3)? No, we can still fully specify the relations as above: *dominates*(a, d), *dominates*(a, b), and *dominates*(b, c). This is so because there is no formal significance to the fact that arrows cross in one case and not in the other (assuming, again, that (2) and (3) are non-planar graphs), or that the distance between symbols (in terms of arrow length and distribution on the printed page) varies. Given an alphabet as specified above, and the binary relation *dominates*, we can see that these two diagrams correspond to the *same* formal object, since they contain the same expressions and relations. In technical terms, the graphs in (2) and (3) are *isomorphic*. This formal object is what we call, for the case of structural descriptions for natural language sentences, an *L-tree*. The drawings, with or without crossing lines, are just *diagrams* of L-trees. Our focus in this book will be on the properties of the mathematical objects as ways to formalise relations between expressions, with diagrams serving a marginal purpose.

At this point, we can get into some foundational linguistic issues. Specifically, what is the status of *L-trees* in generative grammar? Answering this question will define the background for our inquiry and the niche that our theory is to fill. To provide an appropriate answer, we need to consider some core concepts of formal language theory: the study of stringsets and the grammars that generate them was a core part of the birth of generative grammar (Chomsky, 1956, 1959). In a generative grammar as formalised in the mid-to-late 1950s, sentences are *generated* by a system of ordered rules of the form X → Y ('rewrite X as Y') operating over a set of symbols (an *alphabet*); such a system is customarily referred to as a *grammar*[1] (we will use *Gr* for *grammar* to avoid confusion with *G* for *graph* throughout the book). The role of a generative grammar, in the classical view, is to recursively enumerate all and only the set of grammatical sentences of a language (e.g., Chomsky, 1959: 137; Chomsky & Miller, 1963: 283; Chomsky, 1965: 31, 60, 202; see Langendoen & Postal, 1984 for critical discussion): this is what it means to *generate* a set (Post, 1943, 1944). In this context, a *language* is defined as

1 See Post (1943: 203, ff.) for a presentation of a deterministic system based on rewriting rules which can generate Context-Free languages.

> *a set (finite or infinite) of sentences, each finite in length and constructed out of a finite set of elements.* (Chomsky, 1957: 13)

The set of sentences (terminal strings) that a grammar can generate defines that grammar's *weak generative capacity*. The set of *distinct structural descriptions* for terminal strings recursively enumerated by a grammar *Gr* defines the *strong generative capacity* of *Gr* (Chomsky, 1965: 60). We distinguish, then, between string languages and tree languages. Now we can formulate the question in more technical terms: are structural descriptions L-trees or diagrams of L-trees?

Let us begin our analysis by considering the status of L-trees in the kind of rewriting-based system just presented. Frameworks that operate with rules that replace symbols by (possibly null) sequences of symbols (including all versions of Phrase Structure Grammars) are called 'proof-theoretic', 'generative-enumerative', or 'procedural' (Postal, 2010; Pullum & Scholtz, 2001, also Pullum, 2013, 2019; Müller, 2020: §14 for discussion). The sequential, ordered application of these rules of inference (more commonly known as *transition rules* or *productions*) produces a sequence of steps, from an initial designated symbol (which we call *axiom*) to strings composed only of terminal symbols (see Hopcroft & Ullman, 1969: Chapter 2). A *derivation* in *Gr* is the ordered sequence of strings from the axiom to a terminal sequence obtained by applying the rules of *Gr* stepwise: these rules rewrite the symbols on the left-hand side of the rule as the symbols on the right-hand side sequentially (i.e., one symbol at a time, from left to right; this is usually referred to as a *traffic convention*). A derivation ends when a string of only terminal symbols is produced, since no further rewriting is possible. Originally, a *derivation* of a string S was defined as

> … *a sequence* $D = (S_1, \ldots, S_t)$ *of strings, where* $S_1 \in \Sigma$ [Σ being the set of initial—possibly unary—strings of the grammar] *and for each* $i < t$, S_{i+1} *follows from* S_i (Chomsky, 1956: 117)

where

> *a string β* ***follows from*** *a string α*
> *if* $\alpha = Z^\frown X_i^\frown W$ *and* $\beta = Z^\frown Y_i^\frown W$, *for some* $i \leq m$ [where *m* is the last line of a derivation] (Op. Cit.)

For example, let us consider the following Context-Free (CF) Phrase Structure Grammar (PSG):

(4) **Alphabet:** {S, A, B, P, Q, *a*, *b*, *c*}
Axiom: S
Rules: S → A, B
A → *a*, P
P → *c*
B → Q
Q → *b*
Derivation: S
AB
*a*PB
*ac*B
*ac*Q
acb

Note that neither the definition of *derivation* nor the formulation of the grammar in (4) involves a specification of nodes and edges: (4) does not look like the definition of a graph, and indeed it isn't. Rather, it defines an ordered sequence of strings, from an axiom (S) to a terminal string (*acb*). However, given a derivation like the one above, a *tree* can be constructed by relating nodes in any line to nodes in the line immediately above, such that the binary relation between symbols *rewrites as* (or, conversely, *follows from*) becomes the graph-theoretic 2-place relation between nodes *mother of*, or simply *immediately dominates* (see Zwicky & Isard, 1963 and McCawley, 1968: 245 for discussion). Each phrase structure rule can only access a single line of the derivation, the last one, and replace a single symbol of that line by a non-null string (Chomsky, 1959: 143; Lees, 1976: 30). Following this mechanism, we can construct the L-tree corresponding to the derivation in (4) as in (5):

(5)

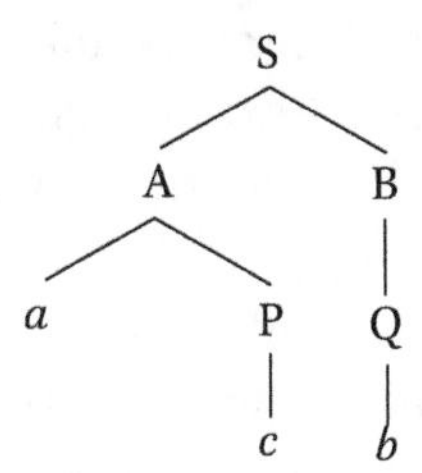

FIGURE 1.4 Tree diagram for derivation of a CFG

In this case, the elements that constitue *intermediate symbols* (A, B, P, Q, S) are part of the alphabet of the grammar just like the *terminal symbols* (*a, b, c*). A rewrite system like (4) does not offer the possibility of assigning a structural description to a string *acb* that relies only on relations defined between these elements directly: intermediate symbols are part of the formal system. The object in (5), not taken as a drawing but as a specification of relations between elements in a formal construct, satisfies the basic conditions to be considered a mathematical object known as a *graph*: a graph is, in the most general sense, a set of nodes connected by edges (see **Chapter 2** for formal definitions). L-trees (the formal objects, not the diagrams), then, are types of *graphs*.

In this monograph we will pursue the idea that the tools made available to us by *graph theory* allow us to formalise aspects of the structure of natural language sentences in a way that mappings from strings to strings and *set theory* do not. As an example, given (5), we can aim at formulating a theory that does not focus on the procedural derivation of terminal strings from an axiom, but on the formulation of declarative constraints over what well-formed graphs look like. We may, for instance, formulate the following statement as a well-formedness condition for a tree language L_T:

> Let x, y, z be indexed categories of the alphabet of L, and $<$ be the binary predicate *dominates*. Then,
> $\forall(x, y, z)\ [((x < y) \wedge (y < z)) \rightarrow (x < z)]$
> $\forall(x, y)\ [(x < y), x \neq y]$

In plain English, the relation *dominates* (not *immediately dominates*) is transitive and irreflexive. Note that this is a statement that may be true or false for a given structure, not a rule that maps an input string into an output string (or a pair of syntactic objects into a set): we may construct a structure for which the statement is *false*. In this case, we would say that the graph is ill-formed (see Rogers, 1997 for a detailed presentation of a declarative system that defines a Context-Free grammar; Oehrle, 2000 for a perspective closer to formal logic; McCawley, 1982 for a semi-formal axiomatisation of a Context-Free grammar with multidominance, also Huck, 1984). We will come back to the distinction between procedural and declarative theories in **Section 1.3**.

It is important to point out that despite the preeminence of string rewriting or set formation systems, graphs feature quite prominently in syntactic analysis since at least the 50s, with Lucien Tesnière's (1959) work on Dependency Grammars. Generative grammar has made use of graphs since the beginning of the 60's, in its transformational and non-transformational varieties. A work as early in transformational generative grammar as Bach (1964: 71) already formulates conditions on phrase markers (P-markers, a structural description of a string; diagrams of P-markers are what we have called L-trees) in terms of their 'topological' [*sic*] properties, making reference to graph theoretic notions and defining specific requirements pertaining to the connections between nodes:

> *A proper P marker (when represented in tree form) is a topological structure of lines and nodes conforming to the general requirement that a unique path be traceable from the termination of every branch to the point of origin of the whole tree (or for that matter from any node to any other node)*

This perspective allowed for a formalisation of conditions over structural dependencies in terms of legitimate relations in *graphs* (see Zwicky & Isard, 1967; McCawley, 1968, 1981a, b; Morin & O'Miley, 1969; Kuroda, 1976), in a departure from rewriting systems and the focus on stringsets that was characteristic of early generative grammar. It is crucial to note, however, that the requirement specified in Bach's fragment that there be unique, unambiguous paths joining nodes in a structural description (with the additional assumption, implicit at this point, that any node in such a path must be visited *only once*) is a staple of the theory of phrase structure in transformational generative grammar, and would define the format of structrural descriptions for decades to come. This requirement has also been adopted in some non-transformational frameworks: we refer here to the tree diagrams of syntactic structures used in Generalised Phrase Structure Grammar (GPSG), Lexical Functional Grammar (LFG), and related theories. In the specific case of LFG, for instance, this implies that there are formal properties that differ between levels of description, *f*-structure and *c*-structure (the former is a set of attribute-value pairs; the latter is a modified version of X-bar theory). Specifically, whereas functional structures allow for a single value to be shared by more than one attribute, constituent structures do not allow for a node to have more than one mother (Bresnan, 2001: 48; the requirement is implicit in the discussion of *c*-structure in Dalrymple et al., 2019: Chapter 3): this is a restriction over allowable structures that becomes essential in the use of a formalism in the description of natural language sentences.

Formal grammars of the type in (4) and their applicability to the description of natural language syntax have been the object of analysis since the very early days of generative grammar, and a big part of the early generative literature was devoted to proving the limitations and inadequacy of pure PSGs for generating structural descriptions for all and only the grammatical strings in a given natural language (Chomsky, 1955a, b, 1956, 1959; Postal, 1964, among many others). The representation of natural language sentences displaying *discontinuity* was often referred to as one of these limitations (see, e.g., Brame, 1978: Chapter 1), but more generally the original argument for transformations pertained to rules whose formulation required reference to the 'derivational history' of a string (Chomsky, 1957): phrase structure rules (PSR) can only make reference to the last line of a derivation, which defines the relation *follows from*. This is a problem if we want to relate expressions that either (i) are introduced at different derivational points, (ii) derive from the conjunction of well-formed strings in a language L to form another well-formed string, or (iii) require us to reorder or delete terms from a well-formed string to form another. To address the shortcomings of pure context-free PSGs, Harris (1957) and Chomsky (1955a, b) introduced a further type of rule in the design of grammars for natural languages: *transformational rules*.[2] As in PSRs, transformations are functions from inputs to outputs, but the nature of their inputs and outputs is completely different from PSRs. These do not map symbols onto symbols (or strings onto strings; a 'symbol' may be thought of as a unary string), but rather *L-trees onto L-trees* (in what follows we will usually omit the *L-* qualification when talking about trees, presupposing it). This allowed the theory to relate an entire sentence to another: e.g., passives to actives, interrogatives to declaratives, conjunctions and clausal complementations to sets of sentences (Chomsky, 1955a: 480–481; Chomsky, 1955b: 146, ff.). The characterisation of transformational rules in Lees (1976) is quite explicit:

> *A transformational rule may be thought of as an ordered triplet [T, B, E] consisting of an IC* [Immediate Constituency] *derivation tree T, a particular analysis of bracketing B of the last line of T, and an elementary transformation E indicating how the elements of B are to be converted so as to yield a new, derived tree T'* (Lees, 1976: 36)

Specifically, transformations map a *structural description* into a *structural change*, specifying the positions of terms before and after the application of the transformation. If we go back to the 1960s, in the so-called Standard Theory (Chomsky, 1965 and much related work) each of the elements (or 'terms') of B (the bracketed analysis of the terminal string of T) is assigned a positive natural number (an 'integer'), which corresponds to its place in the sequence, such that B is an ordered sequence ($X_1, X_2, X_3, \ldots, X_n$). Then,

---

2 We must highlight the fact that Harrisian transformations and Chomskyan transformations are quite different. Harris' model was based on a set of irreducible sentence types (kernels) which were related, often by reversible operations, to expanded sequences (and the whole system was based on the notion of co-occurrence restrictions and distributional classes). Chomskyan transformations, as should become clear from the discussion, are defined on different grounds and play a different role in the theory of syntax.

> *If the structural index of a transformation has n terms,* a_1, a_2, a_n, *it is a* ***reordering transformation*** *if its structural change has any* a_i *as its* k^{th} *term, or if* a_i *is adjoined to its* k^{th} *term, where* $i \neq k$ (Ross, 1967: 427. Emphasis in the original)

Consider, for example, the formulation of the transformation that maps a declarative sentence into a polar interrogative one:

(6) **Interrogative:**

$$\text{S[tructural]D[escription]}: \left\{ \begin{array}{c} \text{(a) } NP, C, VP \\ \text{(b) } NP, C + M, X \\ \text{(c) } NP, C + have, X \\ \text{(d) } NP, C + be, X \end{array} \right\}$$

S[tructural]C[hange]: $X_1–X_2–X_3 \rightarrow X_2–X_1–X_3$
[where C = Pres/Past and agreement; and M = modal auxiliary] (Chomsky, 1964: 227)

We need to note three major aspects of this model: first, transformations do not refer to any aspect of trees: a transformational rule as formulated above maps an array of terms into another array: *John has come* becomes *Has John come?* when used as the input for the transformation in (6). Second, intermediate nodes are used as *variables* over strings in the input of the transformation: in this context, they are unavoidable. Finally, there is no mention of the *grammatical relations* between expressions before and after the application of transformations: consider for example the—quite iconic—case of *Passivisation* (Chomsky, 1957: 112):

(7) **Passive**-optional
Structural analysis [= SD]: NP-Aux-V-NP
Structural change: $X_1–X_2–X_3–X_4 \rightarrow X_4–X_2 + be + \text{-}en–X_3–by + X_1$

The generative model, enriched with transformational rules, allowed not only to derive passive sentences from active sentences, but also to do so without mentioning directly the change in grammatical relations that a transformation such as Passivisation involves (and still does; see e.g. Collins, 2005; Geraci, 2020. See also Perlmutter & Postal, 1983a for discussion about the role of grammatical relations in the definition of Passivisation; we will come back to passives in **Section 13.1**): if passives are derivationally related to actives, we need to capture the fact that the direct object of the active becomes the subject of the passive and the subject of the active becomes a *by*-phrase in the passive, which gets detransitivised. Unlike *Interrogative formation*, for instance, which does not imply a change in grammatical relations, *Passivisation* does. However, the generative formulation of these processes obscures this crucial difference. We will come back to this point below. We have three important ingredients for our argumentation: first, the distinction between stepwise application of production rules vs. formulation of declarative constraints (see **Section 1.3**). Second, the distinction between trees used as notational variants or illustrations for other kinds of formal objects (e.g., unordered sets of syntactic terms) vs. trees used as graph-theoretic objects. And third, the distinction between 'transformations' that change the distribution of grammatical relations among arguments (such as *Passivisation*) vs. transformations that do not (such as *Interrogative formation*). These three distinctions constitute the backbone of our work. The background for these three ingredients is the possibility of dispensing with intermediate nodes (including phrasal categories) in the syntactic description of natural language sentences.

Derivations in classical transformational generative grammar instantiate a more general Immediate Constituency (IC) view of natural language structure derived from its post-Bloomfieldian origins (e.g. Wells, 1947; see Schmerling, 1983b for discussion) and the more general view of a natural language as a recursively-enumerable set of strings and grammar as a procedural, proof-theoretical mechanism (see Langendoen & Postal, 1984; Pullum & Scholtz, 2001; Pullum, 2007, 2019 for critical discussion). We will come back to IC several times throughout this book. IC approaches are based, as we will see in detail below, on a notion of contiguity between expressions that form a grammatical unit (a constituent): these sequences of expressions are obtained by a heuristic procedure based on segmentation and substitution (Harris, 1957), such that a syntactic category is a set of (possibly unary) sequences with overlapping distributions. Contiguity is particularly important for those theories that define constituency in terms of surface properties of sentences alone, without reference to an underlying level where contiguity can be enforced (Dalrymple et al., 2019: 93). In this context, one of the roles of transformations in generative grammar (perhaps their most notable one from a descriptive point of view) is precisely to impose underlying contiguity on superficially discontinuous dependencies (such as *verb-direct object* in a partial interrogative like *what did John buy __?*, where the direct object *what* appears detached from its governor, the transitive verb *buy*, in terms of linear order). The basic idea is that some constructions in natural languages involve a *mismatch* between word order and constituency, and that this mismatch needs to be corrected by proposing a mechanism that takes a deeper level of representation where contiguity holds and reorders some of its parts: this is part of what a transformation like (6), and its more modern variants, does. The elimination of mismatches by positing underlying contiguity is accomplished at the cost of multiplying the levels of representation in the grammar to at least two: one in which grammatical functions and thematic roles are assigned and which constitutes the *input* for the transformation (Deep Structure / D-Structure, depending on the model) and one featuring displaced constituents and enriched with empty categories indicating the positions from which displacement took place (plus displaying other properties that do not concern us now), the *output* of the transformation (Surface Structure / S-Structure).

The inclusion of transformations in the grammar also required the development of indexing mechanisms to keep track of reordered constituents: terms are assigned referential indices, which are maintained through the application of transformations. We need to look at reordering transformations closely: their initial appeal is evident in providing descriptions for sentences that show an apparent mismatch between underlying constituency and superficial linear order, but we need to consider the formal properties of the objects that transformations affect. In early generative grammar, they were arrays of indexed variables over constituents (X_1–X_2, etc.); later on, with the evolution of generative theory and the advent of very general rules (*Affect-α*, *Move-α* in GB and Minimalism; see Lasnik & Saito, 1992), the specification of inputs and outputs for transformations linked to particular constructions was no longer deemed necessary (Chomsky, 1981: 18, for example, claims that '*in the syntax there is the single rule Move-α that constitutes the transformational component*'). This was so because the architecture of GB was focused on filters over representations and how these very general rules interacted with them rather than conditions over derivations: all transformations became 'optional', with deviant structures being filtered out by the diverse 'modules' of the grammar (Theta Theory, Control Theory, Bounding Theory, Case Theory, Binding Theory, the Projection Principle, …). The development of the theory of *displacement-as-movement* in the 70's and 80's gave rise to *traces* to mark the position *from where* a syntactic object had been displaced by a transformational rule (Fiengo, 1977); later on, for intra-theoretical reasons (specifically, a condition that prevents the addition of elements in a derivation that do not belong to the lexical array selected to construct that derivation), *traces* were replaced by *copies* (Chomsky, 1995: 202).[3]

---

3 Strictly speaking, *trace theory* and *copy theory* are not always very different. The version of *trace theory* in Fiengo (1977: 44–45) decomposes *movement* as follows:

Let us look at transformations in terms of their basic formal properties, as input-output mappings. In mathematics, a mapping that preserves relations within selected structure is called an *homomorphism* (see Hefferon, 2014: 181 for a more formal definition of *homomorphism*; we will come back to this issue below). We can ask, then, whether reordering transformations are *homomorphisms*: what does 'preserving relations' actually mean in the context of transformational generative grammar? What relations are there that can be preserved or modified by the application of a transformational rule?

1.2 Transformations and the Preservation of Relations

In the most recent generative framework (the Minimalist Program, MP henceforth), in a departure from the top-down rewriting systems that we saw in the previous section, derivations operate from the bottom-up via a strictly binary, stepwise operation called *Merge*. This operation, in its original formulation, takes syntactic objects X, Y of arbitrary complexity, and forms a set {Z, {X, Y}}, where Z is the *label* of {X, Y}, identical to either X or Y (Chomsky, 1995: 223). Some recent developments, like Epstein et al. (2015), Collins (2017), Chomsky (2020, 2021) dispense with the label altogether in set-formation, and therefore Merge(X, Y) = {X, Y} (only the unordered set containing X and Y), in a further departure from classical formal language theory, phrase structure grammars, and immediate constituent theories. Even when labels are eliminated from the theory, the input of Merge is always a pair of syntactic objects, which means that any already assembled set needs to be taken as a unit for purposes of further applications of Merge: in tree digrams, this is represented by a branching node. This is, in some way, the mirror image of structure building with PSRs:

... movement of NP_i *to position* NP_j *(where A and B are the contents of these nodes) in (30) yields (31) as a derived constituent structure.*

(30) ... NP_j ... NP_i ...
 | |
 A *B*

(31) ... NP_i ... NP_i ...
 | |
 B *e*

On this view, ***NP_i and its contents are copied at position NP_j****, deleting* NP_j *and A, and the identity element e is inserted as the contents of (in this case the righthand)* NP_i*, deleting B under identity.* (our highlighting).

instead of rewriting non-terminals top-down, from an axiom to a terminal string ('expansion-based' grammars), the grammar assembles sets bottom-up ('composition-based' grammars). Intermediate nodes are present in Minimalist tree diagrams, regardless of whether they are labelled: this makes translations between sets and graphs problematic. *Merge* is an operation of set formation, not graph formation; this has far-reaching consequences for the formulation of operations over the outputs of this generative procedure. Generative grammar's departure from graph-theoretic assumptions can be found explicitly as early as Chomsky (1982: 14–15):

> *we can ask whether D-structures, S-structures, etc., have the properties of tree structures. Insofar as they are determined by X-bar theory, this will be the case. But there are other factors that enter into determining their properties. Reanalysis and restructuring processes, for instance, may yield phrase markers that cannot be* ***represented as tree structures.*** [...] *Furthermore,* ***X-bar theory can be constructed so that it does not require that phrase markers have tree properties.*** (our highlighting)

It is clear from this passage that the mainstream development of the theory of phrase structure stayed away from graph-theoretic commitments; however, in early GB the relation between phrase structure and set theory was not as strong as it became in Minimalism under Merge. Consider, for example, the following structure:

(8)

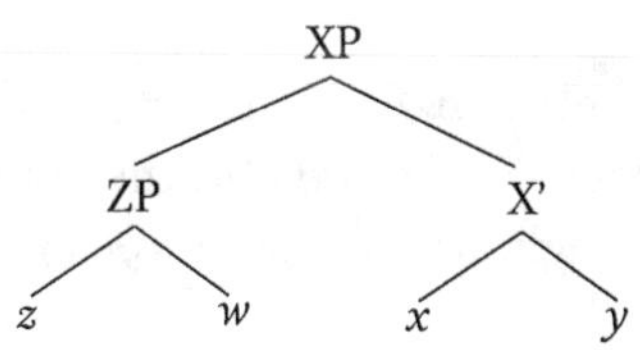

FIGURE 1.5 X-bar theoretic tree structure, from Chomsky (1995)

The fragment from Chomsky (1995: 226) below help us clarify the nature of the relation between the outputs of Merge and tree diagrams:

> *Here* [in (8)] *ZP* = {*z*, {*z*, *w*}}, *X'* = {*x*, {*x*, *y*}}, *XP* = {*x*, {*ZP*, *X'*}}; *more accurately, the tree with ZP as root corresponds to* {*z*, {*z*, *w* }}, *and so on, the labels of the roots having no status, unlike standard phrase markers. Note that w and y are both minimal and maximal; z and x are minimal only.*

In the set-theoretic representation, the status of ZP, X', and XP is unclear: are they sets, members of sets, or proxies for sets (see also Seely, 2006: 189 for related discussion about labels as 'names' for sets, focused on the syntactic 'inertness' of labels)? For example, if ZP = $\{z, \{z, w\}\}$ and XP = $\{x$, {ZP, X'}}, can we replace ZP in XP by $\{z, \{z, w\}\}$? If we can, what is the formal status of nodes like ZP or X'? Chomsky says they have 'no status', but then it is not clear why they are used at all. This is a very important difference between a set-theoretic and a graph-theoretic interpretation of a diagram (8): in a graph, every node and every edge count, since they are part of its formal definition.[4]

Let us take a closer look at reordering under generative assumptions. In what follows, we will use both the set-theoretic notation preferred in Epstein et al. (2015), Collins (2017), Chomsky (2020, 2021), and others as well as the more familiar tree diagrams to make our point (bearing in mind that in the context of Minimalist syntax trees are used as *diagrams* of sets, not as L-trees); we will also label every output of Merge, following a more classical approach. Suppose that we have a phrase marker in which objects X and Y are in a local relation, as represented in (9) in both tree diagram and set-theoretic notation:

(9) 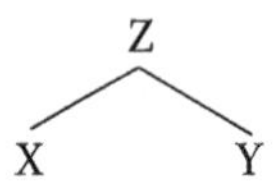

{Z, {X, Y}}

FIGURE 1.6 Minimal labelled tree generated by External Merge

Now suppose that there is some relation *R* between X and Y: for instance, X theta-marks (assigns a theta role to) Y in Z, where Z is the *label* of the object {X, Y}.[5] That relation needs to be maintained throughout the derivation, or recon-

4 Collins & Stabler (2016: 48–49) consider a graph-theoretic translation of Merge, where in a tree like (9) the node Z would be a set, with arcs pointing to its elements. Therefore, Z = {X, Y}, with arcs $e\langle\{X, Y\}, X\rangle$ and $e\langle\{X, Y\}, Y\rangle$. Note that X in {X, Y} cannot be the X in the head of the arc, nor can in the Y in the head of an arc be the same node as Y in {X, Y}: there is a pervasive problem with category types and category tokens (or 'categories' and 'segments', using Minimalist terms). Collins & Stabler do not elaborate on the definition of graph.

5 In 'classical' Minimalism,

> *Applied to two objects α and β, Merge forms the new object K, eliminating α and β.* [...] *K must therefore at least (and we assume at most) be of the form $\{\gamma, \{\alpha, \beta\}\}$, where γ identifies the type to which K belongs, indicating its relevant properties.* (Chomsky, 1995: 223)

We will not discuss the so-called *Labelling Algorithm* that has been proposed under stricter set-theoretic assumptions: given {X, Y}, this algorithm determines the label of the object by

structed at the level of semantic interpretation if it is disrupted by a reordering or deletion rule. We have seen some problems with the latter option, so we would like to give some general strategies for exploring the former. Let us now introduce a further element in the derivation, call it W, which requires a local relation with Y in order to satisfy some formal requirement (which one in particular is not relevant for the present argument). W is external to {Z, {X, Y}}, following a monotonically cumulative approach to derivational dynamics which Chomsky (1995: 190) encodes in the so-called 'Extension Condition': the introduction of new elements always targets the *root* of the tree. In (8), the root is Z, which means that the operation that introduces W in the derivation targets Z and forms a new syntactic object {W, {Z}}, labelled U. This is diagrammed in (10):

(10)

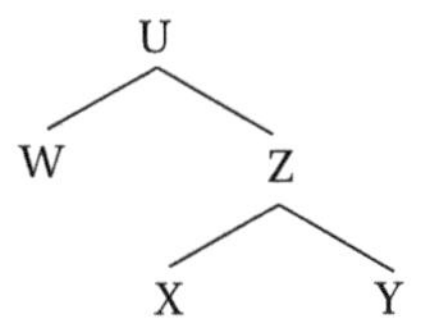

{U, {W, {Z, {X, Y}}}}

FIGURE 1.7 Extended labelled tree generated by External Merge

What happens if a local configuration is required between W and Y (as could happen if, for instance, Y satisfies a criterial feature on W), and the relevant relation *cannot* hold if X intervenes (i.e., if the scope domain of X includes Y but excludes W)? The grammar must somehow create a configuration in which there is a syntactic object which immediately contains W and Y, but not X. As we have seen, transformational generative grammar deals with these situations by *reordering* syntactic objects, by means of the application of rules which map trees onto (derived) trees: these rules are called *transformations*. Recall that the 'trees' we are talking about are not graph-theoretic objects, but simply diagrams for sequences of mappings from strings to strings ordered in terms of the relation *follows from* (we refer the reader also to Lees' 1976 quotation, above). This *reordering* delivered by transformations, which yields non-adjacent dependencies between syntactic objects, is usually referred to as *displacement*, and it can be implemented in a theory of the grammar in different ways. A *displacement-as-movement* approach, which has defined transformational generative grammar since its inception, can either (a) *move* Y to a higher position in the (fea-

searching for an accessible head (Chomsky, 2013 and much subsequent work). Some work goes even further and proposes to eliminate labels altogether (e.g., Collins, 2002, 2017; Seely, 2006).

ture) checking domain of W (thus extending U), outside the scope of X, leaving a co-indexed trace behind (as is the case in so-called *trace theory*); or (b) *copy* Y and *re-introduce* the copy of Y in the derivation where appropriate (the so-called *Copy Theory of Movement*, or *Copy+Re-Merge theory*: Chomsky, 2000; Uriagereka, 2002; Nunes, 2004 and much related work). These two options are diagrammed below:

(11) a. b.

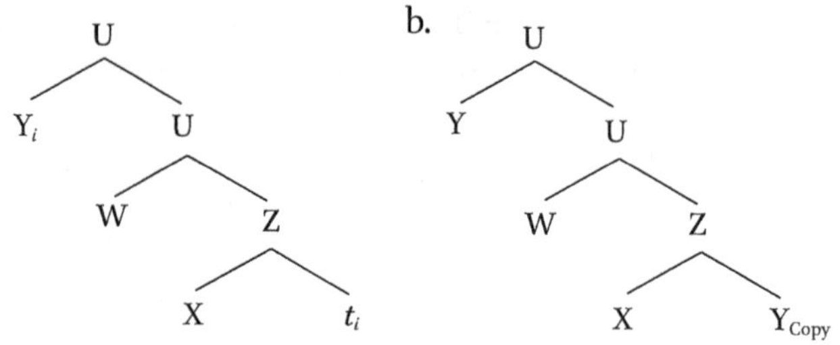

$\{U, \{Y_i, \{U, \{W, \{Z, \{X, t_i\}\}\}\}\}\}$ $\{U, \{Y, \{U, \{W, \{Z, \{X, Y_{Copy}\}\}\}\}\}\}$

FIGURE 1.8 Trace and Copy theories of movement

We need to look at these configurations, and the framework in which they arise, more closely. Chomsky (1995: 225) presents the conception of the syntactic computation in the Minimalist Program as a procedure that maps an Array of lexical items to phonological and semantic representations: a function from a Lexical Array to representations (π, λ) in the 'interface' levels of Phonetic Form and Logical Form respectively. In this context, the operation *Merge* manipulates elements present in what is called the Numeration: this is a set of pairs (LI, *i*), where LI is a Lexical Item taken from the Lexicon and *i* is an integer (known here as an *index*) indicating how many times LI is used in the computation. In the 'classical' version of the Minimalist Program (as presented in Chomsky, 1995), the syntactic computation can only have access to elements in the Numeration, without requiring further access to the Lexicon (see also Chomsky, 2000). Elements in the Numeration are used to create complex syntactic objects via Merge: Merge takes two elements and combines them. However, this only gives us structures which are equivalent to the 'base component' of earlier models: Merge must be allowed to probe not only the Numeration in order to yield more complex structures (in what is called External Merge) but also the syntactic object already created, take an object from there, copy it, and re-Merge it (what is called Internal Merge). Internal Merge, in other words, is what delivers displacement.

The implementation of the property of displacement in Minimalism is not uncontroversial, however. If movement is implemented by means of traces and co-indexing, does the numeration contain *traces*, then? In principle, no. Chomsky (1995, 2000)—correctly, in our opinion—observes that the inclusion of *traces* in a derivation violates the so-called *Inclusiveness Condition* (which bans the inclusion in the derivation of elements that are not present in the Numeration), and therefore, that they are to be replaced by *copies* (as in (11b)). However, as argued in Krivochen (2015b) and Collins & Groat (2018)—among others—, the operation Copy also introduces new elements in a derivation, provided that the copies are not present in the Numeration, as they are *created* as part of the Internal Merge operation (e.g., Nunes, 2004: 89). Therefore, copies also violate the Inclusiveness Condition if this condition is to be understood strictly: information cannot be deleted or lost (see Lasnik & Uriagereka's, 2005 'conservation principles'), but it should also be impossible to add information (in the form of syntactic terminals, for instance) that is not present in the Numeration, including distributional specifications for copies.[6] If an element of arbitrary complexity is Copied and then (internally) Merged, there is no reason to believe that element was present in the Numeration (unless we assume a massive amount of looking ahead that allows the Lexical Array / Numeration to see what is going to be the output of the derivation and thus have all needed elements ready beforehand, crucially including copies), it is therefore treated as a whole new element: after all, until we get to the interface levels (where phonological and semantic interpretation takes place), we have no possibility of establishing a referential connection with an object already introduced in the derivation, except under special stipulations which depart from the simplest scenario and thus require independent theoretical and—more importantly—empirical justification. Conversely, if copies were indeed present in the NUM, the operation Copy would be superfluous, since there would be nothing to copy: all usable elements for the interface levels would be already predicted (somehow) in the NUM (incidentally, making the syntactic component also superfluous and requiring massive amounts of *look ahead*: the computation needs to be aware of derivational steps that have not been

6 Stroik & Putnam (2013: 20) express a similar concern:

> *To "copy X" is not merely a single act of making a facsimile. It is actually a complex three-part act: it involves (i) making a facsimile of X, (ii) leaving X in its original domain D1, and (iii) placing the facsimile in a new domain D2. So, to make a copy of a painting, one must reproduce the painting somewhere (on a canvas, on film, etc.), and to make a copy of a computer file, one must reproduce the file somewhere in the computer (at least in temporary memory).*

The nature and internal dynamics of such 'temporary memory' are not addressed in syntactic works in MGG, although appeal to copies and re-Merges is ubiquitous. More recently, MGG has appealed to *workspaces*, which are in general left undefined. See Krivochen (2023b) for extensive discussion.

developed yet). It must be noted that copies are *not* identical elements: obviously, they are distinct in their syntactic contexts (the structural relations they establish with their neighbours, what each copy is *mother of*, *daughter of*, *sister of* …), but also in their internal composition.

We may clarify this latter point further before proceeding to the former. LIs in Minimalism are assumed to be bundles of features (phonological, semantic, formal), which come in two variants: *interpretable* features can be read by the systems in charge of semantic and phonological interpretation (PF and LF), whereas *uninterpretable* features need to be discharged in the course of the derivation to prevent it from *crashing* (Chomsky, 1995: 232, ff.; Epstein & Seely, 2002). This discharging requires a local relation between at least two categories sharing the feature that is relevant to the case at hand (valued in one instance, unvalued in the other), and if such a local relation does not hold in a tree, a term may be moved in order to create such a configuration (see Chomsky, 2000: 100–101). It is important to note that the theory of formal features in Minimalism goes well beyond a theory of features as used in descriptive morphology or phonology (things like [+/- voiced]; [+/- velar]; [1st / 2nd / 3rd person], etc. See Corvett, 2012 for a detailed surview; Chomsky & Halle, 1968 constitutes a prime example of the use of features in generative phonology): a formal feature can be something as abstract and intra-theoretical as the EPP feature (a head endowed with which triggers the movement of an XP to its Specifier position; Adger & Svenonius, 2011 take the EPP to be a 'second order feature', that is, a feature of a feature), Edge Features (which would make an object 'mergeable'; Chomsky, 2008), Münchausen features (a head endowed with which motivates its own movement; Fanselow, 2003), etc. (see Panagiotidis, 2021 for extensive discussion about what a Minimalist theory of features should do).

In the light of the Copy theory of movement, syntactic objects move up a tree by checking / valuating and discharging features that form lexical items and cannot be interpreted by the semantic or the morphophonological systems (so-called 'uninterpretable' or 'formal' features; see Chomsky, 1995: 276, ff.). This means that in any configuration, no two copies of a syntactic object are defined by the same feature specifications if operations are indeed driven by the necessity to value/check/delete/erase features. Note that the Minimalist derivational apparatus not only multiplies entities, but also requires an additional mechanism to establish relations after displacement rules have applied. This is so because, among other reasons, identity between occurrences of syntactic objects in distinct contexts generated by transformations is not possible in a system that admits feature checking relations as the triggers for transformations, since the feature matrix of an element varies as the derivation unfolds and copies are (Internally) Merged in places in which they can valuate and erase uninterpretable features. The systems in charge of phonological and semantic interpretation (the 'performance systems' in Chomsky, 1995 and much subsequent work) or interface levels with those systems *cannot* establish a dependency between two syntactic objects α and β (e.g., β is a copy of α and only one of them should be pronounced) where they are defined as in (12) (see Epstein & Seely, 2002 for additional discussion):

(12) $\alpha = \{i\text{-}F_1, i\text{-}F_2, u\text{-}F_3, u\text{-}F_4\}$
$\beta = \{i\text{-}F_1, i\text{-}F_2, \sout{u\text{-}F_3}, \sout{u\text{-}F_4}\}$
Where i = interpretable, u = uninterpretable, and strikethrough indicates checked/deleted features.

because there is nothing inherent to α and/or β that suggests they are linked. Any link or relation should be encoded as a diacritic: an additional element in the syntactic representation. Furthermore, as argued in Gärtner (2002: 87), under strict set-theoretic assumptions identity between copies in displacement is also untenable. Suppose that we have the set (13a) created by External Merge, and map it to (13b) via Internal Merge:

(13) a. {X, {Z, {X, {X, Y}}}}
b. {X, {Y, {X, {Z, {X, {X, Y}}}}}}

If *identity* between syntactic objects is defined in a way that includes the specification of set membership for an object, then we can ask whether Y belongs to the set {X, Y} (i.e., if Y is a term of X). Extensionally, it is not an easy question to answer: there are two Ys, one of which does belong to the set {X, Y}, but another one which does not.

Having argued that copies cannot be *identical* (see also Gärtner, 2002: 87, ff.; 2021 for detailed discussion), which entails that Internal Merge necessarily multiplies entities, we must consider the status of relations: if a term is either reordered (as in classical transformational grammar) or copied and re-Merged (as in Minimalism), the relations that object established with other expressions in a structural description are disrupted. A new mechanism or level of representation needs to be invoked to *reconstruct* dependencies from a pre-reordering representation (see Lechner, 1998; Barss, 2001) which delivers underlying contiguity between reordered terms and other objects these are interpreted in relation with or with which they establish a syntactic relation (see e.g. Postal, 1964: 67, ff.; for discussion, Blevins & Sag, 2013: 206, 207).

Above we asked, to what extent does Internal Merge preserve relations within selected structure, and what specific relations are preserved? That depends on the relations: the outputs of transformations are, just like their inputs, strictly *binary* structures; furthermore, movement of heads always targets heads and movement of phrases always targets specifiers (i.e., positions that can only be occupied by phrases) (Chomsky, 1986: 4).[7] This latter condition is particularly Emonds (2007: 347) formulates a Minimalist approach to *structure preservation*:

> *Movement preserves the categories of all its landing sites, but can also have the effect of specifying them further:*
>
> *Definition of Structure-Preservation. A structure-preserving transformational operation is one in which α substitutes for β, where α cannot be specified for a feature differently than β.*

Note that the history of structure preservation, from its original PSG-dependent formulation to the most recent reformulation in terms of featural composition, does not make reference to either properties of trees or grammatical relations. In the original formulation, the configurational information that is preserved depends on whether it is expressable in terms of a PSG (that is: reordering rules cannot insert an element in a position where a PSG could not); in the Minimalist version, there is no configurational information: only the specification that featural compositions cannot differ. This does not rule out the possibility that certain feature structures are restricted to specific configurational positions (e.g., criterial features appear in specifiers of certain functional heads), but in such case configurational information is secondary: relations between objects are restricted to their featural definition. We need to note that early MGG did give grammatical functions an important role to play: part of the argument in Katz & Postal (1964) that transformations cannot affect meaning is that *Passivisation* and *wh-movement* do not modify grammatical relations (1964: 33, ff.): grammatical relations are established at the

---

7 Emonds' (1970: ii) original formulation of the *structure preservation* principle states that *A phrase node X in a tree T can be moved, copied, or inserted into a new position in T, according to the structural change of a transformation whose structural description T satisfies, only if at least one of two conditions is satisfied:* [...] *(ii) The new position of X is a position in which a phrase structure rule, motivated independently of the transformation in question, can generate the category X. (A transformation having such an effect is a* ***structure-preserving transformation****)* [Highlight in the original]
Emonds (2007: 344, ff.) highlights that the original version of structure preservation must be distinguished from Chomsky's, which is a '*weakened*' [*sic*] version.

level of underlying phrase markers, and they correlate '*only with the features of underlying P-markers*' (Katz & Postal, 1964: 33.).

Transformational grammar followed closely on the structuralist conception of clause structure in terms of immediate constituency (see, e.g., Wells, 1947; Harris, 1957), while expanding on the formal work on phrase structure and dominance configurations. In this conception, structural relations like *containment* are primitives, and other relations between expressions, like grammatical functions, are defined in terms of structural configurations: a direct object is the syntactic term that First Merges with a V (see also Chomsky, 1965: 69 for a more classical view, based on strings produced by phrase structure rules). As such, in contemporary generative grammar, grammatical functions are epiphenomena. But that conception of sentence structure, based on configurational relations as primitives and grammatical functions as relational notions, was not unchallenged: as Perlmutter (1983: ix) points out, '*there are significant generalizations, both cross-linguistic and language-internal that can be captured in terms of grammatical relations but not in terms of phrase structure configurations*'. This entails a view of the grammar that is built from grammatical relations up: syntactic structure is built around primitive grammatical relations. Relational Grammar and Lexical Functional Grammar are examples of such views: however, where LFG addresses the relation between grammatical functions and hierarchical relations in constituent structure by proposing two parallel levels (*f*-structure and *c*-structure respectively) related by a mapping function from *c*-structure to *f*-structure (Bresnan et al., 2016: § 4.4), and RG allows for *advancement* and *demotion* operations such that syntactic objects may change their grammatical function in subsequent *strata*, we want to keep the definition of hierarchical relations and grammatical function in a single (graph-theoretic) syntactic representation. The present work essentially agrees with Perlmutter's insight, and explores an alternative to constituency-based frameworks (including all frameworks based on phrase-structure grammars), both transformational and non-transformational. Crucially, once grammatical functions are taken to be primitives of the theory, the strict phrase structure configurations which they are read off in transformational generative grammar can themselves be revised: in particular, we will challenge the idea that syntactic structure is obtained by means of recursive combinatorics yielding unordered binary sets all the way up. Specifically, we argue against *structural uniformity*, where syntactic structure obeys *a priori* constraints that are independent from the properties of the lexical predicates and arguments that appear in specific constructions. In this work we propose a radically different view with respect to procedural syntax in terms of the role of syntax in the grammar and the kind of formal objects that we can use to assign descriptively adequate structural descriptions to natural language strings while minimising the primitives (entities and operations) of the theory.

Perhaps the main inspiration for the line of inquiry that we adopt here comes from the following passage from McCawley (1982):

> *I will assume that the deepest relevant syntactic structures are ordered continuous trees and will investigate the possibility of discontinuity arising in the course of derivations through* ***movement transformations that alter left-to-right ordering without altering constituency*** (McCawley, 1982: 94)

McCawley presents a fascinating perspective: there are rules of the grammar that only alter word order, but not constituency: structural relations are preserved, but linear order is not. As we have just seen, this contrasts with the perspective taken in MGG, where both structural relations and linear order are disrupted by reordering transformations (and where linear order is sometimes seen as a function of structural relations, in particular *c(onstituent)-command*); this crucially includes Internal Merge alongside previous incarnations of reordering transformations (GB's general *Move-α*, construction-specific rules such as those outlined in Chomsky, 1957: 112, etc.). The possibility of having linear order divorced from constituency is essential to the theory of grammar that we are going to develop in this work, and we can go even further if we give up the requirement of structures being *trees* while keeping strict *order* as a basic admissibility condition for structural descriptions (the specifics of which we will come back to below). In this sense, the approach sketched in the passage from McCawley's quotation is an excellent transition between traditional derivational PSGs and the declarative, constraint-based, derivation-less approach that we will develop in this work.

1.3 Declarative vs. Procedural Syntax

When comparing families of syntactic theories, we need to establish what we want syntax to do for us, and what a theory of the grammar looks like: the aims and tools made available by each kind of theory vary greatly, and the choice should be a matter of empirical adequacy. In making these comparisons, it is useful to refer to approaches that are (perhaps) already familiar to the reader. In transformational generative grammar, particularly in its more recent incarnations, the syntactic component of the grammar is a mechanism of unordered set formation, which combines two syntactic terms stepwise (as in, e.g., Chomsky 1995: 243, 2020, 2021; Epstein et al, 2015; Collins, 2017 and much related work). In these views, the generative operation Merge applies to syntactic terms X and Y in a workspace WS and forms an unordered set {X, Y}, mapping WS to WS' and removing the individual elements X and Y from WS'. In earlier incarnations of the theory, the goal of a generative grammar was to recursively enumerate well-formed structural descriptions corresponding to expressions (for discussion, see Langendoen & Postal, 1984: Chapter 1; Pullum, 2019): to *generate* a language (as a set of strings) is to recursively enumerate all and only well-formed expressions of that language. The notions of *strong* and *weak* generative power in Chomsky (1965) and related work are defined with those goals in mind. Later, Chomsky (1995: 148, fn. 1) would claim that '*I have always understood a generative grammar to be nothing more than an explicit grammar.*' However, as noted by McCawley (1988), there has been a change in the way in which 'generative' was interpreted in the 80's (and, we could add, the 90's, 00's, and 10's) with respect to the early days of the theory: the goal now does not seem to be to provide a mechanism to recursively enumerate structural descriptions; rather, 'explanation' in linguistic theory is to be found (at least according to Chomsky) in a theory about learnability and evolvability, but systematic grammatical description is currently seen as a secondary goal (e.g., Chomsky, 2020: 15, 21, 32).

In the view to be developed in this monograph, a characterisation of 'syntax' is that of a *system of expressions and relations* (see e.g. Epstein, 1999: 317–318; Postal, 2010: 4–5; Schmerling, 2018a: 27, ff.). These expressions are nodes in graphs, of the kind informally introduced above (see **Chapter 2** for formal definitions). Specifically, we want to formulate a system that formalises connections between nodes and makes the most out of the smallest possible number of nodes by allowing a single node to establish multiple relations with another node (for example, anticipating material to be dealt with in **Section 6.2.1**, a single syntactic object can be assigned grammatical functions *subject* and *object* by the same predicate in reflexive constructions). What are these nodes? Recall that above we have questioned the formal role of intermediate nodes in generative structural descriptions: both transformational and non-transformational approaches use non-terminal symbols to refer to sequences of symbols that can be part of the input for a rule. This is an heirloom from the rewriting systems core that underlies current transformational generative grammar as well as some non-transformational approaches (e.g., the level of *c*-structure in LFG is modelled after a CFG; see e.g. Dalrymple et al., 2019: 97, ff.). Having a theory not based on node rewriting at all gives us the possibility of considering the formal and empirical consequences of eliminating intermediate nodes in structural descriptions, keeping only basic expressions of the language and specifications of the relations that hold between them (including grammatical functions). In this monograph we explore the consequences of following a radical hypothesis: there are no other nodes in a structural description than those that correspond to overt expressions (in **Chapter 2** we will add the requirement that expressions to be represented in structural descriptions must have semantic values). Pursuing this idea to its limits will reveal if and when it results too restrictive, and if we need to admit the existence of phonologically null basic expressions (see **Section 14.5**). The centrality of the concept of 'basic expression' in our theory calls for a definition, which we borrow from Schmerling (2018a: 16):

> *The* ***basic expressions*** *of L are simply the expressions that are not derived by any operation—they can be thought of as comprising the language's lexicon. If L has a non-empty set of operations (as any natural language does), the outputs of those operations are* ***derived expressions****.* (highlighting ours)

This 'maximise relations, minimise expressions' scenario departs from usual assumptions in MGG, where syntactic operations aim at keeping the relations that each node establishes at a minimum (a single mother and two daugthers) by means of introducing new nodes and assuming an indexing mechanism (as exemplified in (6) above). In this way, a number of second-order conditions over structure building and mapping that arise as a consequence of certain theoretical commitments are given up (including projection-based endocentricity, binarity, the Single Mother Condition SMC, among others).

The aim of an adequate theory of grammar, in the present approach, is to fully specify the relations between nodes in an abstract representation assigned to complex expressions of a language (what we will refer to as the *structural description* of these expressions), and formulate necessary conditions on the admissibility of local structural descriptions. Put simply, we aim to characterise the full set of syntactic relations between semantically interpretable (basic and derived) expressions in the structural description of a natural language sentence. The theory of grammar in this view, then, minimally incorporates: (i) a definition of basic structures, (ii) a recursive definition of complex structures, and (iii) a set of constraints over well-formed graphs. In this sense, and following the distinction made in Pullum & Scholz (2001) and subsequent works, the theory we will sketch here instantiates a *declarative* (sometimes also called *model-theoretic* or *constraint-based*) approach to syntax rather than a *procedural* (sometimes also called *proof-theoretic*) one, at least in general lines. We can unpack this distinction in some detail (see also Pollard, 1997; Pullum, 2013, 2019; Müller, 2020: Chapter 14 for discussion):

- *Model theory*, on which declarative theories are couched, is concerned with finding interpretations for well-formed formulae (WFF henceforth) which make such formulae true: if an interpretation *I* makes a WFF *S* true, we say that *I* is a *model* of *S* (or, alternatively, that *S satisfies I*. We will prefer this latter wording). Model theory is concerned with expressions, not with sets of expressions: an expression, but not a set of expressions, can be a model. Grammars, in this view, consist of finite sets of 'admissibility conditions', in the sense of 'what an expression must look like in order to satisfy the grammar' (Pullum, 2007: 1–2; see also Pollard, 1997: 2). Constraints can be thought of as functions from expressions or pieces of structure to truth values. Constraint satisfaction, furthermore, is *discrete*: either a structure or an expression satisfies an admissibility condition (or the conjunction of conditions) or it doesn't (Haider, 2019). Examples of *model-theoretic syntax* (or *declarative syntax*) are Johnson & Postal's (1980) Arc Pair Grammar, Perlmutter & Postal's (1983a, b) Relational Grammar, Postal's (2010) Metagraph grammar, Rogers' (1997, 2003) 'derivation-less' context-free PSG, McCawley's (1968, 1981b) interpretation of PSRs as node admissibility conditions (see also Sag et al., 1985: 172), most versions of parallel architectures like LFG and HPSG (in general lines, although in implementation and 'grammar engineering' issues of the *order* in which lexical rules apply may arise thus giving rise to derivational restrictions; see e.g. King, 2016 for LFG), and (most versions of) Optimality Theory, which are based on ranked constraints applied to competing (syntactic, phonological) representations (the device in charge of producing these representations, however, can be procedural; e.g. Broekhuis & Woodford, 2013: 125).
- *Proof theory*, on which procedural theories are grounded, is concerned with the enumeration of WFF by means of recursive operations: the meaning of 'generative' in Post's (1943, 1944) work is related to the recursive enumeration of members of a set (here, the relevant set is the set of all and only well-formed sentences in a NL). More often than not, these operations are combinatoric, and based on the syntactic rather than the semantic side of logic: a grammar, in this view, is a set of rules that recursively enumerates the set of well-formed sentences which constitutes a language L (see e.g., Chomsky, 1955, 1956, 1959). Procedural models of syntax are based on the stepwise application of rules to either produce complex objects from combining irreducible objects (*combinatoric procedural theories*, such as Minimalism or Combinatory Categorial Grammar) or expanding symbols until a terminal string is obtained (*expansion-based procedural theories*, such as CF-

PSGs), which translates in the central role of *derivations* and the emergence of *rule ordering* as an explanatory mechanism.[8] Examples of what Pullum calls *proof-theoretic* models (also called *procedural* or *derivational* models, since structure is obtained by means of the ordered application of operations) are all versions of transformational generative grammar (including Minimalist Grammars; e.g. Stabler, 2011), Combinatory Categorial Grammar (Steedman & Baldridge, 2011; Stedman, 2019), and most versions of Tree Adjoining Grammars (Joshi, 1985 and much related work).

Pullum (2007: 2) summarises the tenets of what he calls *model-theoretic syntax* as follows:

> *Grammar, on the MTS* [Model Theoretic Syntax] *view, is about what structure expressions have. It is not about devising a sequence of operations that would permit the construction of the entire set of all and only those structures that are grammatical.*

Model-theoretic syntax is also sometimes referred to as *constraint-based syntax*, although it depends on what exactly is defined as a *model*: some formalisations of LFG and HPSG may be seen as *constraint-based* but not *model-theoretic* (Müller, 2020: xvii). If a model is defined as a set of constraint over well-formed expressions, then the terms *constraint-based* and *model-theoretic* cover the same ground; in this monograph we will use the terms 'model-theoretic', 'constraint-based', and 'declarative' interchangeably. Similarly, 'procedural' and 'derivational' will also be taken to define the same class of grammars.

An important question pertains to the nature of the relation between procedural and declarative syntax: to what extent can they co-exist as part of the same theory? Given that their basic assumptions about what syntax is and how it works are opposite, unification does not seem possible (a point emphasised in Pullum's work, but see Andrews, 2021: 20 for a different perspective and also **Section 14.6**). However, this does not mean that, for the sake of exposition, terminology cannot be borrowed from one or the other (see also Postal, 2010: 6). Our view incorporates descriptive elements from

8 Partee (1973: 510) identifies a 'bottom-up' (combinatory-based) perspective in formal logic, and a 'top-down' perspective in (then current) generative syntax (through phrase structure rules), but says that '*CF-rules can be equally well interpreted as starting at the bottom (with the lexical units) and applying to build up larger and larger phrases*'. In her view, the distinction between *composition-based* and *expansion-based* does not seem to define distinct 'proof classes' (in the sense of Andrews, 2021).

both meta-theories while maintaining logical consistency: we formulate rules as statements about the structure of expressions, rather than as production functions (therefore, aligning with declarative frameworks). We will not 'build' graphs stepwise, or recursively enumerate sentences or structures. This does not mean that this framework does not recognise basic and derived structures: it does (just as pure Categorial Grammar recognises basic and derived expressions). This is different from saying that derived structures are derivationally related to basic structures via composition or expansion. Derived structures are such because their characterisation takes the form of the satisfaction of the conjunction of constraints applied to local structures and their interconnections.

We will borrow some concepts and terms from the early days of MGG that should be familiar to a wide range of linguists, including 'transformations' (as descriptive devices), and notions like Binding (and the principles of Binding Theory; as in Chomsky, 1981: 188 and much subsequent work), Equi(valent NP deletion), extraction (and conditions over extractions; see Postal, 1998 for an important antecedent of our own view), among others. Also, the empirical insights obtained throughout the history of MGG (e.g., Ross' 1967 island constraints, or Rosenbaum's 1967 work on complementation) will feature prominently in our description of English constructions. Methodologically, however, we will mostly build on insights from declarative syntactic frameworks (specifically, Arc Pair Grammar, Metagraph Grammar, Relational Grammar, and Dependency Grammar): in particular, the idea that the grammar is a set of local admissibility conditions that define well-formed graphs (which describe the structure of expressions of a language, in our case, English and Spanish) and what kinds of relations (syntactic dependencies) can be established between nodes in those graphs. The crucial point to bear in mind is that whatever we borrow from procedural syntax will be intended to have purely *descriptive* or *expository* value: there are no derivations in our approach. The theory presented here clearly contrasts with the modern Chomskyan perspective of linguistic structure as being built step by step by means of bottom-up discrete recursive combinatorics: the operation Merge (in all of its incarnations, from Chomsky, 1995 to Chomsky, 2020, 2021) conserves terminal distinctness and unambiguous command paths throughout a derivation.

This shift in perspective forces us to reinterpret the role of 'transformations', which we will keep as merely descriptive devices, pretty much in the same way Postal (2010) uses 'Passivisation', 'Raising' or such terms: we will take *transformations* to have a *descriptive* rather than an *explanatory* value, our goal indeed being to *describe* the phenomena under discussion here. We use rules that make reference to a particular segmentation of an English sequence and a schematisation of its structure and how to operate over that structure to generate, in the formal sense, a sequence that is also grammatical in English, without making claims about psychological reality or universality. What matters is that rules and constraints are not formulated in derivational terms: there is thus no notion of *rule ordering* (Ringen, 1972; Koutsoudas, 1972) or indeed of time (either 'real'—processing—or proof-theoretic—counted as steps in a proof—); there are also *no derivations*. This does not mean, however, that rules and constraints do not interact: as a matter of fact they do, but in a different way from the way they interact in classical transformational generative grammar:

> *Taking any rule R as an implication of the form 'A materially implies B', R applies to any structure S if and only if R's antecedent A is satisfied by S. That determines that S is well-formed only if it satisfies B as well* (Postal, 2010: 7)

Declarative constraints can be ordered, although not in an input-output relation: this is important in order to organise the constraints that apply to basic and derived structures in our model. A model-theoretic approach has no derivations: under such assumptions, the grammar is not a finite way to recursively enumerate well-formed strings. In the present view, which builds on the pioneering work of Stanley (1967), McCawley (1968, 1981b), Zwicky & Isard (1967) and others, *the grammar is a finite set of admissibility conditions over relations between nodes in graphs*,[9] which are structural descriptions of natural language sentences (we will refer to sentences, as well as to grammatically relevant parts of sentences, as 'expressions' of the language). This view entails a departure from the classical approach to PSRs where these are mappings of strings onto strings (following Chomsky, 1959).

---

9 It is interesting to note that our view also contrasts with that of Lasnik & Kuppin (1977), who define a *reduced phrase marker* by means of a set of admissibility conditions *over* (sets of) *strings* (rather than *graphs*, as in the present view), thus the importance of the incorporation of *precedence* relations between monostrings alongside *domination* (Lasnik & Kuppin, 1977: 176–177; see also Lasnik & Uriagereka, 2022). Thus, they define a *reduced phrase marker* (RPM) as follows:

℘ is an RPM if there exist A and z such that
A ∈ ℘ and z ∈ ℘; and if $\{\psi, \varphi\} \subseteq$ ℘, [where $\varphi = x\mathrm{A}z$; A a nonterminal, and ψ a string of terminal symbols]
either ψ dominates φ *in ℘* [where φ *dominates* ψ in ℘ if $\psi = x\chi z$, $\chi \neq \emptyset$, $\chi \neq \mathrm{A}$].
or φ dominates ψ *in ℘*
or ψ precedes φ *in ℘*
or φ precedes ψ *in ℘* (Lasnik & Kuppin, 1977: 177).

Some historical context is in order. McCawley (1968) is often credited with providing a re-interpretation of phrase structure rules as node admissibility conditions (NAC). Let us flesh this out. Consider the PSR A → BC. Then,

> *the base component is a set of node admissibility conditions, for example, the condition that a node is admissible if it is labeled A and directly dominates two nodes, the first labeled B and the second labeled C.* (McCawley, 1968: 247; see also 1981b: 184)

This view has been taken up mostly in the context of Generalised Phrase Structure Grammar (GPSG) and (to some extent) its descendent Head-driven Phrase Structure Grammar (HPSG). For example, Gazdar (1981) formulates a non-transformational approach where

> *A node labelled S in a tree is admitted by the rule* [S → NP, VP] *if and only if that node immediately and exhaustively dominates two nodes, the left one labelled NP and the right one labelled VP. A tree is analysed by the grammar if and only if every non-terminal node is admitted by a rule of the grammar. Under this interpretation, then, phrase structure rules are well-formedness conditions on trees* (Gazdar, 1982: 137)

Similarly, Sag et al. (1985: 127) specify that given a rule like A → BCD,

> *This rule specifies part of the conditions that must hold of a structure rooted in A: namely, that it consist of exactly three daughters whose categories are B, C and D, respectively. However, it does not in itself say anything about the linear order in which B, C and D must occur under A.*

These views are compatible with our conception of what the grammar is (a set of restrictions over well-formed local graphs), and contrast with the approach that takes the grammar to be a set of rules that map strings to strings or combine syntactic terms stepwise.

We will see that there is much to be said about the formal properties of *trees*, and that even within the set of theories that may think of PSRs as node admissibility conditions there is considerable variation, inasmuch as there are several non-mutually reducible or intertranslatable theories whose only common feature is this non-traditional reading of PSRs (for example, McCawley's view allows for transformations in addition to PSRs, whereas Gazdar's does not; a perspective similar to Gazdar's is to be found also in Pollard & Sag, 1994). At this point, once the background for our inquiry has been established, we need to introduce the mathematical notion of *graph* in more formal detail: we will define what a *graph* is, and how it interacts with earlier conceptions of phrase structure (of the kind we have reviewed so far), the empirical limitations of which motivate the present revision of the theory of the grammar.

1.4 On Graphs and Phrase Markers: First- and Second-Order Conditions on Structural Representations

Mathematical definitions pertaining to graph-theoretic concepts will be given in **Section 2.1**, but some preliminary discussion is necessary to set the stage. Suppose that we have a graph G; this will be a set of nodes connected by edges. Let v_i and v_j be two (not necessarily distinct) nodes in G: a v_i-v_j *walk* in G is a finite ordered alternating sequence of nodes and edges (with no repetitions) that begins in v_i and ends in v_j. In an object like (14)

(14)

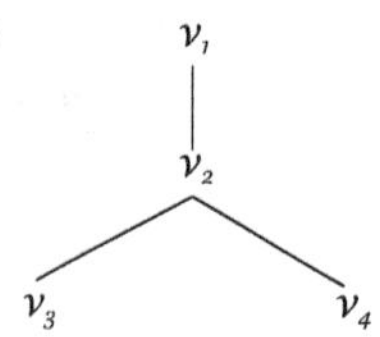

FIGURE 1.9 Sample undirected graph

we can define a walk W = v_1, v_2, v_3, a walk W' = v_3, v_2, v_4, a walk W" = v_4, v_2, v_3, etc. Because there is only one walk connecting any two nodes, (14) is technically a *tree*. In the walks just defined, we have not repeated nodes or edges, and the initial node is distinct from the final node: this is called an *open walk*. The tree in (14) is also not directed: edges are two-way roads (we can go from v_3 to v_2 or viceversa). We can make edges into one-way roads, which we indicate with arrows:

(15)

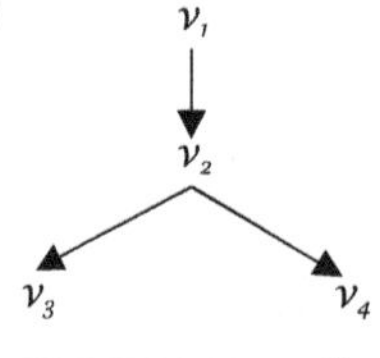

FIGURE 1.10 Sample directed graph

Now there is no walk *from* v_3 to v_2 or to v_4 or to v_1, since there is an edge *to* v_3 but none *from* it. We say, then, that v_3 has *indegree* 1 and *outdegree* 0. (15)

is still a tree, although there are ordered pairs of nodes for which there is no walk connecting them (e.g., $\langle v_4, v_1 \rangle$). In graph theory, a directed rooted tree is occasionally called an *arborescence*. We are still dealing with a tree, since for any two nodes connected by a walk, there is a *unique* such walk.

Crucially for our purposes, further conditions can be imposed over walks. For example, we might require that

- each edge be walked on only once, or
- that each vertex be visited only once, or
- that both of these conditions hold, or
- that neither of them hold

It is important to bear in mind that choosing between these alternatives has empirical consequences for the analysis of syntactic relations, so the choice must be made carefully. As a matter of fact, as we will see shortly, some basic properties of MGG's structural descriptions that we have mentioned above (in particular, the multiplication of nodes in accounts of displacement) can be readily captured in terms of conditions that—implicitly or explicitly—have been imposed over L-trees. Making those conditions explicit will allow us to present our own theory in a way that connects better with a framework already familiar to the reader.

Given this scenario, in particular the choice between conditions to be imposed on walks, we have two possibilities when formalising the theory of grammar in the form of a set of conditions over well-formed graphs:

(16) a. Tend towards maximising relations: use the smallest number of nodes you can, connect them as much as possible (at the cost of giving up 'unambiguous paths' in c-command relations)
b. Tend towards maximising unambiguous paths: keep connections at a bare minimum (at the cost of introducing extra nodes)

(16b) is the option chosen by most works on generative grammar since GB, in particular Minimalism[10] (where it follows from conditions on strict bin-

10 Although we do have to note that already Chomsky & Miller (1963) and Katz & Postal (1964) (see also Chomsky, 1955a, b) assume that Generalised Transformations apply to a *pair* of objects:

The basic recursive devices in the grammar are the generalized transformations that produce a string from a ***pair*** *of underlying strings* (Chomsky and Miller, 1963: 304. Our highlighting)

The recursive power [of a generative grammar] *resides in Generalized Transformations, i.e., those which operate on a set of P-markers* [phrase markers] (***probably always two***) *to produce a single new derived P-marker* (...) (Katz & Postal, 1964: 12. Our highlighting)

ary branching or, more recently, the axioms of set theory), as well as Tree Adjoining Grammars, Lexical Functional Grammar (insofar as the constituent structure component of LFG is essentially a *sui generis* version of X-bar theory but without the axioms of binary branching, endocentricity, or projection; see Dalrymple et al., 2019: Chapter 3; Lowe & Lovestrand, 2020: §§2.1, 2.2), Generalised Phrase Structure Grammar (Gazdar, 1981, 1982), and—to a *much* lesser degree—Relational Grammar and Arc Pair Grammar. The base component of a generative grammar[11] is a context-free grammar of the type [Σ, F] (rewrite a possibly unary sequence of non-terminal or 'intermediate' symbols as a—possibly null—sequence of terminal and/or non-terminal symbols; see e.g., McCawley, 1968 for discussion), further restricted by axioms imposing binarity as a condition over the format of rules (Kayne, 1984; Chomsky, 1986 et seq.) such that F is *always* a string of *two* symbols. This is the kind of system we illustrated in a very simplified manner in (4) above. In such a system, the following conditions thus hold:

a. Every node is dominated by only one node which is distinct from itself (the Single Mother Condition; see **Section 1.6**)
b. Every branching node dominates two other nodes: a branching node has at most (and in the strongest versions of the theory, *exactly*) two daughters distinct from each other and from itself (the binarity axiom in X-bar theory)

Let us analyse in some detail how these conditions, alongside restrictions on walks, determine the format of structural descriptions in MGG. If we define the *neighbourhood set* (or simply 'neighbourhood') of a node as the set of nodes it is directly connected to, a terminal (i.e., nonbranching) node in an X-bar phrase structure tree *t* has a neighbour set of 1, and any nonterminal (i.e., branching) node *n* has a neighbour set of 3 (two daughters and a mother), apart from the root node *S*, whose neighbour set is 2 (two daughters, but no mother). In purely formal terms, the lack of a mother node (i.e., a node being *undominated* in a graph) is the definition of 'root'. Alternatively, we can think of the root of a phrase structure tree in terms of rewriting rules that define derivations as sequences of strings (as in Chomsky, 1956), as a symbol that *only* occurs at the

---

The roots of binarity, which is at the heart of unambiguous c-command paths (Kayne, 1984, 1994) and more recently also labelling (Chomsky, 2013, 2015), are thus much older than X-bar theory (Chomsky, 1970b; Jackendoff, 1977; Stowell, 1981).

11 Or, more specifically, the *constituent structure* subcomponent of the base, since the Lexicon (i.e., the alphabet of allowed terminal and non-terminal symbols) was also part of the base component in the *Aspects* model and later incarnations of transformational theory.

left-hand side of the transition function 'rewrite as', and thus does not *follow from* any other symbol, in the technical sense.[12]

In transformational generative grammar, labelled nodes that correspond to full clauses (S for *Sentence* in the Standard Theory and its developments; CP for *Complementiser Phrase* after Chomsky, 1986) are sometimes also referred to as 'root' nodes, despite the fact that they may be dominated by some other node (as in the case of embedded clauses; but see Emonds, 1970: 5, ff. who claims that non-finite embedded clauses are VPs, not Ss). Derivationally, when a clause is completed, these S/CP nodes are indeed roots of their respective sub-trees. Emonds' (1970: 8) is a useful definition to bear in mind, quite representative of its time:

> *a root will mean either the highest S in a tree, an S immediately dominated by the highest S, or the reported S in direct discourse.*

Consider now the requirements in (a) and (b) above: (a) the SMC and (b) the binarity axiom. Contemporary generative grammar goes to great lengths to justify them as constraints over allowable structures given by Universal Grammar (see Kayne, 1994, 2018; Chomsky, 1995, and much subsequent work). In this context, phrase markers may grow by introducing nodes in the form of non-terminals (projections, phrasal levels) or empty categories (phonologically null terminals). If, following McCawley (1968, 1981b), we interpret phrase structure rules as descriptions of local trees, the graphs generated by a context-free phrase structure grammar that follows (a) and (b) above are always connected, (single-)rooted, and labelled binary branching trees.

A crucial structural relation between nodes in a tree in transformational and non-transformational models alike—as long as they are based on PSGs—is so-called *command*, originally defined by Langacker (1969) in the context of a discussion about *pronominalisation* (to which we will return below):

> *A node A commands a node B if (1) neither A nor B dominates the other; and (2) the S node that most immediately dominates A also dominates B* (Langacker, 1969: 167)

Let us consider again the phrase structure tree in (5), above (repeated here):

12 Note that in this case recursion is not a property of the base component of the grammar but of the transformational component; this was an important feature of the pre-*Aspects* generative theory.

(5)

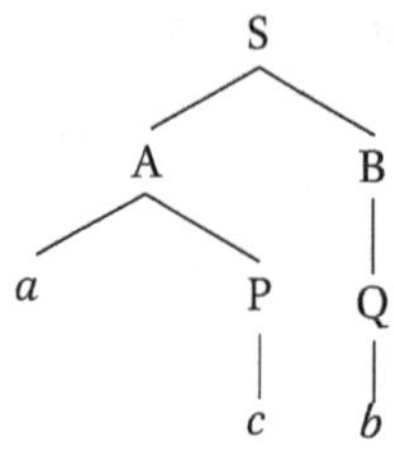

The set C of *command* relations in (5) is C = {(A, B), (B, A), (*a*, P), (P, *a*), (*a*, *c*), (*c*, *a*), (A, Q), (Q, A), (A, *b*), (*b*, A)}: the relation is transitive, total, and symmetric. *Command* was later reformulated by Reinhart (1976) as *c-command* in terms which are still widely used (see Epstein, 1999 for a strictly derivational definition):

> *Node A c(onstituent)-commands node B if neither A nor B dominates the other and the first branching node which dominates A dominates B* (Reinhart, 1976: 32).

Note that the weaker condition imposed by Langacker pertaining to the presence of an *S* node has been strengthened to *the first branching node*: the determination of c-command relations can be now done more locally. Furthermore, in strictly (left-branching) binary-branching phrase markers where each generation is of the form {terminal, nonterminal} (in other words: {head, non-head}, as required in Chomsky, 2009, 2013; Epstein et al., 2015, for labelling reasons and Kayne, 1994, 2018; Uriagereka, 2002, 2012 for linearisation reasons) *c-command* relations between terminals and non-terminals are always *asymmetric*: a head will always asymmetrically c-command the heads further down the structure. We say that a node X *asymmetrically c-commands* a node Y *iff* X c-commands Y but Y does not c-command X. In (5), for example, *a* c-commands P and asymmetrically c-commands everything that P dominates. In a head-complement relation, the head will always asymmetrically c-command all daughters of its complement. The relation between *c-command* and linearisation, as well as the role of *c-command* in defining crucial syntactic and semantic relations (e.g., labelling and Agree as examples of the former; quantifier scope and binding as examples of the latter) makes *c-command* a central relation in MGG. However, conditions derived from the *a priori* requirement that *c-command* relations from terminals to non-terminals be uniquely defined impose heavy restrictions on the format of allowable structural descriptions, which—we will argue on empirical grounds—turn out to be too restrictive.

This view of structure building, while common to transformational and some non-transformational models alike is perhaps best exemplified in Kayne (1984), where explicit conditions pertaining to *unambiguous paths* in c-command relations between terminals and non-terminals are introduced into the grammar. Let us examine this approach in some detail, given the place that graphs play in it. The relevant definitions in Kayne's work are the following:

> *Let a path P (in a phrase structure tree T) be a sequence of nodes* (A_0 ... A_i, A_{i+1}, ... A_n) *such that:*
> a. $\forall i, j \; 0 \leq i, j \leq n, A_i = A_j \rightarrow i = j$
> b. $\forall i, 0 \leq i < n$, A_i *immediately dominates* A_{i+1} *or* A_{i+1} *immediately dominates* A_i (Kayne, 1984: 131–132)

Condition (a) amounts to saying that a walk in T must be a sequence of *distinct* nodes; there are no loops and the walk does not double back on itself. In principle, this is consistent with both the graph-theoretic notion of *path*, which requires that no node be visited more than once, as well as with that of *trail*, which requires that no edge be walked on more than once (but a node may be visited more than once, if we are coming via different edges). Since Kayne uses the term 'path', however, we will use it as well when referring to his approach; we will see that there are indeed reasons to claim that walks in MGG trees must be *paths* rather than *trails*, motivated by the definition of *c-command*. Condition (b) means that a path in Kayne's sense is a sequence of adjacent nodes. Let us now proceed to the definition of 'unambiguous path'

> *An unambiguous path in T is a path* $P = (A_0, ... A_i, A_{i+1}, ... A_n)$ *such that:*
> $\forall i, 0 \leq i < n$
> a. *If* A_i *immediately dominates* A_{i+1}, *then* A_i *immediately dominates no node in T other than* A_{i+1}, *with the permissible exception of* A_{i-1}
> b. *If* A_i *is immediately dominated by* A_{i+1}, *then* A_i *is immediately dominated by no node in T other than* A_i, *with the permissible exception of* A_{i-1} (Kayne, 1984: 132)

Kayne's conditions forbid (i) upward branching (the SMC), (ii) non-binary branching, and—when labels are introduced—(iii) discontinuity. In this context, the '*permissible exception of* A_{i-1}' in (b) is only added '*for symmetry*' [*sic*], because it never comes into play in the analysis of structural descriptions (the SMC is respected throughout).

In Kayne's vision, representative of the GB/MP tradition, terminals need to be nodes with degree 1 (they have a mother, and no daughters), and nonterminals are nodes with degree 3 (having two daughters and a mother), except for the root which is of degree 2 (as seen above). Moreover, in the more recent version of the antisymmetric of phrase structure theory (Kayne, 1994, 2018) and under Bare Phrase Structure (BPS) assumptions (Chomsky, 1994), one of the two most deeply embedded nodes in a tree T must be an empty category: either a trace product of *Move-α* or a base-generated empty node *e* (or, under more recent assumptions, the most embedded category must be allowed to Self-Merge, creating an artificial structural asymmetry; e.g. Guimarães, 2000; Lasnik & Uriagereka, 2022: 104); in any case, it must be a node that does not receive morphophonological exponent and thus is not linearised with respect to the rest of nodes in the tree under consideration. The alternative is to add structure; at least a non-terminal node dominating the most embedded terminal (as in (17c) below) or a projection that can be the target of movement of one of the nodes involved in the point of symmetry (as in (17b) below), in that way the point of symmetry is dissolved (Kayne, 1994: 9–10; Moro, 2000; Moro & Roberts, 2020; Johnson, 2020). We can exemplify these situations in (17):

(17) a. b. c.

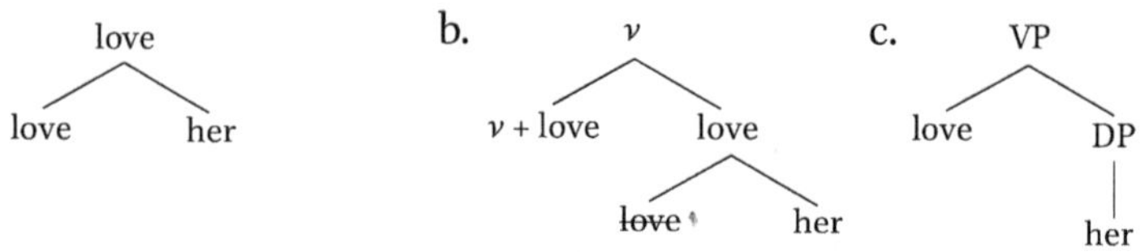

a. Point of symmetry between two terminals, *love* and *her*

b. Asymmetric c-command: <*v* + love, her>

c. Asymmetric c-command: <love, her>

FIGURE 1.11 Examples of symmetry and asymmetry in phrase structure

These conditions over structural representations arise in almost all cases (except in the system explored in Moro & Roberts, 2020) from the purported necessity of phrase markers to comply with the so-called Linear Correspondence Axiom (LCA) as a form to get strings out of hierarchical structures:

> *Linear Correspondence Axiom: d(A) is a linear ordering of T.* [for A a set of non-terminals, T a set of terminals, *d* a terminal-to-nonterminal relation] (Kayne, 1994: 6).

Here, *d* is the syntactic relation *asymmetric c-command*, applying to the set of nonterminal nodes in a tree (Reinhart's definition allows for sister nodes to c-command each other, that is *symmetric* c-command). In (17b) and (17c), the LCA delivers a string *love her* by mapping asymmetric c-command into precedence. (17a) cannot be linearised because *d*(A) is undefined: terminal symbols *love* and *her* c-command each other. The LCA is assumed to be the principle of Universal Grammar (UG) that establishes how to linearise syntactic structure for externalisation purposes (Kayne, 1994; Uriagereka, 2012; see also Moro, 2000 for a more radical version of the theory): that is to say, how to convert trees into terminal strings. In turn, this imposes very strict requirements on possible phrase structure trees, since only those that are linearisable via the LCA are allowed.

As observed in Uriagereka (2002, 2012), the LCA can be interpreted in two ways: a 'weak' interpretation and a 'strong' interpretation. On the 'weak' interpretation, the LCA states that '*when x* [asymmetrically] *c-commands y, x precedes y*' (Uriagereka, 2012: 56), and on the strong interpretations, that *x* precedes *y* ***iff*** *x* [asymmetrically] c-commands *y*, with c-command being both *necessary* and *sufficient*. The importance of the LCA in the present discusson is that it imposes restrictions over what is a well-formed structural description: these restrictions are not based on syntactic or semantic dependencies, but on the possibility of linearising terminals. Note, for example, that if one of the two most deeply embedded nodes was not phonologically empty (as in (16b)), because of the axioms of binarity and the SMC, there would be a structural configuration with two nodes in a *symmetric c-command relation*, which cannot be linearised by the LCA: this is known as a *symmetry point*, and must be avoided at all costs. For example, in Moro's (2000) view, displacement is motivated in natural languages precisely by the need to eliminate symmetry points by moving one of the nodes; a revision of this proposal in Moro & Roberts (2020) makes the ban on symmetric structures depend on the so-called 'labelling algorithm' of Chomsky (2013, 2015, 2020): points of symmetry apparently cannot be labelled by 'minimal search'. The resulting model of phrase structure allowed in an LCA-compliant syntax is one in which *each and every object* (read: each and every node in a tree) *is of one of the following three types*:[13]

---

13 As David Medeiros (p.c.) has pointed out to us, there is nothing in the LCA that prevents a fourth scenario (*d*) *A nonterminal node dominating a single terminal.* However, if the LCA applies to structures generated via Merge (Chomsky, 1995; 2000, 2020), unary branching is banned on independent grounds (at least in the most orthodox works): Merge is by definition binary (Chomsky, 1995: 226; 2000: 81, and much related work even going back to conditions over generalised transformations in Chomsky, 1955a, b; see also Collins & Stabler, 2016, *Definition 13*). Boeckx (2012: 56) mentions a 'plausible' [*sic*] *Anti-Identity* output condition *[xx], which yields the same results as forcing Merge to apply to two distinct terminals (also Collins, 2022). This condition follows directly from an addressing system of the sort introduced in **Chapter 2** below: identically indexed nodes necessarily contract (see also Krivochen, forthcoming a). However, see Lasnik & Uriagereka (2022) for an LCA-inspired cyclic proposal in which nouns can *self-Merge* (based on Guimarães, 2000).

a. A terminal node
b. A nonterminal node dominating two terminals (one of which must then be a *trace* in the sense of Fiengo, 1977[14] or a non-pronounced *copy* in the sense of Chomsky, 2000)
c. A nonterminal node dominating a terminal and a nonterminal

There is some discussion about (b). In Moro & Roberts (2020), for instance, a point of symmetry like that in (b) may be generated (as in Chomsky, 1994), but it is unstable (thus, illegible at the interfaces) and one of the terminals needs to move, thus dissolving the symmetry. In other approaches (Kayne, 2018), symmetry points are not generated at all. In all cases, though, the linearisation mechanism depends crucially on the existence of non-terminal nodes that allow for a definition of antisymmetric c-command.

What about a nonterminal node dominating two nonterminals? This option, which arises in the case of merging specifiers and adjuncts (consider the case of a DP subject which would be generated as a *v*P specifier; set-theoretically, {DP, *v*P}), is explicitly rejected in Kayne (2018: 11): '*Merge never constructs a set consisting of two syntactic objects each of which is a phrase*': a head is Merged *separately* with its complement and then with its specifier (i.e., specifiers would never Merge with phrases). Put differently, a syntactic object of the form {XP, YP} is never generated by Merge (see also Collins, 2017: 50–53; Kayne, 2018). Uriagereka's implementation of Multiple Spell-Out also imposes a ban on complex XP-YP configurations, although of a different kind: one of these is Spelled-Out independently of the other and only then Merged to the rest of the structure. When an XP is Spelled-Out, it becomes internally opaque, for all intents and purposes a syntactic terminal. Therefore, there is no symmetry point in XP-YP if XP is Spelled-Out before Merge with YP, since XP becomes '*something akin to a word*' (Uriagereka, 2002: 49): a syntactic object with no internal structure to be linearised. In both of these approaches, the LCA-offending configuration is not even produced by the generative procedure. This is a more radical approach than having the configuration arise via (External) Merge and then break it down via movement (Internal Merge).

It is worth noting that the LCA, as a linearisation mechanism that restricts local syntactic configurations, is not a privative feature of Move- or Copy-based

14 In a theory that models *displacement* via *movement* of constituents, the notion of 'trace' is defined as follows:

> *Let us call* ***the position from which movement occurs*** *[...] the* ***trace*** *of the node that moves, and let us define proper binding as a relation that holds between a node and its trace only if the node precedes its trace* (Fiengo, 1977: 45. Our highlighting).

Minimalism: works such as Johnson (2016, 2020), Guimarães (2004), Citko (2005), and Citko & Gračanin-Yuksek (2021) advance a Re-Merge approach which yields diagrams where the SMC is not respected (thus allowing for *multidominance*: a node may have more than one mother) and where the LCA plays a prominent role. In these works, linearisation of syntactic structures takes place in strictly binary-branching trees which are intended to represent set-theoretic constructs. It is thus possible to break down applications of the LCA from whole trees to *pairs* of syntactic objects (a classical 'divide and conquer' computational approach): this move allows Johnson to keep the LCA as a linearisation mechanism and have syntactic objects remerge in the derivation (see Stroik & Putnam, 2013 for extensive discussion of potential problems with the Re-Merge view of displacement, depending on exactly how it is implemented). The extent to which 'multidominance' in Minimalist approaches like Johnson's or Citko's is an artefact of diagrams (that is, of pictures or diagrams of L-trees rather than of L-trees as graphs) is not clear: these proposals are not based on *graph theory*, but simply extend top-down PSGs or bottom-up (set-formation) recursive combinatorics; the formalisations follow suit. For example, the recent work of Citko & Gračanin-Yuksek (2021) explicitly assumes set-theoretic Merge as the foundation of structure building. It is interesting to note that Johnson's multidominance approach defines linearisation in terms of *paths* (his 2016: 14 proposal combines elements of Chomsky's 1995 and Kayne's 1984, 1994, 2018 versions of phrase structure: all usual X-bar constructs, like *heads* and *projections*, are still present), despite the fact that, graph-theoretically, multidominance in undirected graphs yields *trails* (walks where a single node is visited more than once but no edge is repeated). It is thus difficult to evaluate such a proposal as different in some significant way from one based on indexing (*trace theory*) or identification (*copy theory*) insofar as all are based on the basic axiom that phrase markers include intermediate symbols, which yield asymmetric c-command relations between terminal symbols.

1.5 Structural Uniformity (and Two Ways to Fix It)

The theory of phrase structure that follows from the requirement that (non-root) nonterminal nodes be of degree 3 (one mother and two daughters) and terminals be of degree 1 (one mother), plus the additional axiom that walks in phrase markers define *paths* between expressions (in the graph-theoretical sense) yields an *a priori* uniform template for structural descriptions. In this section we will examine some empirical problems that arise in this context. The current MGG view of phrase structure implies *structural uniformity*: the computational complexity of linguistic dependencies is invariable. Fukui & Narita (2013: 20) express this view clearly:

> *considerations of binding, quantifier scope, coordination, and various other phenomena seem to lend support to the* ***universal binary branching*** *hypothesis. However, we do not know why human language is structured that way.* [...] *it is likely that theories of labeling and linearization play major roles in this binarity restriction.* [our highlighting]

Crucially, the existence of empirical phenomena that are indeed amenable to a binary-branching analysis does *not* preclude the existence of phenomena for which a binary-branching approach is inadequate. However, binary branching as a model of structural uniformity does imply rejecting *a priori* the possibility that other configurations are available.

A consequence of adopting a proof-theoretic stance is that, because the grammar is a set of production rules that recursively enumerate structural descriptions, the question arises of where in the Chomsky Hierarchy the syntactic component of the grammar is located: are natural languages finite-state? Context-free? Context-sensitive? Turing-complete? Several answers have been proposed in the literature, from the claim that natural languages are regular stringsets (e.g., Reich, 1969; Kornai, 1985) to Turing completeness (e.g., Watumull, 2012) and almost everything in between (e.g., Gazdar, 1981, 1982 for a strictly CF model; Joshi, 1985; Shieber, 1985 for arguments in favour of mild-CS; see Stabler, 2013 for an overview, also Manaster Ramer & Zadrozny, 1990, who present in a very clear way a set of criteria to determine the expressive power of a formalism). However, the computational argument leaves some empirical issues left to solve. We argued in past works that the question as we posed it above may be too restrictive: descriptively adequate structural descriptions for natural language strings (which we take to represent structural and semantic dependencies between expressions), need not be confined to a single level of the Chomsky Hierarchy.

Note that in addressing this issue we take strings to be given (as classical generative grammar did): the grammar assigns structural descriptions to strings, it does not 'produce' strings itself. The function from a set of strings to a set of structural descriptions constitutes the explanandum of syntactic theory (Haider, 1996). At times, restricting the generative engine to a specific level in the Chomsky Hierarchy (by establishing, for instance, that structural descriptions have always the same format) assigns too much structure to substrings: this is the case of CFGs and simple iterative patterns, non-scopal adjunction, and certain kinds of symmetric coordination (Lasnik, 2011; Lasnik &

Uriagereka, 2012, 2022; Krivochen, 2015a, 2016, 2021a). At times, that restriction falls short and configurational information may be ignored (as is the case with CFGs and crossing dependencies in Dutch and Swiss German, see Joshi, 1985; Shieber, 1985 and much related work). In sum: sticking to a single level of the Hierarchy for the sake of structural uniformity may backfire when analysing natural languages. Let us briefly illustrate the kind of problem that arises when a uniform *a priori* template for structural descriptions is adopted (such as X-bar theory or the so-called cartographic enterprise; Rizzi, 1997; Cinque, 1999, 2004). Consider, as a case in point, the following string (see Krivochen, 2021a for a more detailed discussion and a derivational solution to the structural uniformity conundrum):

(18) Some fake fake news

An expansion-based system (e.g., a version of X-bar theory such as that presented in Chomsky, 1986) or a monotonic combinatoric system (such as Merge) can only assign a single structural description to (18): that which goes along the lines of (18) (specific labels are mostly inconsequential; the reader may prefer to replace N' by some array of functional phrases FPs. See Scott, 2002 for a specific implementation of the strictly binary approach to adjective stacking):

(19)

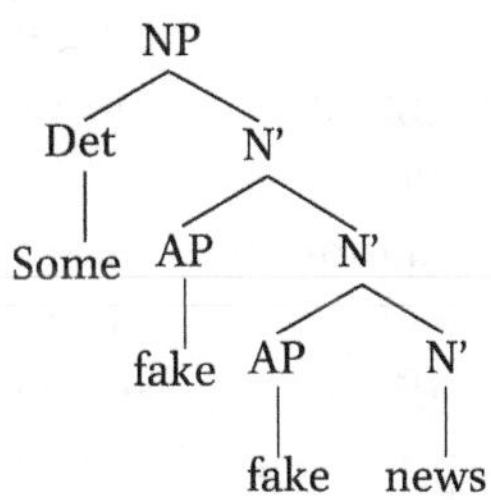

FIGURE 1.12 Strictly binary branching structure for 'fake fake news'

We are focusing on the relation between the adjectives and the adjectives and the noun; issues like the NP/DP debate are orthogonal to our current argument. It is important to emphasise that, if *c-command* relations are mapped to *scope* relations at the level of Logical Form, such that the *scope* of a node A is the set of nodes that A c-commands at LF (Ladusaw, 1980; May, 1985), then because the higher AP asymmetrically c-commands N' (which in turn contains a predicative structure {AP, N}, we assume that the AP is the specifier of the N it modifies based on the usual constituency tests) the only possible interpretation for (18) allowed by a theory that pursues structural uniformity is, roughly, (20) (see e.g. Cinque, 2010: 118):

(20) Some news which is fake as fake news (i.e., truthful news)

In (20), the semantic value of the second *fake* is applied to the semantic value of *news*; then, the semantic value of the first *fake* is applied to the semantic value of *fake*(*news*): *fake*(*fake*(*news*)). But crucially, that is not the only interpretation for (18): a non-scopal, iterative interpretation, where there is no hierarchical relation between the adjectives, is also available:

(21) Some news which sounds very fake (i.e., iteration intensifies the meaning of 'fake')

In this case, we are in the presence of so-called *intensive reduplication*. The meaning of intensive reduplication is reminiscent of the "rhetorical accent" identified by Stanley Newman in his classic work on English stress (Newman, 1946; see also Schmerling, 2018b). The same reading could have been obtained in spoken language by means of vowel lengthening:

(22) Some faaaaaake news ('Some fa·ke news', in Newman's notation)

It is clear that the structural representation in (19) cannot be an *adequate* structural description for (18), insofar as it is unable to account for both interpretations. There is no scope between both instances of 'fake' in the interpretation (21), which means that there cannot be a c-command relation between them: the structure must be different, which also affects the availability of syntactic operations. In the interpretation (21) there is no monotonic application of the semantic value of *fake* to the semantic value of *news*. But it is not all about semantics: syntax is crucially involved in the distinction between (20) and (21). In the case of the iterative reading of *fake fake news*, for instance, it is not possible to cleft only one of the adjectives (**fake is what the fake news was*), which is unexpected under the assumption that the structural description for that string goes only along the lines of [$_{DP}$ some [$_{AP}$ fake [$_{AP}$ fake [$_{NP}$ news]]]]. A structural description that captures the interpretation in (21) and its structural properties must then be *flat* (arguably, finite-state), but only *locally so*: we still want to keep a scope relation between the quantifier and the noun for purposes of functional application, which translates into the requirement that the quantifier c-commands the noun. That is, we want a structural description like (19') for the reading of intensive reduplication (where the substring under A is closed under Kleene star, alternation, and concatenation: the hallmarks of regular expressions), alongside (19) for the scopal interpretation:

(19')

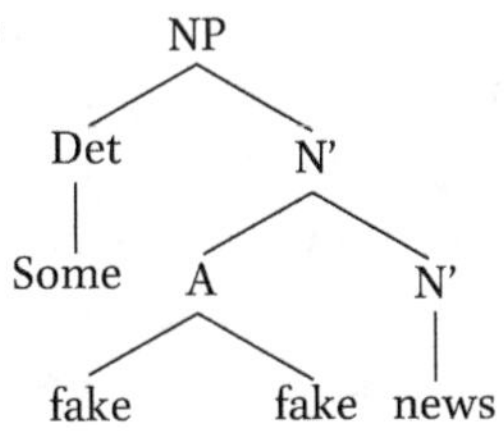

FIGURE 1.13 Locally flat structure for 'fake fake news'

Our point is not notational or typographical: an empirically adequate grammar must distinguish between the two readings of (18), and assign a distinct structural description to each. For the case of intensive reduplication (*fake fake* as 'very fake'), the grammar simply cannot assign the sequence of *fake* internal hierarchical structure: this sequence is best analysed as a FS substring. This accounts for the fact that if we have more than two repetitions of *fake*, the only possible reading is the intensive one, not the scopal one: an even number of *fake* does not make a *true*. The core of our claim, from a derivational viewpoint, is that *some fake fake news* is the result of combining a CF structure with a FS one, and that for purposes of grammatical analysis, it must be possible to express the distinction between these two sub-structures. This is what in past works we have referred to as *mixed computation*, and we will see more instances of sentences that combine dependencies of distinct computational complexity in the course of this monograph.

The problem that we faced can be summarised as follows: an asymmetric relation of the sort delivered by a CFG or Merge is locally adequate (e.g. for the relation between the quantifier and the noun), but not globally so. It *necessarily* assigns too complex a structure for a substring of (18): the iteration of adjectives. This is not a novel point, at least as far as the descriptive observations go: this is already recognised as a problem in Chomsky & Miller (1963), and the issue has been addressed more recently in works such as Lasnik (2011), Lasnik & Uriagereka (2012, 2022), Schmerling (2018b), Krivochen (2015a, 2018, 2021a). The descriptive limitations of PSGs, which are not restricted to iteration, have been observed before, and even the idea of formal systems with mixed types of rules goes back to at least Joshi (1969).[15] However, structural uniformity—now under the guise of so-called Third Factor explanations—has generally

15 In the study of complex dynamical systems, 'mixed computation' is not as controversial as it appears to be in syntax: Binder & Ellis (2016: 4) exemplify physical systems that correspond to grammars of type 1, 2, and 3. In the analysis of the Feigenbaum bifurcation

prevailed based on arguments from parsimony and elegance, at least in MGG analyses. Chomsky's solution to the 'too much structure' conundrum was of course to go beyond phrase structure rules and incorporate a transformational component to grammars. So instead of restricting the power of the grammar, locally (such that substrings could be assigned structural descriptions of distinct computational complexity depending on their semantic properties, as we proposed), the power of the grammar was globally increased. There has been nothing in the MGG metatheory that leads us to prefer a completely underspecified phrase structure building engine over a transformational model in which $\Sigma \rightarrow F$ rules are supplemented with mapping operations from trees to trees (Lees, 1976). Competing analyses have not generally fared better: in LFG, for instance, modifiers are assigned the non-governable grammatical function ADJ(unct), whose value is always an unordered set at *f*-structure (Dalrymple et al., 2019: Chapter 13). Taken at face value, the standard LFG analysis (but see Andrews, 2018 for a modification of the *f*-structural analysis of modifiers) falls short of empirical adequacy for the opposite reasons that the MGG analysis does: it imposes uniform flatness (thus being unable to account for the scopal reading).

This is essentially an argument that combines aspects of weak and strong generative capacity. The success of a theory of grammar is not just measured in terms of generating a stringset or putting each string in correspondence with a structural description, but also requiring that the structure assigned to each string be *adequate*: it must represent the semantic relations between units in each string without imposing extra structure in the form of intermediate or empty nodes (phrasal categories, traces ...). The empirical inadequacy of *a priori* uniform proof-theoretic grammatical systems (as they impose 'too much structure' for intensive reduplication in a case like (18)), which are based on the theory of computable functions developed, among many others, by Alan Turing (1936) can be either avoided or circumvented. It can be avoided by radically changing the perspective and adopting a declarative approach; then define local admissibility conditions that do not require to assume structural uniformity. It is also possible to focus on showing that a proof can be devised that a given string is a well-formed expression of the language, regardless of whether it is a phrase-structural constituent (Müller, 2020: 513). It can be circumvented by

---

diagram (defined by the recurrence equation $x_{(n+1)} = rx_n(1\text{-}x_n)$, plotting a series of values for x_n as a function of the parameter r), for example, mixed computation becomes necessary to characterise different regions of the logistic map.

allowing the grammar, conceived of as proof-theoretic, to oscillate between different levels in the Chomsky Hierarchy when assigning structural descriptions to natural language strings in local domains (e.g., lexicalised elementary trees, as in Lexicalised Tree Adjoining Grammars). This latter option, which entails rejecting the uniformly monotonic nature of syntactic computation that is at the heart of generative grammar while keeping a derivational perspective in which generalised transformations play a crucial role (see **Section 4.3.1**), has been explored in past works.

In the aforementioned works we proposed (considering the analysis of adjectival iteration, symmetric and asymmetric coordination, and chains of auxiliary verbs in English and Spanish) that a descriptively adequate grammar may be able to assign different substrings in a natural language sentence local structural descriptions of distinct computational complexity. We have applied *mixed computation* to the analysis of a range of linguistic constructions in English and Spanish (Krivochen, 2015a, 2016a, 2018, 2021a; Bravo et al., 2015; Krivochen & García Fernández, 2019, 2020; Krivochen & Schmerling, 2016a). 'Mixed computation' means that

> *phrase markers, as structural descriptions of strings, are not computationally uniform but mixed. By saying that a system is 'computationally mixed', we mean that the structural descriptions assigned to strings in L* [...], *need not all be formally identical.* [...] *a computationally mixed system assigns a substring the simplest structural description that captures and represents the formal and semantic relations between syntactic objects.* (Krivochen & Schmerling, 2016a: 36)

The mixed computation proposal comes as an answer to a computational and descriptive conundrum: structural descriptions must be *minimally appropriate*, that is, they must assign no more structure to substrings in L than strictly needed to capture constituency and semantic interpretation. But PSGs are in a very specific sense *procrustean*, only allowing for a single kind of computational dependency in structural descriptions for natural language strings: that allowable in Context-Free languages, nothing more and—crucially—nothing less. Because of this, it has long been recognised that they can be too powerful, as we noted in the analysis of (18) above:

> *a constituent-structure grammar* ***necessarily*** *imposes* ***too rich an analysis*** *on sentences because of features inherent in the way P-markers are defined for such sentences.* (Chomsky, 1963: 298. Highlighting ours)

But they can also fall short when assigning structural descriptions to natural language strings, for there are strings in natural languages the structural descriptions for which display crossing dependencies that cannot be captured by means of a context-free PSG (we will see some examples in **Chapter 4**). What can be done about this? A possibility that has just begun to be explored is that linguistic computation is an *oscillatory process*, which moves up and down the Chomsky Hierarchy in local syntactic domains. What can we do to assign an adequate structural description to iteration that does not contain additional, unnecessary structure? Lasnik and Uriagereka propose that

> [...] *what we need should be, as it were,* ***'dynamic flatness'****. But this is the sort of concept that sounds incomprehensible in a classical computational view, while making sense to those for whom syntactic computations are psychologically real.* (Lasnik & Uriagereka, 2012: 21. Highlighting ours)

> *In a manner of speaking, what we really want to do is* ***move down*** *the* [Chomsky] *hierarchy. Finite-state Markov processes give flat objects, as they impose no structure. But* ***that is not quite the answer either****. While it would work fine for coordination of terminal symbols, phrases can also be coordinated, and, again, with no upper bound.* (Lasnik, 2011: 361. Highlighting ours)

The *mixed computation* approach pursued in our past works is a way to address the issue of PSGs being procrustean while remaining within the space of proof-theoretic grammars: local syntactic domains are defined in terms of computationally uniform dependencies. The iteration of adjectives in *fake fake news*, in the reduplication reading, defines a regular (finite-state) domain, which gets inserted in a wider context-free syntactic context. In this way, we 'move down the hierarchy' locally, with generalised transformations (substitution and adjunction) taking care of the composition of local domains each of which is, internally, computationally uniform.

An alternative to mixed computation is to explore a different grammatical space, based on expressions and constraints over well-formed graphs where relations between expressions are established. In this vein, the present monograph pursues a line of thought that is an even more radical departure from structurally uniform Immediate Constituency tenets than the works referred to above (e.g., McCawley, 1968, 1982; Huck, 1984; Lasnik & Uriagereka, 2012; *inter alios*), while building on their insights. In contrast to transformationally-enriched phrase structure grammars, we attempt to eliminate empty nodes (including traces and copies), indices, restrictions on the degree of vertices, and the requirement of graphs *minimising* connections (*as per* the SMC and strict binarity requirements), thus proposing a theory aligned with tendency (16a), *maximise connections*. We formulated (16a) and (16b) as 'tendencies', because, from a procedural standpoint, derivations can locally oscillate between those extremes: this is another way of capturing the computationally mixed properties of natural language structures (and it is likely that different languages exploit these tendencies differently; we will not deal with these issues in this work, but they are certainly worth pursuing). This view of grammar as a dynamical oscillatory process (Saddy & Uriagereka, 2004; Saddy, 2018; Krivochen, 2018) as interpreted here yields graphs that can be evaluated as maximising or minimising connections *only within local domains*, and restricted by the selectional properties of lexical predicates. That means that a model that embraces *n*-ary branching, discontinuity, and multidominance does *not* necessarily reject binarity and monotonicity altogether: these are properties of phrase markers *at a local level*, whereas mixed computation is a *global* property of a process that assigns structural descriptions to natural language strings.

How can a graph-theoretic model help in the analysis of these cases? We have not yet introduced the formal machinery that will support our analyses, but we can give the reader an idea of the kind of analysis we can provide. Above we highlighted the possibility of doing away with intermediate nodes: having structural descriptions with no VPs, NPs, etc. This entails, from a post-Bloomfieldian viewpoint, no constituency: a theory of grammar that followed this path is not IC-based. However, it can be consistent in its own right. Suppose that each basic expression corresponds to a node in a directed graph. Let us consider a case of iteration where the only possible interpretation is intensive reduplication:

(23) An **old old** man {entered, broke the vase, ran, gave Mary a book ...}

Let us focus on the reduplicated adjectives and their relation to the noun. If each basic expression corresponds to a node, then we would have two nodes: *old* and *man*. Furthermore, *old* modifies *man*: we can indicate that by connecting these nodes by means of a directed edge *from old to man*: we are building a digraph. The first one of those expressions is iterated, though. How can we represent that without adding too much structure? If we do not specify otherwise, in a walk through a graph we can 'visit' the same node more than once; actually, as many times as we want. In iteration, that is what we do: visit the node corresponding to a single expression more than once. Let us diagram the dependencies:

(24)

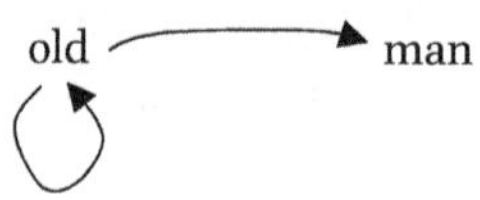

FIGURE 1.14 Finite-state transition diagram

Simplifying much of the discussion in **Chapter 2**, (24) is a graph where *old* dominates *man*, and receives an interpretation where the semantic value of *old* (whatever that is, more on this in **Section 2.2**) is applied to the semantic value of *man* (whatever that is). The existence of an edge from *old* to *old* represents iteration: there is no scope between the occurrences of *old* in *an old old man*. A graph like (24) interpreted as a finite-state transition diagram generates the regular language L = {*old* man*}. Note that this case of iteration differs from *fake fake news* in only allowing for a single interpretation, the non-scopal one.

What happens if we have the iteration of a category, but not of specific lexical items? Consider in this respect an example like (25):

(25) A black, old, heavy book

Here we have a sequence of adjectives all of which modify the noun and not each other. Following the same basic principle as above, where predicates dominate their arguments, we need to have *book* dominated by as many nodes as we have adjectives in our string. The corresponding graph is diagrammed in (26) (see Osborne, 2019: 210, ff. for a related DG perspective):

(26)

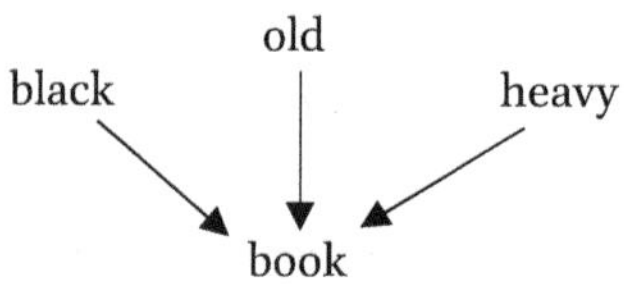

FIGURE 1.15 Graph-theoretic description for adjectival stacking

Intuitively (in terms to be refined in **Section 2.3**), the semantic interpretation of (26) should correspond to the standard approach to intersective adjectives: the book is black and old and heavy, and the denotation of (24) is the intersection of the sets of books, of black things, of old things, and of heavy things. However, we need to go beyond intuition: we will come back to the issue of semantic interpretation in more detail in **Chapter 2**. We must underline the fact that the structural description for (25) does not, unlike standard approaches to adjectival modification based on X-bar theory (e.g., Cinque, 2010; Scott, 2002), impose a hierarchy between the adjectives (e.g., in terms of c-command or dominance): this is precisely what we want to achieve from a syntactic perspective.

A graph like (26) has a node, which corresponds to the expression *book*, dominated by three nodes (and there could be more, of course). This implies a departure from a basic axiom of generative grammar: the Single Mother Condition. We have mentioned it in the preceeding sections, but given its centrality to the generative theory of syntax (transformational and non-transformational) we need to examine this condition in more detail.

1.6 You Only Have One Mother

Recall our distinction between theories of syntax that aim at maximising connections between a very limited number of nodes versus those which tolerate a multiplication of nodes if that means that every node will be dominated by exactly one node distinct from itself. As we have highlighted before, formalisms that position themselves at the (16b) end of the spectrum (which aims at *maximising unambiguous paths*), which include transformational generative grammar as well as LFG, GPSG (and other PSG-based formalisms) and some versions of DG, assume—explicitly or implicitly—what Sampson (1975: 1) calls the 'Single Mother Condition', already familiar to us but which we can now define formally:

> *D is a set of* NODES, *α is a function from D into a vocabulary of symbols (if $\alpha(d) = s$ we say that s is the* LABEL *of d), and δ is a partial function from D into strings over D* [...]
>
> *(iii) for any d, d' in the domain of δ, if $\delta(d) = e_1, e_2, \ldots, e_n$, $\delta(d') = e'_1, e'_2, \ldots e'_{n'}$, and $e_i = e'_{i'}$, for some $1 \leq i \leq n$, $1 \leq i' \leq n'$, then $d = d'$ and $i = i'$.* ***That is, nodes may not branch upwards.*** *We shall call property (iii) the* SINGLE MOTHER CONDITION *(*SMC*).* (our highlighting)

Gärtner (2002: 99) provides a simpler formal definition of the SMC, based on the relation of immediate dominance (ID). Let N be the set of nodes allowed in a PS tree. Then,

> $(\forall x, y, z \in N)\ [x\text{ID}z \wedge y\text{ID}z \rightarrow x = y]$ (see also op. cit., p. 121)

We saw how the requirement of unambiguous paths leads to the SMC in transformational generative grammar. In combination with the axiom of binary branching, this gives us the template that will apply to all structural descriptions: every non-terminal node with immediately dominate at most two nodes, and every node (terminal or not) will be dominated by at most one node. As we saw, the SMC is not an exclusive feature of MGG: for LFG, Bresnan et al. (2016: 46) claim that *c*-structures, which are phrase structure trees characterised by context-free rules, also obey the SMC (see also Westcoat, 2005 for an application of an SMC-based argument to the analysis of *lexical sharing*. In Westcoat's approach, *c*-structures are trees in the graph-theoretic sense); GPSG's trees are also modelled upon CFGs, thus the SMC applies (see Gazdar, 1982; Pollard & Sag, 1987: 55, ff.); and Dependency Grammar trees also follow the SMC (at least in some versions: e.g. Osborne, 2019, but not Tesnière, 1959). If we look at CFG from the perspective of formal language theory, assuming Chomsky's (1956, 1959) idea that derivations are sequences of strings, then the SMC emerges more or less naturally when we translate sequences of strings into trees (McCawley, 1968: 245): if the input for a phrase structure rule is a string $x_1x_2x_3 \dots x_n$ and its output is a string $a_1a_2a_3 \dots a_n$, and assuming that each line of a derivation results from the previous line by replacing a single symbol by a non-null string (Chomsky, 1959: 143), then we can determine the common elements between the input string and output string, and establish a dominance relation between the symbols that have changed (Postal, 1964: 11). In this context, where trees are seen as derivative from sets of strings, it is not possible to have any node dominated by more than a single other node: there is no obvious way to identify symbols that appear in different strings. For example, it is not possible in the following toy grammar to construct a tree where there is a symbol *a* (an expression of category *a*) dominated at the same time by two distinct nodes: we would need to have *two distinct* (*tokens of*) *a's*:

(27) S → AB
A → *a*C
C → *c*
B → *b*PQ
P → *a*
Q → *q*

The corresponding derivation is the following:

(28) S
AB
*a*CB
*ac*B

*acb*PQ
*acba*Q
acbaq

The terminal string *acbaq* contains two *a's*, one preceded by a word boundary and followed by *c*, the other preceded by *b* and followed by *q*. This is not an artificial situation. Consider the derivation for a monotransitive clause in Chomsky (1957: 27):

(29) Sentence → NP + VP
NP → Det + N
VP → Verb + NP

Evidently, there is not a single NP node dominated by both S and VP: NP is not a single variable used twice in a proof (compare with the standard interpretation of logical proofs). Rather, there are *two* distinct NP nodes, each of which is the result of applying a different rule at a different derivational point (we know this even without having to worry about lexical insertion). If trees are looked at simply as ways to diagram relations between strings, multidominated nodes are an impossibility. The idea that derivations involve lexical item tokens is maintained in Minimalism (e.g., Uriagereka, 2008: 16; Chomsky, 2001), which has shifted the focus from strings (as in the Standard Theory; see McCawley, 1981a: 165) to sets.

What about theories not based on Immediate Constituency? After all, tree diagrams are not an exclusive tool of generative or IC models: do other approaches that use trees (which fall under the category *Item-and-Process* in Hockett, 1954; Schmerling, 1983a) consider their graph-theoretic dimension? With respect to Categorial Grammars, Partee (1974)—among others—argues that traditional (i.e., non-combinatory) CGs have the strong generative power of CFGs: it is thus possible to use a 'pure' [*sic*] CG (i.e., Ajdukiewicz / Bar-Hillel style) as a model for the *base component* of a transformational grammar (Partee, 1974: 516–517), insofar as their generative power is equivalent. Let us briefly exemplify this point, because CGs differ from the previous frameworks we have mentioned in non-trivial ways, and we will make use of some insights from Ajdukiewicz-style CG in our exploration of English syntax. In a CG, expressions of the language are assigned to an indexed category, basic or derived. Let X and Y be part of the inventory of basic categories. Then, if an expression of the language is assigned to the category X/Y, this means that X/Y must be concatenated with an expression of category Y to get an expression of category X. For example, we can define the category of 'saturated verb phrases' as the set of expressions that must combine with expressions of category 'noun phrase' to form expressions of category 'sentence': this can be translated into CG notation as the assignment of expressions that we call 'saturated (or 'intransitive') verb phrases' to the category S/NP (see e.g. Partee, 1975: 214–147; Schmerling, 2018a: 151, ff.). This can be diagrammed in an *analysis tree* as follows:

(30)

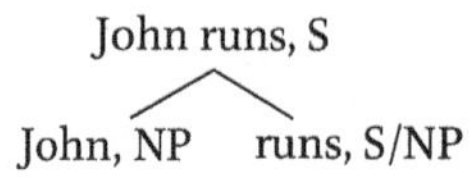

FIGURE 1.16 Categorial Grammar analysis of 'John runs'

The general CF format of Ajdukiewicz-style CG rules is translated into Chomsky Normal Form as follows (let + symbolise concatenation):

(31) $c \rightarrow (c/c_1 \ldots c_n) + c_1 + \ldots + c_n$ (Lewis, 1972)
(e.g., S → NP + S/NP; Partee, 1975: 214)

Things may not be all that simple, though: not all variants of CG have the same generative power; crucially, Montague grammar (which in principle is heavily based on Ajdukiewicz's CG) goes beyond strict CF power if we consider rules of quantification like his rules S14–S16 in PTQ (see Hamblin, 1973: 43; also Van Benthem, 1988 and Buszkowski, 1988 for detailed discussion about the generative power of different variants of CG, including Ajdukiewicz-style and Lambek calculi). These rules yield what is informally known as *quantifying-in*, the replacement of a variable for a term in a scope position (see **Chapter 10** for some additional discussion). In Ajdukiewicz-Montague CGs we could be tempted to say that the SMC is respected: no node in a tree is dominated by more than one node. However, that would entail a misinterpretation of the role of tree diagrams in CG because trees are, strictly speaking, diagrams of a proofs that an expression belongs to a category in the algebra; however, nodes in an analysis tree are not related by *dominance*, *c-command*, or any structural relation familiar to PSGs. Nodes in a CG tree contain three elements: (i) an expression of the language (basic or derived), (ii) the indexed category that corresponds to that expression, and (iii) the rule that has applied to yield that expression (see **Section 5.1.1**). Therefore, the root node in a CG analysis tree defines the category to which a complete expression of the language belongs (see Schmerling, 2018a for details): (30) is a diagram of a proof that *John runs* is a well-formed expression of category S. Tree diagrams have a much less significant role in CGs than in PSGs, if any at all; as a matter of fact, in current models of CG, trees are much less used than in 'pure' CG (see e.g., Steedman & Baldridge, 2011; Steedman, 2019). There is nothing in Ajdukiewicz-Montague style CG that is comparable to a PS tree. The SMC holds for diagrams of mappings between strings, but not for *diagrams* of proofs. Again, the difference between *formal objects* and *diagrams* of formal objects becomes crucial: trees have been used to diagram vastly different formal objects.

A condition similar to the SMC was initially assumed by McCawley (1968) in his formalisation of the base component of a transformational generative grammar. As we saw before, the interpretation of PSGs in McCawley's work, which was later developed in GPSG (e.g., Gazdar, 1981; see also Pullum, 2019 for recent discussion) and LFG (e.g., Dalrymple et al., 2019: 139–140) is that of well-formedness conditions over graphs, such that a rewrite rule like A → B*c* is not seen as a mapping between strings #A# and #B⌒*c*# (as it was in Chomsky, 1956, 1959) but rather as a constraint over what a well-formed graph containing nodes A, B, and c looks like: roughly, *a tree T containing node A is a well-formed tree iff A immediately dominates B and c in T* (see also Huck, 1984). In this context, let x and y be nodes, ρ be a 2-place immediate dominance relation between nodes, with ρ^* its transitive closure and λ be a 2-place immediate precedence relation. Then:

> *if* $x\rho y$ *and* $x'\rho y$*, then* $x = x'$
> *for any two nodes* x *and* y*, if* $x \neq y$*, then either* $x\rho^*y$ *or* $y\rho^*x$ *or* $x\lambda y$ *or* $y\lambda x$
> (see also Zwicky & Isard, 1963: 1 A.4; Wall, 1972: 149 for equivalent conditions)

The first condition entails, again, that nodes do not branch upwards: there cannot be two distinct nodes x and x' such that both x and x' dominate a node y. The second entails that dominance and precedence are each partial orders over a tree: note the absence of $x\rho^*x$ and $y\rho^*y$, which are the configurations that would be generated under certain versions of multidominance which produce closed cycles in a structural description.

Sampson (1975), Jacobson (1977), and McCawley (1982) among others argue against strictly SMC-based theories on empirical and conceptual grounds: there are sentences whose analysis requires us to assume that a single constituent occupies more than one position. McCawley (1982), for instance, provides an account of *parenthetical insertion* and *Right Node Raising* (RNR) based on the assumption that a node could, under specific conditions, be dominated by more than one mother node. This allows structure to be shared between constituents without multiplying the nodes: instead of having rightwards movement leaving a trace or a copy (thus creating a new node) or deletion under identity (in which case we start with corresponding nodes in distinct constituents and delete one of them), a McCawlean SMC-free constituent structure for a sentence such as *Tom may be, and everyone is sure that Mary is, a genius* would be (32), where *a genius* is an immediate constituent of two different VPs (see McCawley, 1982: 99; also Levine, 1985):

(32)

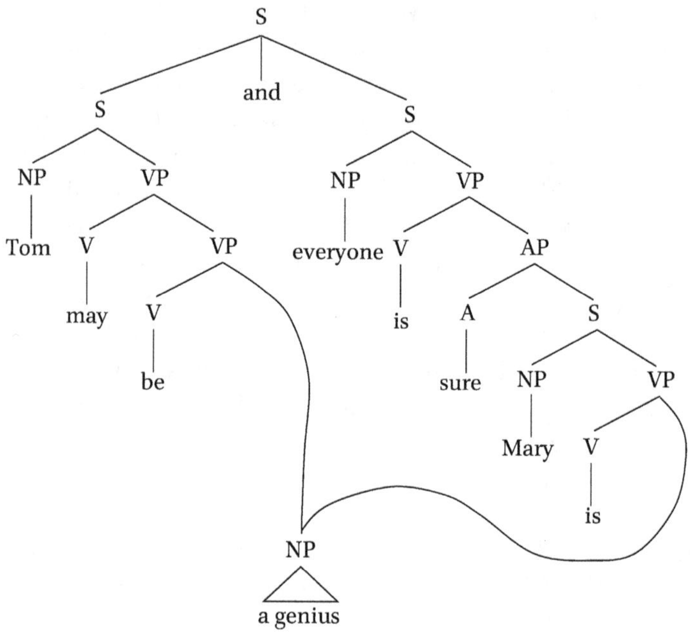

FIGURE 1.17 Analysis of Right Node Raising with multidominance

The point is not merely notational: in (32) there is a single NP node corresponding to the expression *a genius*, which means that a RN raised NP behaves, for all intents and purposes, as if it was in both VP complement positions (Levine, 1985: 496; Wexler & Culicover, 1980: 301). A crucial issue that we will address in this work is to determine under which conditions a single (basic or derived) expression can belong to different local structures (in (32), a single NP is dominated by nodes in distinct sentential structures): this will be the topic of **Section 2.3**. We will come back to rightwards 'extractions' (including RNR and Heavy NP Shift) in **Chapter 14**. For now, it is important to note that grammatical phenomena like RNR and unbounded Across the Board phenomena has been at the core of research in multidominance (Goodall, 1987; Moltmann, 1992, 2017; Johnson, 2016, 2020; de Vries, 2009a; Citko, 2005; see Citko & Gračanin-Yuksek, 2021: Chapter 1 for an overview); however, we will argue that multidominance has a much wider applicability (to anticipate some of the constructions we will analyse, multidominance will feature prominently in our analysis of Binding phenomena, in particular the distribution of reflexives and pronominal expressions, Equi, gapping, and Raising). In this sense, the aforementioned authors do not go as far as explicitly proposing a theory of the kind (16a). The expansion of phrase structure grammars in Peters and Ritchie (1981), called 'Phrase Linking Grammar', also allows for multidominance *locally*, and is thus closer to (16a) than MGG and LFG (see Gärtner, 2002, 2014 for discussion). These works are very important in that they constitute the foundations for our own view (as graph-theoretically oriented theories of grammar), and we will attempt to capture many of their insights in our framework. Other important grammatical frameworks, like Arc Pair Grammar (Johnson & Postal, 1980) and Metagraph Grammar (Postal, 2010: 20) also reject the SMC; these two are particularly important because they constitute formally explicit models with a strong basis on graph theory which are also declarative (see, e.g., Postal, 2010: 4). The approach presented in this monograph was heavily influenced by the theoretical challenges and empirical observations in these works. We share some of their arguments and conclusions, but not all; importantly, our proposal is not equivalent to theirs.

As we have already observed, since at least LSLT (Chomsky, 1955a, b), generative grammar has incorporated a *transformational component* which mediates between kernel sequences (which are derived only by means of phrase structure rules) and morpho-phonological interpretation. The transformational component was incorporated as a way to deal with certain inadequacies of phrase structure grammars, including processes that make reference to relations between separate well-formed sequences (as in the case of conjunction; see Chomsky, 1957: 36, ff.) and the attachment of inflectional affixes to verbal stems, among many others. Transformations operate in different ways and are subjected to different kinds of constraints depending on what they do: given a sequence, they can reorder elements, add elements, delete elements, etc. (see **Section 13.2** for further discussion). We must bear in mind, when considering transformations, that they were formulated within a framework in which *strings*, rather than *graphs*, played a fundamental role due to the background that many early generative grammarians had on *automata theory* and *formal language theory*, but not on *graph theory* (this point is made explicitly in McCawley, 1982: 92; we will deal with exceptions like Zwicky & Isard, 1963 and Morin & O'Malley, 1969 in some detail later). Here, as will become obvious below, we will take *transformations* to have an exclusively *descriptive* value. In this sense, we propose that *transformations*, i.e., mappings from *structural descriptions* to *structural changes* using traditional terms, can be divided in two types:

(33) a. Transformations that introduce new vertices and new edges
b. Transformations that introduce new edges between already existing vertices

And that this division corresponds to McCawley's, in the sense that transformations of the type (33a) also change *grammatical relations* ('relation-changing transformations' in McCawley's terms, RCT henceforth); whereas transformations of type (33b) do not (which we will refer to as 'relation-preserving transformations', RPT henceforth, an obviously derivative term). McCawley (1982: 94) argues that consideration of empirical phenomena provides grounds for distinguishing

> [...] *two essentially different types of transformation that hitherto have been classed together under the single name of movement transformations:* ***transformations that change syntactic relations*** *(not only "grammatical relations" such as "subject of" and "object of", but also relations such as what Pullum (1980)* [Syntactic Relations and Linguistic Universals. *Transactions of the Philological Society*, 78(1). 1–39] *calls "query of", which an item can manifest through its occurrence in some position of syntactic focus), and* ***transformations whose sole syntactic function is to change constituent order.*** (our highlighting)

The present investigation stems from the idea that, in providing empirically adequate accounts for grammatical phenomena in English, these two kinds of processes need to be incorporated to the core of the theory of grammar and appropriately distinguished in formal terms.

In this work, we are not going to be concerned with the linearisation of graphs,[16] because our goal is to provide exhaustive descriptions of expressions and syntactic relations in structural descriptions (thus it would be misleading to use McCawley's term *order-changing transformations* for transformations of the type (33b) since we are not concerned with representing *precedence* in our graphs). Like Tesnière's (1959) dependency *stemmas*, our graphs are intended to represent hierarchical relations only, and while we will introduce a notion of

16 Strictly speaking, the translation of hierarchy into linear order can be taken care of by means of independent linear order statements (for instance, McCawley, 1968 introduces two independent basic relations, ρ 'dominates' and λ 'is to the left of'), which would not be any more or any less stipulative than Kayne's (1994) LCA or other linearisation statements (e.g., Kural's 2005 *tree-traversal* approach; see also Kremers, 2009). We will see some examples of such statements in the discussion of Dependency Grammars, below.

strict ordering as a condition over well-formed graphs, that is not to be understood in *left-to-right* linear terms. By 'constituency'—above—we simply mean 'segmentation of a string and determination of grammatical relations' (but see **Chapter 5** for discussion about the role of grammatical relations in the present model): if grammatical relations are indeed defined in a pre-transformational structure, as in Chomsky (1965: 71), Lees & Klima's binary relation *be in construction with* can be mapped to *be* GF *of*, where GF is a grammatical function or relation ('subject of', 'direct object of', 'adjunct to', etc.). As has been pointed out from a variety of perspectives (perhaps mainly RG and LFG), defining grammatical functions with respect to configurations at the level of constituent structure creates problems for so-called *non configurational languages*: if constituent structure is universal, then so are the configurations where the assignment of grammatical functions takes place. If constituent structure varies across languages, however, the mapping between constituent structure and grammatical functions becomes much harder to establish: it then becomes necessary to assume an underlying, universal level of representation that represents constituent structure and grammatical functions assignment. In transformational generative grammar, this level was Deep Structure, the result of applying phrase structure rules (and some obligatory generalised transformations, such as *substitution*). However, there is no need to resort to base structures (which need to be reconstructed when linearity is disrupted by reordering transformations) if grammatical relations are in fact *primitives* and not derived from a structural template.

Let us clarify this point further: MGG has the X-bar format (or some version of that, more or less representationally oriented depending on the author and the stage in the development of the theory) as a primitive given by UG, and grammatical relations are defined over structural templates generated by X-bar or whatever specific form of PSGs is assumed. In this context, we find definitions of grammatical functions as by-products of phrase structure configuration, like the following:

> *It is necessary only to make explicit the relational character of these notions* [grammatical functions] *by defining "Subject-of," for English, as the relation holding between the NP of a sentence of the form NP⁀Aux⁀VP and the whole sentence, "Object-of" as the relation between the NP of a VP of the form V⁀NP and the whole VP, etc.* (Chomsky, 1965: 69)

In contrast, frameworks like Relational Grammar (and its descendants Arc Pair Grammar and Metagraph Grammar; see Perlmutter & Postal, 1983a, b; Johnson & Postal, 1980; and Postal, 2004a, 2010 respectively) and Lexical Functional

Grammar (Kaplan & Bresnan, 1982; Bresnan et al., 2016) take grammatical relations as primitives, and eliminates derivations as understood in [Σ, F] grammars from the theory (and with them, the need for triggers for derivational operations; in Minimalist terns, that means no formal features, no Agree, no Merge, etc.). Relational Grammar and related formalisms do, however, allow for distinct *strata* in its representations where elements can get *promoted* or *demoted* thus changing their grammatical function: an object -a 2 in RG terms—can become a 1 -a subject—in passive or unaccusative constructions; see e.g. Perlmutter (1978). Promotion/demotion is an RG mechanism we will not make use of: our graphs will represent all relevant syntactic relations in a single object with a single 'stratum' (no Deep/Surface structures, no Initial/Final arcs). In the same vein, we will not invoke parallel modules in the grammar and mapping functions to represent syntactic information (as LFG does, such that both *f*- and *c*-structures capture syntactic relations): syntactic information is 'centralised'. There is still the question of how to link grammatical functions with lexical items in specific syntactic configurations, and that is an essential question that we will return to below. At this point, we need to provide some basic formal definitions, as a way to make our toolset explicit, before getting into the details of grammatical analysis. That is the focus of the next chapter.

CHAPTER 2

Fundamentals of Graph-Theoretic Syntax

We anticipated that, in this work, a *grammar* is a finite set of constraints each of which specifies an aspect of what a structural description for an expression of the language looks like, including conditions on basic and derived structures. The formalism within which our theory is expressed is graph theory (as opposed to, say, set theory): these constraints are formulated over expressions and relations in graphs. Unlike phrase structure rules or derivational operations, constraints are statements that may be true or false for any given structure (such as 'every well-formed tree has a root' or 'if x dominates y in a well-formed tree, then $x \neq y$'): if they are false for a given piece of structure, that structure is ill-formed. A graph of arbitrary complexity (which formalises the syntactic relations between expressions in a sentence) will be well-formed iff the logical conjunction of all declarative constraints that apply to it is true (McCawley, 1968; Gazdar et al., 1985). The grammar, then, looks like a collection of conditions over local structure in terms of legitimate relations between nodes corresponding to expressions in well-formed graphs. Borrowing the prefix from Postal (2010), we can call our graph-theoretic characterisations of expressions and relations *L-graphs*.

2.1 Defining (L-)Graphs

Our focus on constraints over elementary and derived structures entails that the grammar is not procedural, in the sense that there are no derivations, no sequential rule application for expansion or composition (**Section 1.3**). This does not mean, of course, that there are only elementary structures: a constraint-based model may also characterise complex structures formed by several identified elementary structures, just like a proof that a derived expression belongs to a language in pure Categorial Grammar involves the application of syntactic rules of the kind *'if … then …'* to (two or more) basic or derived expressions (e.g., Montague, 1973; Schmerling, 1983a: 4). The analysis of natural language sentences requires the identification of basic and derived expressions and relations between these, and the formulation of admissibility constraints. When the structural analysis of a sentence identifies more than one basic, or 'elementary', structure, well-formedness is still checked locally (McCawley, 1968; Oehrle, 2000). The hallmark of declarative syntax is that the

 | DOI:10.1163/9789004542310_003

formation of complex structures is not modelled in terms of input-output pairs, mapping (sets of) structures to structures. Rather, we may formulate local constraints, or define local structures, and identify in the structural description of a natural language sentence a combination of elementary structures to which these local constraints apply. In this context, the conditions that the structural descriptions of expressions must satisfy need to be made explicit. This chapter presents some basic definitions that will constitute the foundations of our formalism, complemented by conditions over possible dependencies within graphs to be explored in detail in the subsequent chapters:[1]

1. A graph G is a set G = (V, E), where V is a set of vertices (or 'nodes', a more common term in syntax) and E is a set of edges: $v \in V$ is a vertex (or 'node': we will use *vertex* and *node* interchangeably), and $e \in E$ is an edge. There is also an *incidence function i* that associates an edge to pairs of (possibly distinct) nodes. An edge *e* joining vertices *a* and *b* is notated $e = (a, b)$ or $e(a, b)$ (and represented as a line joining *a* and *b*), and *a* and *b* are said to be *adjacent.*
2. We distinguish vertices by using numerical subscripts: when talking about the vertices of graph G, we notate them as $v_1, v_2, \ldots, v_n$. The subscripts *i, j, k*, ... as in $v_i, v_j, v_k, \ldots$ will also sometimes be used to denote arbitrary distinct vertices. Throughout the book, we will use 'vertices' and 'nodes' indistinctly.
3. Edges specify the *directionality* of the connection: $e\langle v_1, v_2\rangle$ where $\langle v_1, v_2\rangle$ is an ordered pair of vertices. Thus, $e\langle v_1, v_2\rangle \neq e\langle v_2, v_1\rangle$ (since there is an order over the pair v_1 and v_2), and the edge is sometimes called an *arc* (in plain English, arcs are one-way roads). In $e\langle v_1, v_2\rangle$, we call v_1 the *head* of the arc and v_2 its *tail.* In diagrams, arcs are represented by arrows. Going back to **Definition 1**, if the incidence function i: $E \to V \times V$ defined on G maps edges to *ordered pairs* of nodes, the graph G is called a *digraph* (short for *directed graph*).

For purposes of linguistic description and analysis, the choice of directed graphs is motivated by the need to capture *predication.* As we will see in detail in **Section 2.2**, there is always a directed edge from a predicate to each of its arguments (in contrast to MGG, RG, and APG, but as in Dependency Grammars): this will help us formulate a compositional theory of semantic interpretation for graph theoretic structural descriptions.

---

1 Introductions to graph theory in which some of the concepts we use here are defined in more technical detail are van Steen (2010), Wilson (1996), Ore (1990), and Gould (1988). Zwicky & Isard (1963) and Huck (1984: Appendix) are linguistically oriented presentations of basic graph-theoretic notions.

4. The *neighbour set* of v_i is the set of adjacent vertices to v_i, usually notated $N(v_i)$ (where N is the number of adjacent vertices to v_i), and the *degree* of v_i is the number of *edges* connected to it.

For example, a vertex v with degree 2 and neighbourhood set $2(v)$ has two edges connected to it, and two vertices which are adjacent. There is no necessary correspondence between *degree* and *neighbourhood set*, since a vertex v_1 may be connected to v_2 by two distinct edges (in which case the neighbourhood set of v_1 would be 1 and its degree, 2) or, under certain conditions, there may be a loop from v_1 to itself. We can exemplify these notions in the following graph:

(34)

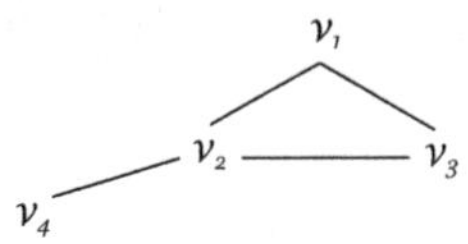

FIGURE 2.1 Sample graph with a closed walk

We have identified nodes with capital letters: the set $V = \{v_1, v_2, v_3, v_4\}$. v_1 and v_2, v_1 and v_3, v_2 and v_3, and v_2 and v_4 are *adjacent vertices*. v_1 and v_3 have a *neighbourhood set* with cardinality 2, v_2's neighbourhood set has cardinality 3, and v_4's has cardinality 1. The *degree* of v_1 and v_3 is 2, the degree of v_2 is 3, and the degree of v_4 is 1. In a directed graph it is necessary to also define the notions of *indegree* and *outdegree*: the *outdegree* of a node is the number of edges that 'leave' that node (i.e., the number of edges that go *from* that node *to* another). The *indegree* of a node is the number of edges that 'enter' that node. If we make (32) into a digraph,

(35)

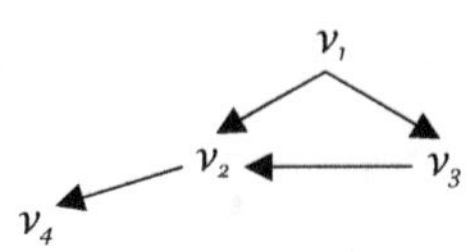

FIGURE 2.2 Sample digraph

we see that v_1 has *indegree* 0 and *outdegree* 2, v_2 has *indegree* 2 and *outdegree* 1, v_3 has *indegree* 1 and *outdegree* 1, and v_4 has *indegree* 1 and *outdegree* 0.

5. Let ρ be a binary asymmetric relation 'immediately dominates': $\rho(v_1, v_2)$ means 'v_1 immediately dominates v_2', where v_1 dominates v_2 iff there is an arc with head v_1 and tail v_2. In other words, v_1 *immediately dominates* v_2 in $e\langle v_1, v_2\rangle$. We will frequently abbreviate $\rho(v_1, v_2)$ as $\langle v_1, v_2\rangle$. We will use ρ^* to denote the transitive closure of 'immediately dominates' (call it *dominate**).

The reason why we restrict ρ and ρ* to be *binary* relations is strictly because they are defined in terms of the existence of an edge between nodes, and edges, by definition, link *two* nodes. This follows from the definition of *edge*, and thus from the choice of graph theory as the framework for our inquiry. Dominance is a *transitive, antisymmetric* relation between nodes. Importantly, *dominance* is not a *total* relation: this will be motivated by an analysis of the dependencies between expressions in structural descriptions for natural language sentences. Importantly, *immediate dominance* is a primitive relation in this work, not one defined in terms of other structural relations; this is because immediate dominance is the most basic relation we can define with two vertices and a directed edge (see Gärtner, 2002: 121–122 for discussion).

6. The *ρ-domain* of a node v is the finite set $\rho = \{v_1, v_2, ..., v_n\}$ that v dominates either directly or transitively.

Considering **Definitions 5** and **6**, we will speak of the *context* of a node v_n to refer to the (finite) set of nodes that v_n is immediately connected to (its *neighbourhood*): the union of the set of nodes that dominate v_n and the set of nodes which v_n dominates.

7. Let τ be an *n*-ary relation 'sister of'. τ is used as an abbreviation for:

(36) $\tau(v_1, v_2, ... v_n)$ *iff* $\exists(v_i) \in G, v_i\rho(v_1, v_2, ... v_n)$

Here we do not require that sisters do *not* dominate each other, which seems counter-intuitive from the perspective of tree-based syntax. Thus, for (37) below (where directed edges have been marked with arrows strictly for illustrative purposes)

(37)

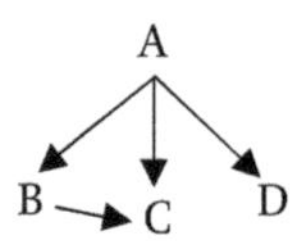

FIGURE 2.3 Sisterhood in a digraph

we have the (unordered) set of dominance relations in (37'):

(37') $\rho = \{\langle A, B\rangle, \langle A, C\rangle, \langle A, D\rangle, \langle B, C\rangle\}$

And, since ρ(A, B), ρ(A, C), and ρ(A, D) hold, we say that B, C, and D are sisters because they share the mother node A, regardless of the fact that ⟨B, C⟩ ∈ ρ. In a slightly different case, illustrated in (38),

(38)

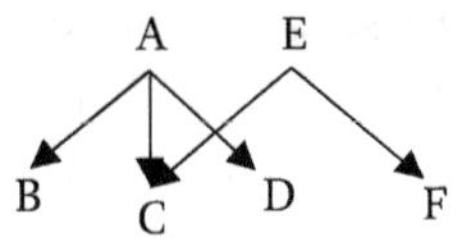

FIGURE 2.4 Multi-rooted digraph

nodes B, C, and D are sisters by virtue of sharing the mother A, and C and F are sisters by virtue of sharing the mother E. We might notate this using a subscript on τ, such that τ_A(B, C, D) is the set of nodes that are sisters because they share the mother node A.

8. The set of all dominance relations in a graph G is called the *ρ-set* of G. Then, we say that (37') is the *ρ-set* of (37). We will see in **Chapter 5** that there are linguistic reasons to revise the format of ρ-sets with respect to (37'): instead of *unordered* sets of dominance relations, we will provide empirical arguments to adopt a definition of ρ-sets as *ordered sets* of dominance relations.

9. $\exists(v_i)\ \forall(v_x),\ v_i, v_x \in G \rightarrow v_0\rho^*v_x$ (see also Peters & Ritchie, 1981: 6; Gärtner, 2014: 2; Huck, 1984: 65, ff. and references therein). We call v_i the *root* of G.

Definition 9 simply states that a graph will have a(t least one) root: a node v_i in G that dominates all other nodes in G (such that for any arbitrary node v_x, v_i will transitively dominate v_x). Trees may only have a single root (if they are rooted at all). All binary trees are rooted.

Peters & Ritchie (1981: 6) provide a further condition pertaining to roots. Let N be the set of nodes in a graph and I a binary relation on N (immediate dominance). Then,

I^{-1} is a partial function defined just at members of N-{r}

Definition 9 and its consequences are especially important, and thus merit further discussion, to be expanded on throughout the monograph. This definition does *not* imply that there is always a *unique* root in a graph (cf. Zwicky & Isard, 1963: 1), only that there is always *at least one*. As a matter of fact, the graph in (34) above has two roots (A and E), that is, two nodes that dominate others but are not dominated themselves. (34) is an example of what Morin & O'Malley (1969: 182) call a *vine*: a directed, acyclic, labelled, possibly multi-rooted graph.[2]

2 We must note that Morin and O'Malley's work was primarily focused on the inadequacies of single-rooted trees headed by a single performative verb in an early Generative Semantics framework (that is, they argue against representations of the kind assumed by Ross, 1970b); their concern was more *semantic* than *syntactic* (in the contemporary use of these terms). The consequences that their formal framework has for the theory of syntax are as far-reaching

However, in the definition of local domains in linguistic structural descriptions we will encounter the need to impose restrictions over how big local graphs can be, what kinds of linguistic units they correspond to, and how they can be combined. For now, we will simply stipulate (but this stipulation will be derived from linguistic conditions shortly) that local structural domains are single-rooted graphs. In what follows we will use the term *arbor* (plural: *arbores*) to refer to *single-rooted, directed graphs*. This means that if multi-rooted graphs arise, they must do so by means of combining single-rooted local graphs: in (34) there would be two local *arbores*, one with root A and another with root E, whose combination yields a multi-rooted graph. Complex structures, including multi-rooted graphs, are obtained by putting *arbores* together in specific ways: in generative grammar, widely construed, operations that relate local structures are the familiar generalised transformations *conjoining*, *embedding*, and *adjunction* (Fillmore, 1963; Joshi, 1985). As much as a proper definition of the building blocks of syntax (single-rooted local graphs), the formulation of conditions over graph composition (i.e., the combination of local graphs) and the dependencies that are allowed for in derived structures ('derived graphs') is a main focus of this monograph.

If the *arbor* is the building block of syntax, it will be fundamental to establish exactly what it may contain, how to restrict its size, and how we can identify several building blocks in a complex structure. This is not a point where graph theory (the formalism) can deliver the answer: we need to consider properties of the linguistic expressions that are nodes in the graphs (grammatical theory). We will come back to this issue in **Section 2.2**.

Let us continue with the definitions.

10. Graphs are *ordered*: a (possibly unary) set of relations imposes an order over vertices.

We know that graphs are sets of nodes and edges. Saying that a graph is *ordered* means that there is one or more binary relation(s) over the elements of its set of nodes and edges. Recall that edges are directed: these impose an order over vertices in terms of the relation *dominates*. Orders may be *total* or *partial*: a set is *totally* ordered *iff* for every pair (x, y) that belongs to the set, there is a relation R such that either $R\langle x, y\rangle$ or $R\langle y, x\rangle$; otherwise it is *partially* ordered. The notion of *ordering* is essential to the formulation of a *traversal*: a way to 'visit'

as they are usually ignored (a point also made in Postal, 2010: 394). For example, Morin and O'Malley characterised structures that anticipate Parallel Merge (Citko, 2005; Citko & Gračanin-Yuksek, 2021) and explicitly defined Multidominant structures as generalisations over the notion of a tree (Morin & O'Malley, 1969: 182).

the vertices of a graph.[3] If we define a walk through the graph, this traversal must be an ordered, alternating sequence of nodes and edges. We said above (in **Section 1.4**) that the concept of *walk* is neutral with respect to whether vertices or edges can be visited more than once (i.e., whether a given vertex v can be ordered with respect to itself so that it transitively dominates itself): now we need to specify what kinds of relations we allow in our walks. As we briefly saw in **Chapter 1**, walks can be either *trails* or *paths*. The notion of *trail* is weaker than that of *path*: a *path* is a walk in which no *vertices* or *edges* are repeated, whereas a *trail* is a walk in which no *edges* are repeated, but *vertices* can be (Gould, 1988: 9; Wilson, 1996: 26; Van Steen, 2010: 37, 61). In Van Steen's words,

> *In a **closed walk**, $v_0 = v_k$, a **trail** is a walk in which all edges are distinct; a **path** is a trail in which also all vertices are distinct. [...] A **directed trail** is a directed walk in which all arcs are distinct; a **directed path** is a directed trail in which all vertices are also distinct.*

Put differently, a *trail* is a walk where no edge is repeated, whereas a *path* is a walk where no vertex is repeated. Linguistic considerations (in particular related to anaphoric and pronominal binding in **Chapters 6** and 7 and filler-gap dependencies in **Chapter 9**) will motivate a local approach to the definition of walks in graphs, relativising the *path-trail* distinction to single-rooted graphs. In this context, we can introduce another definition pertaining to *order*:

11. A vertex v_1 is *dominance-ordered* (*d-ordered*) with respect to v_2 *iff* either $\rho(v_1, v_2)$ or $\rho(v_2, v_1)$ or $\rho^*(v_1, v_2)$ or $\rho^*(v_2, v_1)$. If v_1 is *d*-ordered with respect to v_2, either v_1 is in the ρ-domain of v_2 or v_2 is in the ρ-domain of v_1.

We will see (in **Chapter 5**) that *d-order* does not exhaust the orders imposed over our graphs: in particular, a further source of order needs to be invoked to capture relations between arguments and the determination of grammatical

3 For trees, traversals take the form of three distinct recursive methods (see Kural, 2005; Kremers, 2009 for applications of tree traversal algorithms to the linearisation of tree structures; also Sarkar & Joshi, 1997: 614–615 define a traversal but for purposes of parsing, not linearisation. Medeiros, 2021 combines both). If we take a simple branching node as an example:

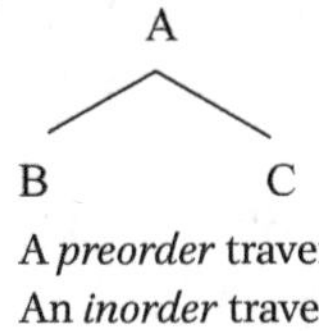

A *preorder* traversal defines the sequence: A B C (root, left, right)
An *inorder* traversal defines the sequence: B A C (left, root, right)
A *postorder* traversal defines the sequence B C A (left, right, root).

functions assigned to each argument (subject, object, oblique …). This contrasts with the situation in MGG, where non-terminal nodes allow for the definition of a total ordering between arguments based on c-command: subjects always c-command objects in the underlying Specifier-Head-Complement (SVO) universal structure proposed in the antisymmetric tradition (Kayne, 1994).

As an important ingredient of our grammatical theory we will also need a definition of *irreducible graph*:

> *A connected graph on three or more vertices is irreducible if it has no leaves, and if each vertex has a unique neighbor* [*sic*] *set. A connected graph on one or two vertices is also said to be irreducible, and a disconnected graph is irreducible if each of its connected components is irreducible* (Koyama et al., 2007: 35)

This does *not* imply, though, that irreducible graphs are *complete* (or *strongly connected*): a graph G is *complete* iff every vertex is adjacent to every other vertex (that is, if there is an edge between any and every pair of vertices, see Ore, 1990: 7; a related usage is that in Gould, 1998: 10, who uses the term *strongly connected* instead of *complete*). Irreducible graphs can, however, be cyclic without being complete. It is in this respect that we find a first major difference between tree-based syntax and the kind of formalism we are advancing here: we do not require for our structural descriptions to be acyclic (cf. e.g. Huck, 1984: 64), but acyclicity may emerge depending on the relations between expressions in a specific piece of data in a natural language. Our graphs may not always end up being *trees*.

12. A directed graph G is *weakly connected* if replacing directed edges with non-directed edges yields a *complete* graph. It is *semi-connected* if for any v_1, v_2, G contains either a walk from v_1 to v_2 or a walk from v_2 to v_1.

It should be apparent that deciding whether graphs corresponding to structural descriptions for natural language sentences are *connected*, *weakly connected*, or *semi-connected* has deep consequences for the kinds of dependencies that can be established between vertices in these graphs. The choice should not be made *a priori*, but rather should be motivated by the analysis of linguistic data. We will come back to this issue repeatedly throughout our argument.

The definitions we have introduced so far in this book use only very basic set-theoretic notation. For the phenomena analysed here, we will see that this notation is sufficient. Unlike some previous graph-theoretic or geometrical analyses of syntax (e.g., Kracht, 2001; Beim Graben & Gerth, 2012; Kural, 2005; McKinney-Bock & Vergnaud, 2014; Gärtner, 2002, 2014, to different extents) we do *not* attempt to encode GB/MP-type derivations or representations in graphs, because—as we have seen—we depart from most basic assumptions in that framework pertaining not only to the nature of syntactic structure (monotonicity, the SMC, movement, etc.), but also the architecture of the grammar (including properties like the algorithmic character of structure building, an autonomous generative syntax and interpretative semantic and morphophonological components, and the relations between these components proposed in the so-called *Y-model*; see e.g. Chomsky, 1981, 1995; Hauser et al., 2002; Siloussar, 2014). We will thus not define notions that are crucial for GB/MP approaches to syntax, like *trace* or *copy* (the former also assumed in Simpler Syntax and some versions of LFG).

The framework presented here is orthogonal to MGG in more ways than one. In the typology presented in Hockett (1954) and Schmerling (1983b), Generative Grammar is an Item-and-Arrangement (IA) framework, in which constituent structure is built by means of discrete recursive combinatorics: an IA theory specifies atomic pieces and ways (procedures) to put those pieces together, for example, step-wise binary Merge in the work derived from Chomsky (1995). The label IA includes all forms of immediate constituent analysis derived from the post-Bloomfieldian tradition. Moreover, MP-style formalisms follow two additional conditions: the *No-Tampering Condition* NTC (operations over syntactic objects X and Y leave X and Y unchanged; Chomsky, 2007: 8) and the *Extension Condition* EC (which prevents counter-cyclic operations; Chomsky, 1995: 190; Kitahara, 1997).

IA frameworks, of the kind just sketched, contrast with Item and Process (IP) grammars. Schmerling defines *language* in IP terms as follows:

> *A language L is a system consisting of the following:*
>
> 1) *an algebra consisting in a non-empty set A of expressions and a (possibly empty) indexed set of operations defined over A; A is the smallest set containing as members all the basic expressions of L and closed under the operations,*
> 2) *a Lexicon, or set of basic expressions indexed by a set of category indices* [for example, the set of integers], *and*
> 3) *a set of syntactic rules that recursively assign any derived expressions of L to indexed syntactic categories. An n-place rule is a triple whose first member is the index of an n-place operation, whose second member is an n-place sequence of category indices (those of the inputs to the rule), and whose third member is the index of the output category*

> *of the rule.* (2) *and* (3) *constitute a recursive definition of the subset of A that is syntactically well formed* (Schmerling, 1983a: 395)

As Schmerling emphasises, '*it is up to an empirical science of linguistics to fill in this framework with a theory of natural language expressions and a theory of natural language categories*' (1983b: 4). Spelling out such an empirical theory of natural language syntax is our task for the remainder of this monograph. Importantly, the formalism in which the theory is expressed does not in and of itself dictate the content of this theory. For instance, graphs can be used to represent diagrams corresponding to structural descriptions in IA or IP: Gärtner (2002, 2014), while adopting aspects of the graph-theoretic Phrase Linking Grammar, remains committed to immediate constituency and derivations. Huck (1984), in contrast, departs from derivational assumptions and adopts a Categorial Grammar perspective.

It will become clear in the chapters devoted to grammatical analysis that our focus on the analysis of specific phenomena in two languages (English and Spanish), the centrality of syntax as an abstract system of expressions and relations, and the interplay that we assume between syntax and semantics, separate the quantitatively-inspired network-approaches from our more narrowly grammatical inquiry (see also the **Appendix**). This means that we are not concerned in this monograph with typological or psychological aspects of language (acquisition or use). Our methodology obeys the requirements of explicit grammatical analysis; a graph in the present theory is *not* designed to be a representation of what humans 'do' (or even what a computarised 'parser' would do) when interpreting linguistic stimuli, however *interpreting* is defined. The grammar, in the present approach, is not a theory of competence or knowledge, it is an explicit characterisation of objects and relations in the grammar of natural language (see also Postal, 2010: 3). Using Haspelmath's (2021) terms, this monograph is primarily a study of *p(articular)-linguistics*: a theoretically informed language-particular description, which he advances is no 'less theoretical' than so-called *general linguistics* (*g-linguistics*). We agree with Haspelmath. This is not to say that there are no relations with the other uses of graphs mentioned above; only that at this point in the development of the theory, it is too soon to tell.

We have hailed the fact that our graph-theoretic grammar allows for a restricted range of multidominance and discontinuity as a major advantage of our graph-theoretic model over (transformational or not) immediate constituent-based PSGs when it comes to empirical analysis (a promise we will do our best to deliver on in **Chapters 5** to **14**). Because we allow for multidominance, discontinuity, multi-rooted representations, *and* indirect loops, both the NTC and the EC are violated. In this sense, our theory would be classified as [-**Ext**] [+**Fin**] in the typology of Frank & Hunter (2021): operations of graph composition do not necessarily apply at the root (thus, the EC does not apply, [-**Ext**]), and syntactic dependencies are locally-bounded as a consequence of lexicalising the grammar (the probing space for any process is finite, [+**Fin**]). At this point, we need to stress that *discontinuity* and *multidominance* must be carefully distinguished and kept apart. Discontinuity can be captured without multidominance (e.g., movement accounts of Right Node Raising, as in Sabbagh, 2007; or the CG treatment of Wrap, as in Bach, 1979, 1980 and applied to non-constituent coordination by Dowty, 1997). In the same way, multidominance structures do not necessarily allow for discontinuity: it is possible to have multidominance theories implemented in IC grammars (e.g., TAGs with links, as in Joshi, 1985; see also the multidominance expansion of GB/MP grammars in Kracht, 2008 or of Minimalist grammars in Gärtner, 2002; Johnson, 2016, 2020; Citko & Gračanin-Yuksek, 2021). In this work, we allow for *both* discontinuity and multidominance and abandon a derivational approach to syntax; this separates the proposal made here from MGG-based graph-theoretic works (e.g., McKinney-Bock, 2013; McKinney-Bock & Vergnaud, 2014; Krivochen, 2023a, b),[4] and puts us closer to APG and Metagraph Grammar (Johnson & Postal, 1980; and Postal, 2004a: 58, ff., 2010 respectively).

Furthermore, it is important to observe that multidominance does not necessarily generate *cyclic graphs*: a multidominated node can have mothers in different sub-graphs (each mother can be the root of said sub-graph, as in our example (38) above, see also (39b) below), in which case the graph is *directed* and *acyclic* (an example of this is the structures generated by Citko's 2005 *Parallel Merge*, if her diagrams are interpreted graph-theoretically[5]). A *directed cyclic* graph can be *rooted* if we have a designated node which is not dominated by any other node (with indegree 0). If there was no encoding of *directionality* in edges, we would have to give up either roots or cyclic graphs (and the requirement of there being a *total order* imposed over graphs). Thus, in the context

4 The present work also differs from the more mathematically oriented works in dealing more with concrete linguistic examples rather than with proving theorems about the structures generated by the system. For the linguist, that might be a relief. For the mathematician, a nuisance. The reader is encouraged, for instance, to compare the presentation here with the rigorous axiomatisation of Copy-Chain and Multidominance structures in Kracht (2001, 2008) or Gärtner (2002).

5 This interpretation is not necessarily intended by Citko, however. As a matter of fact, Citko & Gračanin-Yuksek (2021: Chapter 2) explicitly assume set-theoretic Merge. See also Johnson (2020) for a set-theoretic Minimalist approach to multidominance.

of an application of graph theory to the analysis of natural language grammar, *discontinuity*, *multidominance*, and *cyclicity* must be carefully kept apart.

2.2 Syntactic Composition and Semantic Interpretation

We have defined some basic properties of graphs and walks through them, now we can turn to characterising the nodes that make them up. This is important for two reasons: first, because it will provide us with a way to work rules of semantic interpretation into our graphs, and second (anticipating a point to be made and developed in detail in **Chapter 3**) because the 'size' of local units for purposes of semantic interpretation and the establishment of syntactic dependencies will be determined by the properties of nodes that correspond to *lexical predicates*.[6] We will need to look at our definition of local syntactic domain in more detail, since in **Section 2.1** it was purely configurational (the definition of *arbor* is based only on single-rootedness); in fact, there is a crucial division of labour between configuration and lexicalisation in our approach. For the time being, we will restrict ourselves to the issues of configuration and formal conditions over allowed graphs, leaving examples and analyses for **Chapter 3** onwards. This mode of presentation will allow us to have a relatively 'asceptic' perspective in our definitions, which should make the task of providing a more rigorous formalisation (or finding formal weaknesses) easier in future work. Once we have explained the basic rules, we can start playing.

Definitions 1–12 in **Section 2.1** above apply to local structures as well as complex structures generated by combining local structures: they are declarative constraints that determine what is a well-formed graph. The distinction between irreducible local (or 'basic'), structures and complex, (or 'derived'), structures is fundamental for a theory of grammar: adopting terminology from Tree Adjoining Grammars (TAG) to our current purposes, we may refer to the former as *elementary graphs* and to the latter as *derived graphs*. Note that, considering only the information provided so far, *elementary graphs* are to all intents and purposes equivalent to *arbores*: we will soon refine these two notions so that they are not exactly co-extensive. However, the issue is more general: if the building blocks of grammar are elementary single-rooted graphs,

6 In this sense, as we will emphasise throughout the monograph, our theory of grammar is *lexicalised*; see Joshi & Schabes (1991); Frank (1992, 2002, 2013); XTAG group (2001), and much related work.

the theory requires a way to put these blocks together and form more complex structures. In declarative terms, we need to be able to identify structures that are composed of more than a single elementary building block, if these are to be assigned as structural descriptions for natural language sentences. Before justifying some linguistic requirements on both elementary and derived graphs, we will introduce some formal considerations pertaining to the combination of elementary units: what it is and how it works.

In graph theory, in principle, there is a variety of operations that give us something we could informally call 'structure composition': we can *join* graphs, define their *union*, their *intersection*, and several kinds of *products* (Cartesian product, tensor product, etc.; see, e.g., Bondy & Murty, 2008: §1.4; Gould, 1988: 11–13). Each kind of formal operation delivers different outputs. In this context, what does graph composition look like in our framework? Let us go back to **Definition 9**, which established that every L-graph has at least a root: arbores are single-rooted, whereas complex graphs, which contain more than one arbor, may be multi-rooted. In this context, what is the role of root nodes in defining what the structure of a well-formed complex graph is? In our analysis of graphs that can be decomposed into smaller units (e.g., in the analysis of multi-clausal structures), we will refer to operations of 'graph composition': a complex structure can be broken down into smaller component pieces, each of which must be itself a well-formed graph. Our analysis of graph composition differs from a number of approaches to structure building in that it weakens a condition that has been around in syntactic theory for some decades now (see, e.g., Chomsky, 1955a, b, 1995; Rogers, 2003): the idea that dependencies between clauses occur only *at the root*. In Frank & Hunter's (2021) terms, adherence to this idea defines a theory as [+**Ext**]: they follow some version of Chomsky's (1995: 190) *Extension Condition* (for context, TAGs are [-**Ext**], since structure can grow at the non-root by means of *adjunction*, which we will define shortly).

Departure from [+**Ext**] means that if we link graphs G and G' by means of a common node which we will refer to as a *linking* node, it is *not* necessary that the linking node between G and G' be the root of either graph; all we require is that the relevant nodes in G and G' be *identical*. We will explain what 'identity' entails shortly, but first observe that this approach is distinct from both the definition of *substitution* as a generalised transformation and *adjunction* in the TAG sense. To illustrate this difference, we can define these operations as follows (we will come back to them in detail in the context of linguistic analysis in **Section 4.2**):

> *Substitution*: let T be a tree with root S and T' a tree with a node labelled S in its frontier. *Substitution* inserts T into the frontier of T' at S.

(39)
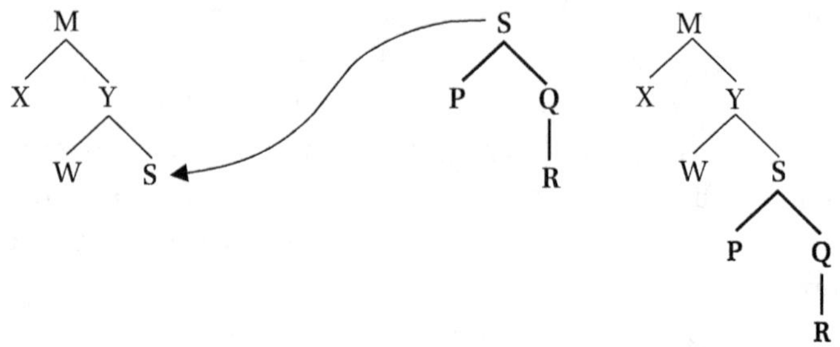

FIGURE 2.5 Tree composition via Substitution

Adjunction: in this case, a piece of structure, called an *auxiliary tree*, is inserted into a designated node (the adjoining site) in an *initial tree*. Informally, the operation *Adjoin* essentially 'rewrites' a node as a tree by 'shoehorning'[7] a piece of structure (i.e., the auxiliary tree) inside another structure (i.e., the initial tree) at a designated intermediate node. The *auxiliary tree* must have two distinguished nodes at its frontier, namely the *root node* and *the foot node* which must carry the same label, and which must be identical to the label of the node that is the target of adjunction in the *initial tree*.

(40)
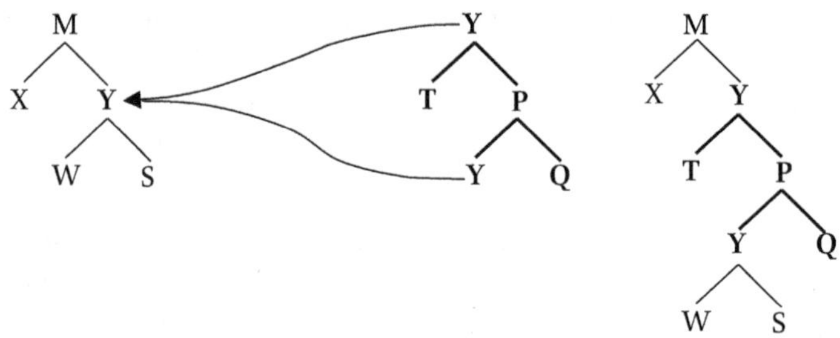

FIGURE 2.6 Tree composition via Adjunction

The notion of *root*—an undominated node—features prominently in the definitions of both substitution and adjunction, and this limits the kinds of relations that can be established between local structures. Bear in mind that whereas the target of adjunction in an *initial tree* may be a non-root, the relevant nodes to pay attention to in the *auxiliary tree* are the root and a node in its frontier, which must bear the same label. We can now turn our attention to the consequences of adopting a different approach from [+**Ext**] theories (an even

7 The term is due to Andrea Padovan (p.c.).

more radical departure from [+**Ext**] than TAGs), which does not require the relations between arbores to be established only at the root (or at the frontier) by definition.

How can we accomplish this? There are several ways, some of which have already been fruitfully explored in the literature. One way, which we will explore in this work, is to assign unique identifiers to nodes, identifiers that can be called regardless of syntactic contexts: a node's unique identifier contains no information about that node's neighbourhood. There are antecedents of such a view, on which we build. Some versions of Tree Adjoining Grammar (e.g., Sarkar & Joshi, 1997; Joshi & Schabes, 1997; see also Han et al., 2010, who employ node contraction but do not explicitly index nodes with addresses), the Multidominance approach in Gärtner (2002, in particular § 3.3.3, 2014; also Peters & Ritchie, 1981), and the phrase structure system in Karttunen & Kay (1985)—to give just a few examples—assume that nodes in an elementary tree are assigned *addresses* (Sarkar & Joshi, 1997 cite Gorn, 1967, whose addressing system is based on assigning unique node labels to nodes in tree data structures as a function of their position), which serve to unambiguously identify corresponding (identical) expressions regardless of immediate dominance relations (what we will refer to as the *context* of a node). Following the notational convention introduced in Krivochen & Padovan (2021), we may use ⦃E⦄ to denote the address of any expression E: a node v_n in a graph G corresponds to a basic expression of the language, and is assigned a unique address that allows us to refer to that node unambiguously. In the context of a theory of natural language grammar, it makes sense to make the set of addresses available in a language L part of the Lexicon of L, together with semantic and phonological information about each expression. The addressing system, as we will emphasise below, is in principle independent from what the content of those addresses is hypothesised to be: it is therefore possible to accept our indexing system and its application to structure sharing but disagree on what addresses point to.

The use of an addressing system is crucial for the framework explored in this monograph because it allows us to formulate the following condition over derived graphs: if nodes v_1 and v_2 on distinct arbores T and T' (such that $v_1 \in$ T and $v_2 \in$ T') are assigned the same uniquely identifying address, then a derived arbor T" that contains T and T' will collapse those nodes into one, call it v_3: $v_3 \in$ T", iff $v_1 = v_2 = v_3$. The composition of arbores is, formally, *graph union*, which delivers what is known in the literature as *structure sharing*. We can illustrate the results of graph union, applying at the root (41a) or at non-root nodes (41b):

(41) a.

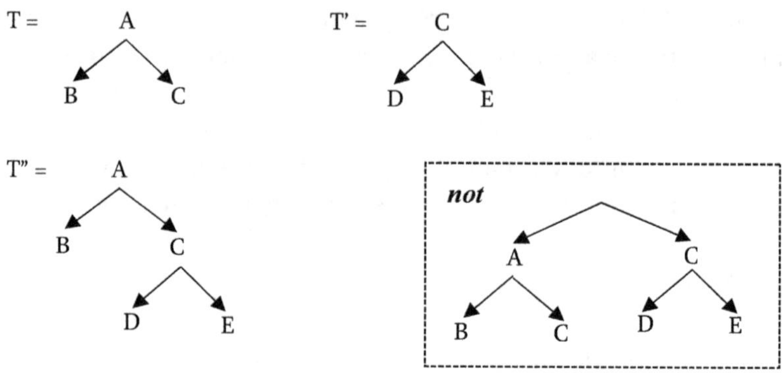

FIGURE 2.7 Single-rooted derived graph

b.

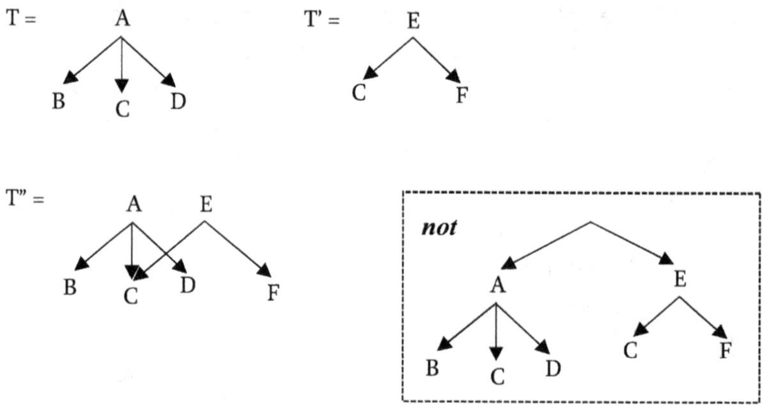

FIGURE 2.8 Multi-rooted derived graph

The number of common addresses between two or more graphs needs not be limited to one: graphs may be linked at any number of common nodes:

c.

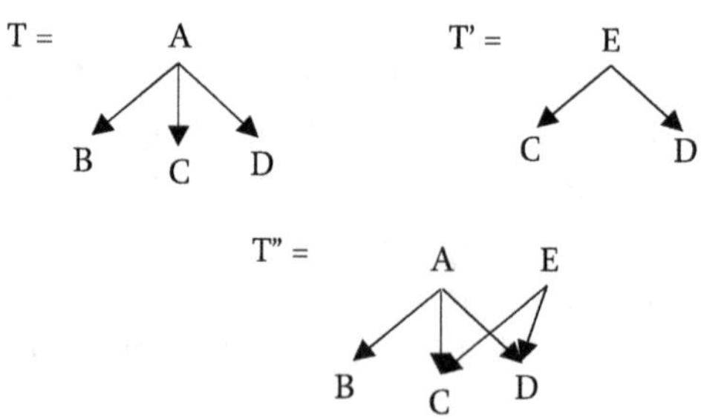

FIGURE 2.9 Multi-rooted derived graph with two shared nodes

Attention must be paid to the fact that in all cases each single-rooted graph is strictly partially ordered: the relation *dominates* is a partial order over a set of nodes in a digraph (since there are pairs that are not ordered by the relation, such as (B, C) in (c)) which is irreflexive, transitive, and antisymmetric. This will become important when we discuss the properties of elementary and derived graphs in terms of whether they are totally or partially ordered (in particular, see **Chapters** 7 and 8). This is an issue independent of graph composition; other than identity between addresses, conditions on graph composition are to be dictated by empirical necessity (which includes a specification of the properties of linguistic expressions that are relevant for the definition of elementary graphs and their composition), not by the graph theoretic formalism.

We can define an order of traversal in graphs such as those illustrated in (41), because they are trees. Tree traversal algorithms apply for binary branching graphs with no loops; in our case we need something more general. Somewhat simplifying matters, given a tree we can define a sequence, starting from the root, that visits all nodes in the graph. For (41a), (41b) and (41c) we can define the following traversal sequences Σ (assuming a *preorder* traversal; see fn. 3 above):

(41') a. $\Sigma_{T} = \langle A, B, C \rangle$
$\Sigma_{T'} = \langle C, D, E \rangle$
b. $\Sigma_{T} = \langle A, B, C, D \rangle$
$\Sigma_{T'} = \langle E, C, F \rangle$
c. $\Sigma_{T} = \langle A, B, C, D \rangle$
$\Sigma_{T'} = \langle E, C, D \rangle$

However, we need to bear in mind that when considering derived graphs things may get more complex because nodes with identical addresses in different elementary graphs have been collapsed into one as part of *graph union*: the union of G_1 and G_2 is the union of their nodes and edges (symbolically, $G_1 \cup G_2 = (V_1 \cup V_2, E_1 \cup E_2)$). When we are dealing with what Morin & O'Malley called vines, *a priori*, we may start a traversal from any root unless an order is imposed over them. Traversals with multi-rooted graphs are common in the analysis of data structures, but in our case the sequence needs to be linguistically motivated. These linguistic considerations, to which we will return in **Chapter 4**, motivate a sequence like that in (41") for the derived graph (41c):

(41") $\Sigma_{T''} = \langle A, B, C, D, E, C, D \rangle$

This sequence entails an order existing between A and E, even though there is no directed edge between them. For the case of trees connected by nodes other than their roots, the term *multitree* is sometimes used. In any kind of tree, however, no direct or indirect loops are allowed: multitrees may be directed, but they are acyclic.[8] It is crucial to bear in mind that the sequences we have just defined are neither walks or ρ-sets: we are not following dominance relations in digraphs. Preorder, inorder, and postorder traversals are defined on undirected trees, and are distinct from ρ-sets. In what follows we will focus on defining the ρ-sets for the sentences under analysis rather than traversals (which have been applied to aspects of tree structure in syntax; e.g. by Kural, 2005; Medeiros, 2021 to account for word order variation).

As we have said, our approach to composition of elementary structures—essentially, graph union—has antecedents in the linguistic literature: for example, Karttunen & Kay (1985) use the term *structure sharing* for a process very similar to this one but restricted to binary trees representing feature structures (structure sharing in this sense has also been extensively used in HPSG, e.g. Pollard & Sag, 1994: 19; Blevins & Sag, 2013: 217, ff. and LFG, e.g. Alsina, 2008; Sells, 2013: 174, ff.). Sarkar & Joshi (1997), in the context of a discussion of Right Node Raising under TAG assumptions, call this process *contraction*, and say that

> *The term* contraction *is taken from the graph-theoretic notion of edge contraction*[9]. *In a graph, when an edge joining two vertices is contracted, the nodes are merged and the new vertex retains edges to the union of the neighbors of the merged vertices* (Sarkar & Joshi, 1997: 612).

In the 'traditional' analysis of coordination in TAGs (see also Han et al., 2010), when coordinands have common arguments (for example, in a case of *gapping* such as (42) below), the operation *Conjoin*,

8 Some authors refer to directed, multi-rooted, (possibly) cyclic graphs as *multigraphs*, by analogy to *multitrees*. However, the term is not common in most introductions to graph theory (e.g., it does not appear in Even & Even, 2012; Wilson, 1996; or van Steen, 2010. Diestel, 2017 does use the term, but with a different meaning), and thus we will not use it.

9 An important caveat: edge contraction, in graph theory, removes an edge and merges the two vertices it previously connected (Gross et al. 2018: 194–195). It does not, however, require that these two vertices be identical in any respect. Graph union (as defined, e.g., in Bondy & Murty, 2008: 29) unifies identical nodes and edges across simple graphs. Note that under strict edge contraction nothing would prevent, in (41a), C and B from being merged in T. This goes, we

> [...] *identifies and merges the shared argument when two elementary trees combine via coordination, yielding a derived tree in which an argument is multiply dominated by two verbal projections.* (Han & Sarkar, 2017: 44)

Our perspective is very close to Sarkar & Joshi's, in the sense that nodes in a graph are assigned addresses that can be referred to at different points in the interpretation of a structure (such that '*corresponding objects in the corresponding expressions have the 'same' addresses*'; Gorn, 1967: 214): as in Sarkar & Joshi (1997) and Karttunen & Kay (1985), this mechanism will prove to be particularly handy when we deal with structural descriptions requiring graph composition (see the definition of *linking* below). However, unlike in the TAG analysis, in our approach (i) structure sharing is not limited to coordination, (ii) we do not make use of a specific operation distinct from other modes of structure composition, and (iii) addresses are context-independent.

For present purposes, the introduction of addresses serves to minimise the number of nodes in elementary and derived graphs. In the latter, this is accomplished by allowing distinct graphs to share common parts of their structure once graph composition operations have applied.[10] Let us give an example: in a sentence such as *John gave Mary a book and Susan a flower*, the coordination superficially affects non-constituents ([*Mary a book*] *and* [*Susan a flower*]). Sarkar & Joshi (1997: 614) propose a derivation of a sentence like that by means of the coordination of elementary trees which do correspond to constituents: structure sharing generates the illusion of non-constituent coordination (see Banik, 2004 for a development of the semantics of coordination under lexicalised TAG assumptions, with phrase structure trees):

---

think, against the spirit of Sarkar & Joshi's analysis, but the use of 'edge contraction' leaves such a possibility open. Our addressing axiom (see below), and the definition of structure composition as graph union deliver the desired results without the linguistic complications introduced by edge contraction, which would allow for distinct expressions to be contracted. See also the **Appendix**.

10 Han & Sarkar (2017) propose an alternative analysis from the perspective of Synchronous TAGs, whereby 'argument sharing' takes place in the semantic representation (using lambda abstraction), not in the syntax. Hoewever, given the fact that the addressing axiom allows us to do structure sharing in the syntax without the need to propose a parallel level of semantic interpretation, we will keep referring to the earlier Sarkar & Joshi (1997) approach.

(42) a.

b.

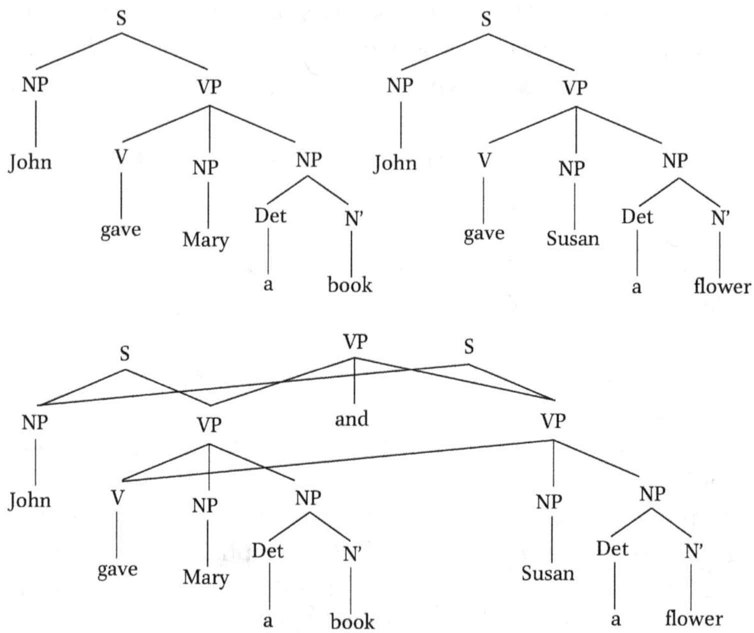

FIGURE 2.10 Analysis of gapping in TAG

In (42b), the subject and the verb are shared between each term of the coordination. Interestingly, even though *gapping* was originally formulated as a deletion transformation (Ross, 1970a), the lexicalised TAG (LTAG) approach, enriched with multidominance, captures the data without deleting any nodes (similarly, Goodall, 1987: Chapter 2 uses a 'three-dimensional' approach that also delivers deletion-less gapping by means of the union of phrase markers; see also Moltmann, 1992). The derived tree in (42b) is the result of composing two elementary trees, each of which corresponds to a term of the coordination, using a coordination schema for the coordination of likes. Tree composition identifies common nodes in the elementary trees by virtue of their addresses (in this case, ⦃give⦄ and ⦃John⦄), and *contracts* these into single nodes, leaving distinct nodes uncontracted. Crucially, this approach delivers deletion effects without deletion transformations: where deletion has multiple nodes as inputs and multiple nodes as outputs (some of which may be replaced by empty symbols, or have their morphophonological exponents erased), structure sharing has multiple nodes as inputs, but only one as an output (see **Section 13.2**). In the theory developed in this book, graph union delivers structure sharing.

The framework in this monograph uses the addressing system as a mechanism that, in combination with the possibility of having multidominance, allows us to cut down the number of nodes not just in *derived* graphs, but also at the level of arbores. As we will see in detail in the following chapters, this approach makes it possible to simplify dependencies that MGG models in terms of *chains*: sets of nodes in distinct structural positions linked by some mechanism (indexing, FormCopy, etc.). The multiplication of nodes in instances of displacement in MGG is related to the fact that, by definition (as noted in **Section 1.2**), the nodes involved in a chain dependency cannot be 'identical', contrary to what is claimed in Chomsky (1995: 231) for Raising to Subject and other cases of so-called 'total reconstruction' (see Sauerland & Elbourne, 2002; also Gärtner, 2002: 87, ff. for discussion). In a system where displacement is triggered by the need to satisfy featural requirements,[11] distinct reconstruction links in a chain will necessarily have a different featural composition depending on what feature has been checked at a particular derivational point. Crucially, because addresses are assigned to nodes in totally ordered digraphs, it is still possible to distinguish between different 'contexts' from which a node is visited in a walk defined in that graph if a node is the tail of distinct arcs: this is a consequence of abandoning the SMC as a condition over well-formed graphs.

It is important to point out that not all symbols in a terminal string are assigned addresses in the structural description of that string, nor is there a direct correspondence between orthographical words in a sentence and nodes in a graph. In phrase structure grammars, every word—or sometimes morpheme—corresponds to a terminal node in a tree: there is no exponent in a string that is not represented in a tree. Dependency grammars often follow the same practice, to the point of having a direct correspondence between words and nodes. We depart from that tradition, in assigning no privileged status to the concept of 'word' (Schmerling, 1983b: 6–7). As in Gorn (1967: 214), only 'object characters' have addresses. Linguistically, 'object characters' are lexical categories (traditionally, V and N are seen as the core categories; see Martin et al., 2019 for algebraic discussion): logically, we have *predicates* and *arguments*. This choice has a long pedigree in logic and (pure) Categorial Grammar: we assign addresses only to *categorematic* expressions, which have a corresponding semantic interpretation that varies (or may vary) across models. Expressions like copulas, complementisers, or case-marking prepositions, have *constant* interpretations across models. These expressions do not have meanings of their own, but every expression in which they occur does have a meaning: following the logical tradition (MacFarlane, 2017; see also Schmerling, 2018a for a

11 Chomsky (2000, 2001) for example, proposes that movement is an 'imperfection' in language design that fulfils the purpose of eliminating the other imperfection of language design: the existence of uninterpretable features.

linguistic perspective from Categorial Grammars), we will refer to them as *syncategorematic*. In first order logic, for example, connectives and parentheses are syncategorematic. In the categorial system explored in Schmerling (2018a), syncategorematic expressions include infinitival *to* in English, the verb *be* in imperatives (as in *Do be careful!*), the indefinite article *a* (see also Montague, 1973), and inflection; García Fernández et al. (2020) identify some 'intermediate elements' in verbal periphrases in Spanish as syncategorematic (e.g., *a* in the analytic future periphrasis ⟨*ir a* + infinitive⟩, or *de* in the modal periphrasis ⟨*deber de* + infinitive⟩), to which Krivochen & Schmerling (2022) add differential object marking *a* in Spanish (as in *ver* *(***a***) Juan* 'to see John', cf. *ver* (****a***) *una película* 'to watch a movie').

What is fundamental to our purposes is that syncategorematic expressions are *not* themselves assigned addresses; we can refine our observation above and say that only expressions to which there corresponds an interpretation are assigned addresses. Importantly, a system of structured data based on expressions pointing to memory locations does not specify what these memory locations actually *contain*: an address only tells the system where to look (e.g., in the Lexicon), but not what it is going to find there. For that, we need a theory of how basic expressions of the language receive an interpretation (and how the interpretations of derived expressions are related to the interpretations of basic expressions); in linguistic terms, we need a theory of *compositional semantics*. In this section we will provide some basic guidelines for what a compositional semantic for graphs could look like, based on the approach presented in Heim & Kratzer (1998) and Dowty et al. (1981), among many others. As highlighted above, everything we have said up until this point may be accepted independently of what follows: what we have called an 'address' could be no different from an arbitrary index in MGG (see e.g. Lasnik & Uriagereka, 1988: 43, ff.; Kural & Tsoulas, 2004 for relevant discussion). In that case, however, the connection between syntactic structure and semantic interpretation would have to be made in some other way, thus complicating the theory with unclear empirical payoffs.

In our graphs, nodes do not correspond to terminal / nonterminal symbols from a typed alphabet or to lexical item tokens: rather, they correspond to basic expressions of the language, and are indexed by addresses which point to their *semantic values* (the notion of 'semantic value' is understood as in Dowty et al., 1981: Chapter 1; Higginbotham, 1985; Heim & Kratzer, 1998). In this context, the *semantic value* of an expression is its denotation. Furthermore, addresses are *unique identifiers*: each address is in a one-to-one correspondence to a semantic value (there is a single address for each elementary semantic value and a elementary single semantic value for each address). Having basic expressions be assigned addresses is at the core of both syntax and semantics in our system. In this context, the definition of the addresses of complex objects, to which correspond derived semantic values, are obtained by means of graph union. This is an important point, which relates to a long tradition in the study of the syntax-semantics interface. What is that 'semantic value'? How does the semantic value of individual basic expressions relate to the meaning of derived expressions? Several answers can in principle be provided: the syntactic framework explored here is compatible with more than one compositional semantic theory, and the choice between competing semantic approaches is partly dictated by the level of granularity at which nodes are defined (basic expressions, features or feature bundles, roots, etc.). This is, we think, a strength of the present framework.

The concerns we have just expressed pertain to what is usually referred to as the Problem of Compositionality, a cornerstone of the interaction between syntax and semantics. Harris (1951: 190) puts the Compositionality Principle in these terms: '(…) *the meaning of each morpheme in the utterance will be defined in such a way that the sum of the meanings of the constituent morphemes is the meaning of the utterance*'. In the formulation of Partee (1984: 153), the Principle of Compositionality says that '*The meaning of an expression is a function of the meanings of its parts and of the way they are syntactically combined*' (Dowty, 2007: 23–24; Bach & Cooper, 1978: 145; Montague, 1973 and much related work; also Kracht, 2013 for a comparison between natural languages and logical languages in terms of compositionality). Much is packed in these definitions, including at least the following underlying assumptions (see Piattelli-Palmarini & Harley, 2004 for discussion):

1. Meaning is a function of the *output* of syntax
2. Furthermore, meaning is a function of the output of *only* the syntax
3. Rules specify modes of combination of syntactically relevant units
4. The grammar has solved the *segmentation* and *classification* problem
5. All intermediate representations in a derivation have a meaning
6. Rules also specify modes of combination of meanings
7. There is some correspondence between two sets of rules, syntactic and semantic

In principle, each of these could be rejected on independent grounds (and most of these have been), and this would give rise to different theories (for an analogous exercise, see McCawley, 1981c). Aside from noting these thorny issues, we will assume that *some* version of Compositionality is linguistically desirable, and provide some preliminary arguments to the effect that what Baker & Jacobson (2007) and Jacobson (2012) call *direct compositionality* can be implemented in the syntactic theory presented in this monograph.

We can ask how to define a function like that referred to in the formulation of the Principle, what the rules of syntactic combination are, exactly how to define what counts as an expression, etc. These are all central questions for a theory of natural language grammar. Therefore, a compositional theory of linguistic meaning must determine both the units that are subject to the Principle of Compositionality, as well as the rules of combination and interpretation: the formulation of the rules of grammar and the elements on which they operate must be explicit. In this section we will sketch some prolegomena to a graph-based theory of grammar that explicitly incorporates a compositional semantic approach.

In the context of this work, we take addresses to point to the semantic value of basic expressions. As anticipated, we will remain mostly agnostic about what these semantic values are, but there seem to be good reasons to assume that semantic values are *intensions*. This choice is not entirely novel, and combines aspects of previous frameworks: as far as intensions go, pure PSGs of the kind explored in Gazdar (1981, 1982) and much subsequent work assume that semantic values are translations of basic expressions into intensional logic (IL). For instance, Gazdar (1981: 156) describes the rules of his context-free system as follows:

> *I take a rule of grammar to be a triple of which the first member is an arbitrary integer (the number of the rule), the second is a PS* [Phrase Structure] *rule, and* ***the third is a semantic rule showing how the intensional logic representation of the expression created by the PS rule is built up from the intensional logic representations of its immediate constituents.*** (our highlighting)

Our choice to have nodes be assigned uniquely identifying addresses, in turn, is based on work on TAG (in particular, Sarkar & Joshi, 1997; see also Joshi & Schabes, 1997, who use addresses to indicate adjunction and substitution sites in derivation trees). In our case, as in TAGs, addresses serve the purpose of uniquely and unambiguously identifying nodes.

Crucially, the choice of what semantic values or 'meanings' are is not dictated by the graph-theoretic formalism: an extensionalist view is also possible, and perhaps the two are not too different in terms of their consequences for the syntactic theory formulated in this work (they do, evidently, differ greatly in other respects). There are, however, some empirical reasons for us to prefer an intensionalist framework over an extensionalist one, motivated by previous work in grammatical analysis. We can summarise some of these reasons. The analysis of English non-progressive *being* in sentences containing proper names such as *I'm being John today* (if I am fulfilling some of John's duties at work, for instance) in Schmerling & Krivochen (2017) is based on the idea that the definite NP immediately following *BE being*—in our case, *John*—picks out a set of properties, but the properties it picks out comprise a proper subset of the property set that it usually denotes. The interpretation of this definite NP, in the aforementioned work, is the translation of this NP into intensional logic. This allowed Krivochen & Schmerling (2017) to define an operator that takes an NP-type extension as input and returns another NP-type extension as output, namely, a contextually determined subset of the set of properties denoted by the input. In other words, by assuming that the semantic value of the expression *John* in *I'm being John today* is a set of properties, we can select a contextually salient subset of those properties (for instance, being a manager). An extensional perspective, according to which the semantic value of *John* is an individual, would be ill-equipped to provide a satisfactory analysis of sentences like the above.

By adopting an intensional approach to the semantics to be paired with our graphs we can distinguish between, say, *John seeks a unicorn* and *John seeks a centaur* (Cooper, 1984: 7): the extension of both NPs *a unicorn* and *a centaur* is the empty set, but their intensions are quite different. In this monograph, we will for the most part just assume an intensionalist view without much further argument, since our focus is syntax, not semantics. The advantages of having addresses point to semantic values include (but are not limited to) the elimination of a distinct (syntactic) level of 'Logical Form' in the grammar where indexing takes place; our syntax is thus 'single-level' (Perlmutter, 1982).

Let us provide some semantic preliminaries. As we noted above, also, there is no specific structural description to capture semantic relations, nor is there a level of representation composed of 'meaning-less' structures (see Jacobson, 2012: 109–110 for discussion): the compositional semantics is read off the elementary relations created in digraphs.

We need to give more details about the nodes themselves, as well as the interpretation of directed edges. In this context, we need to add an axiom to our system of definitions, which is motivated by the fact that we are using graphs as structural descriptions for natural language sentences:

> ***Addressing axiom:***
> *The content of node's address is a semantic value.*

An important part of defining the Lexicon of the system, then, is having a set of semantic values. For predicates, information contained in the Lexicon also includes category, subcategorisation frames (number and syntactic category of required arguments), and thematic structure (the information that in LFG would be specified at the level of *a-structure*): essentially, GB-style lexical entries. The configurational information conveyed in graphs may be enriched, of course. In addition to defining a set of addresses that index the set of nodes, we may specify, for any node, its semantic type: each expression can be assigned a type, basic or derived (see e.g. Han & Sarkar, 2017 for an application of type theory to semantic analyses in TAGs). Summarising, basic types are *e* and *t*, for entities and truth values respectively; derived types are constructed by combining basic types: if A is a type and B is a type, then ⟨A, B⟩ is a type; nothing else is a type. Each derived type denotes a function, specifying domain and range: a type ⟨A, B⟩ is a function with input A and output B. Thus, for example, ⟨*e*, *t*⟩ denotes a function from entities to truth values: this is the type of expressions that must combine with an expression of type *e* (or of type ⟨⟨*e*, *t*⟩, *t*⟩) to output an expression that is assigned a truth value.

Adopting semantic types as a replacement for syntactic categories is not without disadvantages, however. Among them, as Partee (1975: 218) and Bach (1979: 517) observe, is the issue that more than one syntactic category may be assigned the same semantic type (with category splits having a crucial role to play in keeping these categories distinct). There is no one-to-one relation between syntactic category and semantic type. This will become relevant in our treatment of Spanish auxiliary chains: split categories belong to distinct syntactic types but the same semantic type. Thus, as desired, a basic VP such as *trabajar* (lit. 'to work') and the analytic future form *ir a trabajar* (lit. 'to go to work') have distinct syntactic categories but the same semantic type (functions from entities to truth values). Syntactic category provides necessary information related to the satisfaction of subcategorisation properties of predicates and plays an important role in the identification of anchors. So far as we can see, these cannot be completely reduced to type-theoretic annotations.

A more fleshed out semantic approach which annotates each node with its semantic type *in addition to* its syntactic category (assuming some category split mechanism) may use this information to indicate the order of composition in the traversal of the graph (and, derivationally, the order of Merge; see e.g. Krivochen, 2023a, b). If we treat NPs as generalised quantifiers (see **Section 14.4** for discussion), then these can be used to map *n*-ary relations to (*n*-1)-ary ones stepwise (see also Bach & Partee, 1980; Keenan, 2006). The order of composition could, for example, follow the Grammatical Function Hierarchy, since grammatical functions are primitives of the theory (see Dowty, 1982 and

Chapter 5). For present purposes, we will limit ourselves to sketching what a compositional semantics for our digraphs could look like. The importance of syntax for the grammatical architecture must not be underestimated based on the role that semantic values have: the theory presented in this monograph is more than anything a theory of *syntax*. Semantics is at the core, but if we only have a set of semantic values or thematic grids, we cannot do anything with them (other than, presumably, vocatives, interjections, possibly imperatives, and the like). We need to establish relations between those semantic values in order to get compositional outputs: the study of those (syntactic) relations is the main topic of this work.

Above we stated that we will pursue the hypothesis that structural descriptions for natural language expressions are *digraphs*: what determines that relations between nodes are *asymmetric* (i.e., that $e\langle v_1, v_2\rangle \neq e\langle v_2, v_1\rangle$: edges are 'one-way roads') is the semantic properties of expressions corresponding to nodes and how the neighbourhood sets of those nodes are defined. Specifically, we will develop the idea that *predication* imposes restrictions with respect to the directionality of the edges in a graph such that predicates always directly dominate their arguments (thus, predication is read from directed edges, without the need for a dedicated module of predication rules as in Williams, 1980). However, predication alone does not yield anything other than atomic outputs (we can say of a node that it corresponds to a predicate, but nothing else). It is the *connection* between semantic values in directed graphs that yields compositionally interpretable objects, through the establishment of syntactic dependencies. A full development of a semantic theory for our graphs is outside the scope of this work, but some programmatic considerations are in order. For all E (a variable over expressions), let ⟦E⟧ stand for the semantic value of E (Dowty et al., 1980: 19, ff.; Heim & Kratzer, 1998; Von Fintel & Heim, 2011); a compositional semantic theory must at least provide a general way to map syntactic dependencies into semantic relations (the notation for addresses, ⦃ ⦄ is similar to that for semantic values, ⟦ ⟧: this is not accidental since the content of addresses are semantic values and we wanted to highlight their intimate relation). Let us summarise what we have, and what we need:

We have:
A set of graphs G, where G = {E, V}
A set Exp of basic expressions
A set V of vertices (or 'nodes')
A set E of edges
A set A of uniquely identifying addresses
A set S of semantic values (which, as suggested above, are intensions)

If, as proposed in this work, nodes in graphs correspond to basic expressions of the language, we need a function to pair nodes and expressions. This is also true in classical generative models such as Chomsky's (1965, 1970b) insofar as the leaves of phrase structure trees are initially 'dummy symbols' (Chomsky, 1970b: 185); similar considerations apply to more recent 'late insertion' theories. We want to distinguish, for example, the multi-word basic expression *would rather* (Baker, 1970a; Schmerling, 1983b) from the node that corresponds to it in the graph that makes explicit relations between expressions in a sentence like *John would rather not walk*. Each node corresponds to a single basic expression and is assigned a uniquely identifying address, which points to a semantic value: there is a bijective function from ⟨node, address⟩ to semantic values. In this context, we can say that the set of nodes is *indexed* by the set of addresses.

Now we can give a general, semi-formal interpretation rule for connected nodes in elementary graphs (based on Bach, 1979: 516; Heim & Kratzer, 1998: 16; Von Fintel & Heim, 2011: 7; Larson, 2014: 3, among others):

> ***Semantic interpretation rule:***
> A semantic interpretation SI function for L is a function from ⟨N, A⟩ to S. For Exp = v, SI(Exp) = ⟦Exp⟧.
> In an elementary graph, $e\langle v_1, v_2\rangle$ (a directed edge between nodes v_1 and v_2, where v_1 is a functor, and v_2 an argument) is the semantic value of v_1 applied to the semantic value of v_2:
> $e\langle v_1, v_2\rangle$ becomes $[\![v_1]\!]([\![v_2]\!])$
> $\langle e\langle v_1, v_2\rangle, e\langle v_1, v_3\rangle\rangle$ becomes $([\![v_1]\!]([\![v_3]\!]))([\![v_2]\!])$ (v_1 a predicate, v_2, v_3 its arguments)
> $\langle e\langle v_1, v_2\rangle, e\langle v_1, v_3\rangle, e\langle v_1, v_4\rangle\rangle$ becomes $(([\![v_1]\!]([\![v_4]\!]))([\![v_3]\!]))([\![v_2]\!])$ (v_1 a predicate, v_2, v_3, v_4 its arguments)

The semantic sketch we have presented so far—preliminary as it is—works for elementary graphs in roughly the same way that Hein & Kratzer's rules work for phrase structure trees (for derived graphs the semantic type of expressions may become more relevant; we leave this issue aside here), and can also be tied in with Jacobson's (2012) *Type 3* direct compositionality (where syntactic rules may operate over objects as complex as trees, and each syntactic rule is coupled with a semantic interpretation rule without an intermediate level between syntax and interpretation). Let us see the rule in action: recall the graph that we proposed for multiple (non-iterative) intersective adjectival modification in **Chapter 1**, repeated here alongside the arbores that make up the structure:

(43) a. b.

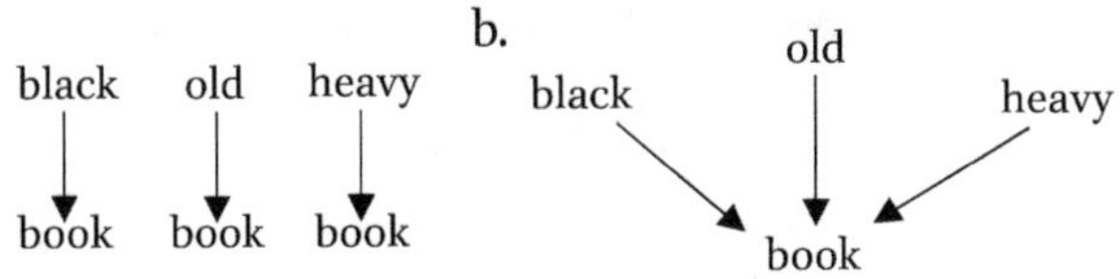

FIGURE 2.11 Intersective adjectival modification

The interpretation assigned to the graph (43b) is the unification[12, 13] of the three modification relations (Shieber, 1986; the same idea is at the core of *structure sharing*): the node with address ⦃book⦄ is common to all three arbores and thus the representation (44a) can be simplified as (44b); note that there is no dominance relation between any of the adjectives:

(44) a. $[\![\text{black}]\!]([\![\text{book}]\!]) \cup [\![\text{old}]\!]([\![\text{book}]\!]) \cup [\![\text{heavy}]\!]([\![\text{book}]\!])$
b. $[\![\text{black, old, heavy}]\!]([\![\text{book}]\!])$

If we have the structure in (43) embedded in a larger object, for example as in (45),

(45) John read the black, old, heavy book

---

12 Unification applies to feature structures in the following way (Shieber, 1986: 14):

In formal terms, we define the unification of two feature structures D' and D" as the most general feature structure D, such that D' ⊆ D and D" ⊆ D. We notate this D = D' ∪ D".

⊆ is used to symbolise a *subsumption* relation between feature structures, in which a feature structure, abbreviated D, contains *part* of the information of another D', such that D' ⊆ D. The concept of *subsumption* is based on that of *dom(D)* (the *domain* of a feature structure, namely, the features it includes, regardless of their mapped values), such that D' ⊆ D *iff* $\forall(x) \mid x \in dom(\text{D}'), x \in dom(\text{D})$. Jackendoff (2011: 276, ex. 10 a, b) provides some examples of Unification in contrast to Merge, which further illustrate our point:

a. *Unification of [V, +past] and [V, 3 sing] = [V, +past, 3 sing]*
(not [[V, +past] [V, 3 sing]], as with Merge)

b. *Unification of [VP V NP] and [V, +past] = [VP [V, +past] NP]*
(not [[V, +past] [VP V NP]], as with Merge)

We can recognise three distinct possible results of the operation, based on Shieber (1986: 15):

- Unification *adds* information (e.g., feature structures are not identical, but compatible. This is the case of our multiple adjective modification)
- Unification *does not add* information (e.g., feature structures are identical)
- Unification *fails* due to conflicting information (e.g., same attributes, different values).

13 If nodes stand for feature graphs (thus making arbores *hypergraphs*), A-N agreement can be implemented in a unification grammar as in Reyle & Rohrer (1988: 6) assuming that Ns have lexically valued agreement features. We leave this possibility aside here.

then we have the following graph as a diagram of the structural description of (45):

(46)

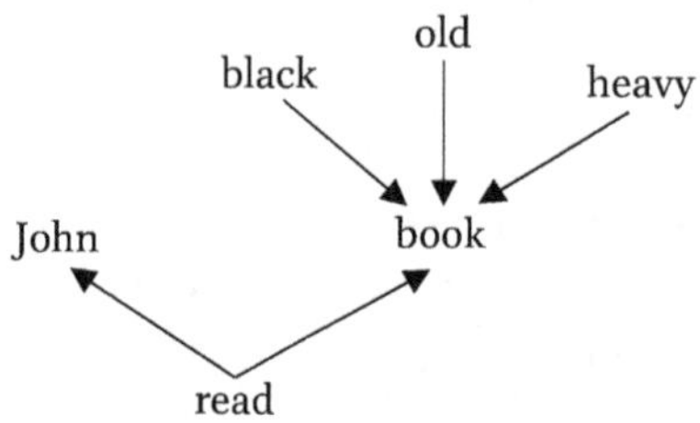

FIGURE 2.12 Graph-theoretic analysis of 'John read the black, old, heavy book'

Anticipating discussion in **Chapter 5**, the arcs e_1⟨read, John⟩ and e_2⟨read, book⟩ are themselves ordered: the set of arcs is itself an ordered set. This is a central point in the argument, insofar as it will allow us to put the order over the set of arcs in correspondence with another set of ordered relations: specifically, the hierarchy of grammatical functions assumed in RG, APG, LFG, and other approaches. This defines an inorder traversal. Importantly, the formalism can accommodate alternative definitions of GFs. For example, assuming a postorder graph traversal, the transitive verb first combines with its object to yield an intransitive verb phrase (an IV), and finally with its subject to form a sentence (see also Dowty, 1982):

(47) (⟦read⟧(⟦black, old, heavy⟧(⟦book⟧)))(⟦John⟧)

The object can be seen as a function from a binary predicate to a unary predicate (Keenan, 2006), whereas the subject is a function from a unary predicate to a sentence. What we have in (47) is the semantic value of the IV applied to the semantic value of *John*.

This proposal is related to certain generative approaches to the semantic interpretation of syntactic structures in which the semantic interpretation are functions of (sub) phrase markers (e.g. Heim & Kratzer, 1998: Chapter 2; also Higginbotham, 1985: 553–554); the differences arise when we consider (a) what the format of structural descriptions is and (b) what kinds of nodes are allowed in those structural descriptions (e.g., are there semantically empty intermediate nodes? Are there syntactically inert semantic operators?). Because edges can only connect two nodes, the process of semantic composition must always proceed in this manner; this is particularly important when we consider that the relation of *dominance* is transitive. This simplified way of translating syntax to semantics follows from our adoption of the DG convention of having predicates directly dominate their arguments: if a node v_1 corresponds to the transitive verb *read* and v_2 to the sortal entity *books*, then the directed edge $e\langle v_1, v_2\rangle$ encodes the fact that the verb selects a direct object. This implies another departure from PSGs, insofar as in a transitive verb phrase the V does not dominate its subject NP: an intermediate node (VP) does (generative arguments in favour of relations being mediated by branching nodes can be found in Neeleman & van de Koot, 2002). Note also that explicitly graph-theoretic frameworks like APG and RG also do not have a directed edge from predicates to arguments, so our interpretation rule could not apply.

There is an immediate difficulty with the interpretation rule above: how can we represent the distinction between subjects and objects, if there is an edge from the V to its subject and from the V to its object? This is a crucial point, since one of the guiding assumptions of the present framework is that grammatical functions are primitives of the theory, and syntactic representations must encode them without adding nodes or diacritics. As anticipated, this will require us to impose an order over arcs, and revise the format of ρ-sets that we have assumed so far. The issue of grammatical functions and how they are related to dominance will be dealt with in detail in **Chapter 5**.

A crucial aspect of our proposal, which we emphasised above, is that nodes in a graph do *not* correspond to 'lexical items' or 'phrases' (or 'constituents', in IA grammars of the kind described in Schmerling, 1983a, b): syntax is based on expressions. In our theory, as already stated, nodes in a graph correspond to indexed *basic expressions* of the language, which are not equivalent to *words* or *phrases*, being defined in a different kind of grammatical system. This aspect sets our proposal apart from other graph-based theories, including RG, APG, and Metagraph Grammar: in these theories, nodes representing 'substantive linguistic elements' are defined in terms of their lexical or phrasal status (although this is often not explicitly said). The choice of what counts as a suitable element to be a *node* in an L-graph (be it a minimally connected tree or a network) impacts on the expressive power and descriptive adequacy of the theory, as we will see in some detail below. However, it is important to note that nothing prevents us from using traditional nomenclature to refer, descriptively, to a set of nodes in a graph. Taking as an example the (preliminary) ρ-set analysis of a sentence like (48),

(48) a. John seeks a unicorn
b. ρ = {(seek, John), (seek, unicorn)}

we may use the proxy 'VP' to refer to the subgraph defined by the directed edge *e*⟨seek, unicorn⟩, as an informal abbreviation for a specification of this sub-graph. Note that in this case, in contrast to MGG's tree representations (but see Graf & De Santo's 2019 use of Dependency Trees in Minimalist Grammars), 'VP' is not part of the formal object: there is no VP node in the graph, nor is VP an address or an address content (see also Seely, 2006: 189 for a related perspective on the role of labels in Minimalist trees). VP, or IV (Intransitive Verb Phrase, a term frequently used in Categorial Grammar to denote an expression that only needs to combine with an NP subject to yield a finite clause; see Schmerling, 1983b, 2018a), can be used as shorthands for a set of nodes and edges in a graph, just like we can use 'S' as a shorthand for a graph that specifies the basic expressions and relations between these in a complete sentence. Evidently, 'sentence' or 'S' is not a graph-theoretic notion, nor is there a node S in a ρ-set (but they are nodes in PS trees). We can use descriptive labels like NP, VP, or S as mnemonic devices, as long as we bear in mind that (a) the theory proposed here is *not* based on IC models, nor is there a notion of immediate constituency in the system, and (b) these intermediate symbols do not correspond to any node in a graph.

Ultimately, the definitions we have provided need to be justified in terms of their usefulness for grammatical analysis, which will be the focus of **Chapters 3** to **13**. We emphasised above the significance of McCawley's distinction between RCT and RPT as the fundamental inspiration for this work; we can now come back to that in somewhat different terms. The core idea that we want to put forth is that MGG's transformations, understood here as *descriptive devices* (as names of constructions), actually *do not change grammatical relations* between elements in a structural description: at most, they *create* new relations, but *without disrupting those already existing* (created by the base component). In order to describe and characterise these relations, locally and globally, we make use of the tools that graph theory puts at our disposition, plus the empirical insights obtained through careful analysis in both transformational and non-transformational theories.

Using classic transformations as names of constructions, we will provide arguments throughout the rest of the monograph (see **Chapter 13** for a summary) in favour of the idea that a majority of 'transformations' change linear order but not such relations as subjecthood or objecthood: these stable dependencies in the description of the structure of a sentence will be at the core of this monograph. A caveat on the status of transformations in this framework is necessary at this point, since we have gone to great lengths to argue for the *declarative* nature of our theory: transformations are taken here simply as descriptive devices, without implying that syntactic objects *actually* move or anything of the sort (there are no *derivations* in the present view, which in and of itself defeats the whole purpose of transformations as mappings). This descriptive view on what transformations are and what they can do was common in the early days of Transformational Generative Grammar.[14]

2.3 Adjacency Matrices and Arcs: More on Allowed Relations

There are several ways to formalise the relations between nodes in a graph. Diagrams with nodes and arrows are useful as illustrations, but they are not formal objects. It is essential to distinguish between *diagrams* of L-trees and L-trees; only the latter are formal entities with properties that can be specified. In the case of the graphs considered in this monograph, we need to be able to specify nodes and connections (direct dominance, indirect dominance, linking, etc.). Connections between nodes in a graph can be formalised by means of a so-called *adjacency matrix* (Van Steen, 2010: 60; Wilson, 1996: 14). Consider a graph G with n vertices: to provide a basis for specifying all possible relations, we adopt an $n \times n$ *adjacency matrix* A(G):

(49)
$$G = \begin{bmatrix} v_{11} & \cdots & v_{1n} \\ \vdots & \ddots & \vdots \\ v_{n1} & \cdots & v_{nn} \end{bmatrix}$$

In A(G), $v_{ij} = 0$ if there is no edge between v_i and v_j. Formally:

(50) $v_{ij} = 0$ *iff* $\forall i, \forall j\, [(i, j) \notin \rho]$

In the theory explored here, the adjacency matrix corresponding to the structural description of an expression is *not symmetric* (see **Definition 4**), due to the fact that graphs are directed: for any two vertices v_i and v_j, $A[v_i, v_j] \neq A[v_j, v_i]$. This means that, for instance, if v_i corresponds to a functor, and v_j to an argument, $A[v_i, v_j]$ will be non-zero, but $A[v_j, v_i]$ will be zero because predicates always dominate their arguments, not the other way around. The main diagonal of A(G) is composed only of zeros if there are no direct loops, if no vertex *directly* dominates itself:

(51) $(v_n, v_n) \notin \rho$

14 For instance, Rogers (1974: 556) lucidly says that:
Transderivational constraints, global and interpretive rules, and transformations, it seems to me, don't **explain** *anything: they describe.*

Assuming that a condition like this holds is rather conventional across the grammatical board, see, e.g., APG's *No Loop Condition* (Johnson & Postal, 1980: 50–51; also Postal, 2010: 12); Dependency Grammar trees by definition also comply with this condition, because dependency is a '*strict mother-daughter relation*' (Osborne, 2008: 1122; also Osborne et al., 2011; Kahane & Mazziotta, 2015; see **Section 4.2** for discussion), with *mother of* being a 2-place relation between *distinct* nodes in the tree such that a node cannot be its own mother. PSGs trivially implement this requirement as well, given the definition of derivations as sequences of strings ordered by the relation *follows from* (Chomsky, 1956). However, we can ask if we really need a restriction like (45) to hold, or if specific conditions in the analysis of natural language sentences may require self-dominance. Recall the examples *some fake fake news* or *an old old man* in **Chapter 1**: we proposed that an adequate syntax for the intensive reduplication reading of *fake fake* or *old old* should be finite-state (Krivochen, 2015a, 2021a; see also Lasnik, 2011; Schmerling, 2018b). If this is correct, then we can *in principle* define a finite-state transition diagram like (52), in which there is a transition from state B to state B with input *b*:

(52)

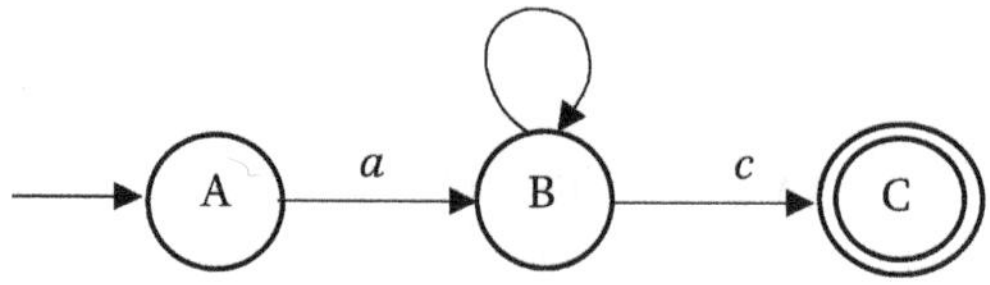

FIGURE 2.13 Finite-state transition diagram

In terms of our example, *a* = some; *b* = fake, and *c* = news (capital letters correspond to states of the finite-state automaton, and to intermediate nodes in a tree representation). However, as has been observed in the literature (see **Section 1.5**) this globally FS structural description is just as inadequate as a uniformly CF phrase structure tree, since it 'flattens' all the structure, not only the fragment that needs to be flattened: there is no way in an FSA to establish a dependency between states A and C, since each state is dependant only on the immediately previous state (the transition between state B and state C, for example, is only dependent on state B and the input *c*: it makes no reference to state A or to previous inputs like *a* or *b*; see also Chomsky, 1957: 19, ff.). This is inadequate because we need to capture scope relations between the quantifier *some* and its argument, as well as between the reduplicated adjective and its argument. Because it allows for the possibility of self-dominance, the graph-theoretic system we are laying out enables us to assign a flat structure just to *fake fake* and nothing more—i.e., to assign a finite state character just to *fake fake* in the larger expression *some fake fake news* in the very specific situation of intensive total reduplication. If we do this, then no additional structure is created in the form of non-terminal nodes at the local level, which effectively provides a solution to Lasnik's problem. If we let *fake fake* be a finite state loop, inserting this locally flat structure for [fake fake] into a CF structure by means of a graph composition operation yields the desired result: a segmentation of the form [some [[fake fake] news]]. At the same time, if each instance of *fake* is an independent node, we have the fully monotonic modification pattern [some [fake [fake [news]]]] 'some truthful news', where the semantic value of the first *fake* modifies the semantic value of *news* and the semantic value of the second *fake* modifies the semantic value of *fake news*.

Immediate self-domination, then, is restricted to intensive iteration, for which a local finite-state description seems to be the most appropriate structure: self-dominance gives us closure under Kleene star, concatenation, and alternation (the latter trivially, since a single expression is being iterated). But a node can also dominate itself when a walk includes two non-subsequent visits to that node. This is *transitive dominance*, the relation *dominate**: $(v_n, v_n) \in \rho^*$ but $(v_n, v_n) \notin \rho$.

Note that if v_n dominates some other node v_m which in turn dominates v_n then v_n is not necessarily multidominated: v_n does not have more than one mother. We will see (in particular in **Chapters** 7 and 9) that this condition allows for a simplification of the *chain formation* mechanism that is central to transformational accounts of 'displacement' phenomena and binding (Chomsky, 1981: 331, 333; 1995; see also Kracht, 2001; Gärtner, 2002 for formal approaches to chain mechanisms in derivational Minimalist systems). In transformational grammar (both GB and MP), *chains* play a fundamental role in the analysis of dependencies modelled in terms of *co-indexing* (anaphora, pronominal reference, filler-gap dependencies, etc.). This means that allowing for multidominance or not in a theory of grammar and specifying exactly which kind of multidominance the system may allow for will have great impact on the empirical adequacy of the theory as this choice will determine the possible treatment that binding and displacement will receive as well as the formal properties of the structural descriptions assigned to sentences featuring these phenomena. For example, some proposals generate trees where it is not possible to define a total order, whereas others are fully compatible with a total order; Kracht's and Gärtner's approaches differ in terms of the kind of multidominance structures they allow for.

We need to characterise the allowed relations in a graph in some more detail: we can do that by considering very simple cases of very local relations between nodes and edges. In this context, it is particularly useful to consider the array of relations allowed, for example, in Arc Pair Grammar and Metagraph grammar. The reader familiar with APG will note that the *Bicircuit* relation (decomposable into Branch(A, B) ^ Branch(B, A) for any A, B: A is a Branch of B iff A's tail node is identical to B's head node) is allowed in the theory exposed here. The relation *Parallel* (where two arcs share both their heads and tails; Johnson & Postal, 1980: 41; Postal, 2010: 12–13) will also become relevant, in particular for our treatment of reflexive anaphora. Let us illustrate some of the relations we have been mentioning, borrowing some graphical tools from APG and MG (see also Harary, Norman & Cartwright, 1965):

(53)

Loop: v_i directly dominates itself. In this monograph, direct loops are used exclusively for intensive iteration.

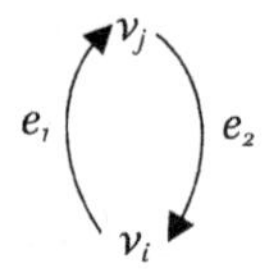

Bicircuit: edges e_1 and e_2 form a bicircuit iff $e_1<v_i, v_j> \wedge e_2<v_j, v_i>$. In this configuration, v_i dominates v_j and v_j dominates v_i. Note that this entails that both v_i and v_j dominate themselves *transitively*.

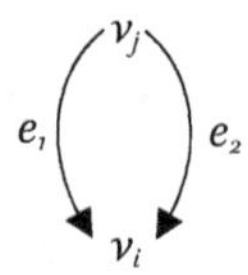

Parallel: edges e_1 and e_2 are parallel iff $e_1<v_i, v_j> \wedge e_2<v_i, v_j>$. In this configuration, v_i dominates v_j, establishing two different grammatical relations with it (see **Chapter 5** for details about these relations). Sometimes these are referred to as *multiedges* (e.g., Gross & Yellen, 2014: 3)

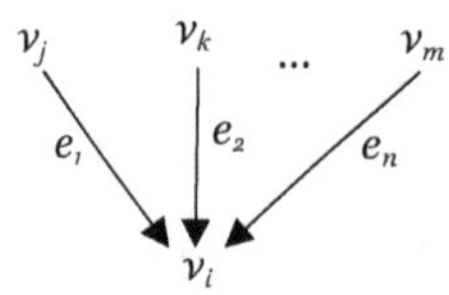

Kiss (Postal, 2010: 13): edges e_1 e_2, …e_n *kiss* iff $e_1<v_j, v_i> \wedge e_2<v_k, v_i> \wedge e_n<v_m, v_i>$. In this configuration, we say that v_i has more than one *mother node*, or that it is *multidominated*. Our account of non-iterative adjectival modification (*old, black, heavy book*) made use of *kissing arcs*.

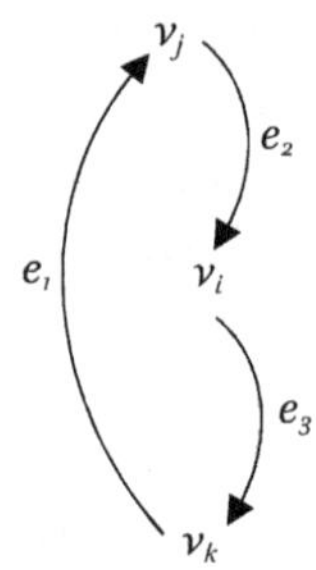

Circuit (Postal, 2010: 12-13): a sequence of edges e_1 e_2, …e_n forms a circuit *iff* for all i such that $1 \leq i \leq n$, Branch(e_{i+1}, e_i) and Branch(e_1, e_n). As in *bicircuit*, every node in a circuit transitively dominates itself.

FIGURE 2.14 Summary of arc relations

We must note that in APG and MG these are relations holding *between edges* (*arcs* in their terms), not between nodes; this makes a great difference when defining conditions over 'long-distance dependencies' and the composition of elementary graphs. To this effect, we have included the kind of relation that holds between nodes in the annotations in addition to the relations between edges (the latter of which are crucial in APG and MG, but secondary here). We allow for less primitive relations than APG, but those we allow are *n*-ary and defined over a *single* level of 'representation': there are no *strata, deep* and *surface structures*, or (more generally) *pre-* and *post-transformational* representations, which are still very much strong in Minimalism (a structural description *before* Agree or Internal Merge may violate a *bare output condition* that the structural description *after* the application of those operations may not). Our graphs are syntactic connections defined over semantically characterised vertices, without—for now—having a representation for linearity or the morpho-phonological exponents of vertices. In this respect, the reader might find it useful to check out work on linearisation of Dependency trees, such as Kahane & Lareau (2016); also the axiomatisation of precedence relations for Metagraph Grammar in Postal (2010: 26). For a Minimalist perspective on linearisation applied to graphs, see the mechanism proposed in Kural (2005), who implements an explicit tree-traversal algorithm in binary-branching trees. More recently, Medeiros (2021: § 9) formulates a two-tier system based on tree-traversal algorithms with the aim of capturing possible and impossible word orders cross-linguistically (in particular, Medeiros provides accounts of Greenberg's Universal 20 and the so-called Final-Over-Final Constraint).

The relations between expressions in graph representations are subjected to the usual locality conditions, which were initially formulated as constraints over dependencies across variables (*à la* Ross, 1967; see also McCawley, 1998: Chapter 15 for a summary of Ross' constraints and Müller, 2011: Chapter 1 for a more recent survey of locality conditions in generative grammar). The empirical insights obtained from the vast research on island phenomena need to be captured. Locality conditions are a fundamental research area in syntactic theory across frameworks, be them transformational (with singulary transformations: Chomsky, 2001; Branan & Erlewine, 2022; with generalised transformations: Kroch & Joshi, 1987; Frank, 2006), or not (e.g. Borsley & Crysmann, 2021; Putnam & Chaves, 2021; Falk, 2009; Kaplan & Zaenen, 1995). In the present framework, it is the formulation of these constraints what needs to be revised: in contrast to the theory in which Ross' constraints were formulated, conditions that make reference to rules are not possible in our framework. This means that conditions of the form '*A rule R cannot apply if* ...' (see, e.g., Chomsky's 1977: 101 formulation of the *Superiority Condition*) must be either reformulated as conditions over expressions or walks or eliminated from the theory if their effects can be captured in other terms (for instance, as part of the definition of elementary graphs). For it to make a contribution to syntactic theory, it is necessary that the theory presented in this monograph is able to provide analyses of phenomena captured by constraints such as the Coordinate Structure Constraint, the Complex NP Constraint, the Right Roof Constraint, etc. This is a self-imposed goal, given the remarkable empirical robustness of some of Ross' constraints (see Postal, 1998 for extensive discussion, in particular of the CSC; also McCawley, 1998: Chapter 15 for examples and a more general discussion and abundant exemplification of Ross' constraints). However, these constraints must be understood (or, better, reformulated) as constraints over dependencies in elementary and derived graphs, rather than as filters over rules. For example, we have a set of nodes, and we want to know if a walk between any two is legitimate (in the most general form, for any A, B nodes in a graph, we want to answer the question 'can I go from A to B?') and if it is, what kind of relation holds between A and B: walking that walk is locally interpreting the structure, and if a walk is illegitimate in terms of conditions over well-formed graphs, so is the interpretation of the relevant graph. It is important to highlight that the constraint-based graph-theoretic syntax we argue for here builds on empirical and formal insights that go back to the early days of the generative enterprise.

We have now enough information to situate this work in the wider context of linguistic theories in terms of what the grammar *is* and what it *does*. In the present conception, the grammar does *not* generate a set of surface structures (as in the Standard Theory-Extended Standard Theory-Revised Extended Standard Theory), a set of derivations (as in Generative Semantics or classical, 1995-style Minimalist Program), a set of form-meaning pairings (as in Categorial Grammar), or a set of constituent structure-functional structure pairings (as in the simplest LFG architecture). *The grammar as understood in the present work defines a set of well-formedness conditions over relations between nodes in local graphs*. It is thus constraint-based. These graphs will not be constructed from atomic elements by means of the application of stepwise operations (such as Merge), or grow by means of replacing symbols by sequences of symbols (as in the formalisation of rewrite rules in Chomsky, 1956, 1959). Elementary graphs represent syntactic relations between categorematic basic expressions of the language (not units of orthography, but of grammar) in the neighbourhood of a single lexical predicate; in this sense, the present proposal may be described as a *lexicalised grammar*. The focus on expressions, and the absence of derivations, implies a further departure from the fundamental MGG assumption that the units of grammar are bundles of features, semantic, phonological, and formal (Chomsky, 1995; Epstein & Seely, 2002; Adger & Svenonius, 2011, among many others). Formal features in classical Minimalist syntax serve the purpose of triggering syntactic operations, in some cases both Merge and Move apply to satisfy some featural requirement (this is particularly prevalent in Minimalist Grammars, where all operations are triggered by features; e.g., Michaelis, 2001; Stabler, 2011, 2013).[15] Without derivations (ordered sequences of stages updated by means of Merge, Move, or Agree), there is no justification for formal features and the operations of checking/valuation that ensure that every term in a structure has satisfied all featural requirements.

The previous sections have introduced the basic definitions and assumptions that guide our inquiry. The following sections will be devoted to the analysis of linguistic phenomena using the notions introduced thus far. We will focus on data that have proven difficult to analyse in Immediate Constituent-based theories, with SMC-respecting structural descriptions. The first phenomenon that we will turn our attention to (and which will give us the chance to refine our preliminary definitions and introduce further conditions on allowed relations in elementary and derived graphs) is a particularly challenging one for immediate constituent-based PSG: relations between expressions that are not linearly adjacent. Specifically, the next chapter will look at *discontinuous constituents*.

---

15 What Frank & Hunter (2021) call 'everyday Minimalism' differs from Minimalist Grammars in having 'free', 'blind' Merge (Chomsky, 2021); however, as Guinsburg (2016) observes, that makes 'everyday Minimalism' incompatible with the foundations of most work in Minimalist Grammars and creates other computational problems.

CHAPTER 3

A Proof of Concept: Discontinuous Constituents

Chapters 1 and **2** introduced the fundamentals of the formal framework within which our inquiry will be conducted: we provided general definitions that will allow us to formulate the conditions that structural descriptions assigned to specific natural language sentences must satisfy. This chapter, and the ones that follow, are dedicated to the application of the theory. We will start the empirical side of the monograph by dealing with the issue of *discontinuous constituents* and how it is captured in a graph-theoretical framework that, strictly speaking, has no notion of 'constituent' as these are defined in PSGs. We argue (not a novel observation) that 'discontinuity' is, in at least some interpretations of the term, a consequence of immediate constituent analyses where structural relations are based on contiguity between expressions (in turn, derived from the heavy focus on configurational languages, in particular English). If relations are maximised in irreducible graphs, there is no need for readjustment rules in cases of linear discontinuity between what the PSG tradition would consider 'constituents', because there is no mismatch between precedence relations between symbols and constituency relations (e.g., the binary relation *is-a*). We propose that the argument for a graph-theoretical approach to discontinuity is two-fold: theoretical advantages over PSGs in terms of the machinery needed to provide a description of the data, as well as the possibility of accounting for grammatical phenomena that IC models often struggle with (although there are IC treatments of discontinuity in the Harrisian tradition). In a sense, it is an argument from weak generative capacity (see Manaster Ramer & Savitch, 1997 for discussion about the theoretical relevance of arguments based on the weak generative capacity of grammars; Shieber, 1985 is an excellent example of a computational claim based on the grammatical analysis of stringsets).

In post-Bloomfieldian structuralism, from which early generative grammar borrowed heavily, constituents at different levels of analysis were defined by a process of segmentation and substitution (Harris, 1951: 269, 279), as illustrated by Harris in this example:

> *we determine by means of substitution what is the status of the given stretch in respect to the utterance (or to the succession of utterances in the speech):* e.g. *given the stretch* ***gentlemanly****, we determine that it is a case of A* [adverb] *from the fact that it is replaceable by* ***fine, narrow-minded****,* etc. *in* ***He's a—fellow****,* etc. (Harris, 1951: 279)

 | DOI:10.1163/9789004542310_004

The aim of IC analyses is to define an exhaustive partition of a sentence into segments that are assigned to distributionally defined classes. These distributional classes, because they are defined based on a segmentation of a string into substrings, depend on an (underlying) relation of contiguity: to use Harris' terms, a 'stretch' (which is individuated by substitution) is a sequence of linearly contiguous 'morphemes' (again, Harris' term). In this context, we can see why discontituity is a fundamental problem in frameworks that derive historically and/or methodologically from post-Bloomfieldian/Harrisian structuralism: discontinuity entails a mismatch between the superficial order of terms and their (underlying) contiguity. In general terms, then, 'discontinuous constituency' refers to a semantic or syntactic dependency holding between expressions that are not linearly adjacent. So defined, discontinuity can be used to refer to filler-gap dependencies like *what did Mary say __?*, which relate an operator and a variable, as well as constructions where there is no operator at all, as in *heat the soup up*, where *the soup* appears between the verb *heat* and the particle *up* and many other constructions in between (Right Node Raising, Relative Clause Extraposition, expletive-associate relations, comparative and degree clauses, Parenthetical Insertion, etc.). Some cases are more controversial than others: if there is no *gap*, there is no discontinuity in L-extractions (see e.g. Gazdar, 1982; Huck, 1984: Chapter 4; and **Chapters** 8 and 10 below). We thus need to specify what *discontinuity* will be used to refer to in the context of this chapter. First, a definition of *constituency* in phrase structure trees can be used to set the scene. In classical PSGs, rewriting rules operate over members of the alphabet of the grammar and ultimately produce terminal strings. The sequence of steps from the initial symbol (the 'axiom') and a terminal string is a *derivation* of that string.[1] In defining the derivation of a sentence, it is possible to also define the categories to which terminal sub-strings belong by following the mapping between strings. Therefore, if we have a symbol NP that dominates any other node or set thereof, it means that whatever set of nodes (of the sets V_N or V_T) is properly contained in the NP can appear within a string whose distributional (and semantic, but these have a very marginal role in IC heuristics) properties are that of an N. Put differently: if the string *a student of linguistics* is an NP, and *of linguistics* is a PP properly contained in the NP, then it means that *of linguistics* is a substring within a string that behaves, as a whole (distributionally and semantically), like any other string **assigned to**

1 As a reminder,

A string β follows from a string α if $\alpha = Z^\frown X_i{}^\frown W$ *and* $\beta = Z^\frown Y_i{}^\frown W$ [...]
A derivation of the string S_t *is a sequence* $D = (S_1, \ldots, S_t)$ *where* $S_1 \in \Sigma$ [the alphabet] *and for each* $i < t$, S_{i+1} *follows from* S_i (Chomsky, 1956: 117).

the category NP (**dominated by** NP in a PS tree). The PP is a **constituent of** the NP. In terms of relations between nodes in PS trees (which are elements of the alphabet of the grammar), we can define the notion of 'constituent' as follows:

> Let *s* be a string of arbitrary length. Then, *s* is a **constituent** in a phrase structure tree T iff *s* is *exhaustively transitively dominated* by a single symbol $\alpha \in V_N$ in T.

In turn, **exhaustive transitive domination** in a phrase structure tree entails that:

I. We refer to *all* symbols in *s*, and
II. All the parent nodes of all symbols in *s* (that is: the set of all symbols for which the relation **is-a** is defined for members of *s*), and
III. There is no symbol that transitively excludes a symbol in *s* if we follow dominance relations

Problems arise when either (i) the *exhaustivity* condition is not respected or (ii) the order between the elements of *s* is non-canonical. Note that linear adjacency between the expressions that make up a constituent is a consequence of the requirement for exhaustive dominance, which is itself a consequence of the way in which phrase structure trees are related to sequences of strings (as discussed in **Chapter 1**). The notion of *discontinuity* has been part of IC analyses since pre-generative times; this can be seen clearly in Wells' (1947) definition of *discontinuous sequence*:

> *A DISCONTINUOUS SEQUENCE IS A CONSTITUENT IF IN SOME ENVIRONMENT THE CORRESPONDING CONTINUOUS SEQUENCE OCCURS AS A CONSTITUENT IN A CONSTRUCTION SEMANTICALLY HARMONIOUS WITH THE CONSTRUCTIONS IN WHICH THE GIVEN DISCONTINUOUS SEQUENCE OCCURS* (Wells, 1947: 104. Capitals in the original)

Wells' proposal for the revision of the IC system includes multiple immediate constituents depending on a single node, which translates into *n*-ary branching for tree representations (see also Hale, 1983: 7). Ojeda (2005) presents the issue in the following terms:

> *Let A, B, C, D be four constituents of some word or phrase. A is said to be discontinuous if and only if (i) B is a constituent of A, (ii) C is a constituent of A, (iii) D is not a constituent of A, and (iv) D is linearly ordered between B and C.*

Note that 'discontinuity' pertains to linear order between constituents (adjacency and precedence) primarily: 'syntax' is affected only insofar as syntactic structure determines linear order. Having linear order between terminal symbols be completely determined by syntactic configuration is a prominent property of the antisymmetric enterprise in Minimalist syntax because relations of asymmetric c-command between nodes in a phrase structure tree completely determines the order of terminals in terms of precedence (see e.g. Kayne, 1994, 2018, 2022; Moro, 2000; Johnson, 2016: Lecture 2; Johnson, 2020; de Vries, 2009a), but it is also not unheard of in other frameworks. For example, Kahane & Lareau (2016) propose a Dependency Grammar approach to the relation between linear order and syntactic structure, such that governor-dependant relations are subject to a linearisation rule that converts linear precedence links in dependency trees. Linear dependency links determine the position of dependants with respect to their governors in terms of 'same' and 'opposite' directions in a left-to-right parsing model. GPSG (Gazdar et al., 1985) separates immediate dominance from immediate precedence: a rule like S → NP VP defines immediate dominance relations, and needs to be accompanied by a *linear precedence statement* (such as 'NP < VP') if only one possible order of symbols is allowed (the absence of a linear precedence statement licenses structures in which nodes occur in either order, not unordered structures; Blevins & Sag, 2013: 213). Postal (2010: 25–26) incorporates a relation *Linearly Precedes* as part of the definition of a Metagraph, as a relation *Edge* × *Edge*; such that *1 arcs* precede *2 arcs* and these precede *3 arcs* if we are dealing with neighbouring arcs (which leaves open the possibility of discontinuity, crucially; see also **Section 4.3**). Precedence between *nodes* is recursively defined upon the relation proposed for arcs (see also Zwicky & Isard, 1963: 2):

> **Definition:** *Node Precedes*
> *Node a **node precedes** b if and only if a is the head of an arc A and b is the head of an arc B such that A linearly precedes B.* (Postal, 2010: 26. Highlighting in the original)

Overall, our view on linear precedence may be seen as close to Postal's, although we are careful in not defining the conditions over which linear precedence relations are established in this work because they fall outside our scope: our aim is to provide a map of syntactic-semantic objects, not a function to map graphs to strings. All in all, we agree with Chomsky (2020) in that precedence between terminals is not a syntactic relation. The goal of our theory of syntax, as emphasised throughout this work, is to provide a specification of expressions and relations in NL sentences: strings are fed to the grammar, not produced by it. We will see in detail how our proposal differs from DG, APG and Metagraph Grammar, and how it can accommodate for different patterns of discontinuous dependencies.

To begin our discussion, we build on the quotation from Ojeda (2005), above. In the framework explored here, considering the definitions provided in **Chapter 2**, Ojeda's definition would be represented as follows:

(54) $\rho^{(*)} = \{(A, B), (A, C)\}$
$\tau = \{(B, C), (B, D), (D, C)\}$
$e = \langle A, B\rangle, \langle B, C\rangle, \langle B, D\rangle, \langle D, C\rangle$

Challenges to the traditional idea of phrase structure come in the form of strings which satisfy constituency tests but which are interrupted by a (possibly unary) string that is excluded from the constituency relation. For instance, languages where verbs and their objects behave like units for syntactic purposes but which may display VSO order (such as Irish; see Blevins, 1990): the substrings that make up the structural unit VP are not strictly contiguous, or languages with free word order in which adjectives may appear linearly non-adjacent to their modified nouns, but where sequences Adj+N pass the relevant constituency tests (such as Warlpiri; see (108d-e) below). The extent to which such cases are problematic depends entirely on whether linear contiguity is a necessary condition for constituenthood, and whether syntactic operations can be formulated that affect segments which are not linearly contiguous.[2]

In this work, syntactic operations ignore linear order of terminals (as a consequence, we also do not assume an underlying level of representation for discontinuous constituents where linear contiguity is respected, be it a Deep Structure, a Lexical Structure, etc.). To illustrate this, consider now the following fragment of a phrase marker, from Ojeda's (2005) discussion of discontinuity (see also Bach, 1979 and Jacobson, 1987 for an analysis of the operation Right Wrap which applied to this case yields V+NP+Prt from an input V+Prt+NP;[3] Carnie, 2010: Chapter 10 offers a number of examples of 'discontinuity' in analogous terms):

---

2 This is a situation that also arises in morphology, see e.g. Harris (1945). Relevant examples include parasynthesis, the lexical entries of auxiliary verbs (which rigidly select the morphology of their complement, and thus have been claimed to include their complement's desinence as part of their entry, see e.g. Chomsky, 1957; García Fernández, 2006), and multiple exponence (see Harris, 2016 for extensive discussion). In this work we will be dedicated to structure above the level of basic expressions (which, by definition, are atomic).

3 Bach (1979: 516) defines Wrap thus:
RWRAP: "Right-Wrap"

(55)

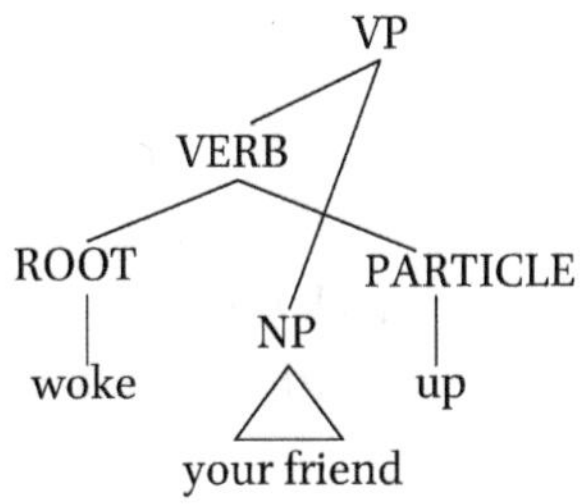

FIGURE 3.1 Discontinuous constituency

The tree diagram in (55) represents Ojeda's structural description for a verb phrase such as *wake your friend up* (cf. Huck, 1984: 28); we will use it as a test subject for our preliminary inquiries into discontinuity (Chen-Main & Joshi, 2010 offer examples of Dependency Grammar-influenced TAG graphs which feature similar discontinuities). The empirical fact to capture is that *wake up* is a unit which is linearly interrupted by the NP. We need to bear in mind that crossing lines have no formal meaning (a point also made by Ojeda) since there is an isomprohic graph to (55) with no crossings: a theory of discontinuity cannot be formulated in terms of *diagrams* of L-trees, but in terms of structural relations. The problems that discontinuity poses for PSGs is not one related to diagrams, but to two axioms that apply to immediate constituent structural descriptions: in Wall's (1972: 148–149) terms, these are the *Exclusivity Condition* and the *Nontangling Condition* (see also Zwicky & Isard, 1963; Gazdar et al., 1985: 49, ff.; Partee et al., 1990: 442–443; Carnie, 2010: 189, ff.; Koopman et al., 2014: 25, among many others):

> THE EXCLUSIVITY CONDITION: *In any well-formed constituent structure tree, for any nodes x and y, x and y stand in the precedence relation P, i.e., either* $(x, y) \in P$ *or* $(y, x) \in P$*, if and only if x and y do not stand in the dominance relation D, i.e., neither* $(x, y) \in D$ *nor* $(y, x) \in D$

(i) *If a is simple, then* RWRAP(a,b) = RCON(a,b) [where RCON = Right Concatenation],
(ii) *If a has the form* $[_{XP}\, X\, W]$*, then* RWRAP(a,b) *is* $X^\frown b^\frown W$.
Note that Bach's definition applies to syntactic phrases rather than phonological material. We depart from Bach in allowing multi-word expressions: xw would be one such expression (corresponding to a single node in a graph). This non-syntactic definition of Wrap will become handy also for the analysis of the behaviour of 'intermediate elements' in Spanish auxiliary verb constructions (see also García Fernández et al., 2020).

THE NONTANGLING CONDITION: In any well-formed constituent structure tree, for any nodes x and y, if x precedes y, then all nodes dominated by x precede all nodes dominated by y

Ojeda observes that the conjunction of these conditions bans a tree like (55) (see also Huck, 1984: Appendix), enforcing, in a sense, planarity. Wall's argument, which is representative of immediate constituent PSGs, requires that constituent structure trees represent both hierarchical grouping (constituency) and linear order between terminals. In such an approach, there is no way to say that *wake up* is a constituent in *wake your friend up*, other than defining the exclusivity and nontangling conditions as conditions over kernel trees and allowing for reordering operations, which move *up* to a position where it is preceded by *your friend*. If we follow this route, the grammar needs two kinds of rules: immediate dominance (ID) rules and left-precedence (LP) rules (see, e.g., Ojeda, 2005: 9–10). We will try to avoid such complications.

Let us describe the relations in (55). Using the tools introduced in **Section 2.1**, we can identify the following dependencies:

(56) ρ = {(VP, VERB), (VP, NP), (VERB, ROOT), (VERB, PARTICLE)}
τ = {(VERB, NP), (ROOT, PARTICLE)}
e = ⟨VP, VERB⟩, ⟨VP, NP⟩, ⟨VERB, ROOT⟩, ⟨VERB, PARTICLE⟩

However, it is not at all clear that in providing a description of expressions and relations in *wake your friend up* we need all that intermediate structure: we should be able to clean it up.

The fact that there is a line crossing over the edge VERB-PARTICLE does not change the structural relations that can be defined in the formal object that (55) diagrams: the NP is still excluded from the object labelled VERB and included in the object labelled VP; in other words, NP and VERB are the two immediate constituents of VP. VERB immediately dominates two pre-terminal nodes ROOT and PARTICLE. In Ojeda's representation, it is *wake up* that takes NP as a complement, not *wake* alone. Finally, and almost as a side note, Ojeda's representation respects the SMC, which we also challenge. An alternative set of relations, after some minimal house cleaning, would be (57):

(57) ρ = {(VP, NP), (VP, ROOT), (VP, PARTICLE), (ROOT, NP)}
τ = {(ROOT, NP), (ROOT, PARTICLE), (NP, PARTICLE)}
e = ⟨VP, ROOT⟩, ⟨VP, NP⟩, ⟨VERB, NP⟩, ⟨VERB, PARTICLE⟩

Note that we have eliminated VERB, and that the NP has two mothers: VP and ROOT. But, do we really need VP as a node in the graph? In the theory explored here, VP is the 'name' that we give a subgraph for ease of reference (that is: the graph 'VP' is an abbreviated way for us to refer the set of nodes ν_{ROOT}, ν_{NP}, and ν_{PARTICLE} and the edges connecting those nodes); 'VP' is *not* a node in that subgraph. Thus, it seems we can dispense with VP as a node. Thus, what we have is the preliminary description in (58), maintaining for the time being Ojeda's labels:[4]

(58) $\rho = \{(\text{ROOT}, \text{NP}), (\text{ROOT}, \text{PARTICLE})\}$
$\tau = \{(\text{NP}, \text{PARTICLE})\}$
$e = \langle \text{ROOT}, \text{NP} \rangle, \langle \text{PARTICLE}, \text{ROOT} \rangle$

The specification of edges requires some justification, and at least one revision. Suppose, for the time being, that ROOT and PARTICLE are both independent basic expressions of the language that are assigned addresses in a graph. Recall our earlier claim that only categorematic expressions are assigned addresses: the particle *up* (as opposed to the preposition *up*) seems to be *syncategorematic* (in LFG it would be treated as a 'non-semantic' preposition, and not assigned a PRED value). Thus, it cannot be a node in the graph: there cannot be an edge *e*⟨ROOT, PARTICLE⟩ or *e*⟨PARTICLE, ROOT⟩. The previous representations are mistaken in assuming that *wake* and *up* are independent expressions. We can eliminate *PARTICLE* as a node, and define *wake-up* as a two-word basic expression of the language, with address ⦃wake-up⦄ and semantic value ⟦wake-up⟧: there is no syntactic dependency between *wake* and *up*. This multi-word expression is what subcategorises for an object *your friend*. This is a point where our approach differs from Dependency Grammars: in general, DGs do not allow for multi-word basic expressions (see **Section 4.2**).

4 The rationale behind the 'tree pruning' undertaken in the paragraphs above, and which we will assume for the rest of this work, is not too different from what underlies LFG's Economy of Expression in *c*-structures:

> *All syntactic phrase structure nodes are optional and are not used unless required by independent principles (completeness, coherence, semantic expressivity).* (Bresnan, 2001: 91)

As in LFG, the restriction assumed here applies to terminals as well as nonterminals: here, a 'terminal' is a node whose ρ-domain is empty, i.e. a node with outdegree 0. A major property of our analysis is that it delivers as few terminals as possible in the description of a sentence, but also allows us to have these terminals as richly connected as possible. It is important to bear in mind that, strictly speaking, *economy of expression* is a condition over structural descriptions. *Economy of expression* is understood in a meta-theoretical sense, as an admissibility condition over which structural descriptions can be proposed within the theory.

The question now is whether we need any intermediate nodes at all: we saw in **Chapter 1** that while they are necessary in formalisms that have roots in Post-rewriting systems, they have no justification given the foundational assumption that graphs define dependencies between basic expressions: intermediate symbols are not expressions of the language, they are expressions of the meta-language. Thus, the use of NP is just as undesirable as the use of VP in the present framework, and the node can be pruned for the same reason. All we have is ⦃friend⦄. An edge *from* the verb *to* the node that corresponds to the expression *your friend* is required to establish a *dependency* (in the technical sense, see Osborne, 2005, 2019) in which a predicate selects its arguments: we thus replace *e*⟨ROOT, NP⟩ with *e*⟨⦃wake-up⦄, ⦃friend⦄⟩. As argued in **Section 2.2**, having directed edges from predicates to their arguments and from modifiers to modified also allows us to define a compositional interpretation for graphs. The revised specification of expressions and relations in the sequence *wake your friend up* is thus (59):

(59) a. ρ = {(wake-up, friend)}
b. ⦃wake-up⦄ ⟶ ⦃friend⦄

FIGURE 3.2 Graph-theoretic analysis of 'wake up your friend'

The word order *wake friend up* compared to *wake up friend* is indicative not of a change in syntactic relations, but of Wrap. Wrap is an example of a relation-preserving process, in which only linear order of phonological material is affected, possibly not a syntactic operation at all. A basic assumption of the present framework, which we have highlighted throughout, is that syntactic configuration and precedence between expressions are distinct systems, and changes in linear order must not be taken to indicate a change in syntactic configuration. Formally, because syntactic relations and linear precedence are divorced, we can define graphs without crossings which are isomorphic to tree diagrams with crossing lines: what matters in our approach is what grammatical relations (in a sense to be made more precise in **Chapter 5**) hold between expressions in these graphs.

Maximising local relations and eliminating movement transformations has other advantages in the analysis of discontinuity. Consider the following examples, featuring restrictive relative clauses (RRC):

(60) a. A man entered who was wearing a black suit (Relative Clause Extraposition, RCE)
b. *What colour did a man enter who was wearing? (violation of the Complex NP Constraint, CNPC; Ross, 1967: 127)

Ojeda (2005)—correctly in our opinion—identifies discontinuity in RCE cases (see also Osborne, 2019: Chapter 8; Huck, 1984: 43, ff.), which ensues the Complex NP Constraint CNPC violation that renders (60b) ungrammatical (this in turn depends on the RC forming a constituent—which is referred to as the *complex NP*—with the N as observed by McCawley, 1998: 451; we will come back to this particular issue in **Chapter 9** below). We need to capture this empirical observation. However, in doing so it would be desirable to ban the RC from being an adjunct to a phonologically null NP or to an unpronounced copy of the subject before raising to Spec-TP to satisfy an EPP property or any such requirement. That is: we want to eliminate possible structures like (61) below, which contains a deleted copy of [a man] or any such transformationally obtained phonologically empty node, even though the terminal string that corresponds to (61) is a well-formed expression of the language:[5]

(61) A man entered ~~a man~~ who was wearing a black suit

A possibility explored in Huck (1984: 45) is to analyse RCE in terms of crossing branches (see also McCawley, 1982, 1998): the relative is a daughter of an NP node, but the arc that dominates it crosses over the VP. There is an extra feature in (61) that Huck's analysis does not consider in depth: the relation between the antecedent and the relative pronoun, which in MGG involves phonologically null elements. Discontinuity alone cannot give us a way to filter such structures, unless a further ban on null copies is introduced. The alternative (looking at the problem from the opposite side) is to assume that *copies are never introduced to begin with*, and thus there is nothing to delete. This, which amounts to adopting a non-transformational standpoint, is the option that we will explore here, since we have *structure sharing* at our disposal. Furthermore, we need to factor in that since we have two clauses each anchored by a lexical predicate, we have also more than a single arbor. Let us make things explicit: we will identify the single-rooted graphs as *arbores* and provide a preliminary specification of the relations of dominance that we find in each (to be revised below):

(62) Arbor 1: [a man_1 entered]
Arbor 2: [a man_2 was wearing a black suit]

5 A transformational analysis would require two movements: first, the NP *a man who was wearing a black suit* moves from Compl-V to Spec-TP. Then, the relative clause is extraposed (or viceversa). If only leftwards movement is allowed (as, e.g., in Kayne, 1994 and much subsequent work), the sequence of movement operations becomes even more complex and produces even more empty nodes.

$\rho_1 = \{(\text{entered}, \text{man}_1)\}$
$\rho_2 = \{(\text{wearing}, \text{man}_2), (\text{was}, \text{wearing}), (\text{wearing}, \text{suit}), (\text{black}, \text{suit})\}$

We can sketch a preliminary analysis with what we have so far. Up to now, we have defined *arbores* based on purely configurational information; namely, (i) the presence of a single root node and (ii) irreducibility (see also Rogers, 2003, whose notion of *local tree* is in principle compatible with our characterisation of *elementary graphs*). But we have given no argument as to why these properties should matter at all in the generation of descriptively adequate structural descriptions for natural language strings: now we will give such arguments, and further clarify the distinction between *arbor* and *elementary graph*. We defined *arbores* in configurational terms, but without specifying what they contain or how big they can get. Limiting the size of local domains in syntax is essential, as it has direct consequences for the theory of locality; our previous discussion was somewhat imprecise in this respect. We will remedy that now.

As we saw in **Section 2.3**, in Lexicalised Tree Adjoining Grammars (LTAGs) the building blocks of grammar are lexically anchored structures: in LTAGs these are called *elementary trees*. By analogy, we refer to the local structural domains that correspond to our irreducible structures as *elementary graphs*. Let us be more precise about what these structures are made of. An *elementary graph* in a structural description is a *single-rooted graph* (i.e., an *arbor*) which contains the following elements (based on García Fernández & Krivochen, 2019a; Krivochen & García Fernández, 2019, 2020):

(63) a. A predicative lexical basic expression *p*
b. Functional modifiers of *p* (e.g., temporal and some aspectual auxiliaries. Cf. Bravo et al. 2015's *functional auxiliaries* and **Section 7.1.1**)
c. Arguments of *p* (e.g., subject, object, oblique …)

The focus of the aforementioned works was the syntax of auxiliary verb constructions, but their implications go beyond the verbal domain. In García Fernández & Krivochen (2019a), following Frank (1992, 2002, 2013) we referred to each structural unit containing elements (63a), (63b), and (63c) as the *extended projection* of *p* (in a related sense to Grimshaw, 2000; Abney, 1987: 57; more on this below). Because each local domain is structured around a single lexical predicate with a relational network, the model of grammar presented here is *lexicalised*. An elementary graph is thus a unit of argument structure (see also Hale & Keyser, 2005: 11; Frank, 2002: 55), in that selectional properties of the lexical predicate (including thematic structure) are satisfied within that predicate's elementary graph (see also Frank, 2013: 240). This idea is intimately related to and indeed builds on the so-called *Condition on Elementary Tree Minimality* (CETM) in LTAG (Joshi & Schabes, 1991; Sarkar & Joshi, 1997; Frank, 1992, 2002, 2006, 2013; XTAG group, 2001):

> *Each elementary tree consists of the extended projection of a single lexical head* (Frank, 1992: 53)

The lexical head of an elementary tree is usually called the 'lexical anchor' of that elementary tree; we will also use the term *anchor* for the lexical predicate that nucleates an elementary graph. More recently, Frank (2002, 2013) elaborates on this perspective, also based on the notion of *extended projection*, slightly reformulating and expanding on the CETM in the following terms:

> *The syntactic heads in an elementary tree and their projections must form an extended projection of a single lexical head.* Frank (2002: 22)

where

> *the extended projection of a lexical head H includes H's immediate projection and the projections of the 'functional shell' that surrounds H.* (Frank, 2013: 239)

Furthermore, TAGs impose a strict locality requirement on syntactic dependencies: this is the *Fundamental TAG Hypothesis*:

> *Every syntactic dependency is expressed locally within a single elementary tree* (Frank, 2013: 233)

The restriction on the size of elementary trees proposed by Frank is essentially what we are going for, provided that the elements in the *extended projection* of a lexical head are the ones in (63a-c): note that under this definition, arguments are part of the extended projection of the predicate that select them. This is a departure from most versions of LTAGs, which make no distinction between lexical predicates and lexical arguments in defining elementary trees: in Frank (2002, 2013) and related works all lexical terminals are anchors regardless of whether they are predicates or not; the versions in Rambow (1993) and Hegarty (1993) have both lexical *and* functional heads be anchors. Let us compare these proposals. For a sentence like *John has run*, the elementary trees proposed by Frank and Hegarty/Rambow are as in (64a) and (64b) respectively:

(64) a.

b.

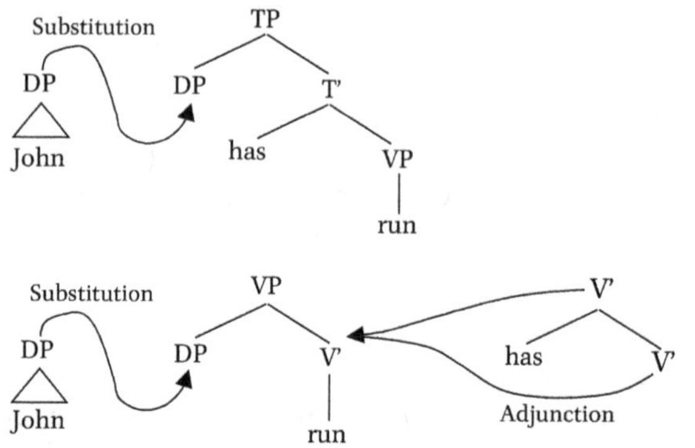

FIGURE 3.3 TAG derivation with Substitution and Adjunction

If we now go back to the definition of *arbor* in **Chapter 2** (as a single-rooted graph), then we can make some adjustments: every elementary graph is an arbor, but there may be arbores that are not elementary graphs: a single-rooted graph with no lexical predicate does not qualify as an elementary graph under (63). This distinction may be relevant for the grammatical analysis of non-predicative constructions, including fragments and interjections, which would be arbores (by virtue of being single-rooted graphs), but not elementary graphs (by virtue of not containing predicates or arguments). Imperatives are possibly also in this category of arbores that are not elementary graphs given their combinatorial restrictions (as noted in Schmerling, 1982, they cannot appear embedded). In what follows, then, we will prefer *elementary graph* when it is important to highlight that the graph we are referring to contains all the elements specified in (63). Otherwise, *arbor* and *elementary graph* will be used indistinctly.

If we allow every head to anchor an elementary graph, the building blocks of syntax become smaller, but there is a tradeoff: there are more composition operations and therefore additional restrictions must be put in place to apply these composition operations *in order*. It does not matter for (64b), since any order of composition (*substitution* > *adjunction* or *adjunction* > *substitution*) yields a grammatical output; however, this is not always the case (an idea that goes back to early generative discussions about rule ordering such as Ringen, 1972 and Koutsoudas & Sanders, 1979; see Krivochen & Padovan, 2021; Padovan, 2021b for discussion). Under Frankian assumptions, for a sentence like *John read an old, black, heavy book* we would need:

(a) An elementary tree anchored by *read*, with substitution sites for two DP arguments
(b) An elementary tree anchored by *book*, with no substitution sites

(c) An elementary tree anchored by *John*, with no substitution sites
(d) An elementary tree anchored by *old*, with root and frontier labelled N'
(e) An elementary tree anchored by *black*, with root and frontier labelled N'
(f) An elementary tree anchored by *heavy*, with root and frontier labelled N'

The ETs anchored by the adjectives adjoin to the ET anchored by *book* at recursive N' nodes, perhaps at the same time (as in tree-local multicomponent TAGs; Frank, 1992; Schuler et al., 2000; Kallmeyer, 2004). This delivers a derived tree corresponding to the string *old, black, heavy book*. This object, of category DP, would undergo substitution targeting a DP substitution site in the complement of the VP headed by *read*. Note that *read* heads a VP, but anchors an ET which is bigger insofar as it contains also the extended projection of V (for concreteness, assume that this extended projection goes up to CP, under standard Minimalist assumptions). The root of that ET will be CP, its anchor will be *read*. Root, anchor, and label are all kept distinct (*read* is not labelled VP, but V): this distinction seems to be crucial in the analysis of the nominal domain (where, under IC assumptions, *old, heavy, black book* is of category NP, not AP despite the N being an argument of A. See, however, Scott, 2002; Bruening, 2020 provides critical discussion). The DP headed by *John* will also undergo substitution, targeting the specifier position of TP. Again, both arguments may be introduced in the same derivational step.

We will see that the determination of what can be anchors has a profound impact on the size of elementary trees, and therefore on the empirical adequacy of the LTAG. With Frank (2013), we propose that the size of elementary graphs (determined by what counts as a lexical predicate in our version of lexicalisation) is a crucial source of cross-linguistic variation. We will come back to this issue in **Section 7.1.1** when looking at relations between auxiliaries in Spanish auxiliary chains.

An important feature of the system presented here is that there is not really a notion of 'projection', because there are no 'heads' (in the X-bar sense) or intermediate non-terminals ('bar-levels', in X-bar parlance) since every arc connects two basic expressions. In this sense, the use of 'projection' here is mostly mnemonic, and not necessarily equivalent to concepts like Abney's (1987: 57) *c-* and *s-projection*. But also, as we have said, our definition of *elementary graph* is more restrictive than that of *elementary tree* in a lexicalised TAG along the lines of Frank or Hegarty/Rambow: only lexical predicates define elementary graphs in our model, whereas *all* lexical heads define elementary trees in a standard LTAG (see also XTAG group, 2001). Thus, we are making *two* claims, one configurational, the other substantive (and linguistically motivated):

I. Arbores are irreducible single-rooted graphs
II. Elementary graphs are arbores defined around lexical predicates: an elementary graph is the smallest set of connected nodes that contains all three kinds of expressions in (63 a–c).

Let us go back to the analysis of discontinuity in the extraposed relative, to make our discussion more concrete. In (62) above, we have *two* single-rooted structures: the one we called Arbor 1 [a man entered] contains an unaccusative predicate *enter*, with aspect and tense marked synthetically and its nominal argument *a man*; the one we called Arbor 2 contains a transitive predicate *wear*, with progressive aspect realised by means of the auxiliary *be -ing*, and both its subject and object *a man* and *a black suit* respectively. We see that each individual sub-graph satisfies the definition of *elementary graph* that we introduced above in (63), and is also compatible with the CETM. Frank (1992) correctly observes that the CETM restricts the size of a single elementary tree, because of the constraints on which 'extended projections' can be built: the scare quotes are required here because extended projections were originally defined in terms of X-bar theory (Grimshaw, 2000: 116, ff.; also Abney, 1987), with endocentricity as a fundamental property. Frank (2013) goes farther, by identifying the size of elementary trees as a source (possibly, 'the' source) of linguistic variation at the syntactic level: if elementary trees have a single lexical anchor, then it depends on what counts as 'lexical' in a particular linguistic system. For example, whereas Frank locates English modals in the same elementary tree as lexical verbs, arguably in Spanish modal auxiliaries are the lexical anchors of their own elementary trees (see Bravo et al., 2015; Krivochen & García Fernández, 2019a, b; we will return to this issue in **Section 6.1.1.** Also important is the fact that the intensional definition of the category 'auxiliary' and 'modal' varies between the Hispanic and English grammatical traditions). Crucially, this means that a sequence *modal + lexical verb* in English and in Spanish would receive different analyses, something we consider an advantage of this theory insofar as there are empirical reasons to argue for these differences (see also Krivochen & Padovan, 2021 for discussion about locality and cyclicity cross-linguistically from a LTAG perspective).

The idea that elementary graphs are 'endocentric' must be taken carefully, for if *endocentricity* is understood as *feature percolation* from a head (that is: the features of a head percolate upwards the projection path, such that XP is a projection of X by virtue of a relevant feature of X—categorial, say—percolating to XP; as in Grimshaw, 2000: def. (3)), this notion clearly *does not apply* to our graphs, seeing as (a) they are not labelled in the PSG sense, and (b) there is really no notion of 'projection' in the present framework. Each elementary graph is defined by the presence of a lexical predicate plus the arguments it subcategorises for and functional modifiers of that predicate; in this sense, we could say that they are 'endocentric'. The notion of 'endocentricity' that is relevant here is that there is an element, contained in the arbor, which determines what else can co-occur in that local structure. That element is, borrowing LTAG terminology, the *lexical anchor* of the elementary graph.

We can now provide a concise definition of *arbor* and *elementary graph* considering the preceding discussion:

> ***Arbor (definition)***: an *arbor* is an irreducible single-rooted graph.
> ***Elementary graph (definition)***: An *arbor* will be called an *elementary graph* iff it contains (i) a single predicative lexical basic expression, (ii) modifiers of that predicative basic expression (which, as in the case of tense and aspect, can be analytically or synthetically expressed) and (iii) arguments of that same predicate.

This is an improved definition, although we will see that there are further considerations to be made which can simplify the formal apparatus.

Clearly, in the description of a complex structure such as (60a) local structures are not disconnected, independent units: there are restrictions over the occurrences of arguments and the dependencies that can be established between occurrences of an object across sub-graphs. We already introduced the basic mechanisms of graph composition in **Chapter 2**, but now we can relate the formal definitions to linguistic conditions. For instance, let $p \in X$, $p' \in Y$, and $p'' \in Z$ be predicative basic expressions in elementary graphs X, Y, and Z. Furthermore, if α is a dependant of p in X, α must be dominated by p in X: $\rho(p, \alpha)$ must hold in X. Now, if we have an argument β which may surface as a syntactic-semantic dependant of more than a single predicate, in our case p, p', p'', then it must be possible for β to be dominated at X, Y, and Z by p, p', and p'' respectively if graph union has produced a derived graph from X, Y, and Z. This is a condition that requires an explicit rejection of the SMC, for β has—in this example—*three mothers*: p, p', and p''. We can illustrate such a configuration with Right Node Raising, as in (65a) or Across the Board *wh*-movement, as in (65b) (we will come back to these in **Chapter 14**):

(65) a. Bill washed, Peter dried, and John broke the dishes
b. What did Bill wash, Peter dry, and John break?

In both (65a) and (65b), p = *wash*, p' = *dry*, and p'' = *break*; in (65a) β = *the dishes*, in (65b) β = *what*. We can provide a diagram for clarity:

(66)

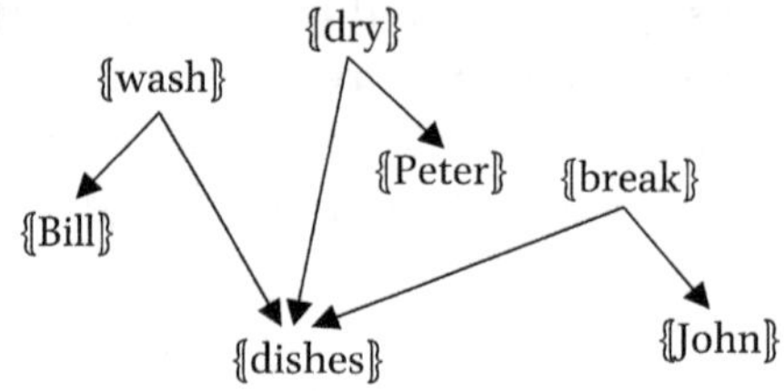

FIGURE 3.4 Structure sharing under RNR

Because there are three lexical predicates, there must be three elementary graphs (which, by virtue of being single-rooted graphs, are also arbores); each of these must contain *two* nominal arguments since all three predicates are monotransitive: each requires a subject and a direct object. When we compose the individual elementary graphs to form a RNR structure, or an Across the Board[6] (ATB) *wh*-interrogative, common addresses are identified as a single node in the derived graph (more details about long-distance *wh*-dependencies are given in **Chapter 10**; rightward extraction is analysed in **Section 14.1**).

Note that we do multiply neither the entities nor the relations: entities are not multiplied because we have just one node β (in (66), *dishes*); or more specifically, the address corresponding to β is called at different points of a traversal through the derived graph (which in this case is a tree); if we define elementary graph composition as *graph union* (such that we consider only the set of nodes in the derived graph that results from the operations of composition), there is no multiplication of β's. The relations are also not multiplied because under a theory that had as many visits to β in the derived graph as there are distinct subcategorising predicates, in our case (with predicates *p*, *p'*, and *p"*) there would still be exactly *three* predicate-argument relations (thus, we are not *joining* the graphs in the technical sense). As with multiple adjectival modi-

6 A rule applies *across the board* if and only if it affects *all* terms of a true coordinated structure (Ross, 1967). Williams (1978: 32) proposes the following generalised definition:
(i) *The structure*

$$\begin{bmatrix} [X_1]_{C_1} \\ \ldots \textit{ and} \\ [X_n]_{C_n} \end{bmatrix}$$

Is a well formed labelled bracketing if $X_1, \ldots, X_n$ *are.*
We will say that a string containing structures defined by [(i)] *is in ATB format.*
Williams' definition anticipates elements of Goodall's (1987) conditions over transformations in terms of *parallel structures*.

fication, we want the semantic value of the derived graph to be the union of the semantic values of the elementary graphs.

The previous paragraph made an essential point in terms of how inter-arboreal dependencies are established; that is, how distinct arbores are related. These considerations must now be applied to the analysis of (60). In the preliminary structural description (62), the expression *a man* serves as a link between both arbores [a man entered] and [a man was wearing a black suit], since the expression *a man*, with address ⦃man⦄ points to the semantic value of the node, and this semantic value is the same in both arbores (the man who entered is the same who was wearing a black suit). In the terms used in Krivochen (2015b), we are dealing with two *tokens* of the same *type*: the expressions *who* and *a man* both have the same semantic value. The crucial issue here is that, as we have defined above, node identifiers are *uniquely identifying pointers*. One way to ensure this rigidity in the relation between addresses and their content is to make their content something invariable across syntactic contexts: the semantic value of the expression that corresponds to the relevant node. We will refer to the relation of distinct arbores by means of a common node as *Linking*. Two or more arbores are *linked* at all common nodes. Linking arbores operates like *Unification* (in the sense of Shieber, 1986; Sag et al., 1986; see also Johnson, 1988: § 2.10 for technical discussion) or *structure sharing*: nodes that have the same semantic value are identified as part of graph union, with no need to resort to an independent indexing mechanism (that's what addresses are for). This approach has antecedents within TAG as well: for example, Vijay-Shanker & Joshi (1991) develop a Unification-based TAG where nodes in elementary trees correspond to feature structures: these feature structures include the definition of a node's relation with other nodes in an elementary tree. Vijay-Shanker & Joshi do not go as far as proposing graph-theoretic structural descriptions with loops and closed cycles, however, and we identify nodes with semantic values, not feature structures. Regardless, the basic idea (namely, that elementary structures that contain nodes with identical addresses, when composed, will collapse those nodes into one) is common to all aforementioned proposals, and to ours.

Configurationally, the fact that a node can be a term of more than a single grammatical relation is captured via multidominance: we make a node a daughter of as many mothers as predicates take it as an argument (see also Postal, 2010: 17 for discussion). In principle, we could be talking about a single predicate and more than one grammatical relation or as many predicates as grammatical relations, always with a single argument: the theory allows for this, therefore, we need to verify whether those relations actually hold in empirical analysis. More than one predicate establishing grammatical relations with a unique argument is a well-known situation: consider *Equi* and *Raising* structures (a similar remark is made in Sampson, 1975, to which we will come back below). But one predicate establishing more than a single grammatical relation with a single argument is not any less common: this corresponds, for instance, to cases of reflexive anaphora (in the sense of Reinhart & Reuland, 1993). We will come back to the analysis of reflexives in **Section 6.2.1** and **Chapter 8** below. It is important to note that the relation *dominance* is not always to be interpreted in predicate-argument terms: that depends on properties of the relevant nodes connected by a directed edge. The theory of grammar cannot, we argue, focus on configuration alone and ignore how the properties of related elements influence configuration; the descriptive importance of *lexically governed* processes and *exceptions* must not be underestimated (see Lakoff, 1965; Gruber, 1965; Dowty, 1978, among many others).

In the light of the preceding discussion, aimed at justifying the elimination of duplicated nodes (copies, in MGG) in derived structures, we need to make some adjustments to our analysis in (49), because—as suggested above—we are accessing *the same node twice*, once per arbor. There is a node v_1 that belongs to arbor 1, assigned address ⦃A⦄, and a node v_2 that belongs to arbor 2, also assigned address ⦃A⦄. This means that v_1 and v_2 have the same semantic value. When arbores 1 and 2, each an elementary structure, are used to derive a composite structure (arbor 3), v_1 and v_2 are identified as one and the same node (call it v_3). This node, however, maintains all dependencies established at the level of elementary structures with other nodes in arbor 1 and arbor 2 (note that the indegree of v_3 will be the union of the indegree of v_1 and v_2). This process is illustrated as follows:

(67)

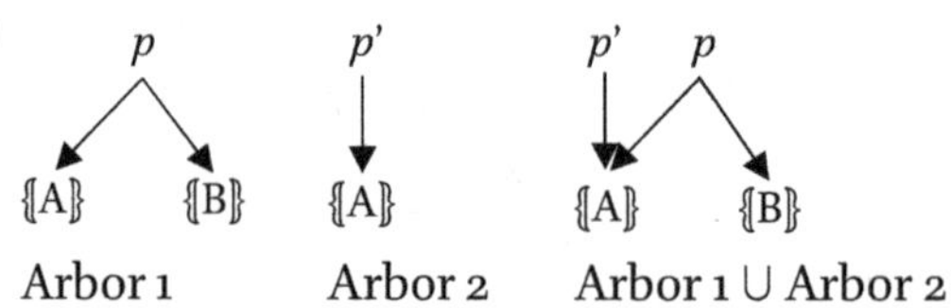

FIGURE 3.5 Structure sharing

The same mechanism applies within a single elementary graph if, for instance, a single expression establishes more than one relation with a predicate within an elementary graph (as in the case of reflexivity, see **Section 6.2.1**).

In this context, we only have one node corresponding to the expression *a man* which is subcategorised for by two distinct predicates. It is a good opportunity to point out that the framework presented here differs from Metagraph grammar in an important sense: given the fact that the content of nodes' addresses are the semantic value of basic expressions, there is no need to resort to *copy* (*pronominal*) *arcs* (Postal, 2010: 40–41) because we do not have multiple terminals for 'a man' and 'who'; in our scenario, there is no need to *copy* or indicate co-reference between distinct nodes (which would be distinct links in a chain in MGG) in any specific way, because there is only *one node* per entity (assuming, as we said before, that nodes are addresses and that addresses are unique identifiers). Thus, we can do better than our preliminary analysis above.

Distinct arbores are related by means of nodes which are the targets for *embedding transformations* (in the sense of Fillmore, 1963) or *substitution/adjunction* (in the sense of Joshi, 1985; Kroch & Joshi, 1985 and related work). Both are *generalised* transformations, insofar as they map sets of structures to structures (*adjunction* can be interpreted as a general case for *substitution*; see Kroch & Joshi, 1985: 11). Before, in **Section 2.3**, we introduced the idea of *linking* distinct arbores, provided that they contain nodes assigned the same address. We can now provide the following semi-formal definition:

(68) ***Linking (definition)***:
*If an elementary graph G contains ⦃v_1⦄ and an elementary graph G' contains ⦃v_1⦄, and G and G' belong to the derived graph G", then G and G' are **linked** at v_1 in G"*

The revised set of relations in sentence (60a) (*A man entered who was wearing a black suit*) would then be (69) (we will come back to the structure of unaccusatives in **Section 5.1**):

(69) Arbor 1: [man entered]
Arbor 2: [man was wearing a black suit]
ρ_1 = {(entered, man)}
ρ_2 = {(wearing, man), (wearing, suit), (black, suit)}

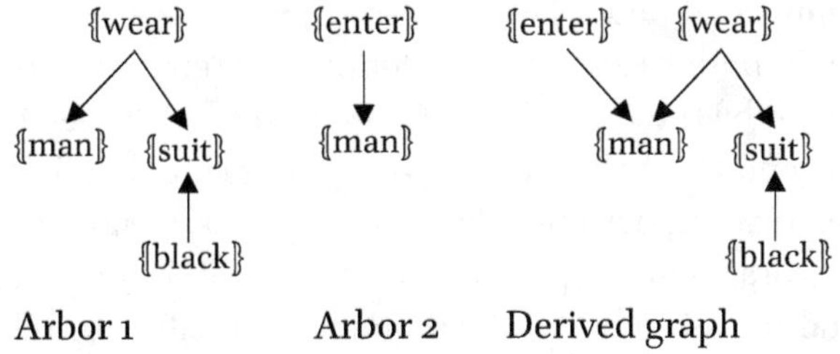

Arbor 1 Arbor 2 Derived graph

FIGURE 3.6 Arbores and derived graph for sentence containing a relative clause

Both arbores are linked at the node ⦃man⦄, which is visited *twice* in the derived graph: once in arbor 1 (as the tail of an arc headed by ⦃enter⦄) and once in arbor 2 (as the tail of an arc headed by ⦃weak⦄) (cf. the analysis of relatives in McKinney-Bock & Vergnaud, 2014). At the level of the derived graph, nothing would change if the arguments defined independent arbores, as in Frank's (2002, 2013) LTAG approach, but under procedural assumptions there are at least two ways to obtain the same output structure (thus, two distinct grammars with different *derivational generative capacity*). In the target sentence we also have the progressive auxiliary *be -ing*, which is a functional modifier of the lexical anchor *wear*. We have omitted it from the definition of the graph since there are additional considerations that apply to auxiliary verb constructions, and which we will come back to in **Chapter** 7. Also, it will be important to define an order over arcs, such that we can distinguish *a man who was wearing a black suit entered* from *a man who entered was wearing a black suit*. The specifics of the order over arcs (which cannot straightforwardly be represented in diagrams) and its linguistic motivation will be given in **Chapter** 5. An important aspect that the analysis reveals, due to its focus on syntactic relations rather than linear precedence, is that the extraposition of the relative clause does not change existing relations: Relative Clause Extraposition is an example of a relation-preserving transformation.

In relating distinct elementary graphs, it is possible to have both hypotactic and paratactic relations, depending on whether linking occurs at the root or not: if the linking node is immediately dominated by the root node in *both* subgraphs G and G', we are in the presence of what Fillmore (1963) referred to as a *conjoining transformation*, a kind of generalised transformation that takes two phrase markers P and P' and yields a new, derived phrase marker with root P" which immediately contains P and P":

$$\left.\begin{matrix} P \\ P' \end{matrix}\right\} \rightarrow P''$$

If *linking* targets an embedded node (i.e., a non-root) we have *embedding*, as in the RCE case above. In both cases, because dominance relations are transitive, anything dominated by the node G and G' are *linked* at is accessible for syntactic relations at G and G'. The notion of *linking* is essential to formulate constraints on possible relations between elementary graphs; it will prove very useful when dealing with issues of *locality* and *opacity*: we will address the question 'under which conditions is a node or sub-graph accessible for relations involving a node outside the first's elementary graph?'. We will see that *linking* as defined above may overgenerate in terms of allowing for connections between graphs to hold if not appropriately restricted: in **Chapter** 7 we will introduce the notion of *self-contained graph* to formalise such a restriction. The theory of locality will be formalised as a set of constraints over cross-arboreal dependencies in combination with the definition of elementary graph.

In the following chapters we will focus on the explicit analysis of various English and Spanish constructions under present assumptions, and in these analyses compare the graph-theoretic approach advanced here with competing syntactic accounts based on their empirical adequacy, the richness of derivations, representations, and well-formedness conditions, and the extent to which additional entities and relations (including empty nodes, metarules, etc.) need to be invoked in each case.

CHAPTER 4

Some Inter-Theoretical Comparisons

Chapter 4 argues that in terms of the complexity of structural descriptions, the approach presented in this book converges with lexicalised TAG s, given the fact that our graphs are themselves lexicalised: we can specify grammatical relations within local graphs which are maintained even after these elementary graphs undergo graph union. The focus will be set on the descriptive adequacy of the structures proposed rather than on mathematical proofs of how restrictive or not the underlying formalism is: frameworks whose underlying formalism is restricted can become unrestricted if, for example, there is no restriction over the introduction of arbitrary features or operations. Substantive aspects of the theory become essential to restrict the formalism: here, what nodes stand for and what constitutes a well-formed elementary graph, and what constraints are imposed through lexicalisation and linking. In this sense, we agree with Müller (2020: 549) when he writes

> *It is not the descriptive language that should constrain the theory but rather the theory contains the restrictions that must hold for the objects in question.*

Similar remarks are found in Shieber (1988: 38, ff.): a theory restricts the formalism, and the formalism expresses analyses (see also Pollard, 1997; Gärtner, 2002: 78, fn. 154). It is not the formalism in which the theory is expressed that imposes restrictions (e.g., a CF grammar, first-order logic, graph theory …), the *linguistic theory* must provide them.

An additional point, more pragmatic than formal, pertains to *how* the theory is used. As also observed by Peters & Ritchie (1973), even though an *Aspects*-style grammar with unbounded deletion *could* generate non-CF languages (in the worst-case scenario, any recursively enumerable language), that excessive power was not used by linguists doing grammatical analysis within the *Aspects* framework (see also Berwick, 1982): an overly powerful expressive formalism does not entail that this excessive power is actually used in practice. In our particular case, the restrictions that apply to both the definition of elementary graph and the operations of graph composition are not part of the formalism (graph theory) or the meta-theory (declarative), but constitute the backbone of a theory of natural language grammar.

What are the limits of our approach? More specifically, we may ask, what is the expressive power of our theory? This is a complex question, which relates

 | DOI:10.1163/9789004542310_005

to the problem of restrictiveness. It is not what the descriptive language *could in principle* do, but what the substantive aspects of the theory *allow* it to do in a specific instance of grammatical analysis. Thus, just like it is not an issue in grammatical description using HPSG or LFG that these formalisms may generate recursively enumerable languages (Stabler, 2013) because the excessive power is just not used in formulating analyses of natural languages (see also Wedekind & Kaplan, 2020: 516–517), we may or may not need to worry about having in graph theory a set of formal devices that is 'too expressive', depending on how we use them in our characterisation of syntactic dependencies. It would be unwise to impose restrictions over the formalism which restrict its usefulness in providing grammatical analyses (again, see Wedekind & Kaplan, 2020: 517; Pollard, 1997 goes as far as claiming that the language in which the theory is expressed should be unconstrained: it is the grammar that imposes restrictions). As pointed out above, this was also noted by Peters & Ritchie in their demonstration that an *Aspects*-style grammar could generate recursively enumerable languages: linguists just did not use the excessive power of unrestricted grammars in their analyses of constructions in natural language. This question is also tricky for a different reason: because we are not recursively enumerating strings, our 'syntax' is, strictly speaking, *not generative*. But there is a sense in which the notion of *generative capacity* (in the sense of Chomsky, 1965: 60; Pollard, 1997 uses the term *generativity*) is meaningful: we want to know what kinds of structural descriptions for natural language strings we can give and what kinds are not allowed. We can use an unrestricted formalism to describe a restricted set of structures: the important thing is that the set of structures (objects + relations) be restricted (finite), and it is here that constraints become essential. In Pullum's (2019: 62) words, '*A grammar becomes in effect a finite library of pictures of local regions of a tree* [read: graph]', where each elementary 'picture' is defined by an admissibility condition. This library is defined by lexicalisation restricting the size of each elementary graph and constraints over relations that we allow within elementary graphs and across elementary graphs. Let us the term *expressive power* to refer to this 'finite library of pictures of graphs', to avoid the 'recursive enumeration' connotations associated with the adjective *generative*. Given our focus on what Haspelmath (2021) calls *p-linguistics*, the restrictions over the expressive power of the formalism come from conditions formulated over the definition of elementary graphs and constraints on relations in derived graphs.

In this context, we can examine the constructions that our theory can provide structural descriptions for and whether such descriptions adequately represent the relations between basic expressions in specific sentences without assigning extra structure or predicting formal relations that do not actually hold (see Graf, 2021 for a discussion of the notion of *weak generative capacity* in the context of computational linguistics that is germane to our perspective; also Manaster Ramer & Savitch, 1997). This is a different, but related, take on restrictiveness: on the one hand, we may be focused on restricting the generative power of the formalism to a specific class of grammars or languages. We may, for example, restrict the form of production rules. However, this alone does not guarantee that the structural descriptions assigned to natural language expressions will not feature unnecessary structure (see **Section 1.5** for discussion). As observed in Chomsky & Miller (1963) and Joshi et al. (1972) among others, phrase structure grammars often assign too much structure to simple expressions, as in the case of non-recursive iterative adjectival modification. In this context, we may leave aside the question of restrictiveness in terms of where in the Chomsky Hierarchy a particular theory is located, and focus on requiring of a theory of syntax to assign no more structure than needed to capture dependencies between expressions.

After this brief introduction, we can be more specific about what this chapter contains. It is devoted to clarifying some aspects of the present proposal that differ from most currently available theories of syntax in terms of how to tackle grammatical analysis. It would make little sense to have yet another theory for 'John loves Mary' and compare the present approach with currently available models of syntax based on these kinds of structures. If there is any merit in this work, we must prove that the system argued for here cannot only provide adequate accounts for well-known phenomena, but also expand the empirical basis of syntactic theory by providing analyses for phenomena that are either underanalysed or which directly elude analysis in one or more approaches. Our main term of comparison will be MGG, given the influential character of analyses formulated in that framework, but non-transformational and non-generative theories will also feature prominently. In particular, we will refer to Arc Pair Grammar and Metagraph Grammar, Dependency Grammar, Lexical-Functional Grammar, and (pure) Categorial Grammar. We will focus on what the atomic elements are in each of these theories, how relations are defined and which relations are allowed, and how different theories provide different analyses for sentences displaying the same (or very similar) formal dependencies.

4.1 Multiple-Gap Relative Constructions

In **Chapter 3** we provided a sketch of an account of so-called *discontinuous constituency*, which is and has been problematic for Item-and-Arrangement-type Immediate Constituent analyses (see Schmerling, 1983a; Wells, 1947 for discussion). But this is hardly enough; further comparison is due and a proper treatment of discontinuity must be provided. To see how our model differs from the usual state of affairs in transformational generative grammar, consider the multiple-gap restrictive relative construction in (70):

(70) A person who people that talk to usually end up fascinated with (from Kayne, 1983: 230)

The GB-MP representation is of (70) (under a head-external analysis of relative clauses; see Demirdache, 1991) goes along the lines of (71) (adapted from Kayne, 1984: 172, adding some phrasal labels to improve readability):

(71)

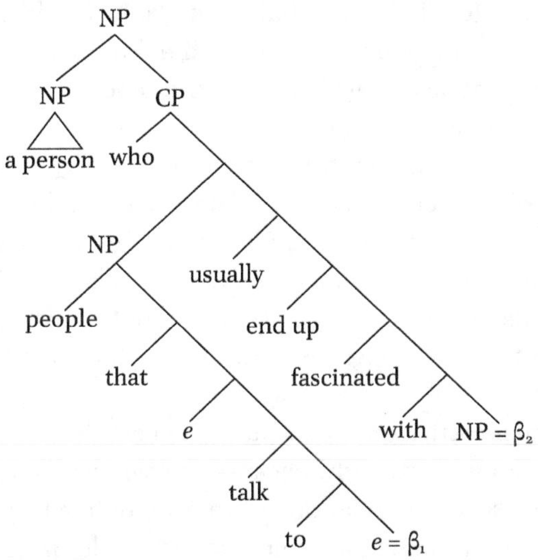

FIGURE 4.1 Multiple gap construction in Kayne (1984)

Where

> *the NP object of* ***with*** *is* (*e*), *so that there are two empty categories* β_1 *and* β_2. *The nodes that belong to* $G_{\beta 1}$, *and* $G_{\beta 2}$ [the government-projections of β_1 and β_2] *have been endowed with '1' and '2', respectively, for the purposes of exposition* (Kayne, 1983: 230)

This structure features multiple co-referent terminals:

(72) a. People = *e*
b. who = β_1 = β_2 = a person

However, despite sharing denotations, those are all *different nodes* in a MGG structural description. This means that extra elements must be introduced in the representation to encode semantic-referential identity: indices (see Lasnik & Uriagereka, 1988: 44, ff. for discussion about the role and ontology of indices). In such a theory, syntactic objects which share indexes receive the same semantic interpretation, in the sense that they denote the same entity. The module of the grammar that is in charge of defining and limiting the possible relations between indexed referential expressions in MGG is known as Binding Theory (Chomsky, 1980, 1981; Reinhart, 1983; Reinhart & Reuland, 1993; see Culicover & Jackendoff, 2005: Chapters 11 and 12 for a different view). Note in (71) that the SMC is respected throughout, which forces this multiplication of nodes in the tree for what is ultimately interpreted as the same entity: coreferential nodes have the same semantic value. All differences pertain to the syntactic context (e.g., 'be the sister of P', 'bear Oblique case', etc.) in which each node occurs, that is, to the local relations (e.g., immediate dominance, theta-marking, etc.) that the relevant node establishes with other nodes.

A second aspect to look at is the definition of *local domain*. In transformational grammar, these are most frequently stipulated in terms of 'government projections', 'bounding nodes', 'barriers', 'phases' (among other notions, usually mutually translatable if not equivalent; see e.g., Boeckx & Grohmann, 2004): they are indicated by the presence of designated functional nodes in the structure (e.g., v^* and C in the standard version of phase theory in Chomsky, 2000, 2008), and independent from lexical properties of the expressions that appear in a structural description.

Finally, in contemporary generative grammar there seems to be no natural way to capture predication domains (mainly, because the syntactic component is autonomous, and thus semantic relations cannot be defined in terms syntactic structure or vice-versa), although some generative developments propose specific phrasal projections for predicates of different types (e.g. Åfarli, 2017; Bowers, 2001 among others). In these latter cases, predication is a property represented in the phrasal skeleton, as a Predicative Phrase dominating a lexical projection (VP, AP, PP). Alternatively, the projection related to predication may be an independently proposed functional category that takes on a new role: for instance, in Hale & Keyser (1993), for example, it is IP dominating VP that makes V into a predicate, since VPs are not predicates in the 'lexical syntax' (Op. Cit., p. 80). Some approaches, like Vogel & Steinbach (1994), challenge Chomsky's (1995) proposal that intermediate projections are 'invisible' (see also Seely, 2006) and assign to these the role of predication. In connection to the previous issue, while at first *phase theory* seemed to be at least partially motivated by 'propositionality' (Chomsky, 2000, 2004), that idea was quickly replaced by a syntax-internal basis for phases (such that phases are the locus of feature checking/valuation; Chomsky, 2008, see also Gallego, 2010: 54, ff.).

The approach proposed here attempts to change both aspects of the MGG approach: the multiplication of nodes and the inclusion of an indexing mechanism and the definition of locality in terms of designated nodes in the structure that act as 'endmarkers' for probing operations and the establishment of syntactic dependencies. Locality is, in the simplest cases, enforced by lexicalising the grammar, and defining elementary graphs as the units where syntactic dependencies are established (essentially, following the Fundamental TAG Hypothesis).

Getting to the analysis, we can define a preliminary set of structural relations for a sentence like (70) above as in (73), which follows rather closely the PSG description in (71) (note that Kayne analyses *end up* as a two-word terminal):

(73) Arbor 1: [people usually end up fascinated]
Arbor 2: [people fascinated with person]
Arbor 2: [people that talk to person]
ρ_1 = {(end up, people), (usually, end up), (end up, fascinated)}
ρ_2 = {(fascinated, people), (fascinated, with), (with, person)}
ρ_3 = {(talk, people), (talk, to), (to, person)}

Following our previous analyses, let us use ⟦person⟧ to denote the semantic value of the expression *person* (which may be a set of properties; see Schmerling & Krivochen, 2017).[1] Similarly, we will use ⟦people⟧ to denote the semantic value of the expression *people*. We can thus get rid of some elements in (62) because we have proceeded as if morpho-phonological form was meaningful for the definition of relations, but we contend it is not. The analysis in (73) has antecedents not only in some versions of the generative analysis that are based on matching (e.g., Chomsky, 1965), but also in other grammatical traditions. Thus, for instance, in his detailed description of Spanish relative clauses, Brucart (1999: 398) argues that the 'propositional value' of the relative clause in *El libro que Luis te regaló es muy interesante* (Eng. 'The book that Luis gifted to you is very interesting') is *Luis te regaló el libro* (Eng. 'Luis gifted you the book'): this

1 We will not deal with the determiner for now (and for indefinite NPs we will see that it makes no difference for future analyses) but in later chapters (mainly **Section 14.4**, but also **Chapter 9**) we will see that some quantified noun phrases require us to have determiner nodes in the graph.

suggests a semantic structure along the lines of *El libro* [*Luis te regaló el libro*] *es muy interesante*. In the case we are considering, the expressions *who* and *person* have the same semantic value, they seem to be simply *tokens* of the same lexical *type* (in the sense of Krivochen, 2015b, 2018) which vary in terms of context (the nodes they are directly connected to in their respective arbores) and morpho-phonological exponent. More practically, there are *not* two distinct nodes for *who* and *person*, but *a single node* with address ⦃person⦄ (pointing to the semantic value ⟦person⟧) which can be the tail of as many arcs as necessary in the derived graph.[2] This analysis of *who* marks an important difference with the raising analysis as presented in Kayne (1994) and Bianchi (1999), for instance, who not only have *who* and *person* as two distinct nodes, but are forced to propose a constituent [$_{DP}$ who [$_{NP}$ person]], which gets dissolved by raising of *person* to the specifier of D solely to comply with the axioms of antisymmetry. Finally, in the analysis above, *end up* dominates the root/anchor of Arbor 2.

A cautionary note is in order: the antecedent and the relative operator are not always coextensive in terms of their graph-theoretic representations. A

---

2 Incidentally, this means that the antecedent is in a way contained within the relative clause. This is not necessarily to be interpreted as supporting the 'raising' analysis (Kayne, 1994; Bianchi, 2000; Bhatt, 2002 and related work), since this approach requires additional syntactic operations of movement and in our opinion makes the wrong constituency predictions (Borsley, 1997; Krivochen, 2022). As we will see in **Chapter 9** our analysis does entail that there is a close articulation between antecedent and relative clause: the antecedent and the *wh*-operator are not two distinct nodes in a derived graph, but only one, which is the tail of multiple arcs. If anything, the present analysis is closer to a version of the 'matching' analysis (Citko, 2001; Salzmann, 2019), but with structure sharing instead of deletion. We need to note immediately that this analysis does not extend automatically to antecedent-less relative clauses (*free* relatives, *transparent* relatives); see Bresnan & Grimshaw (1978), Larson (1987), Wilder (1998), Saddy et al. (2019) for discussion from a variety of viewpoints. It seems to us, however, that there are important connections between the so-called 'grafting' analysis of free relatives (van Riemsdijk, 2000, 2006) and our approach to graph composition (see **Section 8.1**). The main point of contact is that, in our view as in van Riemsdijk's, the embedded predicate AP or NP in a transparent free relative (which determines the distribution of the relative) is a shared element between the local structure corresponding to the relative and the matrix structure:

i. There is [what you might call food$_{\text{Predicate NP}}$] on the table
ii. There is [food] on the table
iii. Mary is [what some consider crazy$_{\text{Predicate AP}}$]
iv. Mary is [a crazy]

If the strings in (ii) and (iv) correspond to an elementary graph and the relative clause to another, these two may be *linked* at the predicative expression (AP/NP). For an analysis of restrictive and appositive relatives under present assumptions, see **Chapter 9** and Krivochen (2022).

well-known observation is that when the relative head is generic or quantified the relative-internal operator (in a restrictive relative) excludes the quantifier (Stockwell et al., 1973: 428, ff.; Brucart, 1981: 100; Bianchi, 1999: 36): thus, we cannot have (74b) as the underlying structure of (74a):

(74) a. All the boys who left early missed the fun
b. All the boys [all the boys left early] missed the fun

In **Section 14.4** we will propose an analysis along the lines of Krivochen (2022), whereby given a generalised quantifier (in the sense of Barwise & Cooper, 1981) only the *set term* may be multidominated, leaving the determiner outside the relation: [All the boys [boys left early] missed the fun]. The issue does not arise with indefinite NPs such as *a person*, which is why the analysis in (73) is for now feasible.

These things considered, the derived graph for (70) can be illustrated as in (75). In what follows we will omit address notation, presupposing it (that is, for any expression E in a diagram of a graph, we will use just E instead of ⦃E⦄):

(75)

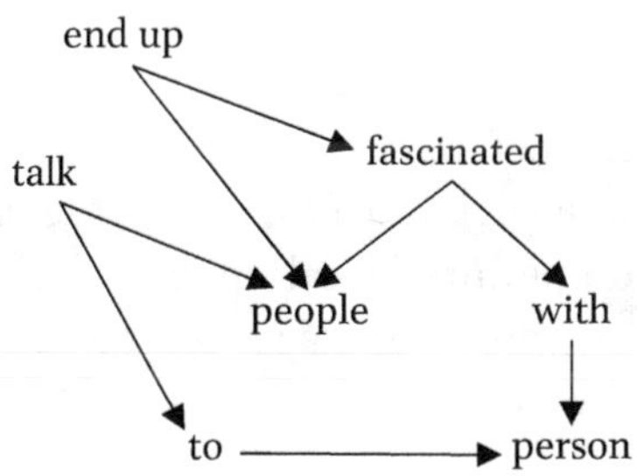

FIGURE 4.2 Graph-theoretic analysis of multiple gap construction

A couple of noteworthy aspects of the analysis: first, we have analysed *end up* as a lexical predicate. This may not be entirely adequate, given the fact that it admits non-thematic subjects (as in *it ended up raining*). In case a better analysis for *end up* was a Raising verb analysis, modifications along the lines of **Section 5.1** would have to be made. Second, both prepositions *with* and *to* are assigned their own nodes: they are categorematic. Unlike 'regime' prepositions, which are invariable (consider, e.g., *rely on* vs. **rely about*) and do not seem to contribute their usual semantic value to the expression in which they occur, both *with* and *to* are contentful, and can be replaced by other prepositions at the cost of meaning change (e.g., *fascinated by* NP instead of *with* NP, or *talk about* NP instead of *to* NP).

Different arbores are *linked* at all common nodes, since all 'ocurrences' of a basic expression are in fact calls to the same semantic value. This makes it possible to build a global compositional interpretation for (75), following the principle that lexical predicates dominate their arguments and functional modifiers dominate the expressions they modify. If those arguments are internally complex (i.e., if an argument is itself an arbor), then the predicate will dominate the *root* of the arbor corresponding to the complex argument(s). Importantly, the node at which arbores are linked may *not* be the root of any of the arbores involved in the process: this is not allowed in the TAG frameworks of Joshi (1985) or Frank (2002) as far as the formation of derived trees is concerned, or the declarative formalism of Rogers (2003). In TAGs, specifically, root-frontier identity is essential in the definition of auxiliary trees. In a departure from TAGs, in the model that we are developing in this monograph graphs need not be linked exclusively through their roots-frontiers, in principle they can be linked *at any node* (note, however, that Sarkar & Joshi, 1997 do allow 'counter-cyclic' linking as a form of structure sharing, but *adjunction* is still a root-to-internal-node operation). The restrictions on linking are actually defined by the grammatical function that a node can be assigned in each elementary graph: importantly, as we will see below, there is not a bijective relation between the set of GF and the set of nodes within a elementary graph.

It is interesting and necessary to briefly compare our representation of a multiple-gap construction with that proposed in Postal (2010: 41) in Metagraph Grammar, since it is also a declarative framework based on graph theory in which grammatical functions (taken to be primitive) play a central role. Specifically, Postal considers the following sentence:

(76) Carla is a woman [who almost every man that __ meets her falls in love with __]

And assigns an arc structure to the bracketed relative clause like (77) (taken from Postal, 2010: 41). Arcs are labelled with the appropriate grammatical relation plus a letter; the letter serves no purpose to the formalism, and only provides us with a way to refer to specific arcs. The non-circled R indicates that the tail of the arc is a relative pronoun:

(77)

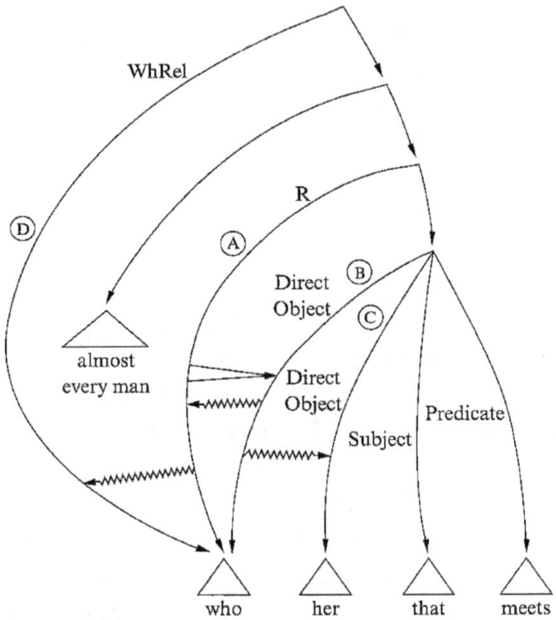

FIGURE 4.3 Multiple gap construction in Metagraph Grammar

Much discussion about the place of resumptive pronouns in the theory of locality aside (but see **Section 14.3** below for some preliminary remarks), we will focus on the differences between the assumptions in APG/MG and our model rather than on explaining every detail of the simplified structure in (77) (see Johnson & Postal, 1980 for extensive discussion). Here we are interested in two aspects that become prominent in the APG analysis:

i) The notion of *pronominal arc* (and *arc antecedence*)

ii) The binary relations *sponsor* and *erase*

The relation *erase* is represented by double arrows (such that arc A erases arc B), and *sponsor* is represented with the wiggly arrow (such that B sponsors A, A sponsors D, and B sponsors C): roughly speaking, APG's *sponsor* and *erase* capture what in Relational Grammar were *promotion* and *demotion* operations. In the analysis of a passive, the initial 2 sponsors the final 1, and the final 1 erases the initial 2 (Blake, 1990: 16). Because erased arcs are not part of the Surface Graph (S-graph) of the sentence (Postal, 2010: 27, 30; Johnson & Postal, 1980: 88–89), arc B does not belong to the S-graph. The notion of S-graph is crucial in APG and MG, and thus it is worth defining it briefly so that the differences between S-graphs and our graphs are fully explicit. In order to do this, first we need to define R(elational)-graphs. R-graphs are finite sets of arcs and nodes which represent all structure of a sentence except that involving *erase* and *sponsor*, and include all grammatical relations between elements; R-graphs need not respect the SMC, do not contain loops, and are single-rooted (Johnson & Postal, 1980: 51, ff). The result of considering *erase* and *sponsor* relations is a *subset* of R-arcs, a S(urface)-graph:

> *An S-graph:*
> *a.* *is connected,*
> *b.* *is rooted, and*
> *c.* *contains no overlapping distinct arcs*
> [...] *S-graphs do not have parallel edges, do satisfy the Single Mother Condition, and have no circuits.* (Postal, 2010: 28)

Furthermore, Johnson & Postal (1980: 90) prove that every S-graph is an R-graph.

It is important to observe that in Postal's representations (R-graphs and S-graphs alike), *who* and *her* are two distinct nodes, with *her* being the tail of a nonreflexive resumptive copy arc. Identity of semantic value aside, the actual morphological exponent of an expression matters more in APG than in our framework. Note also that the complementiser *that* is assigned its own node: this raises interesting questions about the categorematicity of complementisers and other functional categories to which we will return below. Arc B sponsors C and A, and B is deleted; this establishes a relation between A and C called *arc antecede*, which is what in MG (roughly) corresponds to the 'coindexing' mechanism of MGG (Postal, 2010: 34). Note that in order to get correference it is still assumed that the grammar requires a binary relation between arcs, *arc antecede*(A, B), in turn dependant on two other relations: *sponsor*(C, A) and *sponsor*(D, B) and *primitively antecede*(C, D). Postal's conditions and relations are formulated over *arcs* rather than over *nodes* and *walks*, which is already a major difference between his approach and ours. But the point here is that in the graph theoretic proposal we formulate here there is no need to multiply nodes, arcs, or relations: as a matter of fact, correference is one of the major sources of *simplification* in graphs, where 'simpler' is understood in a pre-theoretic way, intuitively as 'having less nodes and/or less edges'. At the same time, **Chapter 5** will focus on the relation between ordered arcs and grammatical functions: an important result of that section is that annotations to arcs (such as Subject, Object, etc.) can be dispensed with if grammatical functions are conceived of as an ordered system of primitive notions (an eminently RG insight) to be put in correspondence with an ordered system of arcs.

In contrast, in our approach expressions *who* and *her* are a single node which links distinct arbores, just like *people* and *person* are in the structural description in (74) above. They denote the same entity, which means that they are assigned the same address; because addresses are unique identifiers, a graph (basic or derived) cannot contain distinct nodes with the same address. If two nodes have the same address, they are identical: one and the same node in the graph. As a consequence, in our model endophoric dependencies are defined in terms of strict ordering between nodes. Provisionally, (we will refine this preliminary characterisation in **Section 6.2.1** and **Chapter 8**) we can say the following:

Identity-ordering condition on pronominalisation:
For any vertices in a graph G v_i, v_j; v_j *can pronominalise* v_i *iff*

a. v_i *and* v_j *denote sortal entities*
b. $⦃v_i⦄ = ⦃v_j⦄$, *and*
c. *the arc of which* $⦃v_i⦄$ *is a tail is ordered before the arc of which* $⦃v_j⦄$ *is a tail in the graph where* v_i *and* v_j *occur*

This condition captures not only the transformational intuition that for A to pronominalise B they have to be coindexed nominals in specific structural relations (Lees & Klima, 1963; Postal, 1969; more recently Hornstein, 2001: Chapter 5; Hornstein & Idsardi, 2014: 14, ff.; Gärtner, 2014; Grohmann, 2003: Chapter 3, among others), but also that pronominalisation is a relation that requires a certain order between the elements it applies to. The issue or arc ordering will become clearer in **Chapter 5**, where it will be linked to the determination of the grammatical function of arguments.

4.2 Dependencies and Rootedness

The next approach we want to mention is Dependecy Grammar. It is important to stress now that, while the internal constitution of each sub-graph may superficially resemble the structures one gets in a Dependency Grammar (Tesnière, 1959; Osborne et al., 2011, 2012; Osborne, 2019; Kahane & Mazziotta, 2015, among many others) in rejecting binarity as a fundamental structure building axiom (see also Krivochen, 2015a, 2018, 2021a) and giving importance to connections defined in terms of sets of nodes which are continuous with respect to dominance (Osborne et al.'s 2012 *catenae* and our *local walks*); there are several important differences, in terms of the structures being characterised and the expressions to which those structures are put in correspondence. We will now go into some of those differences, without being exhaustive: the point is to highlight that the approaches are not notational variants of one another, and that there are non-trivial theoretical and empirical differences that deserve careful analysis.

A basic principle of Dependency Grammar (DG henceforth) is that '*dependents should be grouped together with their head*' (Osborne & Niu, 2017). Furthermore, dependency is a strict *mother-daughter* relation. This is the principle that guarantees that a structural description for a VP like

(78) Walk really fast

captures a segmentation like (79a), and not, say, (79b):

(79) a. [walk [really fast]]
b. [walk really [fast]]

Since *really* is a modifier of *fast*, and *really fast* is, as a unit, a modifier of *walk*, there is a node in the dependency graph that dominates *really fast* and excludes *walk*:

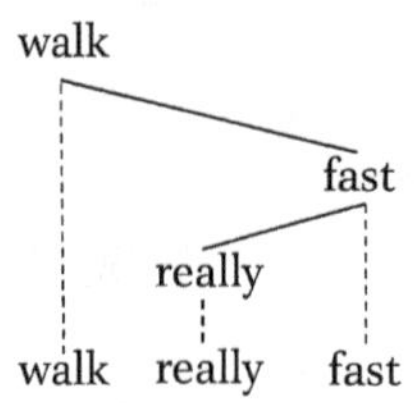

FIGURE 4.4 DG analysis of the verb phrase 'walk really fast'

This tree illustrates one of the most important differences between DG and our approach: in the DG tree, *walk* dominates *fast* despite *fast* being a modifier of *walk*. There is a mismatch between dependency and semantic interpretation, such that the DG tree is not isomorphic to the semantic representation *fast*(*walk*). In our graphs, in contrast, predicates *always* immediately dominate their arguments. This entails that the root of an elementary graph will be the highest-order predicate in that graph, and that root and anchor do not always coincide: for linguistic purposes, in a lexicalised grammar, *anchor* is a more important notion than *root* in that well-formedness conditions over linguistic representations will make reference to the former rather than the latter.

Headedness is essential in a DG, perhaps even more so than in a PSG (which, as Hockett, 1958 and Lyons, 1968 argue, can—and in fact *should*—incorporate exocentric structures in order to be descriptively adequate). But it is not *headedness* we want to question now (despite the fact that there is no real sense in which elementary graphs are 'headed' other than their containing an anchor); rather, we question the segmentation that arises in a DG for a case like (78). We want to capture the fact that there is a connection between *walk* and *fast*, the presence of *really* notwithstanding. There is no need to invoke a notion of *headedness* if all that we are interested in is specifying the semantically significant syntactic relations between expressions in a particular string. The (unordered) ρ-set for (78) looks like (80):

(80) ρ = {(fast, walk), (really, fast)}

Distributionally, paradigmatic choices affect either individual nodes or entire sub-graphs; in principle we do not have any unit in between, although sequences of nodes continuous with respect to dominance (along the lines of Osbornian *catenae*) in elementary and derived graphs play a role in the analysis of syntactic phenomena (such as anaphoric binding or *gapping*).

There are two further aspects with respect to which our theory differs from DG (and most forms of PSGs) and which turn out to be of major importance: first, as in certain forms of CG (e.g., Schmerling, 1983b, 2018a), *nodes need not correspond to words*. There is no fundamental opposition between lexical and phrasal categories at the heart of the formalism: as highlighted above, nodes correspond to basic expressions. In CG terms, basic expressions need not be identified with lexical items (see also Jespersen, 1985 [1937]: 6 for a notational system that similarly assigns atomic symbols to multi-word expressions). Strictly speaking, in PSGs the correspondence between leaves in a tree and words is a consequence of conflating *terminals* with ortographically independent *lexical items*; this is not an inherent property of the formalism. We briefly mentioned multi-word basic expressions above, when dealing with *heat up*, now we can get into some more detail. An example of the kind of linguistic phenomena we have in mind is the behaviour of *would rather* in the following examples:

(81) a. *John wouldn't rather walk
b. John would rather not walk
c. There's no one here who wouldn't rather walk

Schmerling (1983b: 14) proposes that *would rather* is '*a two-word modal*', assigned to the category (FC//IV)/(FC/IV). In Schmerling's system, subject NPs are defined as expressions of category FC/IV: they combine with intransitive verb phrases to form Finite Clauses. An expression of category (FC//IV)/

(FC/IV) is then an expression that combines with an expression of category FC/IV to form a modified expression of category FC//IV. There is evidence in favour of Schmerling's idea that *would rather* is a basic expression (and not the result of a concatenation rule applied to *would* and *rather*). Observe, for example, that negation cannot interrupt the unit unless under the scope of negation itself (as in (81c), noted by Baker, 1970a). A similar paradigm emerges with *would just as soon*, which Schmerling argues (p.c.) is also a multi-word basic expression.

We must emphasise that multi-word basic expressions are *not* the same as idioms (see also Huck, 1988: 255, ff.; Culicover et al., 2017): the interpretation of the sentences in (81), for example, is transparent and compositional; the denomination of *multi-word basic expression* entails that the node in our graph corresponds to more than one ortographic word and that it is this sequence of ortographic words that is affected by rules of the grammar. As observed above, this is not a new idea, it actually predates both PSGs and DGs: Jespersen (1985 [1937]: 6, 25) talks of 'composite verbal expressions' for things like *wait on* in *She waits on us* or *talks with* in *He talks with himself* and assigns *wait on* and *talk with* a single primitive symbol W; similarly, he recognises 'composite prepositions' like *on account of*, and similarly assigns them a single primitive symbol *pp* (1985 [1937]: 6, 32). The empirical applicability of multi-word basic expressions, evidently, goes beyond English grammar: García Fernández & Krivochen (2019b: 37–40) propose a multi-word analysis for Spanish modals that *prima facie* would include a 'preposition' or a 'complementiser' such as the periphrastic future auxiliary ⟨*ir a* + infinitive⟩ or the modal auxiliary ⟨*tener que* + infinitive⟩; that work provides arguments that *a* and *que* in these cases are neither categorematic nor syncategorematic independent elements and that they need to be considered morphologically as part of the auxiliary. In these cases, *tener que* and *ir a* would indeed be multi-word basic expressions (assigned to the category to which auxiliaries belong; see Krivochen & Schmerling, 2022 for a detailed Categorial Grammar analysis of Spanish auxiliaries; García Fernández et al., 2020 for discussion about multi-word basic expressions and syncategorematic expressions in the Spanish auxiliary system).

More fundamentally, the relations that we assume to hold in graphs do not coincide with those assumed in DG, despite some formal similarities between our framework and DGs. As we will see in more detail in **Chapter 5**, in a clause there is always an edge between the lexical verb and its arguments: this is so because the arguments are part of the elementary graph anchored by a lexical predicate. Furthermore, the edge goes *from* the predicate *to* its argument (see also McKinney-Bock & Vergnaud, 2014). Under our assumptions, if the lexical verb is modified by an auxiliary, then the auxiliary dominates the lexical verb (since it modifies the verb, the auxiliary is a functor), but the relation between a lexical predicate and its nominal dependents is always present. Thus, the analysis of a sentence such as

(82) John is reading a book

would involve a single elementary graph, anchored by *read*, and in which the following relations hold:

(83) $\rho = \{(\text{read}, \text{John}), (\text{read}, \text{book}), (\text{be}, \text{read})\}$

In this analysis, the subject and the object are both dependents of the lexical predicate (which is the expression with a subcategorisation frame), not of the auxiliary. This analysis of auxiliary verb constructions contrasts with some DG analyses (e.g., Osborne, 2019), as we can see in the example below:

(84) a.

b.

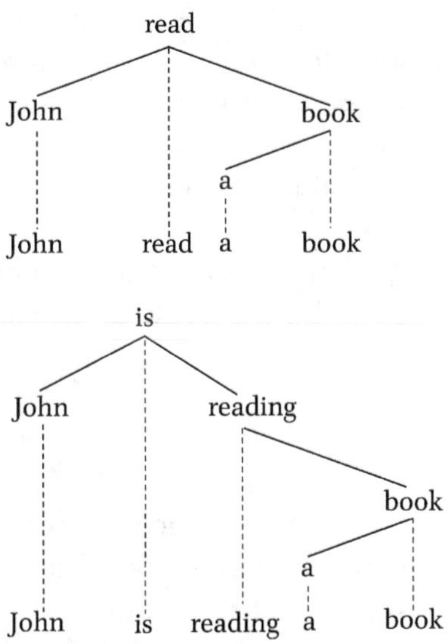

FIGURE 4.5 Dependency Grammar analyses of transitive clause with and without auxiliary verbs

In (84b) there is no direct relation (no dependency) between the lexical verb *read* and the subject *John*: in a DG, '*The subject is always a dependent of the finite verb regardless of whether the finite verb is an auxiliary or content verb*' (Osborne, 2019: 168), with the assignment of grammatical functions and thematic roles being handled by other means (auxiliaries immediately dominate subjects even in versions of DG that assume 'networks' instead of trees, such as Anderson, 2011). This is a crucial point of divergence between (some forms of) DG and the graph-theoretic perspective exposed here: since our grammar is *lexicalised*, the organisation of local domains is centered around lexical predicates, and they dominate all their dependents. As a final note, DG diagrams use terminal words as node labels (Osborne, 2019: 46), but the sense in which *label* is used in this context is not the classical PSG sense (Hopcroft & Ullman, 1969: 19), and no operations are formulated that range over phrasal node variables.

Further restrictions are placed on the relation between words and nodes in DG diagrams which we can look at from a comparative perspective. Specifically (from Osborne, 2005: 253):

I. a. *One wordform per node, and*
   b. *One node per wordform.*

II. *One head per node*

Requirements (I.a.) and (I.b.) are violated in the case of multi-word basic expressions: if there is a single node for *would rather* or *would just as soon* (that is, if nodes correspond to *basic expressions* rather than to *orthographic words*), there is no correspondence between words and nodes. Requirement (II) is trivially void if there is no meaningful notion of 'head' as opposed to 'phrase'.

There is a third requirement formulated by Osborne, which is a more general constraint on the construction of (rooted) trees (see, e.g., van Steen, 2010: 109):

III. *One root node per structure*

The framework explored here violates (III) because our grammar can describe multi-rooted structures as the result of graph composition, which are inadmissible in most versions of DGs and in PSGs (but see **Chapter 12** for a DG account of coordination that makes use of multi-rooted structures), for different reasons. In PSGs, the ban on multi-rootedness is given by definition: the base component of a transformational grammar, as well as the level of *c*-structure in LFG and the basis of GPSG (see Gazdar, 1982 for some early discussion), is a CFG and defines immediate constituent phrase structure trees. Let us repeat some basic information from **Chapter 1** for ease of reference. A CFG is a set $G = (V_N, V_T, P, S)$, where V_N is a set of nonterminal symbols, V_T is a set of terminal symbols, P is a set of production rules (also called 'transition rules') of the form $\Sigma \rightarrow F$ (where Σ

is a possibly unary string in V^+ and F is a possibly unary string in V^*[3]), and S is a starting symbol (also called the 'axiom'), such that $S \in V_N$. Translating PSG derivations into trees (McCawley, 1968), the axiom is a special nonterminal which dominates but is not dominated (since it does not *follow from* any other symbol, in the sense of Chomsky, 1956). Both DG and our graph theoretic approach have the relation *dominate* as an essential device, but it must be noted that if the *root* of a graph is defined *only* as a node that is not dominated by any other node, then it is perfectly possible to have multi-rooted structures as derived graphs. The only requirement for a node v_x to be a root being that there is no node within its arbor which dominates v_x. As we integrate arbores with roots v_x and v_y, graph union may contract nodes that v_x and v_y dominate, keeping v_x and v_y undominated. Summarising, there are local compatibilities between our graphs and Dependency graphs, and between our graphs and PSG s, but the systems are far from equivalent.

Strange though a locally multi-rooted representation might seem to some readers, it is not unheard of (see Morin & O'Miley, 1969 for an early proposal). Multi-rooted structures may arise as intermediate steps in generative grammar: they are however filtered by global representational constraints. Consider, for instance, the *intermediate* representation proposed in Citko (2005: 480) and Citko & Gračanin-Yuksek (2021) for ATB extraction and RNR (which owes much to works like McCawley, 1982):

(85) a. I wonder [what Hansel read and Gretel recommended]
b.

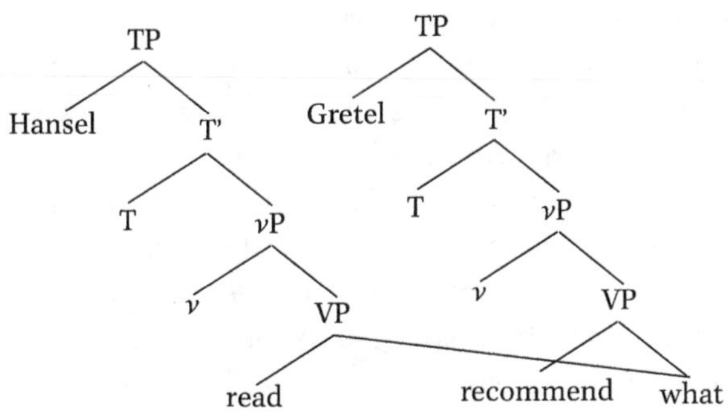

FIGURE 4.6 Intermediate phrase marker under Parallel Merge

3 Hopcroft and Ullman (1969: 1) define that:
If V is an alphabet, then V denotes the set of all sentences composed of symbols of V, including the empty sentence* [notated ε]. *We use V^+ to denote the set V^*–{ε}. Thus, if V = {0, 1}, then V^* = {ε, 0, 1, 00, 01, 10, 11, 000, ...} and V^+ = {0, 1, 00, ...}.*

Insofar as they are undominated, the TP nodes can be characterised as a *root* in each sub-tree corresponding to the structural descriptions of the term of the coordination *Hansel read what* and *Gretel recommended what* since they are undominated in their respective trees. More generally, the output of Parallel Merge is always a double-rooted structure (since its input involves two distinct rooted structures; Citko, 2005: 476), with single-rootedness arising derivationally after the application of further syntactic operations that introduce additional structure:

(86) Parallel Merge (A, B) =

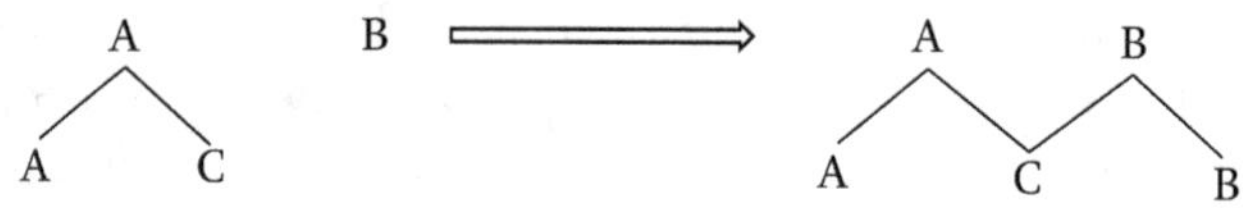

FIGURE 4.7 Parallel Merge

The full derivation proposed by Citko eventually adds more functional structure in the form of a unique Complementiser layer CP that dominates both TP-rooted sub-trees; this yields a single-rooted P-marker with a CP-labelled root. The last step of a derivation (i.e., the step that exhausts the Numeration) cannot be an application of Parallel Merge.

Note, incidentally, that (85b) exemplifies structure sharing, as mentioned above for the ATB extraction out of a coordinated structure, as in *What did Bill wash, Peter dry, and John break?*: in (85) there is a single *what* that is dominated by nodes in both local trees, and which must fulfil the grammatical function *object-of* for two distinct predicates (here, *read* and *recommend*). (85b) is a representation that follows the classical MGG sequence of heads T > *v* > V, including phonologically empty functional (terminal and non-terminal) nodes. Minus non-audible functional structure (nodes like T and *v* and their phrasal projections, TP and *v*P), our *local* description for a sentence like (85a) is not too far from (85b): we have three lexical predicates, *wonder*, *read*, and *recommend*, each of which selects nominal arguments and defines an elementary graph. If we identify all single-rooted local structures, we are left with an analysis like (87):

(87) Arbor 1: [I wonder and]
Arbor 2: [Hansel read *what*]
Arbor 3: [Gretel recommended *what*]
Arbor 4: [read and recommend]
ρ_1 = {(wonder, I), (wonder, and)}
ρ_2 = {(read, Hansel), (read, *what*)}

ρ_3 = {(recommend, Gretel), (recommend, *what*)}
ρ_4 = {(and, read), (and, recommend)}

There is a novel aspect to (87) with respect to our previous analyses, which is the presence of the conjunction *and*: strictly speaking, it is not a lexical predicate, therefore, it cannot define its own elementary graph, but it may root an arbor (see Sarkar & Joshi, 1997 for a TAG approach to coordination that is similar to the one sketched in (87); Perlmutter, 1980: 227 provides a preliminary RG analysis of conjuncts which is in principle compatible with ours but instead of recognising the conjunction as a node, annotates the arcs as Conj arcs). We will come back to the syntax of coordination in some detail in **Chapter 12**, but for now it is important to note that the status of connectives is somewhat anomalous with respect to whether they can define elementary graphs or arbores: they are predicates, but not lexical (they have no relational network, they do not assign grammatical functions). The connective does not play a role in either of the terms of the coordination, it just dominates the root of each term. And, because the coordination is the complement of the verb *wonder*, then *wonder* dominates *and* (and, transitively, everything it dominates as well).

A fundamental difference between our representation and Citko's (see her ex. (13), p. 482) is that in our representation the coordination in (85a) is symmetric—i.e., *paratactic*—rather than asymmetric—i.e., *hypotactic*—(unless Gretel recommended something *after*, or *because* Hansel read it; in which case the asymmetry in semantics would betray an asymmetric phrase marker; see Schmerling, 1975; Krivochen & Schmerling, 2016a for detailed discussion; also **Chapter 11**). In any case, a declarative sentence featuring RNR, like *Hansel read and Gretel recommended that book that's so popular now* would be indeed a multi-rooted structure, linked by the root of the sub-graph that exhaustively dominates *that book that's so popular now* (itself a derived graph): this node is a common expression between the two sub-graphs (as can be seen in (85b) above), thus linking the two elementary graphs headed by *read* and *recommend* respectively. That node is dominated by the verbs in each conjunct, which as we said are the lexical anchors of each elementary graph.

Under MGG assumptions, in a RNR structure single-rootedness can only be achieved by including a designated axiom in the alphabet operated on by phrase structure rules in a top-down model: a recursive rule like (88a) below (a 'flat structure' of the kind argued for in Culicover & Jackendoff, 2005: Chapter 4; Kaplan & Maxwell, 1988; see also the TAG coordination schema in Sarkar & Joshi, 1997: 10; Borsley, 2005 provides empirical arguments against a strictly binary approach to coordination) or an endocentric coordination phrase as in (88b) (see, e.g., Progovac, 1998; Camacho, 2003; Chomsky, 2013):

(88) a. S* → S and S (and S)
b. &P → S &'
&' → and S

Which generate the following structural descriptions, respectively:

(89) a. b.

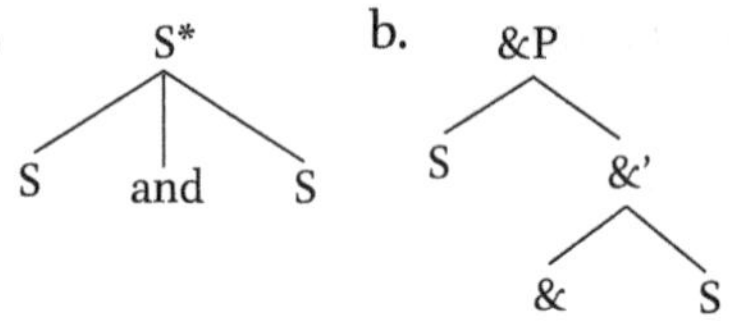

FIGURE 4.8 Structures for coordination

In bottom-up models of structure building (including mainstream work within the MP, but see e.g. Zwart, 2009, 2015), the TPs corresponding to each clause are dominated by a Complementiser layer CP (except Raising and passive clausal complements; see e.g. Chomsky, 2000), and thus extra structure in the form of silent heads must be introduced to comply with *a priori* formal requirements. Our proposal is very different from these. By 'flattening' the structures keeping only what is overtly present in the string, intermediate nodes are eliminated: this can be seen as a version of a rather basic desideratum of *economy of expression* (a fundamental LFG principle governing the level of constituent structure, see Bresnan, 2001: 90, ff.; Dalrymple et al., 2015; also fn. 4 in **Chapter 3**), in that all 'pre-terminal' nodes are eliminated. Recall also that in our system there is no one-to-one correspondence between nodes and words, such that a node can correspond to a multi-word expression (see Schmerling, 2021 for a recent perspective on multi-word basic expressions from a 'pure' categorial perspective). The same requirement (in a slightly stronger form, perhaps) is concisely expressed in Postal (1974: xiv):

> *one should assume the existence of* ***no elements of structure, no levels of structure, and no kinds of representations whose existence is not absolutely necessary*** [...] *I reject all so-called syntactic features, doom markers, other abstract syntactic markers, coding devices, "empty nodes", "doubly filled nodes," and, in short, the entire a priori unlimited set of symbolic elements available in an unconstrained system* (our highlighting)

We agree with Postal's meta-constraints on the theory of grammar, and will continue to assume that a descriptively adequate theory must meet this stronger version of LFG's *Economy of Expression* (or any such constraint limiting the symbols in structural descriptions), insofar as we are trying to define relations between basic and derived expressions (and these are, by definition, overt). The 'absolutely necessary' remark will become relevant later on, in particular in **Chapter 13**. This restriction on the theory raises questions pertaining to the kinds of structural descriptions that can be assigned to expressions, and the expressions that can be proven to be well-formed expressions of the language, in our case, of English. To address the issue of 'unconstrained systems', the following section will deal with the issues of adequately restricting the 'power' of our graphs: by this we mean, informally, that we do not want to be able to describe relations that cannot hold, nor do we want to leave legitimate relations unaccounted for.

4.3 Crossing Dependencies

In the previous sections we focused on the format of structural descriptions; this section approaches the issue from a more substantive perspective. To approximate an answer to this question, we will use Joshi's (1985) discussion of the power of TAGs with links as a reference, and compare the structures that we can characterise with those that would be generated by a TAG. It is worth noting that there are versions of TAG which are closer to MGG than others in some aspects, particularly in accepting binarity as a guiding principle (but not global monotonicity, clearly, which is disrupted by the operations of *substitution* and *adjunction*), the use of traces and empty nodes, and the use of 'bar-levels' in structural descriptions (Kroch & Joshi, 1987; Frank & Kroch, 1995; Frank, 2004, 2006). Because our approach borrows much from Lexicalised TAGs in the definition of local syntactic domains and the way in which that may be composed, the comparison is particularly apt. We must first introduce some basic aspects of derivations in TAGs, to complement our preliminary remarks in **Chapter 2**.

4.3.1 *Generalised Transformations and TAG Derivations*

A TAG is a set G = (IT, AT), where IT is a (possibly unary) set of *initial trees* and AT is a (possibly empty) set of *auxiliary trees*; the trees in the set S = IT ∪ AT are called *elementary trees*. Essentially, a TAG determines how to put basic syntactic blocks together, but does not prescribe much about the internal structure of those blocks. Lexicalisation takes care of much of that. Above, when presenting aspects of lexicalised TAGs we focused on what makes an elementary tree and how big they can get (based on the work of Frank, 1992 et seq.): here we consider how *elementary trees* are put together to yield *derived trees*. Initial trees are single-rooted structures which contain a non-terminal node which is identical to the root of an *auxiliary tree*. *Auxiliary trees* are also single-rooted structures, which contain a node in their frontier which is identical to their root: this allows for *auxiliary trees* to be adjoined to *initial trees* at the nonterminal that corresponds to the root of the AT. Let us illustrate these trees:

Initial Trees: ITs are single-rooted trees, whose 'frontier' is constituted by terminal nodes and/or substitution sites (Joshi & Schabes, 1991: 3). Intermediate nodes (transitively dominated by the root) are non-terminals. ITs will contain a designated intermediate node that bears the same label as the root and frontier of an AT:

(90)

FIGURE 4.9 Initial Tree in TAG

Auxiliary Tree: the root node of the AT and its frontier contain at least a node with the same label (Frank, 2002: 18). Intermediate nodes are, as above, non-terminals:

(91)

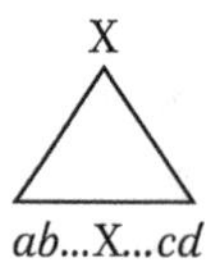

FIGURE 4.10 Auxiliary Tree in TAG

What we are interested in now is the possibility of establishing structural relations between elements belonging to these trees: we want a way to create a syntactic object that contains more than one elementary tree (i.e., a *derived tree*). TAGs introduce two operations for *tree composition*: one is the already familiar *substitution* (see **Section 2.2**), which Frank (2013: 229) defines as follows:

> *an elementary tree β rooted in non-terminal X targets a frontier non-terminal in another elementary tree α that is also labelled X. The result is the insertion of the β at the node labeled X in α.*

The result of substitution, in the case of clausal complementation, is analogous to *tail recursion*: from [$_{S1}$ John said S] and [$_{S2}$ that Mary is brilliant] we get [John said [that Mary is brilliant]], by substitution of S in S_1 by S_2. Substitution can also deliver what in X-bar syntax would be complex (read: phrasal) specifiers: from [$_{S1}$ NP said S] and [$_{NP1}$ John's best friend], we get [$_{S1}$ [$_{NP1}$ John's best friend] said S] by substitution of NP in S_1 by NP_1. Historically (Chomsky, 1955a, b), the formulation of generalised transformations includes a condition which by means of symbol replacement grants unlimited occurrences of *S* within *S*: that is, unlimited embedding by clausal complementation. In classical transformational grammar, where rewriting rules were mappings from strings to strings, *substitution* is defined as a closure property of a set of strings under concatenation. The relevant condition is formulated as follows:

> *Condition 2: if* Z_1, Z_2 [which are strings] ∈ *Gr*(*P*) [the set of kernel sequences generated by the phrase structure component, a component referred to in LSLT as a level 'P'], *then* $Z_1{}^\frown\#^\frown Z_2$ ∈ *Gr*(*P*). (Chomsky, 1955a: 481)

The set of kernel sequences must then include sequences which have undergone *some* generalised transformations, at least conjoining GTs (it is not certain whether embedding GTs would also fall into this category in Chomsky's argument). Chomsky proceeds to argue that, since the set *Gr*(*P*) can include strings with any number of occurrences of #, for # a sentence boundary, then the procedure that generates $Z_1{}^\frown\#^\frown Z_2$ from a pair of underlying strings Z_2, Z_2 (or, rather, a pair of pairs ((Z_1, K_1), (Z_2, K_2)), for K_n an *interpretation* of Z_n; Chomsky, 1955a: 480) is recursive. This procedure is precisely generalised transformations, including both conjoining and embedding GTs (using Fillmore's 1963 terminology). So, in the framework of *Logical Structure of Linguistic Theory* (LSLT), generalised transformations basically replaced # by S, for instance (Chomsky, 1955a: 483):

> Input: #-it-was⌢quite⌢obvious (where ⌢ = linear concatenation, and -= constituent boundary)
> Generalized Transformation: T_{th} (that-clause insertion)
> Output: that⌢S-was⌢quite⌢obvious

Grammars including an operation of *substitution* as defined above are strictly context-free, allowing for *head-tail recursion* (if the placeholder or target of substitution appears in string initial or string final position, respectively) and *center embedding* if the placeholder or target of substitution appears elsewhere (e.g., in a rule like S → *aSb*). In *substitution*, importantly, there is no internal modification to either structure: applications of *substitution* in the LSLT sense (and as borrowed in Chomsky, 1995: 189–190 for the definition of Merge) must 'extend their targets' (Frank & Kroch, 1995: 106) but not tamper with them. As per the interpretation of *structure preservation* in Chomsky (1986), *substitution* targets only specifier or complement positions.

TAGs push the computational power of the grammar up from strictly CF by adding an operation of *adjunction*, which allows us to insert a tree within another tree (pushing some material in the target of adjunction downwards) under specific structural conditions. Joshi (1985: 209) defines *adjunction* as an operation that

> ... *composes an auxiliary tree β with a tree γ. Let γ be a tree with a node labelled X and let β be an auxiliary tree with the root labelled X also.* (*Note that γ must have, by definition, a node—and only one—labelled X on the frontier.*)

We saw *adjunction* at work in **Section 2.2**, we repeat the diagram for the reader's convenience:

(92)

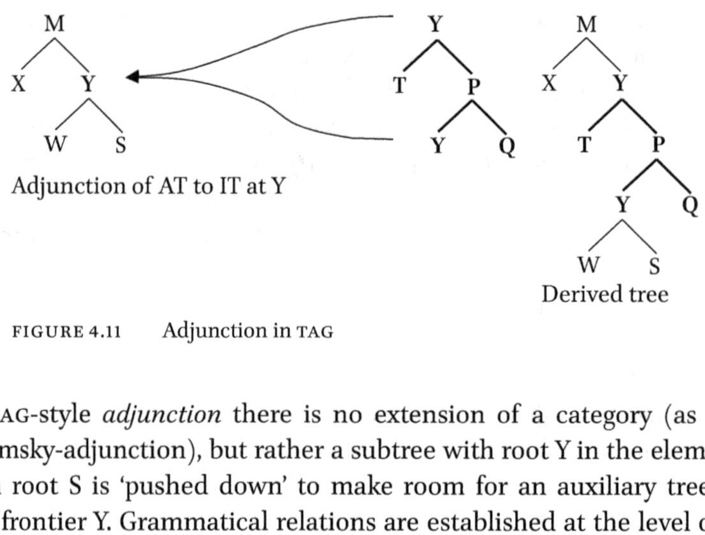

FIGURE 4.11 Adjunction in TAG

In TAG-style *adjunction* there is no extension of a category (as there is in Chomsky-adjunction), but rather a subtree with root Y in the elementary tree with root S is 'pushed down' to make room for an auxiliary tree with root and frontier Y. Grammatical relations are established at the level of elementary trees, and all relations established between elements in S are preserved after adjunction: a node labelled Y immediately dominates W and S in IT, and after adjunction W and S are still immediately dominated by a node labelled Y. Frank (2013: 233, 237) calls these principles the *Fundamental TAG Hypothesis* (repeated from **Chapter 3** above) and the *Non-local Dependency Corollary*:

> ***Fundamental TAG hypothesis:*** *Every syntactic dependency is expressed locally within a single elementary tree*
> ***Non-local Dependency Corollary:*** *Non-local dependencies always reduce to local ones once the recursive structure* [i.e., adjoined auxiliary trees] *is factored out*

TAGs (lexicalised or not) eliminate the early Minimalist distinction between (LSLT-style) *substitution* and (either Chomsky- or sister-) *adjunction* by recognising only two combinatorial operations: *substitution* (replace a designated node in the frontier of a tree with a categorially identical node that is the root of another tree) and (TAG) *adjunction*. As highlighted in Frank & Hunter (2021: 191), TAGs also eliminate the requirement that generalised transformations extend their targets, but capture the empirical consequences of this stipulation in terms of conditions over the positions within an initial tree where adjunction can take place. This depends on the identity between the root and foot nodes of an auxiliary tree and a node internal to an initial tree.

In the treatment of long-distance dependencies, TAGs can be further enhanced by allowing *links* between nodes (we must pay attention to the fact that *link* in a TAG has a very different meaning from *linking* in the present work): links in a TAG play a role similar to that played by traces in a transformational grammar, but there needs to be no 'movement' in a TAG. Specifically,

> *If a node n_1 is linked to a node n_2 then*
> (i) *n_2 c-commands n_1 (i.e., n_2 precedes n_1 and there exists a node m that immediately dominates n_2 and also dominates n_1),*
> (ii) *n_1 dominates a null string* [...] (Joshi, 1985: 214)

Operator-variable relations can be specified within an elementary tree, in terms of links, and since relations within an elementary tree are preserved under adjunction, it is adjunction that gives the impression of long-distance movement, without any Copy or re-Merge actually having taken place (see Kroch, 2001 for numerous examples). We will see an illustration of *links* in the TAG sense in (97) below. TAGs with links can generate *limited* crossing dependencies, by virtue of having elementary trees annotated with links which are preserved under *adjunction*. These *links* characterise binary relations between nodes in an elementary tree.

Links are crucial when evaluating the expressive power of TAGs. A TAG with links can generate the CS language $L = \{a^n, b^n, e, c^n, d^n \mid n \geq 0\}$ *iff a*'s and *b*'s are nested, *c*'s and *d*'s are nested, and *a*'s and *c*'s and *b*'s and *d*'s are cross-serially related (or *vice versa*). The two situations can be diagrammed as follows (see Joshi, 1985: 223):

(93) a. $a_1\ b_1\ a_2\ b_2$ (crossing dependencies between *a* and *b*)
b. $a_1\ b_1\ ...\ c_1\ d_1\ c_2\ d_2\ ...\ a_2\ b_2$ (crossing dependencies between *c* and *d* embedded inside crossing dependencies between *a* and *b*)

For purposes of linguistic description, *a*'s and *b*'s can be thought of as verbs and arguments (Shieber, 1986), or as fillers and gaps (Frank & Hunter, 2021).

Despite their additional power with respect to CFGs, there are CS languages which *cannot* be generated by a TAG with links. These *strictly* CS languages include languages in which *only* cross-serial dependencies are established, like $L = \{a^n, b^n, c^n, e, d^n, f^n \mid n \geq 0\}$, languages that establish crossing dependencies between an unlimited number of categories, and double-copy languages, like $L = \{w\ e\ w\ e\ w \mid w \in \{a, b\}^*\}$ (see also Frank, 2002: 34).

In the present context, because nodes in the graphs we defined are assigned *addresses* which point to memory locations containing *semantic values* of those nodes, the applicability of 'links' in the treatment of long distance dependencies needs to be looked at very carefully: classical TAG linking in phrase structure trees requires *two* nodes (Joshi, 1985; Kroch, 1985), whereas our representations align with multidominance-allowing versions of the theory. TAG-style *links* must not be confused with *traces*: if we maintain the condition that the 'lowest' node in a link must dominate an empty string as in Joshi's definition above, links are only required to represent filler-gap dependencies (in the sense of Postal, 1998; Sag, 2010). Links are not used in TAGs to represent other dependencies between elements in a structural description; in particular, TAG links are restricted to nodes of the same category in addition to the condition that the lowest node dominates an empty string. Because there are no *gaps* (movement-generated null categories) in the present model since there is no movement, there will be no need to resort to TAG-style links (see also the discussion of *wh*-interrogatives on **Chapter 10**).

4.3.2 *Perspectives on Crossing Dependencies*

Syntactic theory usually employs some kind of diacritic or annotation to indicate dependencies between overt elements, in particular, *agreement* and *coreference*. These are dependencies that are not restricted to nodes bearing the same category: as a matter of fact, we have

(94) a. N-V relations
b. N-N relations

Relations of the kind (94b)—the kind that is relevant for *Pronominalisation* in a Lees & Klima-Ross-Langacker view, as well as including e.g., controller-controlee relations—are formulated in terms of having *single N node* multidominated, *not* as a relation between two distinct N nodes which share an index (or some other diacritic indicating identity). This issue in particular will be dealt with in more detail in **Chapter** 7 (recall also the **Identity-ordering condition on pronominalisation** defined above). But (94a) is worth looking at more closely, since it seems to be a different creature. This is the kind of dependencies that arise in the now famous Dutch and Swiss German examples considered in Bresnan et al. (1982), Shieber (1985), Joshi (1985), and much related work:

(95) *Jan Piet Marie zag helpen zwemmen*
Jan Piet Marie saw help swim
'Jan saw Piet help Mary swim' (Dutch; taken from Joshi, 1985: 245)

The dependencies between NPs and VPs in (95) (that is, what NP is the subject of what VP) can be visually represented by means of arrows as in (96):

(96) Jan Piet Marie zag helpen zwemmen

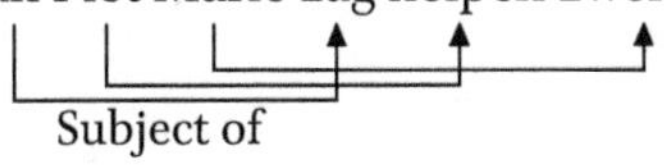

If we use a and b as variables over indexed categories (as in (93)), then the abstract form of (95) is $a_1a_2a_3b_1b_2b_3$: this example displays crossing dependencies (indicated with subscripts). What does an adequate structural description for (95) look like? From the perspective of a TAG with links, the derived structure proposed in Joshi (1985: 249) is the following (links are indicated with dotted lines):

(97)

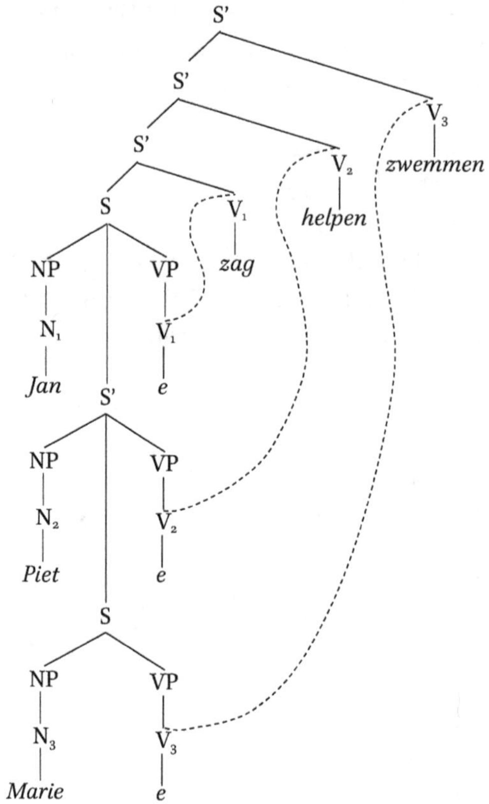

FIGURE 4.12 TAG analysis of crossing dependencies in Dutch

In Joshi's representation, each elementary tree contains a subject and its verbal predicate (rightwards verb extraposition, sometimes known as 'verb raising', applies for reasons independent from the TAG formalism; see Evers, 1975; Kroch & Santorini, 1991). This aspect is particularly important for our own approach. The IT *Marie zwemmen* would be (98), again assuming verb raising:

(98)

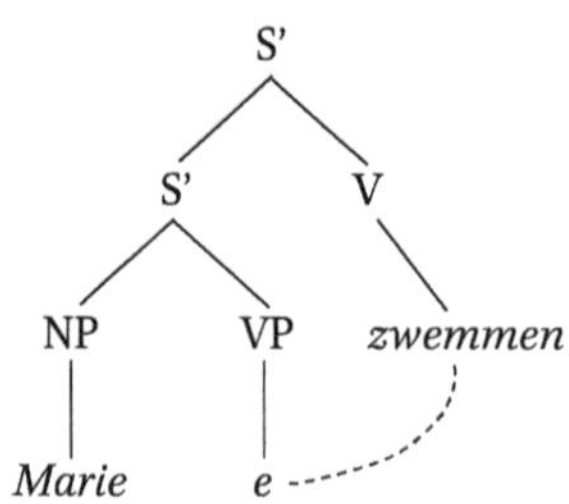

FIGURE 4.13 Elementary tree for Dutch example

The structural description above, adapted from Joshi (1985: 248) is partly based on the discussion of cross-serial dependencies in Dutch in Bresnan et al. (1982). The latter propose the following phrase marker (adapted from Bresnan et al., 1982: 619), in which crucially the subjects do not form a constituent with (or are directly connected to) their corresponding verbal predicates, in contrast to the TAG analysis:

(99)

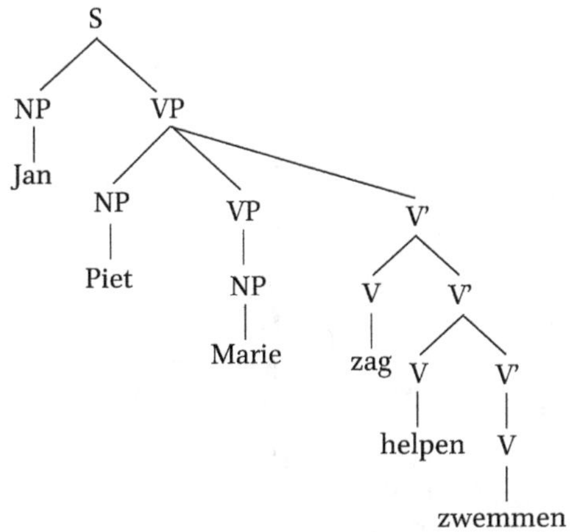

FIGURE 4.14 LFG c-structure analysis of crossing dependencies in Dutch

Despite some differences in the analysis (e.g., note, again, that only in Bresnan et al.'s analysis is there a constituent that contains only the verbs and excludes all the subjects; also there is a constituent that includes the two embedded subjects and exclude the matrix one), both works conclude that there is no need to push the grammar to *full* context sensitivity (i.e., to make the grammar powerful enough that it can generate *all* context-sensitive languages); *weak* context-sensitivity, a.k.a. *mild* context-sensitivity, seems to do the trick. Bresnan et al., unlike Joshi, have a distinct level (*f*-structure) where predicate-argument dependencies are encoded: *c*-structure is linked to *f*-structure by a set of functional equations (that do not concern us here), and is intended to represent constituency, hierarchy, and word order. Joshi's structural description is based on the linear position of elements *as well as* their syntactic dependencies, which is an important difference with our model: recall that the notion of *order* that is essential in the present proposal is not *linear order* (i.e., *precedence*), a point in common with recent Minimalist work (Chomsky, 2020, 2021).

The challenge is set, then. Let us see if we can assign a structural description to (95) that adequately represents syntactic relations in our graph-theoretic model, which although heavily influenced by LTAGs presents some important differences with respect to them (among which the possibility of linking graphs at any node, in principle—limited only by independently motivated aspects of language-specific grammars—must be highlighted). To begin with, we will consider an incorrect, but intuitive, approach. If we grouped in a single arbor everything that a TAG joins by links (doing away with empty terminals, marked *e* in Joshi's diagram), we would get the following:

(100) Arbor 1: [Jan zag helpen]
Arbor 2: [Piet helpen zwemmen]
Arbor 3: [Marie zwemmen]

But things are more complex (the inadequacy of a description along the lines of (100) is indeed prefigured by Joshi). As it is, *and assuming that structural descriptions encode aspects of linear order* (this is a fundamental caveat at this point of the argument), as Joshi does, (100) describes (*weakly generates*, using Chomsky's 1965 terms) the string *Jan zag Piet helpen Marie zwemmen*, which is English-style tail-recursion. This is certainly not what we have in (95). Equally, an approach based on strict center embedding (manipulating only *cyclic monostrings* in the sense of Lasnik & Kuppin, 1977: 176–177) is also inadequate, despite the fact that PSGs can indeed capture dependencies between objects which are 'indefinitely far apart' at the cost of adding 'invisible' structure in the form of nonterminals (see Lasnik, 2011: 356) or feature annotations (Bresnan et al., 1982: 616 argue that such structural descriptions generate an 'artificially restricted' *proper subset* of the relevant structural descriptions, and not particularly linguistically appealing ones):

(101)

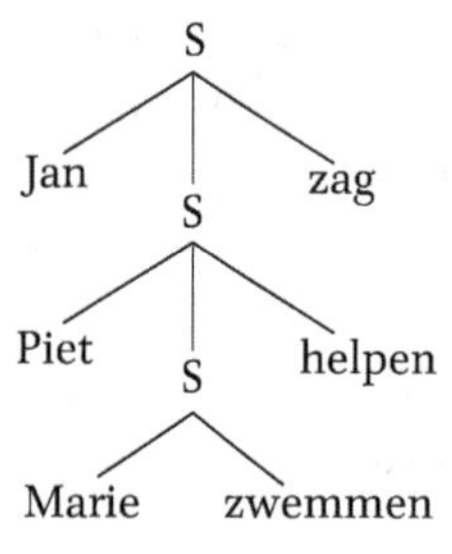

FIGURE 4.15 Center embedding structure for Dutch example

Both verbs *zag* and *helpen* take events (predicate-argument relations; we may call them 'satisfied functions') as complements; their internal argument is a complete event with external and internal argument, both of which are dominated by the relevant verb: this will become obvious in our dominance relations.

Thus, we should have something more along the lines of (101), with the full set of dominance relations in the derived graph made explicit (see also the dependency analysis for an analogous example in Rambow & Joshi, 1994: 13, which corresponds to a 'deep syntactic representation' in Meaning-Text Theory):

(102) ρ = {(zag, Jan), (zag, Piet), (zag, helpen), (helpen, Piet), (helpen, Marie), (helpen, zwemmen), (zwemmen, Marie)}

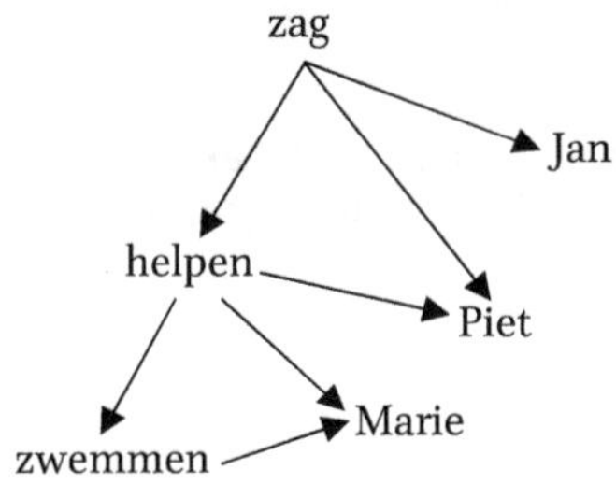

FIGURE 4.16 Graph-theoretic analysis of crossing dependencies in Dutch

The verbal predicates are in a continuous segment with respect to dominance (in Osborne et al.'s 2012 terms, they form a *catena*), and at the same time each verb immediately dominates its arguments. In our structural description we have made no use of empty nodes, intermediate non-terminals (i.e., 'bar levels'), or any MGG-like notion of headedness (in the sense that, for instance, a V is the 'head' of a VP). If there is something like 'heads' in the Dependency Grammar sense, they arise from the consideration of dominance relations, but they are not primitives of the theory (therefore, 'projection' is ruled out; *stricto sensu*, there is nothing to project or intermediate nodes to project to). The relations identified are reminiscent of some work in DG, in that the matrix predicate immediately dominates the predicate it embeds (e.g. Osborne, 2019: 183, also Rambow & Joshi, 1994 and references cited there); however, unlike DGs, in our approach each verb also immediately dominates all of its nominal arguments.

The system described here can also accommodate (provide structural descriptions for) facts from languages where clausal complementation does not yield limited crossing dependencies (thus, mildly context-sensitive), but rather center embedding (thus, strictly context-free). We can give as an example the Turkish counterpart of (95), in (103):

(103)	*Merve*	*Ömer'in*	*Esra'nın*	*yüzmesine*	*yardım*
	Merve.NOM	Ömer.GEN	Esra.GEN	swim.GER.GEN	help.ACC

ettiğini *gördü*
give.REL.ACC see.3SG.PAST
'Merve saw Ömer help Esra swim'

Glossing over morphological complications, the dependencies in (103) are as in (104):

(104) Merve Ömer'in Esra'nın yüzmesine yardım ettiğini gördü

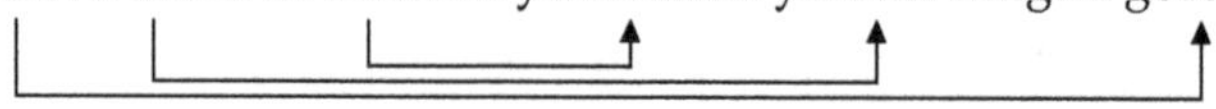

And, of course, the English pattern, in which each verb takes a complement clause to its right (thus displaying only *tail recursion*: the call-function S appears in string-final position):

(105) John saw Peter help Mary swim

The main question that we need to ask ourselves (see also Ojeda, 2005: 8–10) is: do dominance sets vary between Turkish, English, and Dutch? So far as we can see, that depends entirely on what we want structural descriptions to capture. If the goal is to capture (i) syntactic dependencies, (ii) argument structure and grammatical functions, and (iii) word order, then the graphs that correspond to the Dutch, Turkish, and English examples must be different (see Kural, 2005; Kremers, 2009; Medeiros, 2021 for uses of tree traversals to formalise word order alternatives). In Dutch and Turkish, the subject of *see* immediately precedes the subject of *help*; in English there is no adjacency between subjects. If ρ-sets encoded word order, then there must be a directed edge $e\langle Subj_{see}, Subj_{help}\rangle$ in the graphs assigned to Dutch and Turkish, but not English. This is possible (and opens an avenue for further research in terms of cross-linguistic variation), but it is not the path we take here. The questions that we attempt to provide answer to pertain to points (i) and (ii) above; the problem of word order is formulable under current assumptions, but will not be addressed here. As for cross-linguistic variation, we will see that our theory makes very specific predictions about where variation may arise, in line with research on lexicalised TAGs. Given paradigms like the one involving clausal complementation in Dutch (and Swiss German, following Shieber, 1985), Turkish, and English there is a very prominent question that arises in derivational frameworks: how to avoid assigning too much or too little structure (see Krivochen, 2015a, 2016a; 2021a). But this additional structure comes in the form of non-terminal nodes, as is also recognised in Chomsky & Miller (1963), Lasnik (2011), and others: if we eliminate non-terminals (phrasal projections), there is no extra-structure. Admittedly, this is not unlike Alexandre's 'untiying' the Gordian Knot with a sword, *if* the knot was a consequence of Alexandre's own theoretical framework. The challenge, in these early stages of the presentation of our graph-theoretic grammar, is not so much to define the dominance relations in the graphs corresponding to these sentences, insofar as the conditions that we have specified for well-formedness so far have been relatively lax (this will change soon enough). The challenge is to provide, in addition to a restrictive format for graphs, a way to provide an *interpretation* for sentences. We assume a Lexicon where predicates' lexical entries specify the number of arguments they take, but this is not enough. The aforementioned interpretations for structural descriptions must specify, minimally, *who did what* (*to whom*): there must be a way to define grammatical functions, optimally without invoking any additional mechanisms (like annotations, labels, or indices). In the following chapter we will examine the problem of how to specify grammatical functions in dominance sets.

CHAPTER 5

Ordered Relations and Grammatical Functions

It is widely acknowledged that adequate structural descriptions for natural language sentences need to represent, at some level, the grammatical relations that are established between arguments and predicates. At the very least, it must be possible to determine *who did what* (*to whom*) or, much more generally, *what happened*. This much is accepted across the grammatical board. There are differences, however, between theories with respect to the status of grammatical relations in the theory of grammar (primitive vs. derived) and the *locus* at which these relations are determined (whether at a pre-transformational level of Deep Structure, or a level in a parallel architecture such as *f*-structure, Lexical Structure, or Conceptual Semantics, which can map onto other levels by means of interface operations); the descriptive usefulness of grammatical functions for the analysis of natural language sentences (of English or any other language) is however rarely questioned.

This, of course, only means that we must be able to represent grammatical relations somehow. This is not a trivial task: we have said that there is an edge from a predicate to each of its arguments. However, if there are edges from—say—*zag* to both *Jan* and *Piet*, how do we know which one is the subject and which the object? How can we get a proper characterisation of *who did what* (*to whom*)? Let us analyse some alternatives. A first possibility worth looking at is that the difference is marked by means of morphological agreement: subjecthood would be read off agreement morphology. If verb-subject agreement was encoded in an edge *e*⟨Subj, V⟩, there would be no problem: there would be a closed loop between every verb and its subject, but not between every verb and its object in a sentence like (95) above: *Jan Piet Marie zag helpen zwemmen*. That is: we would have a walk Subj-V-Subj but not Obj-V-Obj. This revised picture would yield the dominance relations specified in (106):

(106) ρ = {(zag, Jan), (Jan, zag), (zag, Piet), (zag, helpen), (helpen, Piet), (Piet, helpen); (helpen, Marie), (helpen, zwemmen), (zwemmen, Marie), (Marie, zwemmen)}

However, there seems to be no strong reason to attempt to represent morphological agreement in terms of dominance and define grammatical functions as a secondary effect, given the objectives of the present work. Defining the 'subject' of a clause as the element that agrees morphologically with the verb or that

 | DOI:10.1163/9789004542310_006

is assigned a specific case is—in our opinion—more trouble than it's worth, as the relation between morphological case marks and grammatical functions is not one-to-one (see Perlmutter & Postal, 1983a for extensive discussion).

Another possible course of action is to define grammatical functions configurationally, a classical move adopted, for instance, in Chomsky (1955a: 254; 1965: 69), cited above and repeated here for the reader's convenience:

> *It is necessary only to make explicit the relational character of these notions* [grammatical functions] *by defining "Subject-of," for English, as the relation holding between the NP of a sentence of the form NP⁀Aux⁀VP and the whole sentence, "Object-of" as the relation between the NP of a VP of the form V⁀NP and the whole VP, etc. More generally, we can regard any rewriting rule as defining a set of grammatical functions* (Chomsky, 1965: 69)

In Chomsky's view, grammatical functions are read off syntactic configuration in phrase structure trees. An obvious difficulty in adopting a configurational view of grammatical function along Chomksy's lines arises: there are no phrase structure rules in our system. It would be possible, of course, to eliminate the derivational aspect of the definition of GF and just take configurationality. But there are deeper problems. One of those is that, in PSGs there is no way to represent a single syntactic object with more than a single grammatical function unless additional mechanisms are invoked. Let us limit ourselves to the brief sketch in Chomsky's citation: we have Subject-of and Object-of defined in configurational terms. Let *G* be a PSG with alphabet Σ = {Det, N, NP, V, VP, S} and the following set of rules:

(107) S → NP, VP
NP → Det, N
VP → V, NP

G weakly generates the terminal string Det⁀N⁀V⁀Det⁀N. Here, the first Det⁀N substring is assigned the GF Subject, and the second one is assigned the GF Object. The mapping from structural positions to grammatical functions is possible in so-called *configurational languages*: languages with rigid word order and generally impoverished morphology, such that syntactic configuration conveys a lot of information. This approach is relatively simple and intuitive if the point of departure is IC-PSGs, and it is the idea that was adopted for most of the Standard Theory and its revisions. However, there are several problems with this view (which have been reviewed in detail in the RG/APG/Metagraph literature); here we will simply point towards some of the most notorious. An obvious difficulty is the existence of *non-configurational languages*: languages with free word order and discontinuous constituents where the relation between configuration, word order, and grammatical functions cannot be the same as in a language like English. A classic example is Warlpiri ((108) is taken from Hale, 1983):

(108) a. (Warlpiri, Pama-Nyungan, Australia)
Ngarrka-ngku ka wawirri panti-rni
Man-ERG AUX kangaroo.ABS spear-NONPAST
'The man is spearing the kangaroo'

b. Wawirri ka panti-rni ngarrka-ngku (O-Aux-V-S, grammatical, same meaning)

c. Panti-rni ka ngarrka-ngku wawirri (V-Aux-S-O, grammatical, same meaning)

In a sentence like (108), '*there are no syntactic constraints on the order of words in the sentence, so long as the auxiliary appears in second position*' (Dalrymple, 2001: 65). The first position is always occupied by a constituent, which may or may not be multi-word ((108d-e) are taken from Bresnan et al. 2016: 4):

d. *Kurdu-jarra-rlu wita-jarra-rlu ka-pala maliki*
Child-DUAL-ERG small-DUAL-ERG AUX-3DUAL.SUBJ dog.ABS
wajilipi-nyi
chase-NONPAST

e. *Wita-jarra-rlu ka-pala wajilipi-nyi*
Small-DUAL-ERG AUX-3DUAL.SUBJ chase-NONPAST
kurdu-jarra-rlu maliki
child-DUAL-ERG dog.ABS
'The two small children are chasing the dog.'

Note that the NP 'the two small children' is a discontinuous constituent: the constraint that the auxiliary must be in second position is met in both (108d-e), which means that the string *kurdujarrarlu witajarrarlu* must correspond to a constituent (which can be further analysed as an AP modifying an N head). As emphasised in **Chapter 3**, linear adjacency must therefore be given up as a necessary condition for constituenthood, or an alternative unit of analysis must be defined (such as expressions which are continuous in terms of *dominance* rather than *precedence*; these are Osborne et al.'s *catenae*) Much work in generative grammar and LFG (see Hale, 1983 and Simpson, 1991 respectively, among others) have proposed different constituent structures for Warlpiri and English (for instance, Warlpiri does not seem to have a syntactic object that includes the verb and its direct object but excludes the subject; that is, there is no VP). The configurations in which GFs are assigned do not seem to be universal, which makes it difficult to extend the theory beyond a single language.

But we do not need to go as far as looking at non-configurational languages to find problems with the classical PSG approach. Consider for instance the case of *reflexivity* (to which we will come back in **Section 6.2.1** below): here, a predicate is defined to be *reflexive* if its subject and object are correferential (Reinhart & Reuland, 1993). In this case, a grammar along the lines of Chomsky's proposal must necessarily indicate correference by means of additional devices, diacritics or indices, such that NP-Sister of VP and NP-Sister of V are assigned the same index. As a consequence, the entities in the representation are multiplied in two ways: (i) we have *two* NPs instead of one (because of how the rules are formulated), and (ii) because of this we need to extend the grammar by adding referential diacritics plus a mechanism to keep track of these diacritics. This multiplication, we argue, is neither desirable nor required or motivated empirically. On the empirical side, Johnson & Postal (1980: 16) remark that the Chomskyan view, which is closely related to a rigid view of clause structure (an aspect of the generative theory that, as we have observed several times now, has only became stronger in Minimalism, in particular under antisymmetry theory and cartography) has remained insensitive to difficulties raised by languages that do not conform to the Spec-Head-Compl template allegedly determined by Universal Grammar (which translates to SVO in GF terms, when and if GFs are invoked).[1] In these cases not only entities but also operations have multiplied: reordering transformations must be invoked in order to restore a putative underlying SVO representation (Kayne, 1994: Chapter 4 and much subsequent work). Any internal complexity pertaining to the GFs themselves is also erased: the characterisation in Chomsky's quotation and the work stemming from it does not distinguish Direct from Indirect Objects (or the much richer array of objects proposed in Postal, 2010). Of course, any new GF can be incorporated

1 See, for instance, Haider (2018) for a discussion of the empirical difficulties faced by the Universal (underlying) SVO hypothesis required to make sense of the LCA and how those difficulties have been mostly ignored in MGG. Borsley (1997) also provides indirect argument against the strict antisymmetric approach to syntax, based on the analysis of relative clauses (also Krivochen, 2022).

into the template if GF are defined configurationally, but only at the cost of introducing new nonterminal nodes: for example, we may just add a new XP within the VP that excludes an NP sister of V and dominates an NP, and define Indirect Object in terms of the relation between NP and XP (e.g., an $Agr_{IO}P$, as in Radford, 1997: 242, ff.). But that is an unnecessarily *ad hoc* move, and further articulation in the definition of GF in different languages would require the addition of an unspecified number of nonterminal labelled nodes; in the worst-case scenario, one per GF.[2] This is a situation that we need to avoid.

Are there other possibilities? We contend that there are. Even without adding to the theory levels of representation like *a-structure* for thematic roles (roughly, GB's 'theta grid') or *f*-structure level for grammatical functions (as in LFG; Bresnan, 2001: 19, ff.; Dalrymple, 2001: 8, ff.; see also Williams, 2003 for a theory rich in levels of representation and mapping functions; Culicover & Jackendoff, 2005 also makes use of parallel structures in their grammatical architecture one of which is specifically devoted to grammatical functions) or appealing to intermediate nodes to create specific configurations with respect to which functions are defined (Chomsky, 1965: 71), grammatical functions can be encoded in graphs by adding *annotations* or *labels* to nodes or edges. Examples of the latter strategy are to be found in RG, APG, and Metagraph grammar (Perlmutter & Postal, 1983a, b; Johnson & Postal, 1980; Postal 2010 respectively), where arcs indicate the grammatical relation that is instantiated between the nodes connected by that arc. For instance, a basic (and admittedly incomplete, although in ways that are at present irrelevant) general format for arcs in L-graphs could be (109):

2 It is important to distinguish GFs as conceived of here (following the RG tradition) and the use of 'Subject' as a label in certain forms of cartographic generative grammar. To give just an example, consider the clausal structure proposed in Cardinaletti (2001: 121):
i) [ForceP ... [SubjP ... [AgrSP ... [TP ... [VP]]]]
Following an immediate constituency approach to labelling (based on the relation *follows-from*), we would be forced to interpret (i) as meaning that there is a string that has the distributional properties of a subject, just as any string (transitively exhaustively) dominated by NP has the distributional properties of a nominal. There are two issues with an interpretation of the structure in (i) in terms of immediate constituency (see also Hopcroft & Ullman, 1969 for a view from formal language theory). First, endocentricity pertains to categories, not grammatical functions: the latter are relational notions, two-place predicates: 'Subject' is 'Subject-of(x, y)', whereas categories are 1-place predicates. Second, and perhaps more evidently, such an interpretation entails that a predicate (the VP) is properly contained within the subject, since the SubjP properly contains the VP. However, the predicate of a clause does not behave as a proper constituent of the subject of a clause. The structure in (i) represents a sequence of functional heads that host phrases, and technically defines a one-dimensional *array*.

(109) 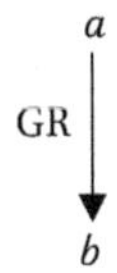

Perlmutter & Postal (1983b: 83)

FIGURE 5.1 Annotated arc in RG

(109) reads 'the linguistic element (or expression) *b* bears the grammatical relation GR to the linguistic element (or expression) *a*'. We must highlight, however, that in a system that includes edges of the kind in (109), an edge is no longer a pair but a triple ⟨*a*, *b*, *p*⟩ (as noted by a reviewer to Postal, 2010, cited in Postal, 2010: 395), where

> *E* ⟨*a*, *b*⟩ *-E still being some binary relation on V* [the set of nodes, including *a* and *b*] *—and p is an edge label (drawn from a set disjoint from V).*

Equivalent proposals for the definition of arcs (the local vertex-edge relations in L-graphs) are made in Johnson & Postal (1980: 10, 37) and Postal (2010: 25–26). Also, Mel'čuk (2003), within a multi-layered model of Dependency Grammar, annotates arcs with integers, such that '*The arc between the predicate and its argument carries the number of the argument*' (2003: 11), where numbers make reference to the place of a specific relation in a hierarchy of grammatical functions.

Along these lines, we could add a specification to each edge, such that the grammatical function is encoded in an index (say, *subj* for Subject, *obj* for Object; we could also have used 1 and 2, in consonance with Relational Grammar practice) and obtain the following representation which revises (106):

(110) e_{subj}⟨zag, Jan⟩, e_{obj}⟨zag, Piet⟩, e_{Vcomp}⟨zag, helpen⟩ …

However, (110) is not quite adequate either, for a number of reasons. To begin with, find this notation cumbersome and not really useful: adding diacritics to the set of tools required by the grammar is only justified if there is no other way to encode grammatical distinctions in the system itself, making use of elements and relations already available (in this sense, we agree with Postal's reviewer). If said distinctions have any real descriptive (or explanatory) value and if our choice of formalism has been fruitful, however, there should be a way to encode them in the grammar itself as a system, rather than as elements of the alphabet. Optimally, the representation of grammatical functions should not involve the multiplication of elements, rules, or principles; it should make use of the basic tools of the formal system already available to us and which have independently proven useful in empirical analysis. This is what we will now try to do.

In **Chapter 2**, we defined the *ρ-set* of a graph G as the full set of dominance relations between nodes in that graph. At that point we remained purposefully agnostic with respect to whether a ρ-set was an *ordered* or an *unordered* set: now, we will eliminate that initial ambiguity and take a stance in order to model grammatical relations in terms of our graph-theoretic approach. We can capture the relevant functional asymmetries (Subject vs. Objects, and within the class of objects, Direct Object vs. Indirect Object) by introducing an *order* into the set of dominance relations following an independently motivated functional hierarchy (see Dalrymple, 2001: 9 for a summary of different approaches to grammatical function hierarchies). Recall that a node v_1 dominates v_2 if and only if there is an edge $e\langle v_1, v_2\rangle$ in a directed graph. Then, we can make the inherent asymmetry of a directed graph into an asset: make that asymmetry correspond to an external hierarchy that can help us in the description of relations without the need to appeal to annotations (to either edges or vertices). We will take RG's hierarchy of grammatical functions (e.g., Perlmutter & Postal, 1983b: 86 in terms of 'nuclear' and 'object' set-defining features; see also Bell, 1983: 148; Keenan & Comrie, 1977) as a reference. Similar hierarchies are found in approaches where grammatical functions are defined at a level of representation distinct from constituent structure, such as Simpler Syntax (Culicover & Jackendoff, 2005: Chapter 6), LFG (Dalrymple et al., 2019: Chapter 2), or HPSG (Müller, 2015: § 2); the idea of having GFs be primitives can be traced as far back as Jespersen (1985 [1937]). RG's definition of *clause* makes this clear:

> *A clause consists of a network of grammatical relations* (Perlmutter & Postal, 1983a: 9)

In this context, primary grammatical functions configure a structured system: they are organised in a *hierarchy*. The RG hierarchy, as presented in Bell (1983: 148) and Blake (1990: 20) is as follows:

(111) 1 (Subject) >> 2 (Direct Object) >> 3 (Indirect Object) >> Obliques

As observed in Blake (1990: 20), some of the RG literature took the hierarchy for granted, but there is extensive work (comparative and not) that supports it (for example, Postal, 2010 is a particularly detailed work dealing with different classes of objects in English). LFG, based on RG, adopts a more fine-graded version of the hierarchy, in which core and non-core primary grammatical functions are distinguished (Bresnan et al., 2016: 97; Dalrymple et al., 2019: 11):

(112) Core: SUBJECT > OBJECT > OBJECT$_\theta$ > Non-core: OBLIQUE > XCOMP, COMP > ADJUNCT

(an XCOMP is a complement clause that is unsaturated: its subject is determined by *functional control*—essentially, *structure sharing*—; OBJECT$_\theta$ and OBLIQUE$_\theta$ are thematically restricted objects—in RG terms, roughly a 2 that can only receive specific thematic roles and a 3, respectively-. Additional primary GFs assumed in LFG include XADJ(UNCT), an open adjunct and VCOMP, the GF of infinitival VPs in ECM contexts. Like RG and APG, LFG assumes a set of secondary, or *overlay* functions (such as TOPIC or FOCUS): these are functions integrated to the clause through its relation to a primary function. An expression cannot bear only an overlay function)

Another declarative formalism, HPSG, defines ARG-ST, the specification of syntactic and semantic properties of arguments of a predicate, as an *ordered list* whose members are also ordered according to Keenan & Comrie's/RG's hierarchy (see e.g. Müller, 2015). Osborne (2019: 8) also presents a structured set of GFs, with a major distinction between Subject and Object and recognising several types of objects (bare—primary and secondary- and propositional). Despite some differences in conceptualisation and formalisation, what we want to stress is that the importance of the GF hierarchy crosses theoretical boundaries.

In RG and APG there is a rich inventory of relational signs (see e.g. Perlmutter & Postal, 1983b: 86), of which we will only consider some. Specifically, and following Bell, we will consider *core* relational signs, and within these, *terms*: adjuncts and *Chômeurs* (in APG and MG, these are elements that have been *demoted* to non-term relations, the prototypical example of Chômeur is the agent in a passive construction) will be left aside. There is a reason for this, and it is that the syntax of some non-core functions, in particular *adjuncts* (which in LFG covers both adjectival and adverbial modification) can be either FS (thus, 'flat') or higher-level (CF, perhaps even mildly CS) depending on a variety of factors, as argued e.g. in Culicover & Jackendoff (2005), Uriagereka (2008), Dalrymple et al. (2019: 215), and Krivochen (2015a, 2016, 2018, 2021a). In contrast, the syntax of *terms* seems to be much more amenable to the classical generative view about strictly hierarchical organisation. 'Adjunct', in generative grammar, is more a statement about the derivational history of a constituent than about its relation to a predicate: how (and *when*; see Stepanov, 2001) it has been introduced in the syntactic structure, and what position in a category it occupies. In the hierarchy in (110), *terms* outrank *non-terms*, and *nuclear* functions outrank *object* functions: 1 is a *nuclear, non-object* function, 2 is a *nuclear, object* function, and 3 is a *non-nuclear, object* function (see also Johnson & Postal, 1980: 198, 250; Postal, 1982: 346 for APG translations of the RG hierarchy). Because the RG hierarchy is remarkably robust when it comes to the empirical analysis of English constructions (see e.g. Postal, 2010) we will maintain not only the *core* functions in our model, but also assume that the hierarchy between them is correct.

Now, how can we capture the GF hierarchy in the ρ-set of a graph? Suppose that we make the ρ-set of a sub-graph an *ordered* set, thus getting the following inequality to hold (square brackets are used to disambiguate the formula, and bear no formal significance):

(113) $[\rho = \langle\langle a, b\rangle, \langle a, c\rangle\rangle] \neq [\rho = \langle\langle a, c\rangle, \langle a, b\rangle\rangle]$

Plainly: an L-graph is an *ordered set of arcs*. We can notate the *ordered* ρ-set of a graph using angular brackets, such that the ordered set of dominance relations between vertices *a, b, c* is as in (114):

(114) $\rho = \langle(a, b), (a, c)\rangle$

This means that the relation (a, b) (an arc from *a* to *b*, abbreviating $e\langle a, b\rangle$) is ordered before the relation (a, c) in an arbor G: we will say that (a, b) *outranks* (a, c) in G. This is so not because of any property of the nodes, but because of the ordering imposed over the relations between them. The sceptical reader may wonder: what have we gained? Quite a bit. If the order between dominance relations (such that a structural description is defined as an ordered set of arcs) is made to correspond with the order between grammatical relations, we have the following relations, semi-formally:[3]

---

3 We must immediately acknowledge that the picture of clause structure that emerges from (115) is necessarily too coarse. It is not a fatal flaw, though, in that the formalism imposes no restrictions over a more detailed partition of GFs. A notion of strict order can, in principle, accommodate the richer array of Object relations recognised and extensively argued for e.g. in Postal (2010). A revised hierarchy, including the richer landscape of Object relations, could initially go along the following lines:
i) 2 (Direct Object) >> 3 (Indirect Object) >> 4 (Subobject) >> 5 (Semiobject)
However, things are rather complicated. The interpretative rule in (115) assumes that there is no Indirect Object unless there is a Direct Object, in line with traditional accounts; that assumption links the hierarchy of grammatical functions with the Case hierarchy, in which

(115) $\rho = \langle (a, b), (a, c), (a, d) \rangle$
Let *a* be a lexical predicate, and *b*, *c*, *d* be arguments of *a*, then *b* is the Subject / 1 of *a*, *c* is the Direct Object / 2 of *a*, and *d* is the Indirect Object / 3 of *a*.

That is: the order of dominance relations corresponds to the order of grammatical relations, such that if an arc $e\langle x, y\rangle$ is ordered before an arc $e\langle x, z\rangle$ (which, recall, we abbreviate as (x, y) and (x, z) respectively) in the description of the elementary graph where expressions *x*, *y*, and *z* appear, the grammatical relation that *y* establishes to the clause anchored by *x* is also ordered before the grammatical relation that *z* establishes to the clause anchored by *x* in the hierarchy of grammatical functions (the two orders are isomorphic, which means that there is a function *f* with an inverse that relates both sets of orders: we can go from one to the other no questions asked). We can express (115) in simpler terms: for *e, e'* arcs in a graph G, if *e* is ordered before *e'* in G, then the expression that is the tail of *e* outranks the expression that is the tail of *e'* in the GF hierarchy. If there is a single NP in a structure, the hierarchy determines that it will be interpreted as the highest possible grammatical relation licensed by V, *modulo* lexical semantics. In (115), *b*, *c*, and *d* are sister nodes by virtue of sharing a mother *a*; the diagram of this graph depicts a 'flat' structure that is common in frameworks that aim at eliminating empty nodes, silent structure, and bar-levels (e.g., Culicover & Jackendoff, 2005: Chapter 4; Postal, 2010; see also Emonds, 2007: Chapter 6 for a perspective much closer to MGG which makes use of locally flat structures). The interpretative rule (115), in combination with the fact that edges are directed, provides a way to define a total order over argument nodes in an elementary graph: the binary, antisymmetric relation *dominance* imposes an order between predicate nodes and argu-

ACC >> DAT. Moreover, it is also not clear whether the relations 4 and 5 are 'as primitive' as the others, or depend on more fine-graded aspects of lexical semantics. For example, consider the following examples that Postal (2010: 73) uses to illustrate the relation 4 (Subobject), with the relevant NP(4) italicised:

ii) Herb neared *the vampire* / Herb wanted *pizza*

Even if the italicised NPs could be otherwise grouped under a certain category label, it seems strange to us that the same relation can be established with two verbs as different as *near* and *want*. As an example, *near* does not require an agentive subject, which suggests that it is not a (garden-variety) transitive V (but rather closer to an unaccusative V, which subcategorise for two internal arguments: a theme and a location):

iii) The train neared the station / *The train wanted pizza

Considerations of this kind prevent us from actively including Postal's extended typology of objects into (113), but further research is required and the question remains fundamentally an empirical one.

ment nodes (such that predicates always dominate their arguments) and (115) imposes an order between arguments in terms of the GF hierarchy (since there is no dominance relation between arguments). In this context, simple templatic arbores for verb typology can be defined as follows (p is a predicate, A an argument):

(116) a. Unergative: $e\langle p, A_1\rangle$ (A_1 = Subj)
b. Unaccusative: $e\langle p, A_1\rangle$ (A_1 = Subj) (we will return to unaccusativity in **Section 5.1**)
c. Monotransitive: $\langle e\langle p, A_1\rangle, e\langle p, A_2\rangle\rangle$ (A_1 = Subj; A_2 = Obj)
d. Ditransitive: $\langle e\langle p, A_1\rangle, e\langle p, A_2\rangle, e\langle p, A_3\rangle\rangle$ (A_1 = Subj; A_2 = Obj; A_3 = Ind Obj/Obl)

In sum, under the simplest assumptions, an n-ary predicate will have outdegree n (see fn. 8 in this chapter). Importantly, however, this configurational template makes no reference to the thematic grid of specific predicates, which is information that presumably is part of predicates' lexical entries (which, again, may be looked at from a variety of perspectives). These entries must specify, minimally, the number of arguments subcategorised (if any), and the semantic/thematic roles associated with those arguments (a point we made already in **Chapter 2**). We are, evidently, leaving aside a great number of issues related to argument structure, and the possible interactions between our syntactic approach and lexicalist vs. decompositionalist approaches (although currently the system seems to lend itself better to interactions with lexicalist approaches).

Alternative approaches to the GF hierarchy are of course possible, and to different extents compatible with our framework. For example, Dowty (1982), while acknowledging the centrality of GF to a number of syntactic processes (e.g., Passivisation, Dative Shift, and Raising) that was characteristic of RG, departs from it in deriving GF from properties of the syntax-semantics interface. Building on previous work in Montague Grammar, Dowty proposes a way of defining GF in terms of order of semantic composition: the first argument to combine with a predicate is the most oblique one, with the Subject being the last GF to be composed. Under Montagovian assumptions about the nature of the syntax-semantics interface, the denotation of a 2-place predicate like *love* is not a function from an ordered pair to a truth value, as it is in classical predicate calculus. Rather, in a sentence like *John loves Mary*, the predicate *love*, of category (FC/NP)/NP (or IV/NP) combines with *Mary* to deliver the expression *loves Mary*, of category FC/NP (or IV). This effectively makes *love Mary* into a 1-place predicate, a function that applies to the denotation of the expression *John* with co-domain *true-false* (if the individual denoted by the expression

John belongs to the set of entities with the property of loving Mary). These combinations are independent from word order, such that the same formal operations apply in, say, SOV, SVO, and VSO languages (Dowty, 1982: 85, ff.). Dowty's system can be conceptualised from the perspective of derivational syntax, given his 'principle of grammatical relations' (1982: 84):

> *A verb that ultimately takes n arguments is always treated as combining by a syntactic rule with exactly one argument to produce a phrase of the same category as a verb of n-1 arguments.*

Derivationally, the order imposed by the GF hierarchy may be linked to the order imposed by stepwise semantic composition (and syntactic combination, Krivochen, 2023a, b). However, since Dowty's framework of choice is Montague grammar, what really matters is the category of the input expressions:

S1: ⟨F1, ⟨IV, T⟩, t⟩ (Subject-Predicate Rule)
S2: ⟨F2, ⟨TV, T⟩, IV⟩ (Verb-Direct Object Rule)

Where T is the category of *terms* (NPs), *t* is the category of truth-value-bearing expressions (sentences), and F_1 and F_2 are rules of concatenation (which also involve case-marking and agreement; Dowty, 1982: 86). These rules are ordered triples: the first member is a syntactic rule (in this case, F_1 is a rule of left-concatenation and agreement and F_2 is a rule of right-concatenation and case marking), the second is the input of the rule, and the third is its output. In this sense, the declarative interpretation of these rules is not unlike Gazdar's (1981) conceptualisation of PSR as admissibility constraints. If, as suggested in **Chapter 2**, we annotated nodes with recursively defined categories, Dowty's system could in principle be implemented in our digraphs without changing the fundamental configurational properties of structural descriptions. Instead of adding category annotations (or type-theoretic annotations), we prefer to go the RG way, and keep GF as primitives.

It is crucial to emphasise, as Postal (1982) does, that the hierarchy in (111) holds for English (and Spanish so far as we can see), but it must not be assumed to be universal: it is the job of comparative grammar to discover whether it is (and since Keenan & Comrie's 1977 seminal work, the application of the hierarchy cross-linguistically has been the object of much empirical research). What *is* universally applicable, however, is the notion of an *ordered set of grammatical relations*, which is an analytical concept.

In this light, let us minimally revise the set of dominance relations proposed in (102) as (117), which is now an ordered set of arcs:

(117) $\rho_{\text{derived}} = \langle$(zag, Jan), (zag, Piet), (zag, helpen), (helpen, Piet), (helpen, Marie), (helpen, zwemmen), (zwemmen, Marie)$\rangle$

The structural description in (117) has *Jan* as the subject of *zag*, and *Piet* as its object. The relation *zag-helpen* (more on which in **Section 6.2**) is not contemplated in the functional hierarchy in (111) because we are not dealing with a V-N relation: the functional hierarchy in (111) and the interpretation of GF assigned to argument from ordered relations in (115) pertain to functions that NPs or constructions with their functional potential (for example, *that*- or *if*-clauses) play in the argument structure of Vs. We can use a classic DG notion to make this somewhat clearer: *valency* (or *valence*), such that a predicative basic expression with valency (or arity) *n* must head at least *n* edges whose tails (which may or may not be distinct) are all assigned distinct GF in its elementary graph G. Tesnière (1959: Chapter 97, § 3) presents the notion with a chemistry analogy as follows:

> *The verb may* [...] *be compared to a sort of atom, susceptible to attracting a greater or lesser number of actants, according to the number of bonds the verb has available to keep them as dependents. The number of bonds a verb has constitutes what we call the verb's valency*

Vs do not count for satisfying the valence of a V, NPs or clauses do: this is easily illustrated in a contrast like **John saw run*$_{\text{V}}$, but *John saw Peter*$_{\text{NP}}$.[4] In *Raising to Object* alternations of monotransitive Vs, it is the syntactic object that corresponds to the whole embedded clause (say, *John saw* [*Peter run*]; the aspectual distinction between the bare infinitive and the present participle—as in *John saw Peter running*—is not relevant here) that satisfies the valence of the V: we have indicated that by establishing a dominance relation between the matrix V and the embedded V (e.g., the edge *e*⟨zag, helpen⟩). However, *helpen* does *not* receive a grammatical function in the minimal sub-graph that contains *zag*: Vs are not assigned GFs. Therefore, even though there are some relations between our analysis and LFG's in terms of the centrality of grammatical functions (in particular, the fact that the embedded clause is explicitly selected by the matrix predicate), the differences are enough to keep the analyses apart (in this specific respect, note that there is no XCOMP or VCOMP function in our theory).

4 In the Standard Theory, it was usual to have clausal complements S dominated by NP nodes (see, e.g., Rosenbaum, 1965; Ross, 1967). This theoretical device captured the same observation that we make in the text.

We will come back to these issues in **Chapter 6**, which deals with the grammar of clausal complement constructions.

A very important condition on well-formed graphs was emphasised in this section: there is always an edge from a lexical predicate and each of the nominal dependents that satisfy its valency (i.e., the dependents that are assigned a grammatical function in a clause that has that lexical predicate as anchor). This requirement separates our theory not only from those versions of MGG in which the maximal projection containing the lexical verb excludes the subject (VP vs. *v*P), but also those versions of Dependency Grammar in which the subject is always a dependent of the finite verb, even if it is not a lexical predicate but an auxiliary.

5.1 A Categorial Excursus on Unaccusatives and Expletives

Before closing this chapter, it is important to complete the descriptive paradigm: it seems intuitively clear how (115) applies to monotransitive and ditransitive constructions, and intransitive unergatives: in RG terms, they all have an initial 1 arc. But we have said nothing so far about intransitive predicates without external arguments: *unaccusatives*. In generative grammar, *unaccusatives* ar sometimes taken to be 1-place predicates that take a prepositional complement: this PP complement relates an argument with thematic role *theme* (the subject) with an argument with thematic role *location* (an oblique-marked NP). For example, the structural description assigned to *arrive* in Hale & Keyser (2002: 196) and Mateu Fontanals (2002: 32) is as follows:

(118)

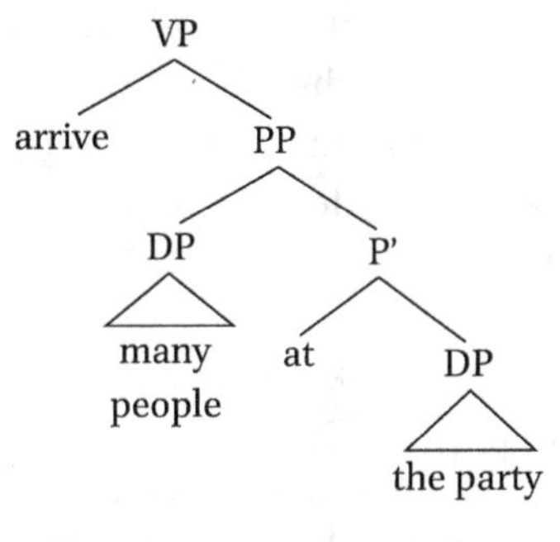

FIGURE 5.2 Tree structure for unaccusative construal

English unaccusatives can appear in (at least) two syntactic configurations: with a pre-verbal definite or indefinite NP subject or with a pre-verbal expletive *there* and an indefinite post-verbal NP with which the verb agrees. Here we take *there-insertion* to be a defining feature of intransitive *unaccusatives*: intransitive verbs with no external argument that do not take an expletive (and which often have a transitive, causative alternation), we will classify as *ergative* (e.g., change of state verbs such as *die, blossom, melt, open, grow, shatter*, etc.; see XTAG, 2001: 83 and references therein). (119) illustrates the relevant behaviour of unaccusatives:

(119) a. $\left\{\begin{array}{c}\textit{A man}\\ \textit{John}\end{array}\right\}$ arrived

b. There arrived $\left\{\begin{array}{c}\textit{a man}\\ {}^{*}\textit{John}\end{array}\right\}$

The question we want to address is, what is the ρ-set that corresponds to (119b)? Is it the same that corresponds to (119a)? If not, how do they differ?

We can be reasonably confident about assigning the representation (120a) to (119a) (but see fn. 8 in this chapter):

(120) a. $\rho = \langle(\text{arrive, man})\rangle$

But (119b) is trickier: *prima facie*, it would seem that we need something like (120b) to represent the dominance relations between basic expressions in the expression:

(120) b. $\rho = \langle(\text{arrive, there}), (\text{arrive, man})\rangle$

where expletive *there* is assigned the GF *subject* by virtue of its position in the ordered ρ-set, and *man* being an object. This entails, crucially, that expletive *there* is an element that is assigned an address, otherwise it would not correspond to a node in the graph. This is an important point to which we will return shortly. Having *man* as an object captures at least an aspect of Perlmutter's (1978: 160) *unaccusative hypothesis*:[5]

5 An important note: under (116), there is no syntactic (configurational) difference between unaccusatives and unergatives: both would be 1-place predicates, with outdegree 1 (minimally, immediately dominating a nominal). This is problematic. A way to keep them distinct is to assume that unaccusative predicates are essentially reflexive in that their 1 and 2 are coindexed. However, unlike true reflexive predicates (which assign *two* thematic roles), there is only *one* thematic role assigned to the 1–2: Theme. Therefore, there is no 'reduction' in the sense of Chierchia (2004) and Reinhart & Siloni (2004) (as this operation requires a predicate that assigns two theta roles as its input). Alternatively, a structure closer to Hale & Keyser's and Mateu's would have the unaccusative predicate as an eventive but non-relational element, dominating a relational head (P). We will use this structure in (126) below:

Certain intransitive clauses have an initial 2 but no initial 1.

However, we need to consider that the *unaccusative hypothesis* was formulated within an RG framework, and that the initial 2 underwent *advancement* to 1, such that at a different *stratum* the 2 becomes a 1 (an initial object becomes a subject). In sentences where expletive *there* co-existed with the postverbal NP (sometimes called its *associate*), RG and APG invoked an *ad hoc* relation: *brother-in-law*. In the APG treatment, the associate would be a Chômeur, and the relation *brother-in-law* holds between a dummy nominal (an expletive) and the NP that it led to be Chômeur (Johnson & Postal, 1980: 631). Agreement between a verb and the associate would be a result of the relation between the expletive and the associate.

Transformational generative grammar models this 'advancement' in terms of *movement*, such that an NP base-generated as the complement of the V moves to the Specifier position of Inflection / Tense to receive Case and satisfy the so-called Extended Projection Principle (EPP) (either as a representational filter at S-structure or as a feature in the I/T head triggering movement), which requires that the Specifier position of the functional category Inflection / Tense (Spec-IP / TP) be filled (Chomsky, 1982, 1995; Belletti, 1988). That movement is how the 2 comes to be a 1. Both approaches are illustrated below:

(121)

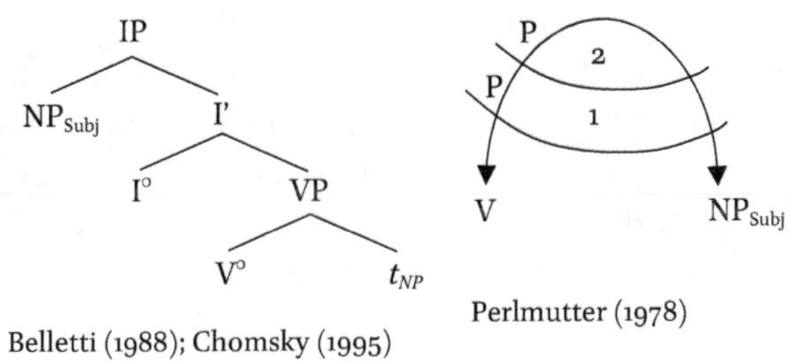

FIGURE 5.3 Generative and RG analyses of unaccusative structures

When a sentence features an overt expletive *there*, the EPP feature in I/T (which requires the specifier of I/T to be filled) is satisfied by insertion of the expletive (Perlmutter does not consider existential sentences with expletive *there* in the 1978 paper, but in Perlmutter & Postal, 1983b: 101, ff. expletives such

---

i) The train arrived at the station
ii) ρ = ⟨(arrive, at), (at, train), (at, station)⟩
We thank Víctor Acedo-Matellán for highlighting the problem of unaccusativity in the present framework.

as *it* and *there* head *dummy arcs* which can be associated to any grammatical function). In other frameworks, however, things are different. For instance, Arc Pair Grammar, RG's successor, takes expletive *es* in a German sentence like

(122) *Es wurde hier getantz*
It become.3Sg.PAST here dance.PART
'Dancing took place here'

to head an arc that *self-erases* (Johnson & Postal, 1980: 228). Strictly speaking, *expletives do not have a grammatical function*; GFs are assigned, rather, to their NP associates if there are any (thus the 'advancement'). Postal (2004a: 65) proposes the existence of *expletive arcs*, with discussion pending on the differences between English expletives *it* and *there* (since they have distinct, non-overlapping distributions). In a lexicalist model such as LFG, expletives do not have a PRED attribute, which means that they have no semantic value (Dalrymple et al., 2019: § 2.5.3): they are syntactically required, but crucially make no semantic contribution to the expression in which they occur. This characterisation is also compatible with ours, as we will see shortly.

We do not have *strata* or *movement / advancement* rules available in our framework, so neither of these proposals is fully compatible with the theory presented here. But there is much to be learnt from them, particularly from RG/APG: expletives do not seem to be assigned a GF. Now, it is crucial to remember that interpretations of nodes in our graphs are the semantic values of *basic expressions* of the language, with the interpretation of derived expressions being obtained by following directed edges. At this point, we need to introduce some aspects of Categorial Grammars, where the concept of *basic* and *derived expression* that we build on comes from.

5.1.1 *Basic and Derived, Categorematic and Syncategorematic Expressions*

In Ajdukiewicz-Montague style Categorial Grammar, expressions belong to indexed categories, which indicate their combinatoric properties: an expression of category X/Y needs to combine via concatenation with an expression of category Y to yield an expression of category X. The set C of categories is recursively defined as follows (see Schmerling, 1983b: 4; 2018a: 28):

C is the smallest set such that:
$X, Y \in$ C
For any $X, Y \in$ C, $X/Y \in$ C

X/Y denotes the category of expressions that denote functions from expressions of category Y to expressions of category X (where both X and Y may be derived categories; so, for example, (X/Y)/(X/Y) is a category, and so is X/(X/Y)).

Formally, we say that the set of categories *indexes* the set of expressions. What we can do now is rephrase our initial question slightly, and ask whether *there* is a basic expression of any indexed category in the algebra of the language. If it is, then there must be a node in the graph that corresponds to the semantic value of *there* (whatever that could be), but not otherwise. The reasoning that follows is loosely based on Krivochen & Schmerling (2022), adapted to the present context.

Let finite clauses be expressions assigned to the category FC. An intransitive verb needs to combine with a noun phrase to form a finite clause; thus, in CG terms we will say that intransitive verbs are expressions of the category FC/NP:[6] in this way, we get *A man arrived* by left-concatenating a basic expression of category NP (*A man*) with a basic expression of category FC/NP. The question now is: what happens with *there arrived*? Well, in and of itself it is not a well-formed expression of category FC: we still need an NP, in this case *right-concatenated* to *there arrived* to form *there arrived a man*. This is important: if we segment the expression as *there arrived* [there [arrived]] and were to assign *there* to the category NP, given *arrived* FC/NP, that would mean that *there arrived* is a well-formed expression of category FC, contrary to fact. This is not the place to discuss the nuances of an appropriate CG analysis, but we do want to highlight two things:

(a) Basic expressions do *not* correspond with orthographical words (i.e.: CG allows for multi-word basic expressions)

(b) CG is flexible enough that the definition of categories in different languages need not be the same. Linguistic variation is encoded in the definition of categories as well as the formal operations; rules of functional application (*slash-elimination*) and functional abstraction (*slash-introduction*) are universal.

Much research pending, we propose the following: expletive *there* is *not* assigned a GF because it is not a basic expression of any indexed category. Rather, unaccusative verbs with expletive subjects (that is, sequences *there + unaccusative V*) are expressions of category FC\NP: they need to right-concatenate with an NP to yield an expression of category FC. We can provide a toy grammar, formulated roughly in Montagovian terms, in order to illustrate our reasoning.

6 We could equally have defined NPs as FC\IV (similarly to Schmerling, 1983b), but for the present purposes we take the verb to be the functor. In CG, the term *functor* designates the constituent of a derived expression whose category is written with a slash (X/Y or X\Y).

This toy grammar includes a specification of the basic and derived categories, the basic expressions, and the formal operations of the language:

(123) **Categories:**
FC (Finite Clause)
IV (Intransitive Verb Phrase)
PN (Proper Noun)
CN (Common Noun)
NP (Noun Phrase)
FC/NP (the category of expressions that left-concatenate with an NP to yield a FC)
FC\NP (the category of expressions that right-concatenate with an NP to yield a FC)
Basic expressions:
IV = {shines, walks, arrives, ...}
PN = {John, Mary, Susan, ...}
CN = {a man, a lamp ...}
Formal operations:
$F_0(\alpha) = \alpha$, for all α.
$F_1(\alpha)$ = the result of concatenating **there** to the left of α, for all α.
$F_2(\alpha, \beta)$ = the result of concatenating α to the right of β, for all α, β.
Basic rules:
S_0. If $\alpha \in P_{IV}$, then $F_0(\alpha) \in P_{FC/NP}$, for all α.
S_1. If $\alpha \in P_{IV}$, then $F_1(\alpha) \in P_{FC\backslash NP}$, for all α.
S_2. If $\alpha \in P_{CN}$, then $F_0(\alpha) \in P_{NP}$, for all α.
S_3. If $\alpha \in P_{FC\backslash NP}$ and $\beta \in P_{NP}$, then $F_2(\alpha, \beta) \in P_{FC}$ for all α, β.

A Montague-style analysis tree for *there arrived a man* (which is the illustration of a proof that *there arrived a man* is a well-formed expression of category FC) would then look like (124), assuming the toy grammar above:

(124)

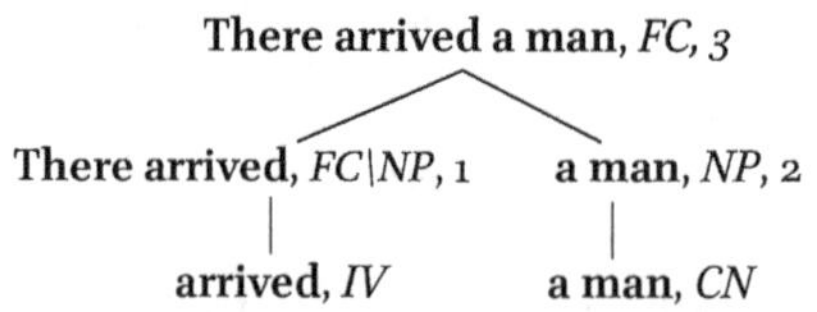

FIGURE 5.4 CG analysis tree for 'There arrived a man'

We have oversimplified the grammar, of course, but we believe our point has emerged relatively unscathed: the operation that introduces *there* yields an expression of the derived category FC\NP, which still needs to combine with an expression of category NP to yield a FC. What this means for us is that there is no node that corresponds to *there* in the ρ-set of *there arrived a man*. The correct structural description for this string is actually (125), ignoring matters of inflection as usual:

(125) There arrived a man
ρ = ⟨(arrive, man)⟩

With Hale & Keyser (2002) and Mateu Fontanals (2002) we take unaccusative verbs to be 1-place predicates which can take either an NP internal argument or a PP internal argument: in the latter case, the theme and the location are both arguments of the P (which determines the kind of figure-ground dynamics that applies to a specific sentence: central coincidence or terminal coincidence). If we add a locative argument to (125), we get (126):

(126) There arrived a man to the party
ρ = ⟨(arrive, to), (to, man), (to, party)⟩

Here, as in Mateu Fontanals (2002: 79, ff.) and related work (see also Harley, 2011), *a man* is an argument of the preposition, and the prepositional structure is the only argument of the V: observe that the preposition has as its first argument the NP *a man* and as its second, *the party*, delivering the semantic *figure-ground* dynamics that we want it to.

The analysis sketched above implies that expletive *there* is a *syncategorematic* element: it has no meaning of its own, but every expression in which it occurs does have a meaning (see Schmerling, 2018a: Appendix A). The distinction between *categorematic* and *syncategorematic* is an ancient one, as the following quotation from Priscian illustrates:

> … *partes igitur orationis sunt secundum dialecticos duae, nomen et uerbum, quia hae solae etiam per se coniunctae plenam faciunt orationem, alias autem partes συγκατηγορήματα, hoc est consignificantia, appellabant* (Priscian, *De Partibus* § 54.5)
>
> [therefore according to the Academic philosophers there are two parts of speech, the noun and the verb, because only these linked together make a complete sentence by themselves; on the other hand, they called the other parts *syncategorematic*, that is, 'meaning jointly'. Translation Susan F. Schmerling & DGK]

Quite more recently, Schmerling (2018a: 44, fn. 10) very clearly summarises the notion of *syncategoremata* as follows (see also MacFarlane, 2017 for a view from formal logic):

> *The designation syncategorematic pertains to material that has no meaning on its own and does not itself belong to any syntactic category but gets its meaning from the words in its context*

Categorematic elements are assigned to syntactic categories and as such can play GF in syntactic construal, but syncategorematic elements cannot; in the present model this means that nodes in a graph correspond only to *categorematic* elements (which can be multi-word expressions, as pointed out above). As desired, in (126) as in (125), the categorematic expression with address ⦃man⦄ is assigned the GF Subject, by (115). *There*, however, does not belong to a category and does not receive a GF, also as desired. There is no semantic value associated to expletives, due to their syncategorematicity: they may be syntactically required, but they are semantically inert (see also Dalrymple et al., 2019: 59, ff. for an LFG view that is germane to ours; Sag et al., 2003: 338 assign 'none' values to expletive *there*'s MODE and INDEX attributes, which make it unsuitable for reference and thematic role assignment). Syncategorematic expressions, as in Schmerling (2018a), would be introduced by a different kind of rule than the ones relating categorematic expressions (say, right/left concatenation).

Still in the spirit of comparing theories, it is crucial to point out that CG analysis trees and our graphs give very different information. Note also that categories in CG are more informative than syntactic labels in PSGs: given the interpretation of the slash-notation introduced above, if we know that an expression is of category X/Y and that one of its constituent expressions is of category Y, we can deduce that the category of the other constituent expression must be X; we do not have that information in category labels like VP or NP. As pointed out by Schmerling (2018a: 28–29) an important emphasis of Ajdukiewicz (1935) is that his CG allows us to *discover* previously unknown categories of expressions; this is referred to as the *heuristic* character of CG. Here, we have taken advantage of that *heuristic* to determine whether a sequence of words constituted a basic expression or a derived expression (the definition of the category is a plus that is not needed for the study of relations), which in the present framework translates into that sequence being a single node or not. In a way, then, pure-CG and the graph-theoretic approach argued for here are complementary: CG analysis trees are proofs that expressions belong to categories of the language; they do not represent dominance or phrasal/non-phrasal objects. Strictly speaking, analysis trees could be replaced by proofs in a different format, for example, the statements that follow:

A man is a basic expression of category CN
A man is a well-formed expression of category NP, by the application of rule S_2

This is exactly equivalent to the right side of the tree in (124) (Montague, 1973: 227). In contrast, our theory is a theory about *graphs*, which must not be confused with *diagrams of graphs*: graphs are sets of nodes and edges, diagrams are graphical/typographical tools which make use of lines and letters or other symbols.

Coming back to the problem of *expressive power*, the fact that we can provide a structural description for cross-serial dependencies (via *linking*) suggests that our graph-theoretic approach can accommodate the core empirical coverage of TAGs. At this point, we need to remember that the expressive power of the formalism (in this case, graph theory) is not as important as what we actually do with it: our system is not unrestricted in the kinds of dependencies it allows or the size of elementary graphs and the relations that can be established between elements in distinct elementary graphs. In particular, as we will see in the next section, there are constraints with respect to inter-arboreal dependencies. These restrictions over relations that can hold between elements in different arbores make our system weakly equivalent to an LTAG with structure sharing (e.g., Kallmeyer, 2004). The following chapters, in particular **Chapters 6, 7, and 8** will focus on formulating filters on well-formed graphs and defining the kinds of structures that they allow us to characterise.

CHAPTER 6

Towards an Analysis of English Predicate Complement Constructions

In this chapter we propose an analysis of non-finite complementation in English from the perspective of graph-theoretic syntax. The classification of so-called 'predicate complement constructions' has been a major concern since the early days of Generative Grammar in both empirical and theoretical fronts: the seminal work of Rosenbaum (1965) distinguishes two kinds of such constructions, classified essentially in terms of VPs taking either NPs dominating Ss or VPs as complements; Perlmutter (1968) and Postal (1974) substantially expand on and modify this proposal. These foundational works, which set the standard for subsequent generative analyses (e.g., Chomsky & Lasnik, 1977; Chomsky, 1981; Kayne, 1981a; Lasnik & Saito, 1991; see Davies & Dubinsky, 2004, 2007; Polinsky, 2013; Landau, 2013 for general overviews and extensive discussion), differentiate between the following kinds of non-finite complements:

(127) a. Raising to Subject
b. Raising to Object
c. Object-controlled Equi
d. Subject-controlled Equi

These terms are used descriptively, without presupposing any theoretical analysis (despite the fact that the nomenclatures that we have used include the names of two transformations, *Raising* and *Equi*(valent) *NP deletion*),[1] as a

1 It is important to point out that we will be dealing exclusively with Exhaustive Control, a subcase of Obligatory Control where there is a strict identity condition between the controller and the subject of the non-finite form, and the controller of the implicit argument is a 'grammatical element' (Landau, 2013: 232). We refer to cases such as (i):

i) Beatrix$_i$ tried/attempted/hoped/decided/managed [PRO$_i$ to kill Bill]

Since Non-Obligatory Control (NOC) has frequently been looked at as a phenomenon at the interface of syntax and pragmatics, its analysis is far less clear in our strictly configurational terms. Relevant examples are like (ii) and (iii):

ii) Potatoes are tastier [after PRO boiling them] (Landau, 2013: 232, ex. (451e))

iii) Clearly, [PRO confessing my crime] was not something they anticipated. (Landau, 2013: 232, ex. (451b))

When NOC allows a bound (endophoric) interpretation, the mechanisms in play may essen-

 | DOI:10.1163/9789004542310_007

matter of fact, *Raising* and *Control* (another term for *Equi*) can be found in works that assume neither *Raising* nor *Control* as syntactic processes or rules (e.g., Osborne, 2019; Börjars et al., 2019). In any case, what we require of an adequate grammatical theory for the English language is that it correctly distinguishes among these structures and assign distinct structural descriptions where necessary. We will proceed in the order indicated in (127), starting from Raising to Subject.

6.1 Raising to Subject

Let us begin by considering (128):

(128) John seems to read the book

One of the first questions we can ask ourselves is '*how many elementary graphs are we dealing with here?*', or, in other words, '*is (128) a monoclausal or a biclausal structure?*'. This entails also asking '*how many subjects are there?*' The issue of monoclausality vs. multi-clausality is a deep one and it may be the case that different languages behave differently in this respect, depending on exactly how 'clause' is defined. The standard view in transformational generative grammar (e.g., Lasnik & Uriagereka, 1988: 9, ff.; Chomsky, 1995: 60, ff.; 111, ff.) is that in (128) the NP *John* receives thematic role from the embedded (sometimes called 'downstairs') predicate and then moves (*raises*) to the matrix (or 'upstairs') clause to receive Case from *seem*. The analysis of these structures is essentially *biclausal*: each verbal predicate *seem* and *read* is contained in separate clauses (dominated by distinct S/CP nodes); the NP *John* originates as the subject of the embedded clause, and moves to the position of subject of the matrix clause. This analysis is also applied to predicative constructions like (129):

(129) John seems sick

From a generative standpoint, in (129) *John* would be base-generated within the projection of the predicative adjective (where it is theta-marked), and move

tially the same as in Obligatory Control. When the interpretation of the implicit argument is not a 'grammatical element', the analysis falls outside the focus of this monograph. This means that partial control and so-called *proxy control* (Doliana & Sundaresan, 2022) are beyond our current scope.

to the superficial pre-verbal position (e.g., Stowell, 1981: 262; Chomsky, 1981: 106): at no point does the subject need to cross clausal boundaries. The structure of (129) along classical transformatonal lines would be:

(130) $[_{IP}$ John$_i$ $[_{I'}$ INFL $[_{VP}$ seems $[_{AP}$ t_i sick$]]]]$

However, things work rather differently between adjectival predicative constructions and infinitival complement constructions from a *semantic* point of view: only in (129) does *seem* behave like an evidential expression, in that to say that 'John seems sick', I need to have some direct perception of John. The same happens, we will see, in the corresponding construction in Spanish: *Juan parece enfermo* (García Fernández & Krivochen, 2019b: § 1.6.1). Let us first deal with examples like (128), and then briefly comment on examples like (129).

Transformational analyses for Raising to Subject have not been uncontested, and the same is the case for non-transformational bi-clausal analyses. To give some examples, Jacobson (1992) and Bresnan (2001) present non-transformational bi-clausal analyses: in LFG, Raising verbs take XCOMP arguments and the subject of the embedded clause is *functionally controlled* by the subject of the matrix clause. Torrego (2002) and Krivochen (2013) present a transformational, monoclausal analysis. Assuming a biclausal analysis entails having *seem* as a semantic predicate capable of taking arguments and projecting a VP, whereas assuming a monoclausal analysis puts *seem* in a class closer to auxiliary verbs (thus, modifiers of other expressions rather than full-fledged predicates themselves). Here, we advance a non-transformational, monoclausal analysis (see also Gazdar, 1982: 152–155; Pollard & Sag, 1994: 133, ff.; Jacobson, 1990). In this sense, it is useful to refer specifically to APG: like RG, APG does not have 'transformations' as such, with Raising constructions involving *immigrant arcs* which presupposes more than one predicate—but not more than one 'clause' in any meaningful sense (see Johnson & Postal, 1980: 134). The reasons for assuming a monoclausal analysis (which translate here into having a single elementary graph) are related to specific lexical properties of predicates, and in a way are essentially *semantic*: Raising verbs are taken to be assimilable to the class of auxiliary verbs, specifically, modals (see Wurmbrand, 1999 for the opposite conclusion). As such, an expression like *seem* in *seems to read* is part of the local domain in which the modifiers and arguments of *read* are expressed (crucially, this also holds in Wurmbrand's view). In our terms, there is only one elementary graph whose lexical anchor is the main verb *read* and of which *seem* is a modifier. The set of relations between expressions in (128) is, then, as follows:

(131) Elementary graph: [John seems to read the book]
$\rho = \langle(\text{seems, read}), (\text{read, John}), (\text{read, book})\rangle$

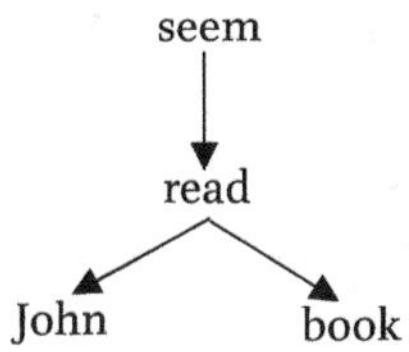

FIGURE 6.1 Graph-theoretic analysis of Raising to Subject

The relations between *read, John,* and *book* illustrate the GF interpretation rule given above in (115). *John* is the subject of *read,* being thematically selected by it but, crucially, there is *no relation* between *John* and the Raising verb *seem* (compare with the DG analysis of the Raising predicate *tend* in Anderson, 2011: 97, where the subject is dominated by both the lexical verb and by *tend*): the subject is not subcategorised by *seem*, but by the lexical verb. One of the main properties of Raising verbs, not imposing restrictions (categorial or semantic) over what we may call their 'surface subject', is straightforwardly accounted for if no syntactic relation holds between *seem* and whatever NP precedes it (the logic underlying this analysis rests in what Larson, 1990: 597 calls a *principle of location*: arguments are generated in structural domains 'headed', or 'anchored' by the predicates that subcategorise them). As we will see shortly, some crucial facts about scope also follow from this assumption.

The treatment just sketched bears some similarities with the so-called 'function composition' analysis of Raising to Subject in Jacobson (1990), in that *seem* does not licence a subject position. Her analysis is more concerned with the semantics than with the syntax, but perhaps because of that the relations between her proposal and ours are particularly interesting to explore. In Jacobson's analysis, a verb like *seem* would be assigned to the category S/S (i.e., a function from propositions to propositions). Combining it with a non-saturated clause, say an expression of category S/NP such as *to be smart* via function composition delivers an expression *seems to be smart,* of category S/NP. The classical TAG analysis (Kroch & Joshi, 1985: 40, ff.; Frank, 2002: 71) shares an important property with Jacobson's analysis, and with ours: Raising verbs license no subject positions

> *The basic contrast between control predicates and raising predicates is that the latter assign no thematic role to their surface subjects. This fact can be captured neatly in a TAG by requiring that raising predicates be inserted in an auxiliary tree with no subject* (Kroch & Joshi, 1985: 44)

However, unlike us, Kroch & Joshi and Frank do assign to *seem* its own elementary tree: *seem* is treated in their work as a lexical head.

The analysis in (131) differs from the classical LFG one, which has Raising predicates subcategorise an open clausal argument (XCOMP) and a non-thematically selected subject, and establish a relation of functional control between matrix and embedded clauses (Bresnan et al., 2016: 305; Dalrymple et al., 2019: 545, ff.). Functional control in LFG is, in essence, structure sharing applied to attribute-value matrices, and in this respect the functional control analysis of Raising in LFG is close to Pollard & Sag's (1994: 136) HPSG treatment (where structure sharing also plays a major role). LFG assumes that Raising predicates do take subjects, but do not theta-mark them. The analysis we propose is monoclausal, with the Raising verb being akin to an auxiliary, not an argument-taking predicate.

How, if at all, does the ρ-set of (129) differ from that assigned to (128), in the light of their semantic differences? This question is important insofar as it pertains to the level of granularity at which our proposal applies: to what extent should graphs (and, more generally, structural descriptions) represent aspects of lexical meaning? A possible answer is simply 'to the extent that those aspects are determined configurationally'. A follow-up question arises immediately: is *evidentiality* a configurational relation or a lexical specification? Here we assume the latter, in the sense that it is not a property of the connections that *seem* establishes with other nodes in a local graph. Evidentiality in English is lexical, not grammatical, and although it may be thought of as related to epistemic modality, Aikhenvald (2004) observes that evidential expressions can co-occur with modals, which makes it difficult to say that evidentiality is a kind of epistemic modality (cf. Palmer, 1986: 51) given the well-known restrictions that English sets on modal expressions. This is important because if *seem* is a non-grammaticalised evidential marker in at least some configurations, then it can be assimilated to modifiers of a predicative expression; in that case, lexical verbs modified by *seem* would define only one arbor. Furthermore, the same form may behave differently in distinct contexts: for example, as we briefly mentioned above, Spanish *parecer* ('seem') behaves like an evidential when it takes a finite complement or a predicative phrase, but not when it takes a non-finite complement (García Fernández & Krivochen, 2019b: 45–49). We will come back to this shortly. Asudeh & Toivonen (2017: §3.1) indeed discuss English Copy-Raising and *flip* verbs (Lakoff, 1965: A-15) as marks of non-grammaticalised evidentiality, which would support an analysis in which Raising predicates are modifiers of the lexical predicate that selects the nominal arguments in a single-rooted elementary graph, as in (131) (see **Chapter 3** above for a brief comparison with Frank's 1992, 2002 *Condition* *on Elementary Tree Minimality*, formulated within an LTAG framework). Recall that in **Chapter 3** we argued that each *elementary graph* in a structural description is a single-rooted graph which contains the following expressions:

a. A predicative lexical basic expression *p*
b. Functional modifiers of *p* (e.g., temporal, aspectual)
c. Arguments of *p* (e.g., subjects, objects)

The crucial thing here is to determine whether *seem* is a predicate that takes nominal arguments or simply modifies another, lexical, predicate. We contend that Raising predicates are (non-lexical) modifiers and do not take nominal arguments (no subjects, no objects): their only argument is the lexical predicate they modify. This lexical predicate may be verbal, adjectival, nominal, or adverbial. Thus, there is no need to assume a multi-arboreal structural description in constructions of the *Raising predicate + lexical predicate* type: nominal arguments are arguments of the lexical predicate, and the Raising predicate is a (modal, evidential) modifier of that predicate, not a lexical predicate itself.

The grammar of *Raising verb + infinitival clause* and that of *Raising verb + predicative NP / AP* should not be automatically conflated. Spanish data can be used to support this conclusion as well: if we consider (132a) and (132b) below, only (132a) allows for the pronominalisation of the predicative phrase; it is not possible to pronominalise the infinitival clause ((132c) completes the paradigm by showing that the restriction against pronominalising a non-finite complement of *parecer* is not limited to copulative constructions):

(132) a. *Juan parece enfermo* → *Juan lo= parece*
J. seem.3Sg.PRES sick → J. CL= seem.3Sg.PRES

b. *Juan parece estar enfermo* → **Juan lo= parece*
J. seems.3Sg.PRES be.INF sick → *J. CL= seem.3Sg.PRES

c. *Juan parece leer el libro* → **Juan lo= parece*
J. seems.3Sg.PRES read.INF the book → *J. CL= seem.3Sg.PRES

Furthermore, as observed above, *parecer* + NP / AP differs from *parecer* + non-finite VP in that the former sequence is interpreted as a result of a perception: I need to have seen Juan or have had some direct perception of his state in order to say (132a) (see also RAE-ASALE, 2009: 2828).

In the light of the preceding discussion, the ρ-set of (129) would be (133):

(133) ρ = ⟨(seems, sick), (sick, John)⟩

As in (129) above, thematic requirements are imposed by the lexical predicate *sick*, which is of course not a finite verb: the anchor of (133) is *sick*. Despite the presence of *seem*, we see no evidence to propose the existence of a new elementary graph, since there is no more than a single *lexical* predicate. Evidentiality notwithstanding, there doesn't seem to be a difference at the level of syntactic dependencies and grammatical functions between (129) and (133), given the fact that ρ-sets are designed to represent relational networks; the differences pertain to aspects of lexical structure and semantics that are at a smaller level of granularity than that we are concerned with here. In all cases, as emphasised above, it appears that English *seem* is not a lexical predicate that can anchor an elementary graph, as it takes no arguments.

An interesting consequence of the lack of direct relation between *seem* and the subject—which we can only sketch at this point—is that cases of so-called *total reconstruction* could be straightforwardly accounted for: these are cases in which movement of the subject seems to be phonological only, since the only copy that gets interpreted is the lower one:

(134) An Austrian is likely to win the medal (Sauerland & Elbourne, 2002: 284)

In cases such as (134), with existentially quantified NPs, the subject takes scope *under* the Raising predicate. Where Sauerland & Elbourne (2002) argue for purely phonological movement (and Chomsky, 1995: 326, ff. distinguishes A'- from A-movement precisely in terms of the availability of reconstruction 'on conceptual grounds'), we argue for the absence of a syntactic dependency between the Raising verb and the clausal subject. It is important to note, however, that the full story is more complex, since universally quantified NPs do not totally reconstruct (and for Raising constructions with existentially quantified NPs, some propose that there are *two* corresponding logical forms: one in which the NP scopes over the Raising predicate, and one in which it scopes below; e.g. May, 1985; see also Lasnik, 2010 for discussion). In that case, the partial multidominance approach to generalised quantifiers in **Section 14.4** could prove particularly useful. Cases of total reconstruction with proper names, such as in a variation of an example by Chomsky (1995: 327) (*John expected to seem to me to be intelligent*) do not require, so far as we can see, any modification to the present account. At this stage in the development of the theory, we must leave this as a promissory note.

We can close this section by noting that the relations in (133) can in principle be extended to other predicates, like *remain* or *be likely* in sentences like (135a) and (135b) respectively:

(135) a. Her bank account remained active after her death
b. Maeve is likely to get that job

When a single Raising predicate is present, the analysis in (131) carries over without issues: *remain* immediately dominates the lexical predicate *active*. However, there are some interesting asymmetries between *seem* and the Raising predicates in (135). In a chain of Raising predicates, *seem* may modify a Raising adjective such as *likely*, as in *John seems to be likely to win the race*: note that here the epistemic contribution of *seem* affects John's likelihood of winning rather than the winning itself. Note, crucially, that the inverse order results in a marginal sentence:

(136) a. ?John is likely to seem to win the race

This restriction is not in place with *remain* + *likely*:

b. Her bank account is likely to remain open after her death
c. Maeve remains likely to get that job

There seem to be some restrictions over *seem* that put it closer to modal auxiliaries (i.e., being mostly restricted to the first position): the combination with the perfect auxiliary *have* is not universally accepted when *seem* is followed by an infinitive (i.e., *has seemed* + infinitive sounds to many speakers worse than *has seemed* + adjective). There may be grounds to invoke a richer semantics in the analysis of these categories: it is possible that Raising adjectives are best analysed in terms of lexical entailments (see Jacobson, 1990 for discussion), but we leave this problem for future work.

6.1.1 *Copy-Raising in a Multidominance Framework*

An analysis of Raising to Subject that allows for multidominance has additional advantages when we look at so-called *Richard*-constructions (Rogers, 1971, 1972, 1974). 'Richard' is a syntactic *copying* transformation (in the sense of Ross, 1967) first formulated by Rogers (1971: 217) and worked out in subsequent publications by him, which relates sentences like (137a) and (137b)

(137) a. it $\left\{\begin{array}{l}\text{seems}\\ \text{looks}\\ \text{sound}\end{array}\right\}$ like Richard is in trouble

b. Richard $\left\{\begin{array}{l}\text{seems}\\ \text{looks}\\ \text{sound}\end{array}\right\}$ like he is in trouble

In Rogers' original formulation, the transformation …

> … *copies the subject NP of the* **like**-*clause in the structure underlying sentences such as those in* [137a] *as subject of the main clause, resulting eventually in sentences such as those in* [137b] (Rogers, 1974: 551)

In more modern terms (e.g., Potsdam & Runner, 2001; Asudeh & Toivonen, 2012), a *Raising to Subject* rule that leaves behind a resumptive pronoun is referred to as *Copy Raising*. It is important to stress that the analyses of Rogers (1972) and more recently Asudeh & Toivonen (2012) and Asudeh (2012) are *biclausal*: in Rogers' paper, the Deep Structure that corresponds to a Richard-sentence contains *two* S nodes; in Asudeh & Toivonen's LFG-based works (following Bresnan, 1982a), Raising constructions are XCOMP, that is, internally unsaturated complements one of whose arguments is filled by functional control (see also Dalrymple et al., 2019: § 15). Crucially for the LFG analysis, Copy-Raising Vs are single-word terminals (i.e., only *look*/*sound*), which take *like* complements (Asudeh & Toivonen, 2012: 325).

In the terms we are working with here, a biclausal analysis would translate into two initial sub-graphs which are linked at the 'copied' node. However, if we take (Copy-)Raising and *flip* verbs to be evidential markers, there seems to be no support for proposing more than a single elementary graph here either: the Raising predicate is a modifier of the lexical verb that assigns grammatical functions to nominal expressions. Our objections to that proposal notwithstanding, there are some aspects of the biclausal analysis that deserve some discussion. In Roger's analysis, one sentential S node corresponds to the perception involved in *flip* Vs, which generates *sui generis* presuppositions, and the other to the object of perception (which is presupposed). The syntactic-semantic structure that Rogers (1972: 310) assigns to the sentence *Harry looks* (*to me*) *like he was drunk* is (minimally adapted) (138):

(138)

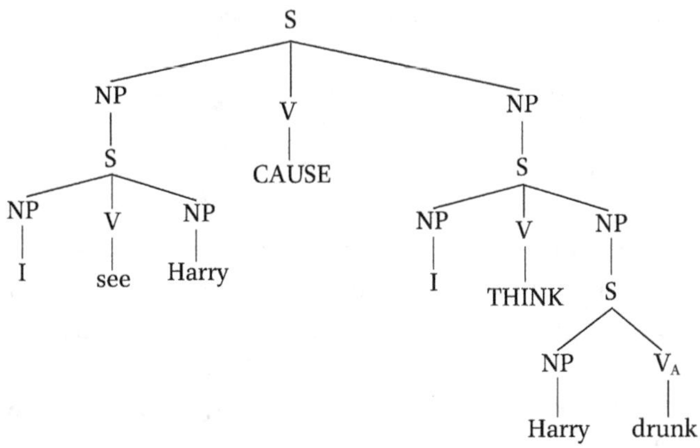

FIGURE 6.2 Structure of a 'Richard-sentence' in Rogers (1972)

If we were to translate Rogers' structure *directly* into a maximally connected graph, there is quite a bit of pruning (in the sense of Ross, 1969b) that we can do. First, we can delete all intermediate (non-terminal) nodes. Then, we can identify identical nodes throughout the structure, such that all occurrences of *I* and of *Harry* are contracted. But there is a more fundamental problem, which is the presence of phonetically empty semantic primitives (as was usual in Generative Semantics; see e.g. McCawley, 1968; Dowty, 1979): here we have the semantic primitives CAUSE and THINK (which must not be confused with the English verbs *cause* and *think*). There is no more justification for those than there is for intermediate phrasal nodes in the present context, which means that a more radical restructuring of the structural description must take place. Before doing that, however, we need to distinguish between different classes of Richard (and Richard-like) constructions, which will require tweaks to the structural descriptions we assign each of them. Initially, we can identify two big classes of Richard-constructions by their syntactic and semantic behaviour: *Pseudo-Richard* and *true Richard* (see also Potsdam & Runner, 2001 for related discussion). Let us characterise them in some detail.

6.1.1.1 *Pseudo*-Richard Constructions

Pseudo-Richard constructions are, predictably, sentences which only look like they are instances of Richard, but have syntactic-semantic properties which set them apart from true Richard constructions. Interestingly, languages which do not feature *true* Richard, like Spanish, do allow *pseudo*-Richard constructions, which suggests that there is further reason to draw the distinction.

A. *Active constructions without internal experiencer:*

(139) Richard acts (*to me) like he's in trouble

(140) *It acts (to me) like Richard's in trouble

Here we have copying without Raising (the matrix verb is a thematic assigner) and without perception either. The structure is thus biclausal, if by 'clause' we understand roughly a unit of predication with a thematic assigner and its arguments. If each of the predicates select nominal arguments, then, by the definition of *extended projection* that we introduced in **Chapter 2**, we should have exactly one elementary graph *per* predicate.

The aspectual interpretation of (139) can be either progressive or habitual, by virtue of *act* being a dynamic verb. Note that the expletive *it* alternation is ungrammatical, cf. (137–140). The ρ-set for (139) is (141):

(141) Arbor 1: [Richard acts like]
Arbor 2: [Richard is in trouble]
$\rho_1 = \langle(\text{act like, Richard})\rangle$
$\rho_2 = \langle(\text{in trouble, Richard})\rangle$

We must also point out that *act like* is, in this analysis, a multi-word lexical predicate[2] (note that it assigns a thematic role to its subject). This will be a staple of our analysis of Richard-style constructions. Interestingly, Rogers' analysis in (141) above does not feature a terminal node for *like*, which suggests that it is not a categorematic expression. Here, the Richard-predicate is not *act* (which by itself does not license Richard) but rather the two-word expression *act like*.

B. *Middle constructions (without internal experiencer):*

(142) This bread cuts (*to me) like it's stale (does not licence Richard; non-factive)

2 Susan F. Schmerling (p.c.) reports that in some Texas varieties it is possible for Richard-predicates to take a *that*-clause as a complement (which also suggests that *like* cannot be a C), at least in the expletive-*it* version:
i) It {looks/seems} like that Richard's in trouble
ii) *Richard {looks/seems} like that he's in trouble
The structure of *V like that S* clauses must be different from that of Richard, in that Raising seems to be blocked. In MGG terms, true Copy Raising may differ from cases such as (i) in

(143) *It cuts (to me) like this bread's stale

Non-Flip Vs do not license an experiencer, that is straightforward enough. In this case, the interpretation is clearly non-agentive, the eventuality denoted by the verb phrase is a state (a property of the grammatical subject). This is the main difference with the previous class of sentences: (142) refers to an individual-level property of the bread. Semantic differences aside, the ρ-set for (142) is (144):

(144) $\rho = \langle(\text{cut like, bread}), (\text{stale, bread})\rangle$

This entails an analysis of *cut like* as a two-word middle voice version of *cut*. A further point to highlight in the ρ-sets (141) and (144) is that, following Schmerling (2018a: 159), finite copula *be* is analysed as *syncategorematic*. Part of the motivation for this is the widely available possibility of omitting copulas in matrix sentences cross-linguistically (e.g., in Latin, Russian, or Turkish); as far as English is concerned, Schmerling argues that copula *be* behaves more like an auxiliary than like a true lexical V. Note also that in MGG analyses of copulative sentences it is not *be* that imposes restrictions on the subject or acts as a true predicate: for example, Longobardi (1985: 211) proposes that copulative sentences do not contain two thematic arguments, but '*always a nonreferential expression with the function of a predicate to be applied to a syntactic subject*'. In the classical GB analysis of predicative copulative sentences, the subject is not generated in the projection of *be*, but in that of the predicative phrase where it is theta-marked (e.g. Stowell, 1981. See also Moro, 2000: Appendix).

C. *Middle constructions* (*with internal experiencer*):

Examples like (145) below, taken from Rogers (1974: 552), differ from those like (142) in that it is possible to have an internal experiencer:

(145) The soup tastes (to me) like Maude has been at the cooking-sherry

We need to underline the fact that the subject position of the lower clause is not occupied by a resumptive pronoun, but by a referential NP: it is *not* possible to claim that (145) has been derived by a *copying* rule (a fact also pointed out by Rogers, 1974: 552). These verbs, which allow for disjoint reference between

---

the presence of a bounding node (a CP headed by *that*) which prevents movement of the downstairs subject.

the subjects of the perception verb and the embedded verb are referred to as *perceptual resemblance verbs* (Asudeh & Toivonen, 2012: 325): for these cases, a biclausal analysis seems to be more strongly supported, since if there are two distinct subjects there must be two distinct lexical predicates. In (145), the lexical predicate *taste* takes a subject *the soup* and the lexical predicate *be* (in its unaccusative alternation) takes a subject *Maude*. If this is the case, then we can try to have the same NP as the subject of both the matrix and the embedded predicate. Let us thus tweak the example, and make both subjects co-referential:

(146) The soup$_i$ tastes like it$_i$'s been left cooking for too long

In (146), the subject of the embedded sentence is not an expletive *it*, but a referential pronoun (bound, with antecedent *the soup*). This is thus still *not* a Richard sentence: we cannot have the expletive-*it* alternation

(147) *It tastes like the soup has been left cooking for too long

Note that verbs that do license the expletive-*it* alternation fall in the class of Richard-verbs: *look, sound, smell*... (at least for some speakers). Cases such as (146) seem to display enough properties of their own to justify classifying them separately. Let us refer to these sentences as *Maude-sentences*. *Maude-sentences* may (but, crucially, do not need to) display at least one of the defining properties of Richard-sentences: (i) the copying procedure that gives us co-referentiality or (ii) the presence of an experiencer *to*-phrase. However, we have seen that there is also at least one defining property of Richard-sentences which Maude-sentences do not display: the *expletive-it* alternation (as shown in (147)). This suggests that we need to make the conditions under which Richard is licensed stricter. This is what we will do next.

6.1.1.2 *True* Richard Constructions

Middle constructions with internal experiencers:

(148) Richard looks/sounds/seems to me like he's in trouble (licenses Richard; non-factive)

(149) It looks/sounds/seems to me like Richard's in trouble

Now we formulate the additional requirements for a construction to be identified as a *true* Richard-construction:

a) It must be possible to have the *expletive-it* alternation (compare (147) vs. (149))
b) The subject of the perceptual resemblance V and the subject of the complement clause of that V *must* be coreferential

Requirement (b) holds for NPs, which can bear referential indexes; but some additional discussion is in order. Consider the following example, from Rogers (1974: 551)—bear in mind that, at this stage in the evolution of the generative theory, expletive *there* was transformationally inserted—:

(150) There looks like there's gonna be a riot

Expletives have no referential indexes (no semantic values; recall that in LFG expletives have no value for the attribute PRED), so requirement (b)—co-reference—cannot be satisfied. This means that expletive *there* cannot have an index or, rather, that it has an index Ø (the empty symbol), as a way of indicating that it has no reference. But there is a unique element in the indexing alphabet that can have no reference in context C, meaning that every instance of *there* in a graph must be 'co-referential' in that they share the prohibition of having a referential index. This constitutes a refinement of the idea, explained above, that expletives are syncategorematic expressions.

More practically: saying that 'the subjects *cannot have disjoint reference*' avoids the problem altogether, because even if we get the expletives *it* and *there* (each of which has a different distribution), they do not (in fact, they *cannot*) have disjoint reference by virtue of there being a single element in the indexing alphabet with no reference in the set C of categories (by the *axiom of extension* in set theory, there is one and only one empty set); and thus we need not invoke null indexes. Note that this generalisation is available to us only because 'Subject' (or 'most prominent grammatical function in the hierarchy') is a primitive of the theory.

Requirement (b) captures Rogers' and Ross' characterisation of Richard as a *copying transformation*. The requirement for co-referentiality comes from the fact that Richard is essentially a *copying* rule. If we copy an NP and leave behind a resumptive pronoun, it will bear the same index as the NP. Without such a requirement, and as Rogers (1974: 551) points out, we could not adequately exclude (151b) as the result of copying applied to (151a), in the idiomatic sense:

(151) a. The shit$_i$ looks like it$_i$'s gonna hit the fan.
b. *The fan$_i$ looks like the shit's gonna hit it$_i$.

(151b) cannot be filtered out unless we assume that Richard is a copying rule: only in this way is the idiomatic unit [the shit hit the fan] preserved. (151a) is grammatical precisely because—in a transformational framework—there is a reordering copying rule (in the sense of Ross, 1967: 427) which displaces the NP *the shit* and leaves behind a copy in the form of a resumptive pronoun. Here, the story is different: there are two elementary graphs (with anchors *look like* and *hit the fan*), linked at a single node with address ⦃shit⦄. Here, *hit the fan* is taken to be a multi-word basic expression; this is consistent with a constructionist approach to idioms. The ρ-set of (151a) is defined in (152):

(152) $\rho_1 = \langle(\text{look like, shit})\rangle$
$\rho_2 = \langle(\text{hit the fan, shit})\rangle$

The core of the observation that Richard is a copy operation is captured by having the 'copied' element be the one that allows for the local arbores to be linked. In order to have a grammatical true Richard construction, the NP that links the graphs must receive the same GF, *subject*, in both elementary graphs. (152b) provides further evidence that Richard as a transformation needs to be limited to *subjects* (a generalisation that would be hard to formulate if GF were not primitive). This follows if we accept the following claims: (i) true Richard-Vs are lexical Vs which take non-finite complements and (ii) an analysis of non-finite complementation as 'multidominance' is along the right lines (note that this would be akin to a configurational version of *functional control*). If we accept (i) and (ii), then the subject is dominated by the Richard-V, with which it agrees in person and number, and by the embedded predicate, where it is thematically interpreted (objects cannot satisfy both these structural relations). Under these conditions, co-referentiality is guaranteed, as desired.

6.2 Raising to Object

We have seen some basic aspects of *Raising to Subject* constructions. Next up in (127) are *Raising to Object* structures. Raising to Object differs from Object Equi in ways that parallel the differences between Raising to Subject and Subject Equi. Within MGG, there has been some debate about what the position of the post-verbal accusative NP is (see Runner, 2006), with Postal (1974), Lasnik & Saito (1991), Koizumi (1995), Lasnik (2001), among many others favouring an overt movement approach (where the subject of the embedded clause raises to a matrix object position), and Chomsky (1973, 1995, 2000) preferring an *in-situ* Case assignment approach (let us call that an Exceptional Case Marking—ECM—analysis), with LF movement to Spec-AgrOP as an innovation in the 1995 version. The core claim in the ECM analysis is that Raising to Object complements are somehow structurally 'defective' with respect to Equi complements: either they undergo S' deletion, or instead of being full CPs (like Equi complement clauses) they are TPs. This structural defectivity allows the matrix predicate to case-mark the subject of the embedded clause (a relation that would be unavailable if there was a CP/S' intervening between the matrix V and the embedded Spec-TP/S).

An important difference between Raising and Equi is that only Raising to Object allows for a syncategorematic subject in the embedded clause, which in some analyses (such as LFG's) suggests that the 'raised' NP is not thematically selected by the matrix predicate. In non-transformational analyses, the difference between Raising to Object and Object Equi is eminently lexical (the lexical entries that follow assume a functional control analysis of Equi, *mutatis mutandis* for an anaphoric control analysis—e.g. Bresnan (1982a), Dalrymple et al. (2019)—, where XCOMP should be replaced by COMP):

(153) a. *believe* V (↑PRED) = 'believe⟨SUBJ, XCOMP⟩OBJ' (Raising to Object V)
b. *tell* V (↑PRED) = 'tell⟨SUBJ, OBJ, XCOMP⟩' (Object Equi V)

The OBJ outside the angled brackets in *believe* means that it is subcategorised for, but not thematically selected. This is an important property of these constructions, and one which at least *prima facie* cannot be reduced to pure configuration: it needs to be lexically specified. Equi verbs, Pollard & Sag (1994: 133) say, '*systematically assign one more thematic role than their raising counterparts*', with Subject Equi verbs assigning one (to their external argument, but not the clausal argument) and Raising to Subject verbs zero, and Object Equi assigning two (external and internal arguments, but not the clausal argument) and Raising to Object verbs, one.

Building on the LFG and HPSG analyses, we will also use structure sharing in the description of Raising to Object: in the simplest cases, we have two elementary graphs which share a node. This is analogous to a proper *Raising* analysis à la Postal (1974), in that the accusative NP is contained in the elementary graph anchored by the matrix predicate. That node is assigned GF *subject* in one of the graphs, and *object* in the other, as seen in the ρ-sets in (154):

(154) The judge believed John to have committed the crime
ρ_1 = ⟨(believe, judge), (believe, John)⟩
ρ_2 = ⟨(commit, John), (commit, crime)⟩
ρ_{derived} = ⟨(believe, judge), (believe, John), (believe, commit), (commit, John), (commit, crime)⟩

The same configurational considerations hold for the analysis of bare infinitives which are complements to perception verbs, such as (155):

(155) a. Mary saw John fall
$\rho_1 = \langle(\text{see, Mary}), (\text{see, John})\rangle$
$\rho_2 = \langle(\text{fall, John})\rangle$
$\rho_{\text{derived}} = \langle(\text{see, Mary}), (\text{see, John}), (\text{see, fall}), (\text{fall, John})\rangle$

We have presented (154) and (155) separately since they do not always receive the same structural analysis in the literature, but they do here (at least in the exemplified configurations). In (154) we have two lexical predicates, anchoring their respective elementary graphs. What we need to focus on is that in the derived graph, the matrix V *see* dominates *both* its nominal object *John* (which we can identify as receiving that GF by the rule in (115); note that under this analysis the GFs of subject and object would not be assigned to a single expression in subsequent derivational steps, but rather in distinct elementary graphs) *and* the embedded V *fall*.[3] As noted above, *fall* itself does not satisfy the valency of *see*, rather, the clause containing *fall* and its subject *John* does: (155) restates the RG analysis, where a Cho arc dominates the root/anchor of the embedded clause. Because *John* is the subject of *fall*, and it is this event (*John fall*) that is the complement of *see*, we have an edge *from* the matrix V *to* the embedded V (the latter, the root of the elementary graph that it anchors). This is an important difference between our treatment of complementation and RG's: in both cases there is an expression that receives distinct grammatical functions from distinct predicates, and edge from the root of the matrix clause to the root of the embedded one, but in RG neither of these roots correspond to expressions (see e.g. Perlmutter, 1980: 224; also Gibson & Raposo, 1986; Scida, 2004: 18 for RG analyses of Romance Raising to Object). Our analysis also differs from DG's (e.g., Osborne, 2019: 183) in that infinitival *to* and copula *be* are taken to be syncateorematic, and there is a direct edge from the matrix predicate to the embedded one (the lexical predicate that imposes selectional and semantic restrictions over co-occurring arguments).

3 A pure 'ECM' analysis is also formulable in this framework: without Raising to Object, the accusative NP does not belong to the elementary graph anchored by the matrix predicate but only to the elementary graph anchored by the embedded predicate. For a sentence such as *they consider John an idiot*, an ECM analysis would be:
i) $\rho_{\text{derived}} = \langle(\text{consider, they}), (\text{consider, idiot}), (\text{idiot, John})\rangle$
We follow Postal (1974) and Lasnik & Saito (1991), among others, in assuming that the accusative NP is a node in the matrix elementary graph.

The elementary graphs anchored by *see* and *fall* are also linked at *John*, which receives a GF in each: *object* in the matrix one, *subject* in the embedded one. We may take note of the fact that the formal properties of structural descriptions for Raising to Object are very similar to those assigned to small clause complements, as in (156):

(156) a. They consider John an idiot
b. Beatrix imagined Bill dead
c. I like my tea strong

Each of these cases involves *two* elementary graphs: there is a matrix verbal predicate (*consider, imagine, like*) and an embedded nominal / adjectival predicate (*an idiot, dead, strong*). The ρ-sets for the sentences in (156) are given in (157), indicating all relations in elementary graphs and derived graphs:

(157) a. ρ_1 = ⟨(consider, they), (consider, John)⟩
ρ_2 = ⟨(idiot, John)⟩
$\rho_{derived}$ = ⟨(consider, they), (consider, John), (consider, idiot), (idiot, John)⟩
b. ρ_1 = ⟨(want, Beatrix), (want, Bill)⟩
ρ_2 = ⟨(dead, Bill)⟩
$\rho_{derived}$ = ⟨(want, Beatrix), (want, Bill), (want, dead), (dead, Bill)⟩
c. ρ_1 = ⟨(like, I), (like, tea)⟩
ρ_2 = ⟨(strong, tea)⟩
$\rho_{derived}$ = ⟨(like, I), (like, tea), (strong, tea)⟩

In a case of secondary predication like (157c), however, the GF *object* of *like* is fulfilled by *tea*, which links the graph that corresponds to the elementary graph defined by *like* with the elementary graph defined by *strong*. In (157a) and (157b), we have added an edge from a predicate to the root of the graph that plays the role of object:[4] this decision is influenced by the analysis of Raising to Object as a rule applying in bi-clausal structures, in which the subject of an embedded clause raises to a VP-internal position in the matrix clause. Still, the *object* of *consider* is not just *John*, and the object of *imagine* is not just

4 In Permutter's (1980: 224) analysis, the root of an embedded clause in Raising to Object is the tail of a Cho(meur) arc that has been demoted from a 2 (this root is not the lexical predicate of that clause in RG, but a distinct, almost 'phrasal' node). A point in common with our analysis, however, is the use of multidominance for the expression that is assigned GF 1 in the embedded clause and 2 in the matrix one.

Bill: *an idiot* and *dead* cannot be considered *adjuncts* by most tests of adjuncthood (e.g., iterability, morphosyntactic recovery under omission, positional freedom), and the predication involved in *an idiot* and *dead* is not independent from the predication in *consider* and *imagine* It may be the case that this specific assumption turns out to be empirically wrong; in that case, we would need to eliminate the edge between the matrix predicate and the embedded predicate. However, there is some support in favour of this approach from the subcategorisation frames assigned to secondary predicates in LFG (e.g., Börjars et al., 2019: 113), and which pertain to properties of the lexical entries of predicates that govern Raising to Object. Recall that subcategorised secondary predicates are assigned the grammatical function XCOMP, which corresponds to non-saturated predicates. Thus, the LFG lexical entry of a verb like *consider* in *consider John foolish* would be

(158) *consider* V (↑PRED) = 'consider⟨SUBJ, XCOMP⟩OBJ'
(↑OBJ) = (↑XCOMP SUBJ)

We can *roughly* identify the PRED value with the semantic value of an expression. It includes whatever *consider* means in some semantic model plus the grammatical functions it subcategorises for: in this case, a subject, an object, and a non-saturated predicative structure (all of which are also semantically selected by the main predicate *consider*). The functional information in the lexical entry also tells us that the *f*-structure that fulfils the function of object of the predicate *consider* also fulfils the function of subject of *consider*'s XCOMP (in this case, the predicative adjective *foolish*). The embedded predicate's lexical entry would be

(159) *foolish* A (↑PRED) = 'foolish⟨SUBJ⟩'

Summarising, we want to capture the following basic insights about the structure of Raising to Object sentences:

i. Sentences featuring *Raising to Object* are minimally *biclausal*, involving at least *two* elementary graphs
ii. The 'raised' NP is shared by both elementary graphs (i.e., the graphs are *linked* at the 'raised' NP)

We follow Rosenbaum (1965) and particularly Postal (1974) (and much work since) in considering the accusative NP in Raising to Object and small clause (secondary predication) structures to be an argument of the embedded predicate, which also establishes a relation with the matrix predicate (cf. Chomsky, 1973: 107, 113, fn. 33); under Raising, this relation is established via movement (e.g., Lasnik, 2001). That insight is also represented in RG, such that an expression bears the relation 1 to the predicate of the embedded clause and the relation 2 to the predicate of the matrix clause (Perlmutter, 1980: 224). In our theory, this relation is represented through an edge from the matrix predicate to the accusative NP, without 'movement' or relational change (no promotion or demotion; see also Postal, 2010). In that sense, it is similar to the lexicalist analysis in Bresnan (1982a). Our analysis of Raising to Object, then, maintains the relation between the embedded predicate and its NP subject, *adding* an edge between the matrix predicate and that NP. Rosenbaum's approach, through Ross' (1969b) rules of *tree pruning*, adequately yields a structure in which there is no bounding node between the main predicate and the embedded predicate, and in which the NP is a 'constituent' of both the matrix and the embedded clauses (receiving a grammatical function in both), which undergo structure sharing. It is important to bear in mind that there is no change of grammatical functions, either in Raising to Subject or to Object: GFs are read off local ordered dominance relations; this allows for a single node to be a subject in one elementary graph and an object in another.

6.2.1 *A Note on Reflexive Anaphora*

The discussion of Raising to Object gives us the opportunity to consider some well-known phenomena under a new light. Specifically, we want to examine some facts pertaining to the interaction between Raising to Object and the licensing of reflexive anaphora. There are two kinds of examples of reflexivity that we are interested in for purposes of the present section, which we illustrate in (160) and (161):

(160) The defendants proved themselves innocent (during the trial) (Lasnik, 2001)

(161) *John believes Mary to like himself

These cases are interesting insofar as they allow us to test some preliminary considerations about bound anaphoric devices, which we will return to in detail in **Chapter 8**. For now, it suffices to say that reflexives must be bound in local domains: this is usually referred to as Principle A of Binding Theory (Chomsky, 1981: §3.2.3). Theories across the grammatical spectrum deal with issues of coreference and the distribution of reflexives, reciprocals, pronouns, and proper names. How do we capture the distribution of reflexives? The first consideration we need to make is that if two expressions are coreferential, they correspond to the same node, given the addressing axiom. This means that the expressions *the defendants* and *themselves* in (160) and *John* and *himself* in (161) correspond to the same addressed node. Chomsky (1981: 188) proposes that anaphors, unlike pronominals, *'have no capacity for 'inherent reference''*: this 'referential defectivity' of anaphors will fall out of our configurational analysis.

The second issue is *locality*: how can we capture the locality requirement imposed on the distribution of reflexives? Anticipating discussion from **Chapter 8**, we will assume that the kind of local domain that in MGG is called a *Governing Category* (Chomsky, 1981) or—better still—a *Complete Functional Complex* (Chomsky, 1986), or a *Minimal Complete Nucleus* in LFG (Dalrymple et al., 2019) here corresponds to *elementary graphs*: these are the domain within which binding restrictions apply. This is to be expected given the condition that all syntactic dependencies be established at the level of elementary graphs (the *Fundamental TAG Hypothesis* seen in **Chapter 3**). Thus, constraints on the distribution of endophoric expressions must be formulated in terms of arc relations in elementary or derived graphs. In order to address this issue, we need to consider what a characterisation of 'reflexivity' looks like in the present context. We will follow Reinhart & Reuland (1993: 662, ff.), and define reflexivity as a property of *predicates* rather than of arguments: a predicate is reflexive if at least two of its arguments are coindexed. In these cases, two or more NPs with the same referential index fulfill distinct grammatical functions and are assigned distinct thematic roles: for example, in *John shaves himself*, *John* and *himself* must co-refer; *John* is the subject/Agent of *shave* and *himself* is the object/Patient. A relational account of anaphoric relations becomes not only possible, but desirable.

The framework we have been exploring gives us a straightforward way to characterise the standard cases of reflexive anaphora, borrowing much from RG and APG. Concretely, we will make use of the relation between edges *'parallel'*, defined in **Chapter 2** and repeated here:

(162) a. 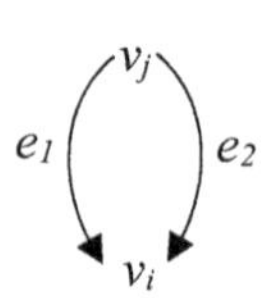

Parallel: edges e_1 and e_2 are parallel iff $e_1\langle v_i, v_j\rangle \wedge e_2\langle v_i, v_j\rangle$. In this configuration, v_i dominates v_j, establishing two different grammatical relations with it.

FIGURE 6.3 Parallel arcs

Let v_j be a dyadic predicate, and v_i a nominal argument of that predicate. The dominance relations in (162a) translate into ordered arc notation as follows:

(162) b. $\rho_{\text{parallel}} = \langle (v_j, v_i), (v_j, v_i) \rangle$

As indicated in the interpretative rule (115), the expression corresponding to v_i is *both* the Subject/1 and the Object/2 of the predicate v_j; we can tell by looking at (162) that the predicate in node v_j is thus *reflexive* in the sense of Reinhart & Reuland (1993). Within our graph theoretic model, then, we can define *reflexivity* in terms of the *parallel* relation between edges, such that a simplex sentence (in the sense of Lees & Klima, 1963) contains a reflexive predicate if and only if the structural description assigned to that sentence features *parallel* arcs[5] (see also Perlmutter, 1980: 210), with a multidominated node (v_i in (162) is the tail of two arcs). In RG this was known as a 'multiattachment' analysis (Berinstein, 1984: 3, ff.) A crucial difference between the RG analysis and ours is that the multiattachment analysis assumes that antecedent and reflexive are two distinct nodes at the final stratum; see Blake (1990: Chapter 3) for discussion and examples. The centrality of the relation *parallel* is due to the fact that this allows a single node to be in two different grammatical relations with another node. In Postal's (2010: 18) terms,

> *one good reason to assume parallel arcs is that for certain types of grammatical relations a single phrase has the possibility of bearing more than one to the same larger constituent.*

This perspective allows us to reduce the number of nodes in a structural description: in our example above, *John* and *himself* do bear distinct grammatical functions and thematic roles. The predicate that heads parallel arcs subcategorises two GF and assigns two theta-roles. If we allow, like Postal, a single expression to bear more than one relation to a predicate, we do not need to assume that *John* and *himself* are distinct nodes in the structure. This advantage holds even if no notion of *immediate constituency* as in the Wellsian-Harrisian tradition is recognised in the formalism, just basic expressions of the language that may be predicates or arguments. The graph-theoretic characterisation of reflexivity in (162) is very close to the RG 'multiattachment' approach, but with some important differences. For example, the structure proposed in Berinstein (1984) for the initial stratum of a sentence like (163a) is (163b) (similarly, Perlmutter, 1980: 209 uses multiattachment for the Italian reflexive sentence *Giorgio si è ucciso*, 'Giorgio killed himself'):

5 The nodes involved need not be the predicate's 1 and 2: binding effects such as those considered in Barss & Lasnik (1986) and Larson (1988) pertain to 2–3 reflexivity.

(163) a. John understands himself
b.

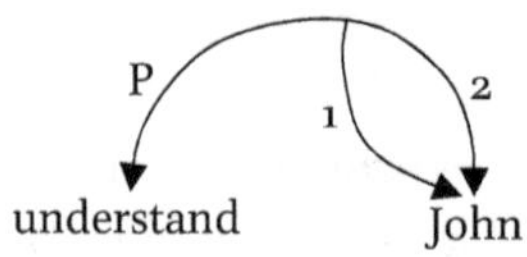

FIGURE 6.4 RG analysis of reflexive predicates

In (163b) we have a single NP that tails the arc 1 and 2 of the predicate P in the initial stratum: this NP is both the subject and the object of the verbal predicate (whose arc is annotated P). There is no direct connection between the predicate and its argument. In our approach, there are only two arcs, which directly connect the predicate and its argument: GFs are determined based on the way those arcs are ordered in the ρ-set of a particular sentence. In any case, our debt to RG and related approaches is hard to overstate. The basic insight, that reflexivity does not involve multiplication of nodes but of edges, is at the core of both approaches.[6]

So far, we have been dealing with monoclausal sentences, in which we have only one lexical predicate. Can we extend now these considerations to more complex structural descriptions, containing more than a single lexical predicate? Here is where (160) becomes relevant. Its complete ρ-set is as follows:

(164) ρ_1 = ⟨(prove, defendants), (prove, defendants), (prove, innocent)⟩
ρ_2 = ⟨(innocent, defendants)⟩
$\rho_{derived}$ = ⟨(prove, defendants), (prove, defendants), (prove, innocent), (innocent, defendants)⟩

The considerations we have made about the relevance of the relation *parallel* to the representation of reflexive predicates extend without the need to resort to further theoretical devices to sentences with more than a single predicate, as in the case of Raising to Object. Note that in (160) *defendants* is both the subject and object of *prove*, here a reflexive predicate, and also the subject of the adjectival predicate *innocent* (which is the lexical anchor of its own elementary

6 Note that a multidominance account of reflexivity, where a reflexive anaphor (in English) is an argument node with indegree greater than 1 in an elementary graph also satisfies the basic conditions in Safir (2014) for D-boundness, in particular the requirement that '*D-bound is the same object in* SEM (*the syntactic input for semantic interpretation*)' without resorting to either indexing or a rich feature system. If nodes in graphs were bundles of features, then feature compatibility would be trivially satisfied. Cross-linguistic variation in the present approach, as in LTAGs, resides in the size of elementary graphs/trees (Krivochen & Padovan, 2021), and thus the size of the syntactic object where D-boundness is defined (see also **Chapter 8**).

graph by virtue of being a lexical predicate). If we had chosen a verbal predicate to be embedded instead of an adjective, the structure would not modify the conditions under which reflexivity is licensed in the simplest monoclausal English cases (see **Chapter 8** for further discussion about Binding Theory within the present framework).

A look at the ρ-set of (161) is now in order. In this case the anaphor is the Object (2) of the embedded predicate, which means that this predicate must be reflexive. Concretely, it means that, in order to license a reflexive pronoun, a configuration like (162) must be an adequate characterisation of the embedded clause: there must be parallel arcs within an elementary graph in order to have reflexivity. Suppose that *himself* in (161) is coindexed with *John* (that is, assume the reading *John$_i$ believes Mary$_j$ to like himself$_i$*). This means that the expression *John* and the expression *himself* correspond to the same node in the graph, with address ⦃John⦄. Example (161) contains two elementary graphs, one for each lexical predicate *believe* and *like*. In that case, what we have in (161) is the following set of relations:

(165) ρ_1 = ⟨(believe, John), (believe, Mary), (believe, like)⟩
ρ_2 = ⟨(like, Mary), (like, John)⟩
ρ_{derived} = ⟨(believe, John), (believe, Mary), (believe, like), (like, Mary), (like, John)⟩

In this analysis we cannot say that the embedded predicate *like* establishes two distinct relations (by means of two distinct edges) with a single expression, by virtue of *Mary* and *John* being distinct nodes. There is no *parallel* relation in the structural description of (161). The reflexive anaphora is not licensed due to the absence of *parallel edges* in the elementary graph corresponding to the embedded clause (*Mary to like himself*), and (161) is thus adequately excluded as a well-formed expression of the language. The same argument can be invoked in filtering out cases such as Chomsky's (1995: 327) **John expected him to seem to me to be intelligent* (ungrammatical in a coindexed reading for *John* and *him*, acceptable in a disjoint reading): because under Raising to Object *him* should be part of the elementary graph of *expect*, if it is coindexed with *John* we should have parallel arcs, yielding an anaphor and not a pronominal. And indeed *John expected himself to seem to me to be intelligent*, where the parallel arc condition is satisfied and the 1 and 2 of *expect* are the same node, is somewhat clunky but grammatical (and the contrast between *him* and *himself* in that context is clear-cut).

To summarise, our treatment of *Raising* has built on the idea that there is no GF change: in *Raising to Subject* structures the raising predicate (be it single-word, like *seem* or multi-word, like *be likely*) can be thought of as a modal modifier of the lexical predicate akin to an auxiliary, and as such—in English—it does not anchor an elementary graph. In *Raising to Object* there are indeed two local graphs, anchored by distinct lexical predicates, but the 'raised' NP does not change GF: defining GFs at the level of elementary graphs allows us to link those graphs at the shared NP while maintaining the GF that it has been assigned in each local structure. The effects are not unlike LFG's analysis of *Raising* as functional control, insofar as all non-contextual properties of the shared NP are identical between the matrix and the embedded clauses. However, in contrast to the LFG analysis, the dependencies are syntactic, not lexical.

6.3 Object-Controlled Equi

Let us now focus on the structure that is next in (127), *object-controlled Equi*. The reason we are dealing with it after Raising to Object is that there are some superficial parallels between these constructions, in the sense that they are, superficially, sequences of the form NP_{Nom} + V_{fin} + NP_{Acc} + V_{inf}. Essentially, there are two main classes of analyses of Equi vs. Raising: (i) the distinction is strictly lexical, with no configurational differences (roughly, the LFG approach), and (ii) the distinction is configurational, in that Equi and Raising sentences receive distinct structural descriptions (roughly, the MGG approach). The lexicalist view may limit itself to specifying the thematic structure of the lexical predicates involved and the cases assigned to each nominal argument with structure sharing applying 'automatically' as part of the definition of the composition of local domains (e.g., via graph union), or it may also include a specification of what is shared.

In this context, it is particularly interesting to consider the proposal in Hornstein (2001, 2003) that the mechanisms of obligatory control should be reduced to the mechanisms of movement. We will return to the so-called Movement Theory of Control (MTC) in the next section, but for the time being, it is worth summarising the main tenets:

- Theta roles are features
- There is no upper limit to the number of theta features an NP can bear
- An NP can take extra theta roles in the course of the derivation: movement to theta-positions is allowed

In Hornstein's approach, object control is a consequence of Shortest Move: the fact that in ditransitive Equi verbs the closest NP to PRO controls it (with *promise* being marked) translates to the derivational economy principle *Shortest*

Move. Hornstein summarises the reasoning behind his approach thus:

> *A theory that has both movement rules and construal rules has two ways of establishing inter nominal dependencies in the grammar*. (Hornstein, 2003: 11)

We agree with this assessment, and indeed so do most non-transformational theories so far as we can tell. The difference is where the explanatory burden is put: in Hornstein's model, movement takes care of everything. In ours, the relevant mechanisms are construal and structure sharing via graph union. Our theory, in and of itself, is not any more or less 'minimal' than Hornstein's: there is simply a different choice about what the way of establishing 'inter-nominal dependencies' is. In the simplest case, structure sharing would take care of both Raising to Object and Object Equi as far as configuration is concerned: the format of the elementary and derived graphs involved would be the same:

(166) a. The judge believed John to have committed the crime
$\rho_1 = \langle$(believe, judge), (believe, John)$\rangle$
$\rho_2 = \langle$(commit, John), (commit, crime)$\rangle$
$\rho_{derived} = \langle$(believe, judge), (believe, John), (believe, commit), (commit, John), (commit, crime)$\rangle$
b. Mary told John to shut up
$\rho_1 = \langle$(tell, Mary), (tell, John)$\rangle$
$\rho_2 = \langle$(shut up, John)$\rangle$
$\rho_{derived} = \langle$(tell, Mary), (tell, John), (shut up, John)$\rangle$

There are several things to note here, which serve as a recapitulation of assumptions that we have been making throughout the analysis of complementation. First, we are dealing with two single-rooted sub-graphs with anchors *believe* and *commit* in (166a) and *order* and *shut up* in (166b), linked at the node with address ⦃John⦄. Note that, unlike in predication analyses such as Bach's (1982) or Chierchia's (1989), in our structures each predicate anchors a distict elementary graph, and control is determined by structure sharing, not order of composition and functional application or lexical entailments (see also Landau, 2013: 48).[7] This aspect of the analysis was prefigured in Sampson's (1975) sketch of an

7 As a matter of fact, the absence of an edge between matrix and embedded predicate precludes a predication analysis, since there is no application of the semantic value of the matrix predicate to the semantic value of the embedded predicate. In Bach (1979), Subject Equi constructions are assimilated to garden-variety transitive structures, being delivered by *right*

analysis of Equi, and versions of the LFG in which (English) Equi and Raising are both analysed via functional control (Börjars et al., 2019 exemplify this analysis), and is consistent with the traditional generative observation that Equi constructions are *biclausal*. Second, the relations 'object of *order*' and 'subject of *shut up*' are defined in different elementary graphs, but there is no need to multiply the nodes because grammatical functions are read off the local ordered ρ-set. Despite the configurational similarities, the thematic structure of Raising to Object and Object Equi verbs are different (a difference that has gotten renewed attention in MGG; see Chomsky, 2021: 24, ff.).

Most, if not all, syntactic analyses identify problems for a unified structural analysis of Raising to Object and Object Equi, which in empirical terms take the form of the well-known asymmetries exemplified in (167) and (168):

(167) a. I expected the doctor to examine Bill ~
b. I expected Bill to be examined by the doctor (Raising to Object)
c. I told the doctor to examine Bill ≠
d. I told Bill to be examined by the doctor (Object Equi)
e. Bill was expected to be examined by the doctor (~ (a))
f. Bill was told to be examined by the doctor (≠ (c))
g. I believe there to be unicorns in the garden (Raising to Object)
h. *I required there to be unicorns in the garden (Object Equi)

Note that in both Raising to Object and Object Equi it is possible to have a post-verbal reflexive, which attests to the objecthood of the accusative NP:

(168) a. Atreus$_i$ believed himself$_i$/*him$_i$ to be a great warrior
b. Atreus$_i$ forced himself$_i$/him$_{*i/j}$ to be a great warrior

---

concatenation (RCON). Equi verbs are assigned to the category V/VP ((*t*/*e*)/(*t*/*e*)), and they compose with the embedded predicate (producing an expression of category TVP, the category of expressions that need to combine with an NP to form an 'intransitive' VP) before composing with any argument:

i) John persuaded Mary to go: ((**persuade'**(**to go'**))(m))(j) (Landau, 2013: 48)

(i) has the semantic value of **persuade** applied to the semantic value of **go**; then, the semantic value of **persuade to go** applied to the object *Mary*, and finally the semantic value of **persuade to go**(m) applied to the subject, *John*. Recasting such an analysis under present assumptions would require an edge between matrix and embedded predicate, and would leave the distinction between Equi and Raising as strictly lexically determined (and the facts in (167–168) unexplained). Note that if we proposed a ρ-set like (ii) to correspond to Bach's (i)

ii) ρ_{derived} = ⟨(persuade, go), (persuade, John), (persuade, Mary)⟩

there would be no expression to saturate the GF Subject subcategorised by *go*, which goes against the definition of elementary graph as a unit of argument structure saturation. There is no obvious way in the system presented in this monograph to have *persuade to go* as a TVP.

Under present assumptions, reflexivity is a symptom of parallel arcs in an elementary graph. In both cases, as a reflexive is licensed, the elementary graph anchored by *believe* must contain parallel arcs: *himself* is a clause-mate of *Atreus* in both cases. The difference is what predicate subcategorises for it. Note also that in both cases a reflexive is possible in the embedded clause:

(169) a. Paul$_j$ believed George$_i$ to only trust himself$_i$
b. Paul$_j$ advised George$_i$ to only trust himself$_i$

These cases suggest that the 'raised' object is also a clause-mate of the embedded reflexive. Again, parallel arcs must be involved under our definition of reflexivity. If parallel arcs are required for reflexivity, and if only categorematic expressions are assigned addresses (see **Section 6.1.1**), then the impossibility of **John believes there to trust himself* derives (at least partly) from the fact that the embedded subject position is occupied by a syncategorematic expression which cannot head parallel arcs. The possibility of having syncategorematic subjects is, we believe, a lexical property, not a configurational one (or, perhaps better put, syncategorematic subjects are a configurational issue only to the extent that configuration is determined by lexical properties). Similarly, the lack of synonymity in local and long-distance Passivisation in Equi (167 (a-d)), if related to the thematic structure of the predicates, is also lexical in nature (but see Schmerling, 1976 for a cautionary tale on the value of arguments from synonymy). Argument structure and theta grids become again essential: the graph-theoretic modelling of relations between expressions is not a theory of the lexicon, and it is in the lexicon where theta grids and subcategorisation frames are specified (part of this information is in what HPSG calls ARG-ST, a list of arguments taken by each predicate, which are ordered following the GF hierarchy, but ARG-ST does not include thematic roles assigned by a predicate, which we do need). In the analyses sketched above, however, lexical specifications are supplemented by a configurational distinction: in Object Equi the two predicates (matrix and embedded) are not contiguous with respect to dominance: the output of graph composition is not an arbor in Object Equi (since there is more than one root), whereas it is in Raising to Object (as the matrix predicate dominates the embedded one).

6.4 Subject-Controlled Equi

We have deliberately left Subject-controlled Equi for last, because of the problems that its analysis presents to syntactic theory: Subject Equi is trickier than it seems. During the days of the Standard Theory and GB, Raising vs. Equi was considered an essential distinction which made use of two different mechanisms: *movement* vs. *deletion* in Equi-based theories; *movement* vs. *lexical insertion* (and indexing) of a null pronoun in GB and other lexicalist theories (see Polinsky, 2013; Landau, 2013: Chapter 2; Davies and Dubinsky, 2004: Chapter 1 for overviews). In GB, the term Equi was mostly abandoned in favour of Control (the relation between the bound null pronoun PRO and the overt NP that gave it reference in OC contexts), since these structures did no longer involve deletion of identical NPs. But more recently in the development of generative syntax, the distinction in the mechanisms involved in Raising and Control has been blurred. As mentioned above, Hornstein (1999, 2003) and much subsequent work has argued that there is a single underlying mechanism, which is Movement, unifying Raising and Control. The differences between these structures would reside only in the fact that NPs acquire further theta-roles in the course of the derivation in Control, but not in Raising structures (because Raising verbs are not theta-assigners); this proposal entails the elimination of PRO and the proliferation of copies/traces as well as features (since, as seen above, theta roles must become features under MTC assumptions). Interestingly, Chomsky's most recent analysis of Raising vs. Control in Chomsky (2021: 21, ff.) seems to revive Equi, in a way (something that he himself notes). Take a sentence such as *John tried to win*. In Chomsky's analysis, Merge generates the following set:

(170) {John, {INFL, {tried, {John, {to, win}}}}}

The crucial question in this context, says Chomsky, is whether the two *John* can be assigned the relation Copy-of (which would entail that they constitute a chain generated by Internal Merge). Given the assumption that derivations are strictly Markovian, without any evidence of previous derivational steps or way to access the derivational history of a structure, the deciding factor is the theta-criterion: each *John* gets a different theta role, which means that the configuration cannot have been generated via Internal Merge. Thus, ⟨John, John⟩ is not a movement chain (contrarily to what would happen in Raising to Subject). The only other alternative is what Chomsky calls a *Markovian gap*, where the lower *John* gets deleted. In this recent analysis, which we may refer to as Neo-Equi, the two subjects (upstairs and downstairs) are introduced via External Merge (as only External Merge can fill a thematic position), and if no relation Copy can be established between them, one of them gets deleted and becomes invisible for other syntactic operations.

Non-transformational frameworks, naturally, reject a movement analysis. HPSG embraced structure sharing in feature structures for Raising, and indexing for Equi (Pollard & Sag, 1994: 134; Sag et al., 2003: 373–374; Polinsky, 2013: 598): Sag et al. argue that cross-linguistic data does not favour a structure sharing analysis of Equi (a similar argument is used in the LFG analyses of Equi as *anaphoric control*, which also involves co-indexing; see e.g. Dalrymple et al., 2019: 561, ff.). But there are structure sharing analyses of Equi too. Recall that Relational Grammar allowed for multidominance (called 'multiattachment') for local reflexives: cross-clausal multiattachment is also possible (Blake, 1990: Chapter 3), and used to deliver structural descriptions for Equi and Raising without multiplying the nodes in structural representations. Along similar lines (at least in terms of the rejection of the SMC in structural descriptions) Sampson (1975: 6) sketched (but did not develop) a multidominance analysis of Equi, which is equivalent to a structure sharing approach:

(171)

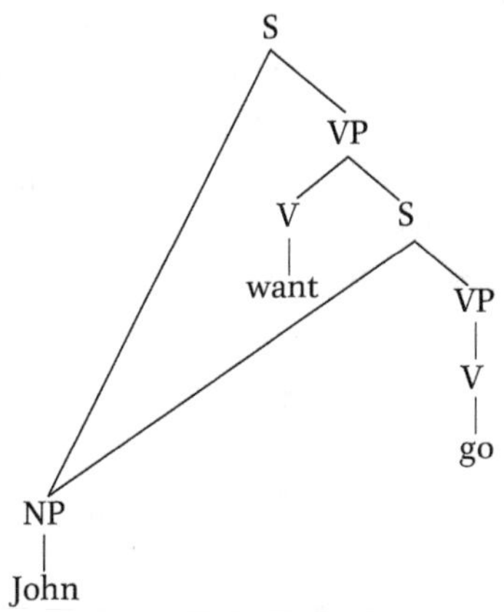

FIGURE 6.5 Multidominance analysis of Subject Equi in Sampson (1975)

As observed in **Section 6.1**, in LFG, Raising Vs take XCOMP ('open') complements: the SUBJ of XCOMP is also an argument of the matrix verb. This is specified in the matrix verb's *f*-structure (Bresnan, 2001; Dalrymple, 2001), and the relevant relation in Raising is actually *functional control*: the subjects of the matrix and embedded Vs need to have the same *f*-structure in Raising to Subject and the object of the matrix V and the subject of the embedded V need to have the same *f*-structure in Subject to Object raising (Dalrymple, 2001: Chapter 12). This is a point of contact between the LFG and HPSG analyses of raising: they both rely on structure sharing.

Some LFG analyses (e.g., Dalrymple, 2001, 2007; Dalrymple et al., 2019) propose that the complement of an Equi verb, unlike that of a raising verb, is *not* open: we don't have an XCOMP but a COMP, with a phonologically empty category PRO in the subject position of the embedded predicate and a relation of *anaphoric control* between the subject or object of the matrix V and PRO (in Subject- and Object-controlled Equi, respectively) where there is no need to have the same *f*-structure. The subject of the matix clause and the subject of the embedded clause are semantically related, but *are not the same syntactic object* (as opposed to the Raising cases). As mentioned above, there are some functional control analyses for English Equi. In contrast to the *anaphoric control* analysis, Falk (2001), Asudeh (2005), Börjars et al. (2019), among others, propose an analysis of Equi which involve *functional control* (and thus total identification of functional structures, comparable to Kartunnen & Kay-style *structure sharing* or RG-style *multiattachment*) and as a consequence needs no PRO subject (see Bresnan, 2001: 297, ff. and Dalrymple, 2001: 325, ff. for accessible introductions to the distinction between *anaphoric* and *functional control*). Functional control is graphically indicated with a line that joins the *f*-structures with identical values: in this case, the subject of the matrix predicate functionally controls the subject of the XCOMP (formally, we have the following functional equation: (↑SUBJ) = (↑XCOMP SUBJ)). As Landau (2013: 59) observes, it is an underived axiom in LFG that the controllee is always a SUBJ, although an argument can be made for control as a syntactic process being sensitive to the GF hierarchy. The *f*-structure for a sentence like *the doctor tried to examine John* under functional control (as in Börjars et al., 2019) would be as below:

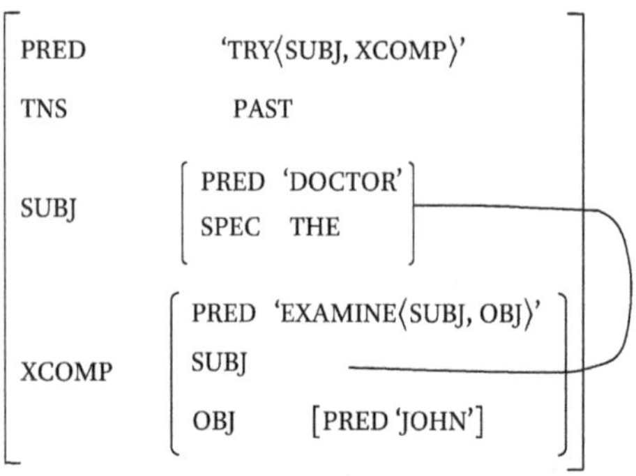

FIGURE 6.6 F-structure for Subject Equi with functional control

Crucially, as Dalrymple (2007) points out, the answer to the question whether an adequate analysis of Equi involves *anaphoric* or *functional* control (and even if the same analysis is valid for all cases) is not given by the theoretical framework of LFG, but by the linguistic facts. This is important because it allows for different languages to follow different strategies when building predicate complement constructions; given the strong empirical focus of the present work this is clearly a view we are sympathetic towards.

The reason we have spent some time with LFG's analysis of Subject Equi is that we need to ask the question whether Raising to Subject and Subject Equi should receive distinct structural descriptions in terms of the number of elementary graphs or the relations defined in the graphs or whether the difference can be encoded in lexical specifications (at least in English; as pointed out above, it is possible and theoretically consistent that other languages may choose to encode the distinction by different means). In the latter case, these specifications would pertain to the interpretation of nodes, but not to the connections they establish with other nodes in a structural description. What these lexical specifications look like depends on the kind of lexicon that we want to have. At the very least, the Lexicon should include information about the *arity* of predicates as well as their thematic grid: how many arguments they take, their syntactic category, and which semantic roles they assign to these arguments. For purposes of this work, the crucial ingredient of the proposal is that there is neither movement nor empty categories involved in either Raising or Equi constructions. The *predication* analyses of Equi were among the first to not have deletion transformations or PRO (in some versions at least; see Landau, 2013: 47, ff. for discussion), but their focus was semantic, not configurational (i.e., control was taken to be a semantic relation, not a syntactic one). Our analysis is not a predication analysis, however: there are two 'clauses' (two elementary graphs) and there are two subjects (as in GB's PRO analysis), but there is also structure sharing, and these two subjects become a single node in the derived graph.

The subcategorisation requirements of Equi and Raising verbs differ in interesting ways. As mentioned above, only Raising predicates allow for syncategorematic arguments:

(172) a. It {seems / *tries} to rain
b. There {seemed /*tried} to appear an image on the screen
c. Paul {believed / *told} there to be no reason to argue
d. Paul {wants / *tells} it to rain tomorrow

The presence of syncategorematic expressions suggests that in Raising sentences there are 'less' arguments than in Equi sentences: syncategorematic expressions are not assigned addresses (as they have no semantic values), and therefore cannot be shared. We want to exclude, then, elementary graphs for structures such as

(173) a. [$_{\text{EG1}}$ It seems]
b. [$_{\text{EG2}}$ it to rain]

The role of the expletive is simply to fill the subject slot in the surface syntax, and cannot participate in syntactic relations such as control or predication. Under the analysis of Raising in **Section 6.1**, there is a single elementary graph in the grammatical version of (172a), in which *seem* is a modifier of *rain*. Furthermore, we need to specify that *rain*'s subject is not referential: that is not part of the syntactic configuration in which *rain* appears, but a lexical property of *rain* which needs to be represented in the lexicon. In the case of (172b), two analyses are in principle available: either *there* is the 1 and *an image* is the 2, or *an image* is the 1, and there is no. 2. The analysis in **Section 5.1** follows this second route, which leaves unaccusativity as a lexical, rather than configurational, property (but see fn. 8 in **Chapter 5**).

Above, in **Section 6.1**, we argued that Raising to Subject structures receive a monoclausal analysis, which translates into their structural description being a single-rooted graph. The argument was, partly, that Raising verbs such as *seem* are (in specific contexts) modifier expressions of lexical predicates rather than lexical predicates themselves, therefore, they are part of the elementary graph defined by that lexical predicate. Can we say the same for Equi verbs? Consider to this effect examples (174a) and (174b):

(174) a. Aerith *tried / wanted* to sell the flowers (Subject Equi)
b. Aerith *was likely / seemed* to sell the flowers (Raising to Subject)

If we follow the line of reasoning for Raising verbs, the question is whether *try* is a modifier of *finish* or a different predicate altogether; if the latter, then we need to propose a multi-rooted structure linked at the node corresponding to the subject NP, which is the subject of both predicates.

Unlike Raising predicates, Equi predicates impose thematic restrictions with respect to their co-occurring arguments; this is a traditional observation and has been at the core of generative analyses of these constructions. As has been pointed out repeatedly in the literature, (175a) is anomalous not because *winter* cannot *arrive*, but because it cannot (literally) *try*:

(175) a. #Winter tried to arrive early this year
b. Winter seemed to arrive early this year

It seems unlikely that the difference between (175a) and (175b) can be reduced to pure configuration: the thematic properties of *try* versus *seem* are resistant to being fully syntactified, in that appeal to lexical specifications is at some point necessary (even if thematic roles were features, they would be specified in the Lexicon, as part of the lexical entry of predicates). In the present context, the lack of semantic selection on the part of *try* means that the subject NP is a nominal dependant of the matrix predicate as much as it is a nominal dependant of the embedded predicate: therefore, we need to conclude that each predicate, the *Equi-governing* V and the embedded V anchor an elementary graph each. In the light of this discussion, then, we can specify the ρ-sets for (174a) and (174b) as follows:

(176) a. $\rho_1 = \langle$(try, Aerith)$\rangle$
$\rho_2 = \langle$(sell, Aerith), (sell, flowers)$\rangle$
b. $\rho = \langle$(be likely, sell), (sell, Aerith), (sell, flowers)$\rangle$

Only (176b), the ρ-set of a *Raising to Subject* sentence, defines a single-rooted elementary structure, with a root *be likely*, which is a multi-word modal epistemic expression (and the highest-order functor in the structure), and a lexical anchor *sell*. (176a), the ρ-set of a Subject Equi sentence receives a different structural description, with two elementary graphs linked at the node with address ⦃Aerith⦄ (see also Sampson, 1975: 6). This is because each elementary graphs is defined by the presence of a single lexical predicate, and we only have two distinct lexical predicates in Equi structures (in (176a), one is anchored by *try* and another by *sell*). The two lexical predicates in an Equi structure are not contiguous with respect to dominance (thus, not a DG *catena*): this is a property common to Subject Equi and Object Equi predicates. This approach delivers a distinct structural description for Equi vs. Raising (as is also proposed in MGG analyses of these constructions) without stipulations about the presence or absence of designated nonterminal nodes.

We can now proceed to the analysis of more complex cases. We can look, for example, at Equi predicates embedded under other Equi predicates. Consider first the following sentence:

(177) Melvin promised to try to win (adapted from Johnson & Postal, 1980: 540, ex. (164))

In this case, we have *three* lexical predicates, *promise*, *try*, and *win*. Two of those take non-finite complements, *promise* and *try*. Let us compare three structural descriptions for (176): the MGG one, the APG/MG one, and ours so as to assess the advantages and disadvantages of each. We begin by considering the structural description assigned to (177) in MGG analyses (based on Chomsky, 2001 and related work, but cf. Chomsky, 2021; see also Polinsky, 2013 for an overview of MGG analyses of control):

(178) a. [$_{CP}$ [$_{TP}$ Melvin$_i$ [$_{T'}$ [$_{\nu P}$ ~~Melvin~~ [$_{\nu'}$ promise+ν [$_{VP}$ ~~promise~~ [$_{CP}$ [$_{TP}$ PRO$_i$ [$_{T'}$ to [$_{\nu P}$ ~~PRO$_i$~~ [$_{\nu'}$ try+ν [$_{VP}$ ~~try~~ [$_{CP}$ [$_{TP}$ PRO$_i$ [$_{T'}$ to [$_{\nu P}$ ~~PRO$_i$~~ [$_{\nu'}$ win+ν [$_{VP}$ ~~win~~]]]]]]]]]]]]]]]]]]

Each control verb takes a CP complement; the full functional structure of the clause is projected. Noteworthy features of the structural description (178a) are the proliferation of copies (an occurrence of NP in *ν*P guarantees theta-marking to the external argument NP, whereas movement of this subject NP to Spec-TP satisfies T's EPP feature; Case assignment can take place at a distance as a by-product of T's phi-feature valuation via Agree, with Null Case being assigned to PRO by non-finite T) and the necessity of an independent indexing mechanism to establish the relation of *control* between *Melvin* and all instances of PRO (such that the entity denoted by *Melvin* is understood to be the 'promiser', the 'trier' and the 'winner'). While we reject non-audible structure and empty categories (two major aspects of the structural description (178a)) as well as movement, this representation is useful to see the monotonically recursive structure of a sequence of predicates with similar lexical specifications insofar as syntactic context goes.

Next up is APG's representation (minimally adapted from Johnson & Postal, 1980: 540):

(178) b.

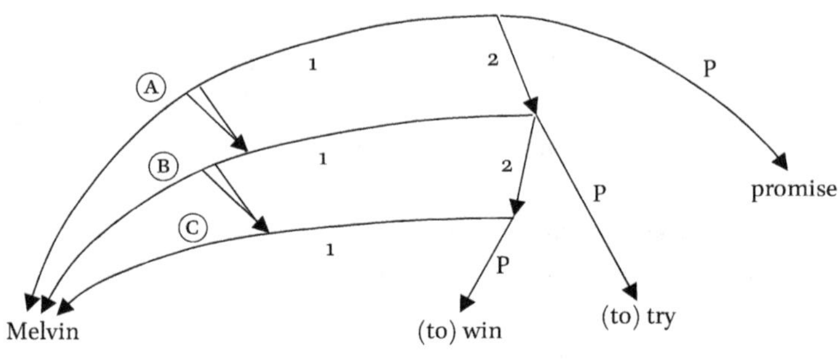

FIGURE 6.7 Arc-Pair Grammar analysis of Subject Equi

This structural description for 'stacked Equi' features three P(redicate) arcs, one per V. *Melvin* is the 1 of all three predicates, and the 2 of *promise* and *try* corresponds to a clause (i.e., a predicate plus its own 1 and 2 arcs). There is no need to resort to PRO or indexing at all in APG; rather, it is arcs that get erased (here marked with double arrows): A erases B, B erases C. It is interesting to see that there are arcs in (178b) which do not join expressions: because there are no arcs between predicates, the formalism is forced to admit that an arc may join two nodes which do not correspond to expressions in the 'superficial form' of the sentence (e.g., the tail of the topmost 2 arc is not an expression, but the head of another arc which has no associated overt exponent). Note that also in APG the predicates are not ordered with respect to dominance: none of *promise*, *try*, or *win* dominates the other(s). Dominance is a partial order over the nodes in this graph.

The conditions we impose over edges, by definition, are stricter: an edge *must* join two and only two nodes, and every node must be justified in terms of *economy of expression*, regardless of their formal convenience. A fundamental assumption in the present work is that nodes must be categorematic expressions: there is no intermediate node that corresponds to a phrasal constituent (e.g., [promise [$_S$ ...]]), since ours is not a theory based on immediate constituency. Our structural descriptions do not resort to non-audible structure in labelled projections and empty categories. We can, of course, use a shorthand to refer to a set of nodes (say, VP), but that will not be a part of the graph: as highlighted before, that shorthand it is just convenient descriptive nomenclature, not a formal object. The structure will have as many roots as lexical predicates; each of those lexical predicates will form its own sub-graph with its functional dependants and nominal dependants, and these sub-graphs will be linked at those nodes assigned the same addresses. We have three lexical predicates, which means we have three elementary graphs:

(179) Elementary graph 1 = [Melvin promise [EG 2]]
Elementary graph 2 = [Melvin try [EG 3]]
Elementary graph 3 = [Melvin win]

The ρ-set for the derived graph for (177) (after linking the three elementary graphs at the common node with address ⦃Melvin⦄) is (178c):

(178) c. $\rho_{\text{derived}} = \langle$(promise, Melvin), (try, Melvin), (win, Melvin)$\rangle$

Let us look at another compex case, such as (180):

(180) Barrett seems to expect Tifa to tell Aerith to try to sell the flowers

(180) combines Raising to Subject, Raising to Object, Subject Equi, and Object Equi: *seem* is a Raising verb, *expect* is a Raising to Object verb, *tell* works here as an Object Equi verb, and *try* is a Subject Equi verb. We admit that (180) is somewhat heavy, but it is grammatical. A laboratory sentence, for sure, but it helps us illustrate the possible dependencies that our grammatical theory allows for. In this case we will go directly to our own proposal, leaving the MGG and APG structural descriptions as an exercise to the reader.

Above we said that Raising predicates do not take their own dependants, but rather inherit those of the lexical predicate they modify.[8] Therefore, being functional modifiers, Raising predicates do not define elementary graph: they are nodes within elementary graph defined by lexical predicates. With this in mind, the elementary graphs and ρ-set of (180) are as follows:

(181) Elementary graph 1 = [Barrett seems to expect [EG 2]]
Elementary graph 2 = [Tifa tell Aerith [EG 3]]
Elementary graph 3 = [Aerith try [EG 4]]
Elementary graph 4 = [Aerith sell the flowers]
$\rho_{derived}$ = ⟨(seem, expect), (expect, Barrett), (expect, Tifa), (expect, tell), (tell, Tifa), (tell, Aerith), (try, Aerith), (sell, Aerith), (sell, flowers)⟩

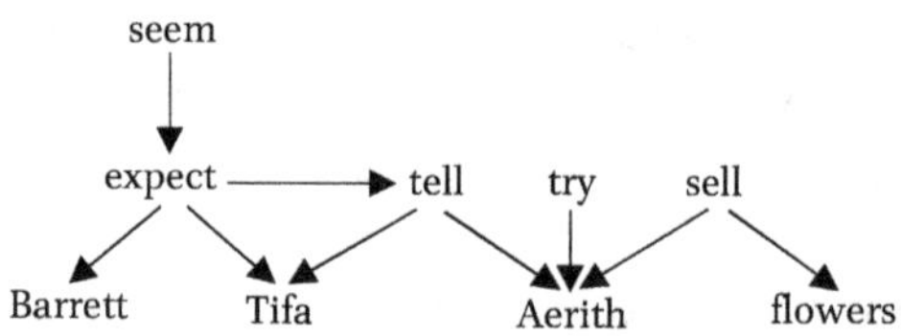

FIGURE 6.8 Graph-theoretic analysis combining Raising to Subject, Raising to Object, Subject Equi, and Object Equi

The derived graph defined in (181) is multi-rooted: there are four lexical predicates *expect*, *tell*, *try*, and *sell*, of which *expect* is modified by *seem* (bear this structural pattern in mind, for we will come back to it in our extension of the present system to Spanish auxiliaries in **Chapter** 7) and *tell* is dominated by *want*, which is a Raising to Object predicate. *Seem*, *try*, and *sell* are roots, by virtue of being undominated. As before, it is important to emphasise that arguments are directly dominated by the lexical predicates that select them, regardless of finiteness (a difference between our approach and DG). The GF of each nominal element in each elementary graph can be read off the ρ-set, assuming (115).

The English cases that we have analysed so far suggest that *Raising to Subject* and *Subject Equi* must receive a different structural analysis, in addition to what may be encoded in lexical entries (arity and thematic structure, both of which are crucial aspects of a theory of the lexicon). However, this conclusion

8 More specifically, we propose that raising predicates are modal auxiliaries; see Krivochen (2013) for an argument based on Spanish. García Fernández & Krivochen (2019) analyse the prototypical Spanish raising verb, *parecer*, and similarly conclude that, when followed by an infinitive, it is indeed an auxiliary.

must not be generalised to other languages automatically: just like the English analysis was motivated empirically (which does not mean it is correct, just that the decisions we have made attempt to capture empirical properties rather than being motivated by intra-theoretical notions), so must be the structural descriptions assigned to Raising to Subject and Subject Equi sentences in other languages. Let us see an example of cross-linguistic variation in this regard. In the Spanish grammatical tradition, the possibility of a verbal predicate to combine with a meteorological V is taken to be an indication of its status as an auxiliary, and thus akin to Raising predicates (see, e.g., García Fernández, 2006; Bravo, 2016a). Thus, there is a contrast between (182a) and (182b):

(182) a. *Parece* *llover*
Seems.3SG.PRES rain.INF
'It seems to rain' (Spanish; Raising V)

b. *#Intenta* *llover*
Tries.3SG.PRES rain.INF
'It tries to rain' (Spanish; Equi V)

As the argument goes, Raising verbs do not impose restrictions on their co-occurring NPs, and nor do auxiliaries (with the possible exception of dynamic modals, possibly some deontic modals, and some aspectual phasals; see e.g. Bosque, 2000; Hacquard, 2010; García Fernández, 2006). Thus, the anomalous character of (183a, b) below is due to the incompatibility between the subject NP and the embedded predicate, not the Raising V or the perfective auxiliary:

(183) a. **La piedra parece correr*
The stone seem.3SG.PRES run.INF
'The stone seems to run' (Spanish; Raising V)

b. **La piedra ha corrido*
The stone have.3SG.PRES run.PAST.PART
'The stone has run' (Spanish; auxiliary V)

The problem with the sentences in (183), analogously to the English case (175a), is not that a stone cannot *seem+V* or *have+V-ed*, but rather that it cannot *run*. This is represented in the theory in terms of a lack of argument structure in Raising V and (most) auxiliaries, they do not assign thematic roles because, *stricto sensu*, they do not select arguments (they 'inherit' them, in the words of Jacobson, 1990). And because meteorological verbs do not take arguments either, auxiliaries and Raising verbs can freely combine with them. Lexical verbs (including Equi Vs) do restrict the class of NPs that they can co-occur with, and thus—so the argument goes—they *cannot* combine with meteorological verbs. Or, rather, they *shouldn't*. However, we do get examples like the following, both of which are perfectly natural and acceptable sentences:

(184) a. *Quiere* *llover*
Want.3SG.PRES rain.INF
Lit. 'It wants to rain'

b. *Parece* *llover*
Seem.3SG.PRES rain.INF
'It seems to rain'

The English equivalent of (184a) would be literally *it wants to rain*, which the theory predicts to be ill-formed due to the incompatibility of *want* and *rain*: in simple words, things that can *want* cannot *rain*. The more natural status of (184a) seems to support a view in which configurational information is supplemented with lexical specifications: (184a) is, in our opinion, better analysed in terms of lexical properties rather than as a garden-variety Equi construction. This *querer* (but not *intentar* in (182b)) is going down the road of grammaticalisation. Of course, this must not be overgeneralised: any run-of-the-mill Equi construction with *querer* receives (in principle) the same ρ-analysis as the English examples; (184a) simply cannot be assigned the same structural description as (185) below, despite superficially featuring the same verb:

(185) *Juan quiere terminar el trabajo pronto*
Juan want.3SG.PRES finish.INF the work soon
'Juan wants to finish the paper soon'

The interpretation of *querer* in (184a) is certainly not one that involves a volitional or even sentient entity; rather, it means something along the lines of 'the conditions are such that it may start raining any moment' (possibly making ⟨*querer* + infinitive⟩ as it appears in (184a) a 'preparatory conditions' modal periphrasis, along the lines of Martín Gómez, 2022). The thematic properties of *querer* in the context of a meteorological verb are not the same as those in other contexts, which seems amenable to a lexical rule treatment.

As for (184b), it has been argued that *parecer* followed by an infinitive is indeed an auxiliary (Krivochen, 2013; García Fernández & Krivochen, 2019b: 45–50). In generative terms, having *parecer* as an auxiliary entails that it would not be generated as the V head of a VP (which would be the structural description assigned to *seem* as a *Raising* verb in MGG and even TAG; see e.g. Postal, 1974; Davies & Dubinsky, 2004: 340–343; Polinsky, 2013; Frank, 2002: 71) but rather as a T/I head of TP/IP. Here, the question is phrased in different terms: does Spanish *parecer* anchor an elementary graph? Recall that the structural description that we have provided for English *seem* when followed by an infinitive takes it to be akin to an auxiliary verb. Can we say the same about Spanish *parecer*?

Let us consider how *parecer* interacts with auxiliary verb constructions. An interesting point is that *parecer* admits a sequence of auxiliaries (an 'auxiliary chain') as a complement, as in (186):

(186) *Juan parece haber estado trabajando hasta tarde*
Juan seem.3SG.PRES have.INF be.PART work.GER until late
(*porque llega a casa después de lo= previsto*)
(because arrive.3SG.PRES to home after of CL= foresee.PART)
'J. seems to have been working until late (because he gets home after schedule)'

It is important to note that I do not need to have a direct perception of John working to say (186), I just need evidence that may be indirect (for example, the fact that he arrives later than expected). Recall that when *parecer* is followed by an infinitival clause, its complement cannot be pronominalised: only when *parecer* is followed by a small clause is this possible (↛ indicates lack of equivalence):

(187) a. *Juan parece {estar enfermo / trabajar}* ↛ *Juan lo= parece*
Juan seems be.INF sick / work.INF ↛ Juan CL= seems
'Juan seems to be sick / Juan seems to work'

b. *Juan parece {enfermo / un tonto}* → *Juan lo= parece*
Juan seems sick / a dumb → Juan CL= seems
'Juan seems sick / Juan seems (to be) a dumb person'

This suggests that the category of the complement is different, with the complement of *parecer* in (187b) being a complete predication structure; above we assumed that this predicate is the lexical anchor of the elementary graph. What is the status of *parecer* in (187a)? Syntactically, in (186) we have a chain of auxiliary verbs (namely, *haber*$_{Perfective}$ + *estar*$_{Progressive}$) and the question is whether *parecer* is part of this chain (that is, if the chain of auxiliaries in (186) is *parece haber estado* or just *haber estado*). If *parecer* is part of the chain, then it seems that it is the highest auxiliary in the chain and, as such (in a matrix finite clause), the one that manifests inflectional features. The corresponding structural description, in this case, involves a single elementary graph: there is only one subject (*Juan*), one lexical predicate (*trabajar*), and only one event (that of working), plus two functional modifiers: perfective ⟨*haber* + participle⟩ and progressive ⟨*estar* + gerund⟩. If, on the contrary, we hold that *parecer* in a sequence *parecer* + infinitive is a lexical verb that takes as its complement a non-finite subordinate clause (as in traditional analyses of English *seem*), then we necessarily have to say that *haber estado trabajando hasta tarde* in (186) is an infinitival complement in a biclausal structure. We can compare the two (simplified to show only relevant phrasal labels) structural descriptions in MGG terms:

(188) a. [$_{TP}$ Juan$_i$ parece tener que estar [$_{VP}$ t_i trabajando hasta tarde]]
b. [$_{TP}$ Juan$_i$ parece [$_{TP}$ t_i tener que estar trabajando hasta tarde]]

Here, as in Krivochen (2013) and García Fernández & Krivochen (2019b), we argue that (188b) is inadequate, and that the correct structure for the Spanish construction ⟨*parecer* + infinitive⟩ is (188a): *parecer* is an auxiliary, and as such it is part of the auxiliary chain. This allows us to keep the relation between *Juan* and *trabajar* a local one, within a single elementary graph as desired (since it is *trabajar* that subcategorises for *Juan*). Furthermore, it also allows us to unify the epistemic semantics of *parecer* with other modal verbs, like *tener que* or *poder*. We will come back to the issue of Spanish auxiliary chains on **Chapter** 7, since the possibilities that Spanish offers us in its auxiliary system (for instance, the fact that modals are not positionally restricted, as they are in English, since they have full inflectional paradigms) will force us to reconsider aspects of cyclicity and locality in auxiliary verb constructions. We will come back to this issue in **Section** 7.1.1.

Let us take stock. Following the discussion in García Fernández & Krivochen (2019b) (see also RAE-ASALE, 2009: § 28.6d, ff.; 37.10n), if *parecer* indeed satisfies the criteria to be considered an auxiliary when followed by an infinitive, then in principle it must belong to an elementary graph defined by the presence of a lexical head: this would be the lexical verb. This statement assumes, crucially, that all auxiliaries are functional modifiers of a lexical head (i.e., that auxiliaries do not modify one another), an assumption that while appropriate for the rigid English auxiliary system, is not necessarily empirically adequate for Spanish. As a matter of fact, in **Chapter** 7 we will argue exactly this: that the class of Spanish auxiliaries is not syntactically homogeneous. This will have deep repercussions on the definition of arbores in sequences of auxiliaries: we will argue that not all languages structure their sequences of auxiliaries in the same way; some display fixed order between auxiliaries and monotonicity in modification relations within the auxiliary seqnence (e.g., English), whereas others allow for more flexibility and combinatory possibilities (e.g., Spanish, Italian). This means, crucially, that the size of elementary graphs (and the order in which composition operations apply, in derivational systems) is subjected to cross-linguistic variation: this is an important point in the analysis of variation in LTAGs (see e.g. Frank, 2013: 238; Krivochen & Padovan, 2021).

6.5 A Note on Raising and Polarity: 'Opacity' Revisited

Raising structures present an interesting interaction with the licensing of polarity items, which we will only briefly comment on here, much research pending. It is well-known that Negative Polarity Items (NPIs) like 'any' or 'ever' need to be licensed by a negative expression, be it an adverb like 'no', 'never', 'seldom', a quantifier like 'none', or a propositional operator like *interrogation*. In the simple cases, *licensing* requires c-command such that an NPI must be within the c-command domain of (the minimal XP containing) a licenser (Horn, 1989, 1997; Ladusaw, 1980; Giannakidou, 2002). Ladusaw (1980: 112) formulates the so-called *Polarity Hypothesis* as follows:

> *A NPI must appear in the scope of a trigger (a downward entailing element). If the trigger appears in the same clause as the NPI, the trigger must precede the NPI*

Alternatively, we may simply say that NPIs are licensed in *downward-entailing contexts* (see Giannakidou, 2002 for discussion).

There are two conditions formulated in Ladusaw's hypothesis, one pertaining to structural order (hierarchical relations) and another to linear order. Here, we are only interested in the former. In the context of this work, we express the structural order condition as follows:

> *A node v_i appears* ***in the scope of*** *a node v_j in G iff:*
> i. *There is a walk w in G such that v_j precedes v_i in w, and*
> ii. *There is no v_k such that v_k is the root of a sub-graph including v_i but excluding v_j*

Condition (i) captures the insight that in Osborne's version of DG is achieved via *catenae*, in that in both cases we are making reference to nodes which are *d-ordered*. May the following contrast suffice as an illustration of the accessibility condition (with licenser and licensee in **bold**):

(189) a. *A man in love was **ever** happy (no trigger; thus, NPI is not licensed)
b. **No** man in love was **ever** happy (**No** as a local, accessible trigger; the NPI is thus licensed)
c. Which man $C_{[+Q]}$ in love was **ever** happy? (the interrogative operator $C_{[+Q]}$ as a local, accessible trigger; the NPI is thus licensed)

We will now analyse some situations in which we get unexpected licensing under standard assumptions about phrase structure and opacity in MGG. These assumptions, in what pertains to the present point, are the following:

I. Non-complements (adjuncts, specifiers) are internally opaque (see, e.g., Chomsky, 1977; Huang, 1982; Uriagereka, 2002, among many others)
II. A negative polarity expression *within an adjunct* cannot legitimate an NPI in object position (as a consequence of (I) plus the additional assumption, which follows from X-bar theory, that material inside an adjunct to XP never c-commands material in object position in XP)

Assumption II is related to the fact that polarity scope relations are restricted to single rooted sub-graphs where a walk can be defined from trigger to NPI (see Sternefeld, 1998a, b; more on this below). This is in a way a consequence of Assumption I: given MGG phrase structure building procedures, all non-complements need to be Chomsky-adjoined (Chomsky, 1955a). Thus, when we consider a *kernel sequence*,[9] they are still not 'there' in the derivation and cannot license an NPI which has been introduced by lexical insertion / External Merge (see also the relative timing between *constituent structure rules* and *conjoining* and *embedding generalised transformations* in Fillmore's 1963 architecture).

The opacity of adjuncts for certain grammatical purposes is not exclusive of derivational frameworks. For example, Dalrymple et al. (2019: 656, ff.) observe—from an LFG standpoint—that under certain conditions there cannot be functional control involving a dependency path that ends in an adjunct.

9 In classical, pre-Aspects generative grammar, *kernel sequences* are strings of terminal nodes derived by means of phrase structure rules exclusively (Chomsky, 1955a: 481). In contrast, *kernel sentences* are those which result from applying *obligatory* transformations to the terminal strings generated by the [Σ, F] grammar—in this case, a PSG in normal form—(Chomsky, 1957: 45).

In a declarative framework such as LFG, it is necessary to define locality constraints in terms of sequences of *f*-structures that separate a filler from a gap. Opacity effects involving adjuncts are characterised not at the level of constituent structure, but at the level of functional structure, with *c*-structures being annotated with functional uncertainty paths (regular expressions that define the path from a filler to a gap).

It is thus crucial to determine what is an adjunct and what is not (which in turn implies a certain phrase structural configuration). In general, the distinction can be captured in IA as well as IP frameworks (which is a nice argument for its extra-theoretical reality). In X-bar syntax, a *complement* is defined as the sister node of the head and daughter of an intermediate projection (Chomsky, 1970b; Stowell, 1981), a configuration defined by the PSR X' → X (YP). In some developments in which specifiers are configurationally assimilated to adjuncts (e.g., Kayne, 1994; Uriagereka, 2002; Chomsky, 2013), the relevant structural opposition is expressed in more general terms, as holding between complements and non-complements. The distinction need not be made in configurational / derivational terms, however. From the point of view of Categorial Grammars, Dowty (2003) formulates the essential features that distinguish *complements* (read: *internal arguments*) from *adjuncts* in an admittedly simplistic manner, but which captures the core aspects of the syntax and semantics of the argument/adjunct distinction in a very elegant way:

> *A constituent Y in a phrase [X Y] (or in [Y X]) is an ADJUNCT if and only if (i) phrase X by itself (without Y) is also a well-formed constituent, and (ii) X (without Y) is of the SAME syntactic category as phrase [X Y]. (X is in this case the HEAD of the phrase [X Y].)* (Dowty, 2003: 34)

The issue here is thus not the position of a phrase with respect to a head in a configurational template (such as X-bar), but rather the definition of the categories of inputs and outputs of a function. In CG terms, this means that for any category X, an adjunct to X will be of category X/X.

> *Then, a constituent Y in [X Y] is a COMPLEMENT if and only if (i) X by itself (without Y) is not well formed, or else (ii) if it is grammatical, then X standing alone not have the same category as in [X Y] (and does not have exactly the same meaning as it has in [X Y])* (Op. cit.)

In this case, if an expression is of category X/Y, it needs to be combined with an expression of category Y to yield an expression of category X: in this configuration, Y is a complement.

In semantic terms,

> *If Y is an adjunct, the meaning of [X Y] has the same kind of meaning (same logical type) as that of X, and Y merely restricts [X Y] to a proper subset of the meaning/denotation of X alone. Where Y is a complement in [X Y], (i) the meaning of X by itself, without Y, is incomplete or incoherent. Else, (ii) X must be understood elliptically* [...] *Also, the same adjunct combined with different heads affects their meaning in the "same" way semantically (e.g. walk slowly vs. write slowly). But the same complement can have more radically different effects with different heads (e.g. manage to leave vs. refuse to leave)* (Dowty, 2003: 34).

After this brief introduction of the argument-adjunct distinction, and given Assumptions I and II above, we can examine some data. Consider now the following examples, in which we will analyse the conditions for NPI licensing:

(190) a. **No** policeman seemed to all of the reporters to have **any** chance of solving the case
b. The police seemed to **none** of the reporters to have **any** chance of solving the case
c. *The police seemed to have **any** chance of solving the case to **none** of the reporters
d. *To **none** of the reporters, the police seemed to have **any** chance of solving the case[10]

Example (190a) satisfies this condition: the NPI (here, **any**) needs to appear within the scope of its trigger (here, **no**) (where scope is defined in terms of c-command; Ladusaw, 1980: 37). The opposite scenario arises in (190c), in which (depending on where we assume the *to*-phrase is adjoined, more on this in a moment) either the NPI c-commands the trigger or the NPI is directly free (i.e., not bound). But it is the grammaticality of (190b) that results of particular interest: the trigger appears within a phrase that does not seem to c-command the NPI; however, the NPI appears to be somehow (non-canonically) licensed. The status of the *to*-phrase in Raising to Subject constructions is in

10 For ease of contrast, compare (190d) with the 'inversion-via-focus' version which, our informants report, is perfectly grammatical:
i) To none of the reporters did the police seem to have any chance of solving the case
We will come back to this contrast below.

general assumed to be that of an adjunct, in phrase structure grammars (and in the 'mixed' PSG/CG approach in Dowty, 2003): note that in general it can be omitted, and when it is materialised its linear position is not as restricted as with complements. Let us illustrate the situation with a *Richard* construction, already familiar to us:

(191) a. (to me) it looks (to me) like Richard is in trouble (to me)
b. (to me) Richard looks (to me) like he's in trouble (to me)

The *to*-phrase in (190a) above presents similar positional freedom:

(190') a. (to all of the reporters) **No** policeman seemed (to all of the reporters) to have **any** chance of solving the case (to all of the reporters)

This is to be expected, since the *to*-phrase does not intervene between the licensor **no** and the licensee **any**. However, when the polarity expression that licenses the NPI occurs within the *to*-phrase, things get more restrictive. Summarising (190b, c, d) in a single example, we get both peripheral positions yielding severely degraded or directly ungrammatical results:

(192) (*to **none** of the reporters) The police seemed (to **none** of the reporters) to have **any** chance of solving the case (*to **none** of the reporters)

Recall that NPIs need to be *licensed* in some way in a downwards-entailing context. In the terms we have been working with here, adapting Ladusaw's condition, *licensing* requires the existence of a walk between licensor and licensee: this entails that the licensee is *accessible* to the licensor. In short, we seem to have elements to formulate the following condition in (193):

(193) ***Licensing (first preliminary formulation)***
An expression corresponding to node v_i may ***license*** *an expression corresponding to node v_j iff $(v_i, v_j) \in \rho^*$*

This definition entails that there must be a *walk* from v_i to v_j (we will drop 'expression' in what follows, presupposing it). While this condition seems to be sensible, it is not enough: a walk between the peripheral occurrences of **none** and the NPI **any** is certainly formulable, but would yield incorrect empirical predictions (we would then expect all options in (192) to be grammatical, contrary to fact). What else is needed, then? It seems that here the property that graphs are rooted and elementary graphs are single-rooted (recall that all elementary graphs are arbores, but not all arbores are elementary graphs) comes in handy: **Definition 10** in **Chapter 2** establishes that graphs contain at least one node which is not dominated by any other node within that subgraph; that is the *root*. In elementary graphs, there is only *one* root; derived graphs may be multi-rooted. This means that the traditional distinction between *root* and *non-root* transformations (e.g., Emonds, 1970 and much subsequent work) can be captured in the present framework. And how exactly does this help? To begin with, we propose that the relevant structure of (192) is (194), using S to denote the root node for convenience:

(194) (*to **none** of the reporters)$_{\text{S-Adjunct}}$ The police seemed (to **none** of the reporters)$_{\text{VP-Adjunct}}$ to have **any** chance of solving the case (*to **none** of the reporters)$_{\text{S-Adjunct}}$

Let us flesh this out: both left and right peripheral topic adjuncts are assumed here to be adjoined to the root (see also Baltin, 1982; Kroch, 2001), whereas the intermediate adjunct is somewhere in the VP, in phrase structure terms. Here, what matters is simply the root vs. non-root distinction, because if the adjunct is a sub-graph, we want to know in which positions the nodes that the sub-graph contains are accessible for operations outside that sub-graph, including licensing. Thus, we seem to need to reformulate the condition on licensing above to incorporate the nuances we have briefly discussed in this paragraph. The result goes along the following lines (with condition (*iii*) to be slightly reformulated below):

(195) ***Licensing (second preliminary formulation)***
Let G and G' be sub-graphs and v_i and v_j be nodes. Then, $v_i \in G$ may license $v_j \in G'$ iff
i. *$(v_i, v_j) \in \rho^*$, and*
ii. *G' is* **not** *adjoined to the root of G, and*
iii. *There is at least a node in address in G that is identical to a node address in G'*

In this reformulation we have incorporated a few corrections with respect to the first version of the definition. First, licensing is a node-to-node relation. Also, we added the condition that the licensee cannot be contained in a sub-graph that is adjoined to the root of the graph containing the licensor: 'left' or 'right' periphery in (193) is a matter of linear order, not of syntactic dependencies. It is essential to bear in mind that *root node* does not mean that we are dealing with a matrix clause, for there are embedded roots in the case of the application of generalised embedding transformations (in the sense of Fillmore, 1963; these could also be understood in terms of *substitution* in Kroch & Joshi, 1985; Frank, 2002: 17; see also Chomsky, 1955a; 1955b: Def. 26). We illustrate an *embedding* generalised transformation in (196) (see also **Section 2.2**):

(196)

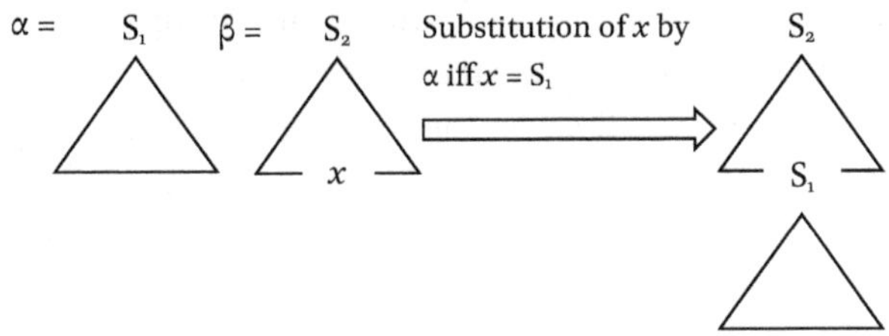

FIGURE 6.9 *Embedding generalised transformation*

S_1 is still the root of the sub-graph α (whose internal structure does not concern us now), which is what matters. Strictly speaking, the diagram above illustrates the *substitution* of *x* by S_1 under identity, because the node *x* is in the frontier of β, and it does not dominate anything else. If *x* dominated some structure, there are additional requirements should the TAG definition of *adjunction* be used (Kroch & Joshi, 1985: 9, ff.; 1987: 111): in this case, the structure is not extended in the same way as it is in Chomsky-Adjunction (such that, for instance, Chomsky-adjoining an XP to VP necessarily extends the VP; see Frank, 2002: 20 for some discussion about differences between TAG-adjunction and Chomsky-adjunction). As we saw in **Chapter 4**, in TAG-style *adjunction*, the structure dominated by *x* in β would be 'pushed downwards' by the adjunction of α, which in turn needs to have a node labelled *x* in its own frontier in order to preserve structural relations.

Now, let us consider a contrast that we mentioned in passing above: that between (190d), repeated here for convenience as (197a), and the new example (197b)

(197) a. *To **none** of the reporters the police seemed to have **any** chance of solving the case
b. To **none** of the reporters did the police seem to have **any** chance of solving the case

We will refer to the process that gives us fronting *without* subject-auxiliary inversion in (197a) as *topicalisation*, and fronting *with* inversion as *focalisation*. This choice of terminology is not innocent: the kind of argument we put forth here contrasts with Rizzi's (1997: 295–296) pertaining to the mechanisms of *topic* and *focus*. In the light of the conditions (195 *i–iii*) above for licensing, we can interpret the contrast in (197), in which an NPI is licensed under fronting + inversion (*focus*) but not under simple fronting (*topic*). Let us discuss this contrast in some more detail.

In his seminal analysis of the structure of the left periphery of the clause, Rizzi (1997: 295) observes that there may be more than a single *topic* per clause, but only one *focus*. He initially considers (but promptly proceeds to reject) the idea that the derivation of these structures differ in that only *topicalisation* involves adjunction, but *focalisation* does not. There are, however, reasons to think that the initial view may be correct for the cases we are analysing. If *topic* involves adjunction, that means that we are dealing with (at least) two distinct sub-trees which are related by the creation of an edge between nodes in the adjoined object and the target of adjunction (in Rogers' 2003 terms, we have a *composite* tree). Furthermore, the target of adjunction in *topicalisation* structures seems to be the root, at least in the cases we have observed here. If this is so, then conditions (195 *i–iii*) are sufficient to filter out NPI licensing from a *topic*. But then what happens with (197b)? Two options are logically possible at this point:

I. *Focalisation* is adjunction, like *topicalisation*, but to a non-root
II. *Focalisation* is not adjunction at all: focalised structures only require a single local graph

Given the fact that we do not have transformations, the choice between hypotheses I and II depends on the answer to the following question: are we in the presence of more than a single derivational space, and thus, more than a single sub-graph? Here, we will argue that we are *not*. Much research pending, the polarity expression in the focalised constituent generates an intrusion effect with *focalisation*, but not with *topicalisation*, such that something that should not be able to license an NPI, *de facto* does. We attribute this intrusion to the possibility that *focus* only involves a single sub-graph to begin with, which would be equivalent to proposing that the specific kind of *focus* that we see in (197b) is *base-generated* (but this does not extend to other kinds of *foci*, see Jiménez Fernández, 2015 for a classification of *foci*, and García Fernández & Krivochen, 2019a for arguments in favour of a multidominance analysis for *verum focus* in Spanish that is compatible with the approach to *focalisation* proposed here, although under different, derivational, assumptions).[11]

---

11 As we have emphasised above, the present framework does not directly pertain to intonation or prosody. However, a tangential piece of evidence could be called upon to strength-

In this chapter we presented a sketch for a treatment of English clausal complement constructions aimed at adequately capturing relevant empirical generalisations while at the same time simplifying the theoretical apparatus. We have also leveraged the analysis of Raising constructions involving a single elementary graph, and looked at aspects of NPI licensing under specific structural conditions. In looking at Raising, some interesting aspects of Spanish Raising to Subject constructions became relevant to the general framework: a predicate like *parecer* is a Raising predicate only in specific contexts (García Fernández & Krivochen, 2019b: 45, ff; see also RAE-ASALE, 2009: § 37.10n). The next section will be devoted to the dynamics of VPs as well, but considering the interaction between auxiliaries and lexical verbs in Spanish. We will also take a look at the way in which dependencies with clitics work across sub-graphs, focusing on the 'transparency' or 'opacity' of parenthetical clauses, which by definition are not derived by the same rule sequence that builds the matrix clause to which they are attached. We will argue that a strictly configurational theory of (non-)monotonicity in the grammar is too restrictive when it comes to predicting possible dependencies across elementary graphs.

---

en the idea that only topics, but not foci, involve more than one sub-graph: in the contrast between (197a) and (197b), only in the former do we have a separate intonational unit for the fronted phrase. This follows directly if we have root-adjunction under a principle like Emonds' (1970: 9): '*a root S immediately dominated by another S is set off by commas*'.

CHAPTER 7

More on Cross-Arboreal Relations: Parentheticals and Clitic Climbing in Spanish

By and large, the theory presented here has been devised as a tool to define explicit maps of dependencies between expressions in English sentences. We have indeed warned the reader that extensions and applications to other languages are not automatic, because the focus of this model is not aprioristic universality, but rather description of specific features of particular natural languages. In the present section we will, however, show that in principle there are problematic aspects of dependencies in Spanish sentences that can be captured using a natural extension of the model that we have sketched so far, an extension that we will then proceed to apply to further English data.

The graph-theoretic approach pursued here is strongly local (recall that all dependencies of a predicate must be expressible within a single irreducible elementary graph; cf. Frank's 2002, 2013: 233 *Fundamental* TAG *Hypothesis*), and at the same time can straightforwardly describe relations across arbores or elementary graphs (that is, relations between nodes which belong to distinct local single-rooted graphs that contain a single lexical predicate). We can take a sub-graph and embed it in another sub-graph, provided that there is a node in the target graph that either dominates or is assigned the same address as a node in the embedded sub-graph. This much is not very different from the mechanism of *substitution / adjunction* in a Tree Adjoining Grammar (Joshi, 1985; Kroch & Joshi, 1985; Frank, 2002, 2013), but because there is no restriction that the target for embedding is the root of a sub-graph if graph composition involves structure sharing, we can speak of *generalised adjunction*, which also sounds rather cool. In principle, any number of graphs can be linked at any node: restrictions over generalised adjunction must be required by the analysis of specific phenomena in specific languages.

We started our inquiry by trying to flesh out the distinction between *relation-changing transformations* and *relation-preserving transformations* (using the terms from McCawley, 1982 introduced in **Chapter 1**): the former modify the distribution of GF (and thus, the configurational relatons between arguments and predicates), whereas the latter only affect linear order (leaving syntactic relations untouched). More generally, we extended the notion of 'relation' such that it applies to local relations in a graph: let *a*, *b* and *c* be nodes in a graph G, and let there be a relation R(*a*, *b*). If we want to establish a connection between

 | DOI:10.1163/9789004542310_008

a and *c*, an RCT would *replace* R(*a*, *b*) by R'(*a*, *c*), in which case we are left with only one relation; whereas an RPT would *add* a new relation R' to the set of relations in G without modifying or disrupting R.

Let us now consider a new situation. Assume that we have a string *abcd* that features the following dominance relations:

(198) $\rho = \{(a, b), (a, c), (a, d), (c, d)\}$

And assume also that there is an embedding operation inserts the substring *efg* at *c*, yielding *ab*[*efg*]*d*. If the insertion of *efg* is implemented as TAG-*adjunction*, then it 'pushes' *d* downwards in the hierarchical structure (Joshi, 1985: 209). Now we have a derived graph, with two elementary graphs: one corresponding to the string *abcd* and another to *efg*; furthermore, we need to assume that either *e*, *f*, or *g* are identical to *c*, in order to respect the conditions on *linking*. Let us assume that $f = c$. A familiar question arises: is *efg* opaque for purposes of operations or dependencies at (i.e., required by an element in) *abd*? That is, can we create an edge from any individual element in the original string to some individual element in the adjoined string? If so, under which conditions?

The answer to this question relates to the more general problem of how to establish relations *across* local syntactic units, a problem that has been at the core of syntactic theorising within both transformational and non-transformational frameworks. In some cases, the problem has been addressed from the perspective of constraints on inputs: if there is a rule *r* that relates syntactic objects, how can we specify the structural description that an expression of the language has to satisfy in order for *r* to apply? For example, *binding principles* can be formulated in such a way, and in general *indexing* mechanisms are approached from this perspective. Standard Theory-style constraints (like the *Tensed S Condition*, below) often adopted the form *rule r cannot apply in configuration C*, which is essentially a condition on the input of *r* (see Müller, 2011 for discussion):

> *Tensed S Condition*:
> *No rule can involve X, Y in the structure*
> $\ldots X \ldots [_{\alpha} \ldots Y \ldots] \ldots$
> *Where α is a tensed sentence* (Chomsky, 1977: 89)

In other cases, it is the *output* of the rule that is under scrutiny. Conditions which are formulated over *outputs* are sometimes referred to as *representational constraints* (e.g., in Müller, 2011), and they apply to the output of *r*, which in turn implies that the structural description for *r* has been met (in other words, that the *input* of the rule was well-formed). An example of such *output* conditions is the *Empty Category Principle* (ECP) as formulated in Chomsky (1981) (see also Lasnik & Saito, 1984):

> $[_\alpha\ e]$ *is properly governed* (Chomsky, 1981: 250)

The ECP was the reason that motivated punctuated (a.k.a. 'successive cyclic') movement in GB, since traces needed to be antecedent-governed and thus *operator-variable* relations needed to be *local* (i.e., within the limits of a cyclic node, S' / NP).

In some cases, it is the rule itself that is subject to a constraint: this is the case with some early economy conditions in Minimalism, including Last Resort:

> *Move F raises F to target K only if F enters into a checking relation with a sublabel of K.* (Chomsky, 1995: 280)

This last situation is the one that will be the least relevant to our argument.

This brief discussion serves as a background for the following questions: What would a condition over dependencies across elementary graphs look like under present assumptions? Does the distinction between RCT and RPT interact with the formulation of these conditions?

We would like to propose, initially, that the insertion of an element in a graph that *only disrupts linear order* without changing constituency does *not* count as defining an arbor. Put differently, the configurations that are allowed under our definition of *licensing* are sensitive to the distinction between RCT and RPT. Thus, we suggest the following preliminary constraints (from Krivochen et al., in preparation):

(199) [*X* ... [... *α* ...] ... *Y*] *yielding R*(*X*, *Y*) *is a legitimate configuration iff*
   i. [... *α* ...] *has been introduced non-monotonically,* **and**
   ii. *The rule introducing* [... *α* ...] *is an order-changing rule* (*in the sense of McCawley, 1982: 94*) [a *relation-preserving transformation* in the terms we use here]*; that is, there is no element in α that either dominates or is dominated by a non-root node in the target of embedding*

We will proceed to exemplify how these licensing conditions work now, focusing our discussion on legitimate relations between argumental *clitics* and the predicates that select them under specific syntactic operations: *clitic climbing* and *parenthetical insertion*.

Let a be a node corresponding to an argumental clitic and let b be its governing V; thus, we have the following relation: R(a, b) = R(Clitic, V). Now we need to ask: what exactly is R? Here we argue that a clitic-governing V structure is modelled in terms of *dominance*, thus, something like *hacerlo* (Lit. do_{INF}+it_{CL}) displays the dominance relation ρ(hacer, *lo*), where the accusative clitic is the direct object of the verb *hacer*. We will leave discussion pertaining to the subject aside for now, let it be a ghost arc PRO_{arb} or some such (as in *hacerlo es impensable* '(for anyone) to do it is unthinkable') simply for expository purposes (but see **Section 14.5**). In RG/APG terms, we would have something along the lines of the following preliminary graphical representation:

(200)

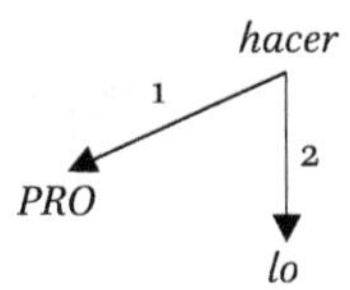

FIGURE 7.1 Graph-theoretic analysis of Spanish infinitive with object clitic

Or, using our ordered notation, which also conveys information about the GF assigned to each nominal ((200) and (200') are completely equivalent):

(200') ρ = ⟨(hacer, PRO), (hacer, *lo*)⟩

These considerations hold at a very local level, and for the syntactic representation of predicates and their selected arguments: recall that elementary graphs are the units of argument structure. Looking at the issue globally, we need to consider V-clitic relations in a wider syntactic context, and constrain the possible dependencies that clitics can establish with otherwise *in abstracto* potential hosts. To this end, we need to make some assumptions explicit. Here it is useful to adopt a mildly derivational standpoint for expository purposes: let us assume for a while that structural descriptions are indeed obtained by procedural means, via the ordered application of syntactic operations (e.g., concatenation). To begin with, we will assume that a clitic is *ordered* (via immediate or transitive dominance) with respect to *all* other nodes within its *elementary graph*, following the definition of *order* in **Chapter 2**, to be substantially expanded on in **Chapter 8**. We assume that it is the ordering imposed over nodes in a graph that allows for a componential interpretation to be assigned to those nodes. If a clitic is not ordered with respect to a given node v, that node v (which may be the root node of an auxiliary tree) does not disrupt any relation within the minimal sub-graph containing the clitic and its *closest* host (where *distance* is measured as the *number of nodes and edges* between two nodes).

In a derivational system, how can we account for the opacity of certain adjoined domains? One possibility is that, by definition, all elements within a derivational cascade (Uriagereka, 2002) are accessible to any other element within the same cascade, which is strictly ordered. If elements within a parenthetical are not accessible to certain elements this means, as a provisional observation, that parentheticals cannot be ordered with respect to the rest of nodes in a structural description. Such a derivation is not uncommon, and indeed has been proposed in the literature: see for instance approaches along the lines of Emonds (1979), McCawley (1982) who introduce parentheticals via post-cyclic rules. Similarly, de Vries (2007), Kluck (2015), and others, propose introducing parentheticals via a kind of Merge called *par Merge* the output of which does not dominate its input (and thus the parenthetical cannot be *d-ordered* with respect to its host; Ott, 2016: 36 refers to this as 'inclusion without dominance', however that could be formalised). Ross (1973) may be seen as a precursor to these analyses, introducing parentheticals via a transformational rule known as *slifting* (we will come back to this issue in more detail in **Chapter 8**). If the material within the parenthetical is not part of a strictly ordered derivational current, then it cannot enter dependencies with nodes in the matrix sentence. In turn, this entails that any node contained in a parenthetical should not count as intervening nodes for clitic climbing, because the clitic would not be ordered with respect to any of the nodes that constitute the parenthetical by virtue of it not being part of the same structural unit as the clitic. This is a strong prediction, and indeed verified empirically:

(201) a. */?? *Los hermanos se= la= dejan a Ana*
The brothers CL.3Sg.DAT= CL.3Sg.ACC= let PRT Ana
preparar e algunas veces
prepare.INF *e* some times
'Her brothers let Ana prepare it sometimes' (example from Emonds, 2007: 200; judgment ours)

b. Los hermanos se la$_i$ dejan, a Ana, preparar e_i algunas veces

(201) can only be acceptable if *a Ana* is introduced by an order changing rule in McCawley's terms (observe that in the acceptable (201b) *a Ana* appears

between commas, as a parenthetical). If this is the case, the clitic is ordered with respect to *dejar* and *rammati*, but *not* with respect to *a Ana*. Therefore, the annotated string (201b) complies with the requirements in (195) over legitimate licensing configurations, and is thus predicted to be grammatical.

However, this is not (it cannot be) the full story: what counts as a 'parenthetical' (in the sense of 'adjoined opaque domain', whose opacity is, in traditional IC-PSG terms, given by the fact that they are neither subcategorised for nor monotonically assembled) is not always clear. Under standard MGG assumptions about the derivation of parentheticals (see e.g. Ott, 2021 for a derivational approach based on the opacity of parentheticals), we would expect that (202b) below would not work but, somewhat surprisingly, it does:

(202) a. *Juan puede, en realidad {tiene que / debe},*
J. may.3Sg.PRES, in reality {has to / must},
hacer=lo
do.INF=CL.3Sg.ACC
'J. may do it, actually he has to / must'

b. Juan puede, en realidad lo {tiene que / debe}, hacer

c. *Juan lo puede, en realidad {tiene que / debe}, hacer

The reason why we would expect (202b) to be ungrammatical under a more or less standard MGG view of what a 'parenthetical' is that, if *en realidad tiene que* is derived in parallel and inserted via some sort of *generalised transformation* (including Late Merge, Pair Merge, and similar proposals; we group together a number of approaches based on the fact that we have insertion of a structure into another structure) into the matrix clause *Juan puede hacerlo*, then a node belonging to this adjoined sub-structure should *not* be a suitable host for the accusative clitic *lo* (the direct object of the verb *hacer* 'to do') for either of two reasons. First, the clitic cannot move *into* the domain derived in parallel because it is opaque for operations triggered outside the parenthetical: see Chomsky (1973) for a Subjacency-inspired view; Uriagereka (2002) and Cerrudo Aguilar (2015) for a perspective based on parallel derivations and Multiple Spell-Out; De Vries (2007, 2009b) presents an analysis of parentheticals as analogous to *amalgams*, as forms of *external re-Merge* (Merge of a syntactic object embedded under a root R with a syntactic object with root R' not embedded in R). In (202a) the parenthetical is not evidently 'anchored', in Kluck's (2015) terms. Second, if the clitic is base-generated inside the parenthetical, it cannot be thematically interpreted as an argument of the lexical verb in the matrix clause (because opacity is a double edged sword: nothing comes in, nothing goes out). It is important to note that we are dealing with two distinct problems: one pertains to how parentheticals are linked to matrix sentences, the other to the internal structure of parentheticals and their opacity. McCawley and Emonds shed light on the former issue, but not directly on the latter (see Potts, 2002 for a movement analysis of *as*-parentheticals in English, which deals with issues of islandhood). Further refinement of the analytical machinery is thus required.

In this context, following McCawley and speaking of 'order-changing rules' seems to be a felicitous choice, in the light of the following contrasts pertaining to the positional freedom of the relevant sub-graphs:

(203) a. Los hermanos se$_i$ la$_j$ dejan, a Ana$_i$, preparar e_j algunas veces
b. Los hermanos se$_i$ la$_j$ dejan preparar e_j algunas veces, a Ana$_i$
c. A Ana$_i$, los hermanos se$_i$ la$_j$ dejan preparar e_j algunas veces
d. Los hermanos, a Ana$_i$, se$_i$ la$_j$ dejan preparar e_j algunas veces

(204) a. Juan puede, en realidad lo tiene que, hacer
b. Juan puede hacerlo, en realidad {*tiene que / debe}
c. Juan lo puede hacer, en realidad {*tiene que / ??debe}
d. *Juan puede hacer, en realidad lo {tiene que / debe}

In contrast to the positional freedom that we see in (203) with respect to the possible sites where the parenthetical *a Ana* can appear in the string, it seems to be the case that the rule that introduces *en realidad (lo) tiene que* into the matrix sentence *Juan puede hacerlo* in (204) is more than simply an order-changing rule: the possible adjunction sites are more restricted.[1] One way of looking at this pays close attention to lexical semantics: the *locus* of parenthetical adjunction depends on whether there is a semantic relation between the adjoined domain and the matrix clause; specifically, we can look at the auxiliaries in the adjoined phrase and the matrix clause, since these are the possible hosts for clitic climbing and since the parenthetical seems to be a 'correction' affecting

1 Note that in (204) we have used two modals: *tener que*, which contains what García Fernández et al. (2020) call an *intermediate element*, and *deber*, which does not (see fn. 5 in this chapter). The reason is that intermediate elements (such as *que*) resist being clause-final, and therefore (204b) is ungrammatical for reasons other than the relation between the clitic and its host. In (204b), a modal with no intermediate element works fine, but (204c) and (204d) are very marginal and ungrammatical respectively regardless of whether the modal has an intermediate element or not.

the modal in the matrix clause. In the specific case of (204), the modal auxiliary in the adjoined clause (*tener que / deber*) is related to that in the matrix clause (*poder*) on a scale, such that *tener que* means deontic obligation, whereas *poder* only means deontic possibility, a weaker notion than obligation (Kratzer, 1991; Brennan, 1993; Bravo, 2016b): whereas possibility quantifies existentially over possible worlds, necessity quantifies universally (Lyons, 1977). In this sense, linear order is relevant: it represents the scale along which the meanings of the modals are ordered, from 'weak' to 'strong'. When there is no scalar relation between the auxiliaries involved (because, say, we have a modal auxiliary *poder* and a phasal aspectual auxiliary *empezar a*, which belong to different semantic scales), the construction ceases to be acceptable:

(205) *??Juan puede, en realidad lo= empieza a,*
J. may.3Sg.PRES, in reality CL.3Sg.ACC= start to.3Sg.PRES,
hacer
do.INF
'Juan may do it, actually, he starts to do it'

But even this slightly more nuanced account cannot be the whole story, even though it does give a partial account of the kinds of auxiliaries that can be adjoined in this manner. We still have unanswered questions, even in the descriptive front. Structurally, why is the parenthetical *a Ana* in (203) not an intervening object between the 'gap' and the coindexed clitic, but the parenthetical containing the auxiliary in (204) *is*? A lexically-oriented proposal can shed some light on where the parenthetical can appear in terms of scalar relations between auxiliaries in the matrix clause and the parenthetical, but it says nothing about syntactic dependencies between objects in the matrix clause and objects within the parenthetical (other than the root). Of the two problems we identified above, the second (the opacity of some units for purposes of relations with objects in other arbores) remains unaddressed; we will now focus on it.

Let us go back to the contrast between (203) and (204): when does an embedded arbor 'get in the way' of links in the matrix one? The possibility we will explore here is that if the adjoined graph is a 'self-contained' unit, *it does not count as intervening* for operations at the target of adjunction. We may rephrase this, appealing to TAG terminology: if an Auxiliary Tree (AT) is a *self-contained unit*, it does not intervene for purposes of operations at the Initial Tree to which the AT is adjoined (Joshi's 1985: 214 expansion of TAGs by means of *links* preserved after *adjunction* yields results that are similar to ours in weak generative capacity). Of course, making what 'self-contained' means explicit is paramount. We define the notion as follows:

(206) ***Self-containment (definition)***
A graph G is a self-contained syntactic object iff $\nexists(v_i)$, $v_i \in G$ *such that*
i. $\langle v_j, v_i \rangle \in \rho^*$ *holds for* $v_j \in G'$ *and* $G' \neq G$, ***and***
ii. v_i *receives a grammatical function* GF *in G*

That is: a graph is self-contained if it does *not* contain any node corresponding to an argument (i.e., a node that establishes with a predicate in G one of the relations in (110)) that is dominated by a node outside that graph (i.e., an argument of a predicate in G', where G' ≠ G). Note that *self-containment* is the notion that we appealed to informally in condition (195 *iii*) for licensing above, which we repeat here for convenience:

> *There is at least one node address in G that is identical to a node address in G'*

In this context, then, the final version of licensing conditions is as follows, which now includes condition (195 *iii*) (in **Chapter 6**) formulated in terms of *self-containment* (a notion that we need to define anyway, for independent reasons):

(207) ***Licensing (final formulation)***
Let G and G' be sub-graphs and v_i *and* v_j *be nodes. Then,* $v_i \in G$ *licenses* $v_j \in G'$ *iff*
i. $\langle v_i, v_j \rangle \in \rho^*$, *and*
ii. *G' is not adjoined to the root of G, and*
iii. ***Neither G nor G' are self-contained***

If there is *no* dominance relation, then the nodes in the 'self-contained' subgraph are *not strictly ordered with respect to the nodes at the target of adjunction*, because there is no walk communicating those. An order defined on the set of nodes in the derived graph will thus *not* be *total*: this accounts for the positional freedom of most free parentheticals (including the one in (203)). We can give the relevant set of *local* dominance relations for the matrix and parenthetical clauses in (204a) as follows (recall that Spanish modals are lexical predicates, and as such anchors of their own EGs):

(208) ρ_{EG1} = ⟨(poder, Juan), (poder, lo), (poder, hacer), (hacer, Juan), (hacer, lo)⟩
ρ_{EG2} = ⟨(tener que, Juan), (tener que, lo), (tener que, hacer), (hacer, Juan), (hacer, lo)⟩

The syntactic object which corresponds to EG 2 in (208), is not 'self-contained' in the sense that it dominates a node that is also dominated by an element in another domain, in this case, the clitic *lo* is dominated by *poder* in elementary graph 1, as well as *hacer*, which belongs to *both* elementary graphs. This follows the lines of the multidominance approach to RNR and other rightwards extractions (McCawley, 1982, 1987, 1998; see also **Section 14.1** below). This last claim requires some unpacking: were we working within a transformational framework, the D-Structure / pre-transformational representation of (204a) would need to look like (204') to get the correct interpretation:

(204') a. Juan puede hacerlo, en realidad tiene que hacerlo (*D-Structure*)

In this case, we are interested in the fact that the clitic is a node which belongs to two subgraphs playing an argumental role in both of them (it is the direct object of the lexical verb *hacer*), but the clitic is part of a bigger sub-graph, which is *not* self-contained: the arc *e*⟨hacer, *lo*⟩. In other words: the sub-graphs are *linked* at the argumental clitic *lo*. That is not the case with the node with address ⦃Ana⦄, which *is the multidominated node itself*: because it is not a node that links two or more graphs by being a proper subset of them, but is an adjoined graph in and of itself, it is thus free to 'move around', changing just the linear order between nodes (but *not*, we stress again, grammatical relations).

7.1 Discontinuity and Clitic Climbing in Spanish Auxiliary Chains[2]

A crucial point is that so-called 'clitic climbing' (discontinuity relations between clitics and their governing verbs) occurs across sequences of verbal predicates under specific conditions (for a transformational account, see Rivas, 1974; for a Relational Grammar analysis which is now classic, see Aissen & Perlmutter, 1983). In this section we will focus on clitic climbing through *auxiliary chains* (in the sense of Bravo et al., 2015; García Fernández et al., 2017 and related works). Here we use the expression *clitic climbing* in a purely descriptive manner, to denote strings in which a clitic's morpho-phonological host is not the lexical predicate that takes it as an argument (without implying that the

2 Much of the data and discussion in this section is adapted and developed from Krivochen & García Fernández (2022). We are grateful to Luis García Fernández for allowing us to use that material here.

clitic has literally *moved* from one position to the other; see Ordóñez, 2012 for a general perspective). Some background on auxiliary chains is necessary before our discussion. As classically used for Spanish, these terms refer to sequences of one or more auxiliary verbs and a non-finite form of a lexical (or "main") verb, giving rise to a single predication and within the limits of a single clause (RAE-ASALE, 2009: § 28.5). Descriptively, an auxiliary chain is any verbal periphrasis in which there are at least two auxiliary verbs; the presence of an auxiliary chain presupposes, of course, the presence of a lexical VP. In previous works we defined an 'auxiliary chain' semi-formally as follows:

> *An auxiliary chain* CH_{AUX} *is a string* $\{\{x^\frown y^\frown z \ldots n\}^\frown VP\}$ *where*
> i) $\{x, y, z \ldots n\} \in$ *Auxiliary Verb*
> ii) $n > 2$
>
> BRAVO ET AL., 2015: 75

Defining the set of Spanish auxiliary verbs is no easy feat. García Fernández (2006) lists approximately 100 auxiliaries, but a revision of the criteria for auxiliarihood in García Fernández & Krivochen (2019b) results in only 60. What counts as an auxiliary depends to a great extent on the criteria considered by different grammatical traditions, and it is not our place to revise those criteria here. Therefore, we will operate with a definition of *auxiliary* that follows the more restricted framework of García Fernández & Krivochen (2019b).

The focus of this section will be set on the interaction between clitic climbing, which we looked at in the previous section, and structures where a single auxiliary takes coordinated VPs as its complement. Consider now the following examples featuring coordinated auxiliaries, coordinated lexical verbs, and clitic climbing, where traces have been added for expository purposes only:

(209) a. *Podrías gustar=le y decir=lo a todo el mundo*
Could.2Sg.COND like.INF=CL.3Sg.DAT and say.INF=CL.3Sg.ACC to all the world
'He/she could like you and you could tell it to everybody'

b. *Le$_i$=podrías ser t_i infiel y decir=se=lo a todo el mundo*
CL.3Sg.DAT=could.2SG.COND be.INF t_i unfaithful and say.INF=CL.3Sg.DAT=CL.3Sg.ACC to all the world
'You could be unfaithful to him/her and tell it to everybody'

c. **Lo*$_i$*=podrías* *ser=le* *infiel* *y*
CL.3Sg.ACC=could.2SG.COND be.INF=CL.3Sg.DAT unfaithful and
decir t_i *a todo el mundo*
say.INF t_i to all the world
(the clitic climbs to the auxiliary from the second term of the coordination; the result is ungrammatical)

(210) a. *Nos=estás* *molestando y* *mirando*
CL.1Pl.ACC=be.AUX.2SG.PRES bother.GER and look.GER
'You are bothering us and looking at us'

b. **Estás* *molestando=nos* *y* *mirando*
Be.AUX.2SG.PRES bother.GER=CL.1Pl.ACC and look.GER
(intended reading: same as (208a))

c. **Estás* *molestando y* *mirándo=nos*
Be.AUX.2SG.PRES bother.GER and look.GER=CL.1Pl.ACC
(intended reading: same as (208a))

Let us describe the data. In (209a) we have the auxiliary *poder* followed by two coordinated infinitives, *gustar* and *decir*, each of which hosts a clitic which corresponds to its internal argument. (209b) features the clitic which depends on the first terms of the coordination having climbed above *poder*, whereas the second term of the coordination has both accusative and dative clitics *in situ*. The contrast between (209c) and (209d) is particularly interesting, since it shows that there is no problem in distributing an *auxiliary* between coordinated infinitives (such that (209c) means *estás molestándolo y estás mirándonos* 'you are bothering him and you are looking at us'), but it is *not* possible to distribute a *clitic* between coordinated terms. As far as the second set of sentences goes, in (210a) we have the progressive auxiliary *estar* followed by two coordinated infinitives but, unlike the examples in (209), in (210) there is only one clitic proclitic to the auxiliary which is nevertheless interpreted distributively with respect to the coordinated infinitives (as can be seen in the English translation). The analysis presented in Krivochen & García Fernández (2022) explains why some elements can be distributed across coordinated lexical verbs in periphrastic constructions while others cannot, from a derivational standpoint. Our goal here is to provide adequate structural descriptions for these sentences which capture their syntactic and semantic properties within a non-derivational framework.

An important constraint to bear in mind is that it is impossible to make two clitics climb from different terms of the coordination. (211) illustrates the result of making a clitic from each term climb, yielding an ungrammatical sentence:[3]

(211) *Se_i=lo_j=podrías ser t_i infiel y
CL.3Sg.DAT=CL.3Sg.ACC=could.2Sg.PRES.COND be t_i unfaithful and
decir t_j *a todo el mundo*
say t_j to all the world

These restrictions have nothing to do with lexically governed processes pertaining to either the auxiliaries or the lexical verbs (it is also worth pointing out that argumental and non-argumental clitics behave exactly the same for all present intents and purposes), since both (212a) and (212b) below are well-formed as individual sentences:

(212) a. Le_i=*podrías ser* t_i *infiel*
CL.3Sg.DAT=could.2SG.COND be.INF t_i unfaithful
'You could be unfaithful to him/her'

b. Lo_i=*podrías decir* t_i *a todo el mundo*
CL.3Sg.ACC=could.2Sg.COND say.INF t_i to all the world
'You could say that to all the world'

The relevant condition that needs to be invoked to account for the ungrammatical cases is, we argue, Ross' (1967) *Coordinate Structure Constraint* (CSC):

> *In a coordinate structure, no conjunct may be chopped, nor may any element contained in a conjunct be chopped out of that conjunct.* (Ross, 1967: 428)

3 It is relevant to note that there is no *a priori* constraint on making clitics climb from different predicates, as seen in (ii):
(i) *Les=deja traer el diccionario al examen*
CL.3Pl.DAT=let.3Sg.Pres bring.INF the dictionary to.the exam
'He/she lets them bring the dictionary to the test'
(ii) *Se=los=deja traer al examen*
CL.DAT=CL.3.Pl.ACC=let.3Sg.Pres bring.INF to.the test
'He/she lets them bring it to the test'
The relevant constraint applying to (211) must thus pertain to the structure of coordination, not to the fact that there is CC from two different predicates. Incidentally, *los* manifests plural number due to the fact that *se* cannot inflect in number; in certain varieties, like ours (River Plate Spanish), that forces the number features of the indirect object migrate to the accusative clitic.

It is necessary to state this more precisely, because the CSC has been understood in at least two different senses, not always yielding equivalent results. As correctly pointed out by Postal (1998: 83),

> *It seems correct to divide Ross's original formulation of the CSC into separate principles. The one I called the Conjunct Constraint* [...] *forbids the extraction of coordinate conjuncts themselves. The other, the CSC, bans (non-ATB) extraction from true conjuncts.*

Here, we use the term 'CSC' to refer strictly to the *second* of Postal's principles: non-ATB extraction from *true* conjuncts is to be filtered out.[4] We will come back to the mechanisms of ATB dependencies in **Section 14.1** in the context of the discussion of RNR, but it is worth introducing some aspects of that discussion here.

Let *R* be a relation between nodes *a, b, c, d* in G and G' such that R(*a, b*) holds in G and R(*c, d*) holds in G', where *a* and *c* are lexical predicates. Then, we say that *R* is an ATB relation in G" *iff b* = *d*. Crucially, *b* and *d* are assigned GFs in G and G' respectively. This formulation is rather clumsy, however. We can simplify things by appealing to the notion of *linking* defined in (69) above, and repeated here:

> *If a graph G contains* ⦃v_1⦄ *and a graph G' contains* ⦃v_1⦄, *and G and G' belong to the derived graph G", then G and G' are* ***linked*** *at* v_1 *in G"*

To capture ATB rule application, we need to further specify (i) that v_1 needs to receive a GF in G and G', and (ii) that G and G' are sub-graphs of G". Furthermore, it is possible that the GFs must be the same, assuming that dependencies across coordinated structures occur under the condition that those structures be parallel (in the sense of Goodall, 1987). We have no need of indexing operations to keep track of what has 'moved' where, and of course the multiplication of copies is also avoided: an issue when considering traditional movement-based approaches to ATB operations is that for *n*-coordinated terms affected, *n*-1 occurrences need to be deleted. Thus, in order to get *what did John buy and Mary break?*, there has to be a way to get two NPs in object position (of *buy* and *break*), transform them into *wh*-phrases, and then move only one to the left periphery, deleting the other. Or, alternatively, leave the object NPs *in situ*,

4 Ross (1967: 176) first defined ATB phenomena as a class of rules that affect (e.g., displace, delete) identical constituents of all the conjuncts of a coordinate structure at once.

lexically insert a *wh*-phrase in the left periphery of the clause, assign the same index to both object NPs and the *wh*-phrase, and delete the NPs. Other options are also possible, but they all involve multiplying indexes, or occurrences, and require *ad hoc* selection and deletion rules. We will return to some of these considerations in **Section 14.1**, for the time being it is worth noting that if we allow for multidominance in maximally connected local graphs the conditions that account for the relevant data involving clitic climbing in coordinated auxiliary verb constructions can be captured without the need to invoke any further principles or operations.

If we see clitic climbing as an operation that creates new edges in addition to the primitive relation between the lexical V that assigns the clitic a GF, this should also apply to climbing through elementary graphs. The only constraint to bear in mind is the CSC in Postal's sense: non-ATB climbing is not allowed. Thus (213a, b), in which clitic climbing does not affect elements in both conjuncts, are correctly excluded (traces are expository only):

(213) a. *Lo$_{\text{ACC}}$ podrías serle$_{\text{DAT}}$ infiel y podrías decir $t_{\text{CL.ACC}}$ a todo el mundo
b. *Se$_{\text{DAT}}$ lo$_{\text{ACC}}$ podrías ser $t_{\text{CL.DAT}}$ infiel y podrías decir $t_{\text{CL.ACC}}$ a todo el mundo

In these examples, as well as in (211) above, the CSC strictly understood is violated, and the results are ungrammatical. So far so good. However, can we say something about how the complex *grammatical* cases are analysed? After all, in the grammatical cases in (209) and (210) clitic climbing is not all that is going on, there is (descriptively) also *gapping* of the auxiliary verb in the sense that the auxiliary is interpreted as affecting both terms of the VP coordination. As pointed out above, the corresponding interpretations for (209a) and (210a) above are (209'a) and (210'a) respectively (the position of the clitic, either proclitic to the finite auxiliary or enclitic to the non-finite verb, makes no interpretative difference):

(209') a. Podrías gustarle y podrías decirlo a todo el mundo
'He/she could like you and you could say that to all the world'

(210') a. Estás molestándonos y estás mirándonos
'You are bothering us and you are looking at us'

The auxiliaries *poder* and *estar* are distributed with respect to the lexical verbs: a structural description in which the auxiliary only occurs in the first conjunct cannot adequately represent the semantic relations between expressions in the sentence, because the auxiliaries modify the verbs in both terms of the coordination (see Krivochen & García Fernández, 2022 for further discussion about the syntax and semantics of these sentences). We should be able to provide an adequate structural description without multiplying the occurrences of the auxiliaries or the clitics. Let us, then, specify the ρ-sets for (209a) and (210a), leaving aside for the time being the structure of coordination (to which we will come back in **Chapter 12**):

(214) ρ_1 = ⟨(poder, *pro*), (poder, gustar), (gustar, *pro*), (gustar, *le*)⟩
ρ_2 = ⟨(poder, *pro*), (poder, decir), (decir, *pro*), (decir, *lo*)⟩

(215) ρ_1 = ⟨(estar, *pro*), (estar, molestar), (molestar, *pro*), (molestar, *nos*)⟩
ρ_2 = ⟨(estar, *pro*), (estar, mirar), (mirar, *pro*), (mirar, *nos*)⟩

Elementary graphs 1 and 2 in (214), where each term of the coordination has its own clitics which are not correferential, are linked at the auxiliary *poder*; in this case there is no need to involve *deletion* in *gapping*: the operation is licensed when two elementary graphs are linked at node that corresponds to a verbal expression (in this case, an auxiliary). It is worth pointing out that, because the clitics are not correferential, climbing is not licensed since it is not possible to make it happen ATB; in other words: because the arbores are *not* linked at the node that corresponds to the clitic's address, however we choose to formulate the operation Clitic Climbing, it cannot possibly affect both terms of the coordination (see also (212)).

The analysis for (215) is slightly more complex, because the arbores are linked at *two* nodes: *estar* and *nos*. We predict, then, that it is possible to apply Clitic Climbing *across the board*, as well as to have *gapping* of the auxiliary. In other words, we predict that (216) below (= (210a)) should be grammatical and furthermore that it should indeed be interpreted as (210'a), with both the auxiliary and the clitic being distributed over the conjuncts:

(216) *Nos=estás* *molestando* *y* *mirando*
CL.1PL.ACC=be.AUX.2SG.PRES bother.GER and look.GER
'You are looking at us and bothering us'

Both of these predictions are borne out, without the need to invoke additional conditions or theoretical tools: the notion of *linking* can naturally capture the requirement for cross-arboreal operations to apply ATB in the relevant cases.

As a summary, we have given arguments to show that our graph-based model can provide adequate descriptions for at least a set of expressions belonging to languages other than English; the present section has focused on some aspects of the dynamics of auxiliaries and clitics in Spanish. From this it must not be inferred that we are giving our conditions the status of universals, or that they apply to any additional phenomena: these are empirical claims, and as such cannot receive an *a priori* answer. Continuing in this line, we will turn to some fundamental issues about the syntax of sequences of auxiliaries, which will reveal the need to allow for distinct structural descriptions for English and Spanish auxiliary chains. This is but an example of a more general argument against universal structural uniformity.

7.1.1 *Two Classes of Auxiliaries in Spanish*

The analysis of examples featuring auxiliaries in the previous section, preliminary as it was, required us to be able to articulate relations within and across elementary graphs: it is not a novel claim that some auxiliaries have a syntactic behaviour similar to Equi and others are closer to Raising verbs (see e.g. Bosque, 2000; García Fernández, 2006 for arguments based on Spanish data). However, the full extent to which the label 'auxiliary' encompasses syntactically heterogeneous objects is rarely addressed. In their analysis of sequences of auxiliaries, Bravo et al. (2015) and subsequent work identified Spanish auxiliaries which could both modify other auxiliaries and be modified themselves (dubbed *lexical auxiliaries*) and auxiliaries which could only modify, but not be modified (dubbed *functional auxiliaries*). This complex articulation of sequences of auxiliary verbs contrasts with the traditional perspective, according to which auxiliaries are always modifiers, but never modified (or, in other terms, always 'auxiliate', but are never 'auxiliated' themselves), found—implicitly or explicitly—in works like Alarcos Llorach (1994), Gómez Torrego (1999), and RAE-ASALE (2009: § 28), among others. The set of auxiliaries that can be modified by another auxiliary includes *modals, phasal aspectual auxiliaries*, and iterative *volver a*; the set of pure modifiers includes *progressive* ⟨*ser* + gerund⟩, *perfective* ⟨*haber* + participle⟩, *temporal-aspectual* ⟨*ir a* + infinitive⟩, and some more. The definition provided in Bravo et al. (2015) and Krivochen & García Fernández (2020: 157) summarises this state of affairs:

> *In a chain of auxiliaries* CH_{AUX} $\{\{x^\frown y^\frown z \ldots n\}^\frown VP\}$
>
> (*i*) LEXICAL AUXILIARY VERBS *can modify other auxiliaries and the lexical VP and be modified themselves by other auxiliaries in a chain.*
>
> (*ii*) FUNCTIONAL AUXILIARY VERBS *can only modify lexical heads (auxiliaries or main verbs), but cannot themselves be modified (i.e., these can only be functors).*

TABLE 1 Lexical and functional auxiliaries in Spanish

Transparent / functional	Opaque / lexical
Progressive ⟨*estar* + gerund⟩ (Eng. *Be V-ing*), perfective ⟨*haber* + participle⟩ (*Eng. Have V-en*), ⟨*ir a* + inf⟩ (Eng. future tense *will*), ⟨*acabar de* + inf⟩ (in its 'recent past' reading; Eng. *Have just -en*)	Phasals (⟨*empezar a / comenzar a* + inf⟩ 'to start'; *terminar de / acabar de* + inf 'to finish'; ⟨*continuar / seguir* + ger⟩ 'to keep -ing'), positionally unrestricted modals (⟨*tener que* + inf⟩ 'to have to'; ⟨*poder* + inf⟩ 'to be able to/ to be allowed to'; ⟨*deber* (*de*) + inf⟩ 'to have to'); scalars (⟨*llegar a* + inf⟩ 'to go as far as to', ⟨*acabar* + ger⟩ 'to finish by -ing'); first-position auxiliaries (⟨*soler* + inf⟩ 'to be accustomed to -ing', ⟨*haber de* + inf⟩ 'to have to'); ⟨*haber que* + inf⟩ 'it is necessary to'; ⟨*tardar en* + inf⟩ 'to take (time) to'; ⟨*volver a* + inf⟩ 'to do sth again'.

Table 1 illustrates (not exhaustively) the members of each class, based on the aforementioned works.

The distinction between lexical and functional auxiliaries is *empirically* motivated (based on the possible relations of modification between members of an auxiliary chain), but it has far-reaching consequences for theories about the format of phrase markers. These aspects have been analysed in detail in previous work from both Phrase Structure grammars and (pure) Categorial Grammar perspectives (Bravo et al., 2015 and Krivochen & Schmerling, 2022, respectively). We can briefly summarise the argument against a strictly monotonic (cartographic or not) structural description for Spanish auxiliary chains here. The difference between lexical and functional auxiliaries is illustrated in (217) below, where lexical and functional auxiliaries are marked as such using L(exical) and F(unctional) subscripts:[5]

---

5 See García Fernández et al. (2020) for extensive discussion and justification of why *que* is grouped with *tener* in the modal *tener que*, *a* with *ir* in *ir a*, etc. Summarising, these elements are not prepositions or complementisers, despite their morphological exponent: García Fernández et al. call them 'intermediate elements'. Intermediate elements in verbal periphrases fall into three categories, based on a number of syntactic tests (e.g., clitic climbing, VP fronting, *wh*-movement, possibility of having complements of more than a single syntactic category):

i. Syncategorematic intermediate elements (the ones in ⟨*terminar / acabar de* + infinitive⟩, ⟨*empezar / comenzar a* + infinitive⟩, ⟨*tardar en* + infinitive⟩, ⟨*deber de* + infinit-

(217) a. *Juan va* a_F *poder*$_L$ *empezar* a_L *trabajar*
Juan go-to.AUX.3SG.PRES be-able-to.INF start.INF work.INF
allí
there
'J. will be able to start working there'

b. *Juan suele*$_L$ *estar*$_F$ *trabajando*
Juan AUX.3SG.PRES.HAB be.AUX.PROG.INF work.GER
'J. is usually working'

c. *Juan está*$_F$ *debiendo*$_L$ *llegar a tiempo*
Juan be.AUX.3SG have-to.GER arrive.INF on time
'J. is$_{Prog}$ under the obligation to arrive on time'

In (217a), what is temporally anchored by the future auxiliary ⟨*ir a* + infinitive⟩ is the deontic obligation denoted by ⟨*poder* + infinitive⟩, not the inchoative aspectual *empezar a* or the lexical verb *trabajar*. The future obligation, in turn, pertains to the start of the event of working; that is, *va a tener que* modifies *empezar a*, which in turn modifies *trabajar*. However, ⟨*poder* + infinitive⟩ and ⟨*ir a* + infinitive⟩ do not modify *trabajar*: we can see this from the lack of entailment relations indicated with ⇏ in (217a'):

---

ive⟩, etc.). Auxiliaries followed by these intermediate elements allow VP fronting (they always get fronted *with the auxiliated V*, there being an independent ban on stranding in Spanish outside of RNR), *wh*-movement and clitic climbing, but only allow non-finite forms as a complement.

ii. Categorematic intermediate elements (the ones in ⟨*empezar / comenzar por* + infinitive⟩, ⟨*terminar / acabar por* + infinitive⟩). The auxiliaries that they appear with allow clitic climbing and (marginally) *wh*-movement, but not VP fronting. Also, they can take more than a single kind of complement (specifically, either non-finite verbs as well as NPs) which suggests a prepositional nature.

iii. Forming multi-word basic expressions (the most grammaticalised ones: e.g., ⟨*ir a* + infinitive⟩, ⟨*tener que* + infinitive⟩, ⟨*haber de* + infinitive⟩). These auxiliaries allow clitic climbing as well as *wh*-movement, but not VP fronting. For all syntactic intents and purposes, these auxiliaries are multi-word basic expressions, and the intermediate elements is best analysed as morphologically part of the auxiliary. This does not prevent them from undergoing (obligatory) Wrap:

(i) ¿Qué tiene Juan que hacer? (via RWRAP(tener que, Juan) = tener⁀Juan⁀que)
'What must Juan do?'

(ii) *¿Qué tiene que Juan hacer?

See **Chapter 3** for related discussion.

(217a') *Juan va a* *poder* *empezar a* *trabajar*
Juan go-to.AUX.3SG.PRES be-able-to.INF start-to.INF work.INF
'Juan will be able to start working'

⇏ *Juan va a* *trabajar*
⇏ *Juan* go-to.AUX.3SG.PRES work.INF
'Juan will work'

⇏ *Juan va a* *empezar a* *trabajar*
⇏ Juan go-to.AUX.3SG.PRES start-to.INF work.INF
'Juan will start working'

⇏ *Juan va a* *poder* *trabajar*
⇏ Juan go-to.AUX.3SG.PRES be-able-to.INF work.INF
'Juan will be able to work'

If the structure assigned to a chain of auxiliary verbs was uniformly monotonic, as seems to be the case of English chains, whose rigid order has been extensively documentes (see, e.g. Chomsky, 1957; Ross, 1969; Quirk et al., 1985; Bjorkmann, 2011; Harwood, 2014; Ramchand & Svenonius, 2014; Ramchand, 2018) then the prediction is that the highest auxiliary (in our example, ⟨*ir a* + infinitive⟩) will take scope over everything to its right, since the scope of a node in a phrase structure tree is the set of nodes that it c-commands (Ladusaw, 1980; May, 1985). But in Spanish there is no independent evidence that *ir a* c-commands the phasal auxiliary *empezar*, nor that it has scope over the lexical verb *trabajar*, as evidenced by the lack of entailment relations in (217'). The temporal auxiliary *va a* localises the modal *poder* in the future (such that John will have the possibility to start to work), but not the actual starting point of the event of working or the event of working itself. In turn, the future possibility pertains to the starting point of the event of working, but not the event of working itself: if we say that John is able to start working we are not saying that he works. This means that the unit *va a poder* modifies the phasal *empezar a* but not (directly) the lexical verb *trabajar*. An adequate segmentation for (216a), therefore, needs to group *va a* with *poder* in a syntactic unit that excludes *empezar a trabajar* and at the same time allows for cumulative modification as we go down the auxiliary chain: something along the lines of [[*va a poder*] [*empezar a* [*trabajar*]]] as opposed to [*va a* [*poder* [*empezar a* [*trabajar*]]]].

Bravo et al. (2015) call lexical auxiliaries *opaque* because, as (217a) illustrates, they do not let temporal and aspectual information from functional auxiliaries like ⟨*ir a* + infinitive⟩ pass through: ⟨*ir a* + infinitive⟩ modifies *only* the lexical auxiliary ⟨*poder* + infinitive⟩, not having scope over anything to its right. The question, from the perspective of syntactic construal, is how this is possible if the structural description of a chain of auxiliary verbs has the form [Aux_1 [Aux_2 [Aux_3 ... [Aux_n [VP]]]]] (as in the references cited above, with some variation pertaining to the specific labels assumed for each auxiliary). Our answer in past works, to be explored here under a slightly different lens, is that Spanish auxiliary chains cannot be assigned such uniform, monotonic structural descriptions. Rather, as proposed in Krivochen & García Fernández (2020: 156),

> *For purposes of modification, the introduction of a lexical auxiliary in the derivation closes a cycle: lexical auxiliaries can, unlike functional auxiliaries, both modify and be modified.*

There is a property of Spanish auxiliaries that makes all the difference: the fact that (with the well-documented exception of ⟨*soler* + inf⟩ and ⟨*haber de* + inf⟩), they have full inflectional paradigms. Unlike English modals, which are restricted to chain-initial position due to the fact that they lack non-finite forms (McCawley, 1975; Pullum & Wilson, 1977), Spanish modals and aspectual phasals may appear anywhere in a chain, provided that there is no semantic incompatibility. (217a) showcases a modal in second position (after temporal ⟨*ir a* + inf⟩), but it is not difficult to create natural sounding examples with modals in third position, for example:

(218) a. *Juan* *ha*$_{Aux1}$ *estado*$_{Aux2}$ *teniendo que*$_{Aux3}$ *trabajar* *todo*
J. have.3Sg.PRES be.PART have-to.GER work.INF all
el fin de semana
the end of week
'John has been under the obligation to work for the whole weekend'

And even more than one modal in the same sentence, given well-known ordering restrictions:

(218) b. (*future* + *modal*$_{deontic}$ + *perfect* + *modal*$_{dynamic}$ + *phasal*)
Juan va a tener que haber podido
J. go-to.AUX.3SG.PRES have-to.INF have.INF be-able-to.PART
terminar de trabajar
finish.INF work.INF
'John will have had to be able to finish working' (if he wants to enjoy the weekend)

c. (*progressive* + $modal_{deontic}$ + $modal_{dynamic}$ + *phasal*)
Juan está teniendo-que poder terminar-de
J. be.3SG.PRES have-to.GER be-able-to.INF finish.INF
trabajar temprano
work.INF early
'John is having to be able to finish working early' (otherwise he can't take care of his baby)

The fact that Spanish allows for modals to be modified by other auxiliaries implies a stark contrast with the situation in English, where the only possible chain is as in (219) (adapted from Ross, 1969: 77):

(219) Boris $will_{Modal}$ $have_{Perfective}$ $been_{Progressive}$ $being_{Passive}$ $examined_{V}$ by the captain

Because of the rigidity of English chains of auxiliaries, phrase-structure and templatic approaches (based on an *a priori* hierarchy) result descriptively adequate, and have enjoyed remarkable success. In Chomsky (1957: 39) the structure of an English auxiliary chain is captured by means of a PSR with optionality (C = agreement features, M = modal auxiliary):

(220) Aux → *C*(*M*) (*have* + *-en*) (*be* + *-ing*) (*be* + *-en*)

The ρ-set of (219), under a classical *Syntactic Structures*-style analysis, can be summarised as in (221):

(221) ρ = ⟨(will, have); (have, been); (been, being); (being, examined)⟩

The lexical anchor of the elementary graph would be the lexical verb, *examined*.[6] If we look at the generative analyses of English auxiliary chains, the general properties of the ρ-set in (221) hold. Furthermore, since there is no possibility to change the order of auxiliaries, proposing a rigid skeleton of phrasal nodes corresponding to each auxiliary makes sense from the point of view of theory, and does not present immediate problems when conducting gram-

6 Note that we have used the modal *will*, which contributes temporal meaning (Huddleston & Pullum, 2002: 180). As we said before, the situation is less clear with root modals, since they have been argued to assign a theta-role to their subject (Bosque, 2000; Hacquard, 2010; but see Wurmbrand, 1999). We will leave this issue aside here.

matical analysis. However, this is not the case for Spanish: as we have seen, different orders are possible, all of which correspond to distinct interpretations. Moreover, there is no evidence to defend the position that some of those orders are in some sense 'derived' from others (e.g., by the application of movement rules).

Let us consider the role of what we have called functional auxiliaries in chains. In (217b) (*Juan suele estar trabajando*), the functional auxiliary *estar* intervenes between the lexical auxiliary *soler* and the main verb. If functional auxiliaries are *transparent* for purposes of modification relations in auxiliary chains—that is, if they let that information through—we predict that the lexical auxiliary modifies the next lexical element namely, the main verb. This prediction indeed holds:

(217b') *Juan suele estar trabajando* ⇒ *Juan suele*
Juan AUX.3SG.HAB be.INF work.GER ⇒ Juan AUX.3SG.HAB
trabajar
work.INF
'Juan is usually working' ⇒ 'Juan usually works'

In (217c) (*Juan está debiendo llegar a tiempo*) the modal auxiliary *deber* appears in a progressive periphrasis, as the complement of the functional auxiliary *estar*. As in (217a), the lexical auxiliary *deber* absorbs the aspectual (imperfective, progressive) modification from this functional auxiliary, so that what is understood progressively is the obligation to arrive on time. The event of arriving *per se* is not so understood.

The reader may wonder if the lack of entailment is a general property of sequences of auxiliaries. That is: does the distinction between *lexical* and *functional* auxiliaries play any role? To illustrate that it indeed does, let us now consider an example where all auxiliaries are *functional*: here, we expect that none of them 'absorbs' the information from the auxiliaries to its left, and therefore all auxiliaries modify the lexical verb without modifying each other:

(222) *El ministro va a haber sido asesinado*
The minister go-to.AUX.3SG.PRES have.INF be.PART murder.PART
'The minister will have been murdered'

In (222) we have two functional auxiliaries: temporal *ir a* and perfective *haber*, plus passive *ser* (which is somewhat anomalous as a functional auxiliary given its role in changing the distribution of grammatical functions between arguments rather than providing event-bound temporal or aspectual information;

see Krivochen & Schmerling, 2022 for discussion). Given our characterisation of functional auxiliaries above, they all modify the lexical verb *asesinar* with no one auxiliary modifying any other. For instance, diathesis is not something that can be aspectually modified, or located in time (so neither *va a* nor *haber* can modify *ser*). None of them can absorb temporal or aspectual information, but rather they cumulatively modify the lexical anchor of the elementary tree. In *va a deber trabajar*, it is the obligation conveyed by the modal *deber* that is located in the future due to *va a*; this future obligation modifies the lexical verb. In *va a haber trabajado*, however, the situation is different: we cannot say that the perfect is located in the future in the same sense insofar as *haber* does not denote an event.

The preceding discussion summarises much previous work, to which the reader is directed for details. There are two crucial points that we want to highlight here before closing this section. The first is that lexical auxiliaries, to the point that they can take functional modifiers, are candidates to anchor elementary graphs. The second is that, since Spanish allows for variable orders in its sequences of auxiliaries, a templatic approach is inadequate; this inadequacy entails that we cannot know what the elementary graphs will contain *a priori*. We need to analyse a specific sentence to find elementary graphs. As an example, let us provide the ρ-set for the auxiliary chains in (217a) and (222) as (223a) and (223b) respectively (although we will return to the passive in **Section 13.1**):

(223) a. ρ_1 = ⟨(va a, poder), (poder, empezar a), (poder, Juan)⟩
ρ_2 = ⟨(empezar a, trabajar), (empezar a, Juan)⟩
ρ_3 = ⟨(trabajar, Juan)⟩
b. ρ = ⟨(va a, asesinado), (haber, asesinado), (sido, asesinado), (asesinado, ministro)⟩

In an auxiliary chain that contains lexical auxiliaries, there will be always more than one elementary graph: one for each lexical auxiliary and one for the lexical verb. However, if an auxiliary chain only contains functional auxiliaries, they all modify the lexical verb, which is the only lexical anchor by virtue of being the only *lexical* predicate in the structure. This is easy to formalise under structure sharing and unification, but monotonic concatenation-based grammars usually predict structural relations that are not there. Recall that anchors in the present work are defined by *two* necessary properties: (i) being predicates and (ii) being lexical categories. In this respect, we contrasted our approach with that of Frank (2002, 2013), where condition (ii) is sufficient (thus, nouns are lexical anchors despite not being predicates), and Hegarty

(1993) and Rambow (1993), where neither condition holds (since functional modifiers are also anchors for elementary trees). We compared these proposals in (64) in **Chapter 3**. As suggested above, also, it is possible that the rich scenario that emerges in the discussion of the definition of elementary trees can be used to analyse cross-linguistic variation, in terms of different languages being more or less restrictive in their requirements to define anchors for elementary trees (Krivochen & Padovan, 2021 provide some examples of this, considering evidence from English, German, and Spanish).

Let us summarise the discussion. We provided arguments that prove that a monotonic approach to sequences of auxiliary verbs run into empirical problems when it comes to appropriately delimiting the domains across which modification takes place. Among other reasons for these limitations, monotonic branching is uniformly to the right or to the left, and semantic relations, mapped from syntactic structure via c-command, are similarly monotonic. As an alternative to the monotonic view of structure building, we proposed that the syntax of chains of auxiliary verbs require us to consider the properties of the specific auxiliaries that appear in a chain when assigning a structural description to it: there cannot be an *a priori* template because there are several possible combinations of auxiliaries, some of which define different elementary graphs and some of which do not. The approach pursued here captures the properties of auxiliary chains that in past works were modelled using a derivational approach. The basic idea is that Spanish auxiliary chains are composed of local elementary graphs, anchored by lexical predicates. The graph-theoretic viewpoint adopted here captures the empirical insights from Bravo et al. (2015), García Fernández et al. (2017), Krivochen & García Fernández (2019, 2020), and related works in a simple manner: *functional auxiliaries* are modifiers of *lexical auxiliaries* or of *lexical verbs*, and therefore belong within the elementary graph defined by the latter (see also Krivochen & Schmerling, 2022 for a perspective from Categorial Grammar). Lexical auxiliaries anchor their own elementary graphs, whereas functional auxiliaries belong in the elementary graphs defined by either lexical auxiliaries or lexical verbs, as functional modifiers of these. This allows us to provide a working definition of *lexical head*, as anticipated above: a lexical head is a head that can modify and be modified.

Such a picture contrasts with the traditional approach to English sequences of auxiliaries (Chomsky, 1957; Ross, 1969; Bach, 1983; Quirk et al., 1985; Cinque, 2004; Ramchand, 2018, to name but a few), where the structure of a sequence of auxiliaries is monotonically growing (one auxiliary head at a time in a binary-branching tree) and is thus structurally uniform. This means that there is a source of linguistic variation to be found in what exactly counts as a lexical anchor and why, and therefore in the size of elementary graphs. Segmentations internal to sequences of auxiliaries are not considered necessary in the analysis of English sequences of auxiliary verbs. An English sentence featuring a chain of auxiliary verbs would involve a single elementary graph, following traditional analyses. This is not a privative property of transformational generative models or cartographic approaches. Falk's (1984) analysis of the English auxiliary system assigns the same *f*-structure to all auxiliaries (they all behave like Raising predicates in subcategorising for ⟨XCOMP⟩SUBJ, an open complement clause and a non-thematically selected subject), and Bresnan's (2001) analysis is also uniform, although she treats all auxiliaries as features at the level of *f*-structure. Falk's revision of his analysis in (2003) mixes the predicate analysis and the feature analysis: *will, would, have, do, be* (progressive) are treated as features, but *may* (both in its epistemic and root readings) is a full-fledged predicate (nothing is said of *ought to, need, must, have to* or passive *be*); this revised analysis is closer to our perspective. Within TAGs, Frank (2013: 253) situates a modal auxiliary (*should*) in the same elementary tree as the lexical verb that anchors the tree: a sentence like *the linguist should finish his book* receives the following analysis:

(224)

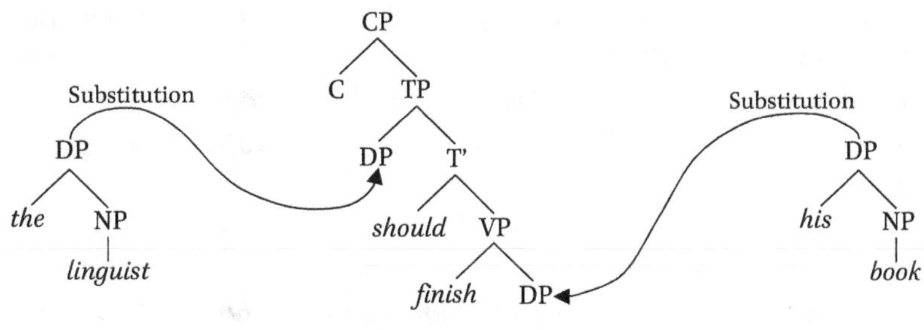

FIGURE 7.2 LTAG derivation for 'The linguist should finish his book'

On the other hand, as we saw in **Chapter 3**, Hegarty (1993) and Rambow (1993) make both functional and lexical heads (including auxiliaries) be anchors of their own elementary trees: in their view, *should* would be an anchor, but so would *finish, linguist,* and *book* by virtue of being heads. In our approach, in contrast, if an arbor contains exactly one lexical predicate, and its functional modifiers as dependants of that lexical predicate, then functional auxiliaries belong in the arbor defined by the lexical head (auxiliary or verb) that they modify. Since Spanish allows for more than a single lexical auxiliary in a chain, it follows that there may be more than a single arbor in a Spanish auxiliary chain, whereas there is only one in an English chain. If we add to this the fact that Spanish modals can be modified by functional auxiliaries (unlike English modals, which have no non-finite forms and thus are restricted to first-position only in finite clauses), then the LTAG representation for a Spanish sentence such as *El lingüista ha debido terminar su libro* ('The linguist has had to finish his book'), following Krivochen & Padovan (2021), would be (225):

(225)

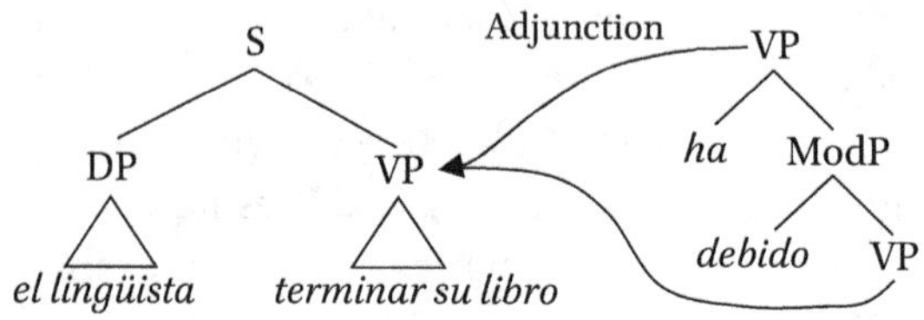

FIGURE 7.3 LTAG derivation for 'El lingüista ha debido terminar su libro'

The elementary trees in (225) allow us to define a syntactic object that includes *ha debido* (the modal modified by the perfect auxiliary) but excludes the lexical verb *terminar*: in this way, the fact that the perfect does not affect the lexical verb can be straightforwardly accounted for, since by the *Fundamental TAG Hypothesis* all dependencies are established at the level of elementary trees. The graph-theoretic approach in this monograph captures the same empirical facts and theoretical insights about the locality of syntactic dependencies (cf. Frank's 1992, 2013 CETM) as the derivational system in a LTAG. Furthermore, it provides a way to frame the issue of cross-linguistic variation. This perspective, as we said before, is germane to the LTAG proposal in Frank (2013: 238) that

> *the set of elementary trees in the grammar of a language is the only locus of cross-linguistic differences, imposing substantial restrictions on the kind of variation that we will find crosslinguistically.*

In a lexicalised grammar, this entails that there is variation with respect to what exactly is a lexical head, and what counts as its extended projection. Let us go back briefly to our argument about Raising *seem* and its Spanish counterpart *parecer*. If *parecer* is an auxiliary, then for us to define an elementary graph we need a lexical predicate that *parecer* can modify: this can be either a lexical verb or a lexical auxiliary (e.g., in *parece tener que hacerlo* 'seems to be under the obligation to do it'; note the order *epistemic* > *deontic*). In this analysis, *parecer* does not behave any differently from auxiliaries like ⟨*tener que* + infinitive⟩ or ⟨*poder* + infinitive⟩ in their epistemic readings (in terms of the structural description it must receive, in the way in which it contributes to the compositional meaning of a sentence that contains it, and in the way it interacts with other auxiliaries following the order *epistemic* > *deontic* > *dynamic*[7]). This contrasts with the behaviour of English *seem*, which English theories of Raising recognise as a lexical verb; *seem* does not fulfil the requirements to be considered a modal auxiliary (or an auxiliary at all). This picture of variation is precisely the kind of scenario predicted in LTAG s.

The consideration of a grammatical phenomenon that can relate expressions in distinct local domains within auxiliary chains, Clitic Climbing, motivated the analysis of the conditions under which nodes that belong to an elementary graph are accessible for relations with nodes in other elementary graphs (in other words: to which extent elementary graphs are 'transparent' and how to restrict graph union): this led us to ask deeper questions about what the theory of locality looka like in the present framework. In doing so, we formulated a further condition over dependencies across elementary graphs (based on the notion of *self-containment*), which we will make use of in the analyses that follow.

7 Spanish allows for sentences where all three kinds of modal auxiliaries appear: in a sequence such as *debe tener que poder hacerlo* ('it must be the case that he/she is under the obligation to be able to do it') ⟨*deber* + infinitive⟩ can only be interpreted as epistemic, ⟨*tener que* + infinitive⟩ only as deontic, and ⟨*poder* + infinitive⟩ only as dynamic, even though each of these auxiliaries allows for different interpretations in other contexts (see Bravo, 2016b, 2017). Interestingly, ⟨*parecer* + infinitive⟩ has the same restrictions: in a sequence of modals, it can only appear in the first position, reserved for epistemics. When *parecer* is not followed by an infinitive its distribution is freer and it can appear after a modal (e.g., *puede parecer inteligente, pero no lo es* 'he/she may seem smart, but he/she is not'), which is consistent with the analysis in García Fernández & Krivochen (2019b), in which only ⟨*parecer* + infinitive⟩ is an auxiliary.

CHAPTER 8

On Unexpected Binding Effects: a Graph-Theoretic Approach to Binding Theory

This chapter focuses on co-reference relations within structural descriptions: the kind of structural and semantic relations that are usually referred to in the syntax-semantics literature as 'binding' (see also **Section 6.2.1**). Giving up the SMC and the binarity requirement has an evident impact on how co-reference relations are defined, and locality conditions holding in relations between antecedents and anaphoric elements (reflexives, reciprocals, and pronouns) must also be redefined. The approach to binding developed in this chapter is only a preliminary sketch and by no means an attempt to capture the entirety of empirical insights of decades of work in binding phenomena. However, we try to show that an approach to structural descriptions in which the neighbourhood set of any given node is maximised has some descriptive empirical advantages over more rigid models of phrase structure which only allow a single word per terminal node and a maximum neighbourhood set of 3 (one parent node and two daughters). This chapter introduces and develops constraints on relations between nodes which have the same address (and thus the same semantic value) in distinct elementary graphs and which are therefore collapsed into a single node in derived graphs (via *graph union*). We refer to these relations as *linking* the elementary graphs. This is a crucial concept in our framework which will be taken up on subsequent chapters.

A crucial point made in **Chapter 7** was that a one-size-fits-all approach to the relations between parentheticals and main clauses cannot derive adequate structural descriptions for all cases. There are instances in which a parenthetical clause *does* intervene in the dynamics of the matrix clause, and there are also instances in which parentheticals are 'invisible' for operations at the target of parenthetical insertion. We proposed that the crucial factor to consider was whether the parenthetical is a *self-contained unit* or not (cf. Kluck's 2015 distinction between *free* and *anchored* parentheticals), in the sense of (205) above. This is a restriction over relations in derived graphs. If two elementary graphs G and G' are *linked* at some node, then *in principle* they are mutually accessible: it is to be expected that a closer inspection may reveal that some node or arc in one of these may not be ordered with respect to some node in the other or some further condition may be violated. In this section we will come back to English data, and present further evidence in favour of a 'mixed'

 DOI:10.1163/9789004542310_009

approach to parentheticals (and other adjoined sub-graphs) in which opacity is not an automatic consequence of adjunction. The data we will consider pertains to pronominal reference and crossover effects, and will provide the basis to rethink the principles of Binding Theory (BT) in terms of ordered relations in digraphs. Before we start, it is useful to remember the principles that constitute so-called Canonical Binding Theory in MGG (Chomsky, 1981: 188; see also Chomsky, 1995: 211 for a Minimalist update to the classical principles):

(A) *An anaphor* [reflexives, reciprocals] *is bound in its governing category* [where the governing category for α is the minimal category that contains α, a governor of α, and an accessible subject for α; in practical terms, the governing category of α is the minimal NP or S node dominating α]

(B) *A pronominal is free in its governing category*

(C) *An R-expression is free*

Binding principles have become a more or less widely agreed upon aspect of syntactic theory, such that even non-transformational theories have versions of BT that resemble MGG's closely (see, e.g., Pollard & Sag, 1994: 254 for HPSG; Bresnan, 2001: 215; and Dalrymple, 2001, §11.2 for LFG). The principles of BT have remained remarkably stable throughout their history, with revisions being more slight tweaks and terminological changes than ground-breaking reformulations (although alternative systems do exist, see e.g. Safir, 2013 for a summary of some such alternatives). Thus, we can find, instead of reference to *governing categories* (which as we just saw played a big role in binding and bounding in GB), a different approach to 'freedom' in Chomsky (1995: 65):

> *an r-expression α must be A-free, that is, not c-commanded by a pronoun in an A-position linked to α in the binding-theoretic sense*

Feature-based Minimalist proposals to binding also exist, such as Antonenko (2012), in which reflexivity is not analysed configurationally or in terms of GFs but as the effect of a feature which determines coreference. The syntactic domain within which binding conditions are establishes is now the *phase*, defined by the presence of designated nodes (transitive *v* and C in the original version of the theory, in Chomsky, 2000, 2001).

Some aspects of BT were analysed already in **Section 6.2.1**, in particular the idea that the distribution of anaphoric pronouns can be formalised in terms of parallel arcs, with a single expression establishing more than one relation with a predicate. In this chapter we will analyse some unexpected binding patterns and then formulate the relevant generalisations that govern the distribution of anaphoric expressions, pronouns, and R-expressions in graph-theoretic terms.

We begin by looking at the following example (taken from Bresnan, 2001: 81), with indexes and gaps added for expository purposes:

(226) The one$_j$ he$_i$ should be spoken to e_i by e_j, for God's sake, is his$_i$ mother$_j$

Here we have an interesting mix between a multiple gap construction like (70) in **Section 4.1** (Kayne's example *A person who people that talk to usually end up fascinated with*) and issues related to the syntax of parentheticals, which were the topic of **Chapter 7**. In Bresnan's example, [for God's sake] is a self-contained unit, and thus has its own syntactic and semantic independence. Using Kluck's (2011, 2015) categories, it is a 'free parenthetical' since it is not related to material in the host clause. However, we can modify the example to show not only that parentheticals can be accessible from the matrix clause they adjoin to if they are *not* self-contained, or 'anchored' in Kluck's terms (i.e., if they contain a node that dominates a node also dominated by a node in the matrix clause or if they contain a node immediately dominated by a node in the matrix clause), but that there are restrictions over what in MGG is usually referred to as the *Spell-Out* of nodes (i.e., their morpho-phonological exponent) which would be unexpected (and remain unexplained) under an 'adjoined sub-graphs are always opaque' theory (cf. Uriagereka, 2002, in the most radical interpretation of Multiple Spell-Out[1]). Let us take a look at the following paradigm, which expands on (226):

(227) a. The one he$_i$ should be spoken to by, for his$_i$ sake, is his$_i$ mother
b. *The one he$_i$ should be spoken to by, for John's$_i$ sake, is his$_i$ mother
c. *The one he$_i$ should be spoken to by, for his$_i$ sake, is John's$_i$ mother
d. The one John$_i$ should be spoken to by, for John's$_i$ sake, is his$_i$ mother

The cases that we are interested in are (227b) and (227c). The traditional MGG literature on Binding Theory (Chomsky, 1981, 1995, and related work; see Safir, 2013; Truswell, 2014 for overviews) would account for the ungrammaticality of (227b) and (227c) in terms of a Principle C violation: the relevant cases would receive structural descriptions in which the R-expression [John] is

1 Specifically, Uriagereka (2002: 53–54) says:

If a noncomplement [this would include specifiers as well as adjuncts. Parentheticals are neither complements nor subcategorised for; therefore, their configurational status should be analogous to that of adjuncts] *is spelled out independently from its head, any extraction from the noncomplement will involve material from something that is* ***not even a syntactic object*** (*or,* ***more radically, not even there***). (Our highlighting).

bound within its governing category, which is a configuration excluded by Principle C (*R-expressions are always free*, where 'free' means 'not c-commanded by a co-referential NP'). That proposal would indeed work for (227c), in which the pronoun [he] and the R-expression [John] co-exist in the same derivational space (specifically, they belong in the same *phase*; e.g. Wurmbrand, 2011: 60; also Uriagereka, 2002). (227b), on the other hand, poses the following problem: in order to blame its ungrammaticality on a Principle C violation, we need to be able to claim that the R-expression is bound, but how can it be? The R-expression is contained within a parenthetical, which cannot be monotonically derived together with the rest of the sentence (either by top-down phrase structure rules or bottom-up Merge), and therefore there should be no c-command relation between the R-expression inside the parenthetical and the pronoun in the matrix clause.[2] Two possible solutions appear:

I. The parenthetical is visible for purposes of dependencies between elements in the matrix clause by the principles of Binding Theory because these apply late in the syntactic computation, *after* adjunction of the parenthetical to the matrix sentence (e.g., at LF; see Lebeaux, 2009)

II. The parenthetical is visible because it is *not* a self-contained unit

Note that only I. requires a multiplication of levels of representation (to at least two: a syntactic level and LF, or, more generally, one where *adjunction* takes place and one where *indexing* takes place). But even if indexing and the computation of reference does indeed take place at a very late stage of the syntactic derivation, it is not clear how to appropriately filter out the cases in which parentheticals are completely opaque, for instance (228):

(228) *What$_i$ did John, who asked for e_i, get a book for his birthday?

In this case, the appositive relative clause is opaque for purposes of extraction (see also Kluck, 2011: § 3.4.1), and the ungrammaticality of (228) could be blamed on a violation of the Complex NP Constraint (Ross, 1967: 127) which bans either leftwards or rightwards extraction of X from a configuration of the kind [$_{NP}$ ... [$_{S}$ ... X]], the Adjunct Condition, which bans extraction from non-subcategorised phrases (Huang, 1982), or some other constraint. However, opacity for extraction and transparency for binding effects are not easy to reconcile under these assumptions: why would indexing have access to a syntactic domain to which operator-variable relations cannot? The problem, evidently,

2 See Kluck (2011: § 3.4.3) for discussion about binding effects in amalgams and transparent free relatives.

is even more pervasive for analyses of pronominalisation as movement. What do we need in order to account for these data? Minimally, we must make explicit the conditions under which a given syntactic domain becomes opaque for extraction: we need to define the conditions under which a node may be dominated by another in elementary graphs G and G', both of which belong to a derived graph G''.

There are, as far as we can see, several ways to achieve this. One, in the spirit of derivational syntax, is to define the relative order between structure composition (parenthetical insertion, substitution, adjunction) and extraction: we can determine whether the one rule feeds or bleeds the other. This is precisely the kind of approach taken in the derivational proposal advanced in Krivochen (2019a) to account for the impossibility of extracting either the result or the direct object out of a resultative construction (either transitive, like *John hammered the metal flat*, ergative, like *the river froze solid*, or unergative, like *Mary shouted herself hoarse*). We can illustrate the relevant phenomena (traces, as usual, are purely expository):

(229) **Wh-movement:**
a. *What did John hammer the metal? (answer: 'flat')
b. *What did the river freeze? (answer: 'solid')
c. *What did Mary shout herself? (answer: 'hoarse')

(230) **Topicalisation in embedded clauses:**
a. *... that / because, the metal$_i$, John hammered t_i flat.
b. *... that / because, the river$_i$, t_i froze solid.
c. *... that / because, herself$_i$, Mary shouted t_i hoarse.

In that work, we proposed that the opacity of a resultative construction was due to the fact that resultative constructions involve at least two syntactic objects (roughly, one for the process, another for the change of state) in parallel derivational spaces, one of which gets 'atomised' by virtue of introduced by adjunction in the other (see Mateu Fontanals, 2002, and the top-down model of Zwart, 2009 for a similar take on atomisation): this process makes the atomised element opaque for operations at the target of adjunction. Furthermore, the opacity of an adjoined domain (in that case, resultative clauses) is a consequence of the combination of two factors: (a) *derivational timing* (relative to the ordering of embedding and singulary transformations) and (b) whether the adjoined sub-graph is a *self-contained unit*. In that work, which advanced a TAG expansion of the syntactic-semantic machinery assumed in Generative Semantics (adding generalised transformations to the operations of *predic-*

ate raising and cyclic *lexicalisation*), we proposed the following constraint on trans-derivational dependencies:

> *Let γ and β be two sub-trees such that γ contains a node X that corresponds to the root of β. A singulary transformation T_S triggered from γ can affect β **iff** T_S is intrinsically ordered*[3] ***after*** *an embedding transformation that adjoins β to γ at X.* (Krivochen, 2019a: 59)

What singulary transformations (e.g., *wh*-movement or topicalisation) *cannot have access to*, we argued, is *elements embedded within* β (i.e., dominated directly or transitively by the root node in β); only β as a whole can be affected by a singulary transformation at γ ordered after *adjunction* of β to γ (similar opacity conditions can be formulated within the Multiple Spell-Out framework of Uriagereka, 2002, 2012). The order of operations *adjunction* > (*sub-*)*extraction* is excluded. Note that the same kind of argument could be made by formulating a principle like Frank's (2013: 233) Fundamental TAG Hypothesis, which we repeat for the reader's convenience:

> ***Fundamental TAG hypothesis:*** *Every syntactic dependency is expressed locally within a single elementary tree*

Because we would be asking the grammar to establish a syntactic dependency *across* elementary trees, the Fundamental TAG hypothesis would rule out the corresponding representations. There would need to be an elementary tree containing a *wh*-operator but not the variable it binds.

In Krivochen (2019a) we committed ourselves to a strongly local derivational model of syntax with an ordered system of rules (building on Fillmore, 1963). Despite the fact that the theory in this monograph is not derivational, some empirical consequences of our previous derivational proposals hold also in the present model. To the extent that such empirical consequences pertain to locality conditions in the establishment of dependencies, they must be captured regardless of the formal foundations of the meta-theory.

---

3 A weaker version of this condition would refer to *extrinsic ordering* (a proposal along these lines is in effect suggested in Lakoff & Ross, 1967 [1976]; McCawley, 1968 opposes it). Ringen (1972: 267) presents the distinction between *intrinsic* and *extrinsic* ordering very clearly:

> *If in a grammar, G, rule X is ordered before rule Y, then X and Y would be* ***extrinsically*** *ordered if G restricts how these rules can apply; that is, if these same rules could apply in the order Y before X in some derivations if not restricted by G. X and Y would be* ***intrinsically*** *ordered if there is only one order in which these rules could ever apply in any derivation; that is, if it would be impossible for these rules to apply in the order Y before X.*

The kind of argument just presented (based on the relative ordering of rules, with some rules creating contexts for other rules to apply and some rules restricting the application of other rules) is not available to us now, since there are no derivations. Here we go further along the line of reasoning that we adopted in Krivochen (2018, 2019a) as far as the emergence of interpretative domains goes: the strongest hypothesis would be that a graph G can be assigned a compositional semantic interpretation *if and only if* it is self-contained. It should be apparent that graphs corresponding to independent simple sentences are indeed self-contained (we will come back to this point in **Section 14.2** below, when analysing *deletion*. In other words, every self-contained basic or derived graph can be assigned a compositional interpretation. We must highlight that above we said ‘graphs’, not ‘elementary graphs’: the reason is that getting a graph G none of whose nodes is dominated by a node outside G may require the composition of more than a single elementary graph. If an interpretation for a graph implies (at least) defining local walks in that graph, we need all relevant nodes to be ordered with respect to one another. Informally, we say that if a node or a set of nodes is not ordered in the graph G that we are walking at some point, then it is not possible to assign a compositional interpretation to that node or set thereof in G. We can formulate a general condition to this effect:

(231) ***Total Order Condition***
A node is strictly ordered with respect to every other node in the ρ-set of the arbor that contains it

Ojeda (1987) formulates a ‘Full Leaf Order’ condition which can be thought of as the ‘linear precedence’ counterpart of our order condition, as follows:

> ***Full Leaf Order:*** *For all leaves u, v,* [either] $u < v$ *or* $v < u$. [where ‘<’ is a binary antisymmetric relation ‘precedes’] (Ojeda, 1987: 258)

In other words, linear precedence is a total relation imposed over terminals (see also Kayne, 1994). The importance of an *order* imposed over a set of nodes in a graph being *strict* and *total* is also emphasised in Sternefeld’s (1998a, b) connectivity-based approach to Binding Theory, which we will come back to below. For clarity, a binary relation $R(x, y)$ defines a *strict order* over a set or class $D(x)$ if and only if:

for all a, b, c,
$R(a, b) \rightarrow D(a) \wedge D(b)$; [irreflexivity]

$\neg(R(a, b) \wedge R(b, a))$; [antisymmetry]
$R(a, b) \wedge R(b, c) \rightarrow R(a, c)$. [transitivity]
R(a, b) may again be written a < b (mod. R). (Krivine, 1971: 7)

The condition in (231) should be completed with the additional requirement that the *ordering* of a graph be both *total* and *strict*: this in turn implies that for any nodes *a, b, …* in G, or any sub-graphs G, G', the order imposed over all nodes involved is *irreflexive, antisymmetric,* and *transitive*. The idea we put forth here is that when nodes are disconnected from a sub-graph (and are thus not ordered with respect to any other node in the relevant sub-graph) they *cannot be assigned a compositional interpretation*. In Ojeda's view, which is possibly the other side of the coin, a node that is not ordered with respect to any other node does not precede or follow any other node, thus it cannot be 'located' in a string. We are concerned with structures rather than strings, but it is intuitive how these conditions can be made compatible.

The conundrum we are facing with the paradigm in (227) above seems to resist traditional accounts. In sum, the problem is that delaying the application of Binding Theory principles derivationally (say, to LF, after the 'overt' syntactic computation and covert operations) does not solve the essential, underlying issue: how come sometimes parentheticals are visible for purposes of operations at the matrix clauses to which they are adjoined, and sometimes they are not? However, if we combine the observations made within the context of BT with the non-monotonic perspectives on computation advanced here and in past works, we can enhance the descriptive and explanatory power of the grammar. In (227), the parenthetical [for John's sake] can form a chain with the other two occurrences of the same *type* (namely, [he] and [his]) because it contains a multidominated node with address ⦃John⦄, such that the following relations hold (we use {} and not ⟨⟩ for this ρ-set because it is *not* an ordered set, it is merely a partial dominance set):

(232) ρ = {(to, John), (John, should), (for, John), (John, mother)}

Because it is *not a self-contained unit*, the sub-graph corresponding to the structural description of *for John's sake* is visible for binding purposes (independently of *when*, in derivational time, BT principles are supposed to apply), and thus the materialisation of the occurrence of ⦃John⦄ in the parenthetical as *John* does trigger a Principle C violation.[4] Equally, we can see parentheticals

4 Kluck (2015: 108) mentions cases where an R-expression within an anchored parenthetical does not trigger a Principle C violation:

(more specifically, appositive relative clauses) and fronted adverbial clauses triggering crossover effects:[5]

(233) a. The President, whose son$_i$ is involved in a collusion scandal, betrayed him$_i$
b. *Who$_i$ did the President, whose son$_i$ is involved in a collusion scandal, betray t_i?

(234) a. Now that President Trump$_i$ has been offered Mexico's help in the wake of Hurricane Harvey, he$_i$ may be accepting assistance from a country full of 'bad hombres'
b. *Who$_i$, now that President Trump$_i$ has been offered Mexico's help in the wake of Hurricane Harvey, t_i may be accepting help from a country full 'bad hombres'?

The ungrammaticality of (234b) is unexpected (and remains unaccounted for) if non-restrictive/appositive relative clauses are assumed to be fully opaque for purposes of operations or dependencies at the matrix clause to which they are adjoined, and this holds regardless of the derivational timing of adjunction to the root (the 'Main Clause Hypothesis' of Emonds, 1979) or the establishment of a discontinuous dependency (McCawley, 1982, based loosely on Wells, 1947). Similarly, with a fronted adverbial clause we have crossover effects that are unexpected if the root-adjoined object is not internally accessible: the NP that causes the crossover effect is embedded inside that clause, and under an MGG approach to binding relations it is not clear how either (233b) or (234b) satisfy the structural conditions for *strong crossover* (see also Postal, 1971):

> *The central cases of the strong crossover configuration arise when Wh Movement has taken place from some A-position A* asymmetrically c-commanding some pronoun P in an A-position, to an Ā-position Ā* that c-commands*

---

i) He$_i$ kissed, Cassanova that Bob$_i$ is, many women (the example and judgment are Kluck's)

The status of (i) is not entirely clear (our informants have judged it as either marginal ? or impossible *), but the effect of Condition C seems to be ameliorated with respect to (ii), for example (which our informants judged as much worse than (i)):

ii) *He$_i$ kissed, because Bob$_i$ is a Cassanova, many women

Principle C seems to be somewhat anomalous in this respect. If we consider crossover, effects do appear:

iii) *Who$_i$ does everyone, including his$_i$ mother, love t_i?

5 We are indebted to Susan Schmerling for her help in constructing these examples.

> *P. In a right-branching language like English, this will arise only if P is "between" Ā* and A*. The movement can thus be said to "cross over" P.* (Sportiche, 1985: 461)

We want to emphasise that, under an MGG view, it is not evidently possible to explain the ungrammaticality of the (b) cases via crossover without additional stipulations, because the NPs *whose son* (in (233b)) and *President Trump* (in (234b)) do *not* c-command the position A*, marked with traces for expository purposes (it is important to observe that in (233b) the *wh*-operator is coindexed not with the possessor, but with the possessed). Since the problem seems to us to be in the theoretical assumptions about phrase structure and configuration rather than on the definition of *crossover*, we will propose in what follows an alternative formulation of Binding conditions within local digraphs rather than defined over phrase structure trees.

In our account of the effects just presented we will combine the insights that follow from theories focused on derivational timing (essentially following Fillmore, 1963 and building on Uriagereka, 2002; Krivochen, 2019a) with the relativisation of the notion of representational opacity: as we saw in the previous section, *a sub-graph is opaque for purposes of establishing relations with objects outside that subgraph if and only if it is a self-contained unit*. Sportiche's definition of the configurations that generate *strong crossover* effects means that *strong crossover* is essentially a violation of Principle C of Binding Theory (Chomsky, 1982: 20), which establishes that referential expressions must be free (i.e., not c-commanded by a coreferential element). We agree with Postal (2004c: 206) in being skeptical about finding any *explanatory* value in Principle C: the Principle does *not* explain why crossover effects appear, requires a category of expressions to be defined in an essentially stipulative manner (R-expressions, a category that includes otherwise heterogeneous elements: variables as well as proper names; see e.g. Higginbotham, 1983) and, descriptively, only captures a subset of the facts (for discussion, see Postal, 2004c; also Koster, 1987: 346, ff.). We want to avoid resorting to a category like R-expressions, or to a negative constraint (such as 'R-expressions cannot be bound').

If the theory exposed here is read with Harrisian glasses (or in the context of an approach like that in Lees and Klima, 1963), then it should be possible to formulate a rule of *pronominalisation* which makes reference to the ordered occurrences of a particular node when defining a walk through a graph.[6] Because

---

6 For instance,

> *An NP_1 may pronominalize an identical NP_2 if NP_1 is to the left of NP_2 under any conditions, or if NP_1 is to the right of NP_2 and NP_2 does not command NP_1* (Bach, 1970: 121)

edges are directed, we can capture both forward and backward pronominalisation (in the former, the antecedent linearly precedes the pronoun; in the latter, the pronoun linearly precedes the antecedent, see Ross, 1969a), both of which are formulated by Ross as *cyclic rules* (in current terminology, they apply within elementary graphs). We have indeed pursued this line of reasoning in **Chapters 5** and **6**. It must be explicitly said, however, that such an extension of the graph approach (that is, including aspects of the morpho-phonological exponents assigned to nodes in the graphs) requires the addition of a memory stack that can keep track of which nodes are visited when, and whether, at any given node, we have visited that node already. Constraints on *forward* and *backwards pronominalisation*, which seem to interact with cyclic conditions (see, e.g., Ross, 1969a: 192), would have to be explicitly formulated, etc.[7] In any case, it is crucial to highlight that the requirement that nodes in an arbor be *strictly ordered* with respect to all other nodes within that arbor is *not* interpreted in terms of string precedence. We will see below that the requirement of *ordering* is sufficient to formulate constraints that can take care of the *crossover* cases we are interested in, given that these constraints interact with the definitions of *linking* and *self-containment* proposed above.

At this point, and considering the argument involving local reflexive anaphora in **Section 6.2.1** the reader may want to generalise these observations, and thus they may ask: is it possible to formulate a theory of Binding in constraint-based, graph-theoretic terms? It is, and, at least partially, it has been done before (under different assumptions from ours, we will come back to this momentarily). Consider, for instance, the definition of *Binding Tree* in Sternefeld (1998a: 158):

> *Given a tree Σ and a node $\alpha \in \Sigma$, the Binding Tree for α is the smallest subtree $T \subseteq \Sigma$ that satisfies the following conditions:*
>
> a. $\alpha \in T$
>
> b. *The root of T is the root of Σ*

Where 'to the left of' can be interpreted as 'visited before in a walk in G'. *Command*, in the context of Bach's discussion, is Langacker's (1969) notion, which is based on the asymmetry of dominance: in our terms, the ρ-domain of a vertex is an *ordered* set, as it is a set of dominance relations.

7 There is also the obvious fact that pronouns can occur in contexts which do not satisfy the structural description for any *pronominalisation*-like rule. Compare:
(i) John$_i$ thinks that he$_i$ is smart
(ii) He is smart
There is no reason to believe that the pronoun 'He' in (ii) has been derived transformationally from a proper NP (or by visiting a node for the nth time). This point, as far as we know, was first explicitly made in the generative literature by Postal (1969: 202).

c. *If $\beta \in T$ and γ is the local trace of β, then $\gamma \in T$* ***unless***
   (i) *α is an R-expression*
   (ii) *β reflexively dominates an adjunct that dominates α, and*
   (iii) *γ is not a reconstruction site*
   [Highlight in the original]

Sternefeld's definition in turn owes much to Barss (1986) and Lebeaux (1994), and is formulated in a late-GB framework. We will build on this definition, but we do need to point out that there are some aspects of that definition that are not translatable to the framework that we are working in. For example, the notion of *reconstruction site* plays no role in the theory sketched here: on the one hand, we allow for a node to be visited more than once in a walk through a graph (we constrain the number of times we walk through each edge, not through each node). On the other, MGG-style *reconstruction* is an essentially derivational operation (since, given a pre-transformational structure and a post-transformational structure, reconstruction involves, well, reconstructing the position of a displaced term to its pre-transformational position at the level of semantic interpretation), and there are no derivations (no ordered sequences of application of production rules) in the framework explored here. Moreover, our nodes do not correspond to lexical items, as we have stressed throughout the monograph: rather, they are addresses that point to semantic values. This last consideration is related to Montague's (1974: 194) approach to the interpretation of basic expressions:

> *a model should assign to a basic expression not a denotation but a* ***denotation function****, that is, a function that maps each infinite sequence of individuals onto a possible denotation of the expression* (Highlight in the original)

In this context, formulating binding principles that make reference to the morpho-phonological form of a node (e.g., to say that something is an anaphora because it contains the morpheme *-self*) does not seem to be too promising either formally or empirically. It is well-known, besides, that the distribution of nominal expressions does not always coincide with what is expected from binding principles: a clear example is the non-anaphoric interpretation of *him/herself* in the sentences that follow:

(235) a. Mary$_i$ complained that [the teacher gave extra help to everyone but herself$_i$] (from Baker, 1995: 64)
b. Himself$_i$ a man of science, John$_i$ had to verify Bill's results (from Krivochen, 2019b: 106)

c. The postdoc$_i$ did all the paperwork herself$_i$ whose salary is not nearly enough (Op. Cit.: 99)

The point is, *anaphors, pronouns, R-expressions* are defined *by* and *for* Binding Theory in GB/MP, and it is not always clear (a) at which level each principle applies, and (b) whether binding primitives are morpho-phonological, semantic, or syntactic (i.e., if an anaphor is always a SELF object, or a semantically reflexive object, or an object that establishes a local syntactic relation with some other object). In the context of this work, things work slightly differently. Importantly, of the three conditions that define a *Binding Tree* in Sternefeld (1998a), only conditions (c (*i*), (*iii*)) make explicit reference to non-configurational aspects: R-expressions in (c-*i*) and reconstruction (which in turn depends on movement) in (c-*iii*). Conditions (a) and (b) are locality conditions, which ensue that the Binding Tree for α is a minimal rooted tree, and condition (c-*ii*) can be thought of as a monotonicity condition: if β binds α in *T* if the walk between α and β does not visit a root *R* that includes α and excludes β; or, to use Sternefeld's condition adapted to current notation, if $(R, \beta) \notin \tau$.

If these considerations are incorporated into the theory, then we can formulate some preliminary generalisations which pertain to the interpretation of 'bound' elements, which captures the basic insight of the traditional BT principles:

(236) ***Binding Graph (definition)***
The Binding Graph of α $G_B(\alpha)$ *is the minimal graph G that contains* α *and a single root R*

Binding graphs are arbores. Note that if α an argument, then the root of the minimal graph that contains α will be either a lexical predicate or a functional auxiliary, and the binding graph for α will be an irreducible elementary graph containing α. In this sense, binding graphs cover similar ground as the redefinition of *govering category* in Chomsky & Lasnik (1995: 102): whereas traditional governing categories included NP and S (categories that licensed subject positions), the version in Chomsky & Lasnik (1995) …

> … *generalizes from NP and IP to complete functional complex* (CFC), *where a* CFC *is a projection containing all grammatical functions compatible with its head.*

This change was motivated by cases where pronominals and anaphors are not in totally complementary distribution: in a sentence such as *They hate* [{*each other's/their} parents*], both an anaphor and a pronoun are possible, and unless Principle A and Principle B are evaluated with respect to different domains, then either *their* will be bound within its governing category, or the reciprocal *each other* will not be. Then, Principle A was relativised to CFC, whereas Principle B may continue to make reference to Governing Categories (Jaeger, 2004). Unusual as it is to find explicit reference to grammatical functions in MGG, the characterisation of a CFC (and thus of governing categories) in these terms brings it closer to out characterisation of elementary graphs, and indeed to a lexicalised grammar.

Then, we can preliminarily define the objects subjected to BT as follows

> *If α is bound within its Binding Graph, call α an 'anaphor' regardless of its morpho-phonological exponent*

> *If α is bound outside its Binding Graph, call α a 'pronominal' regardless of its morpho-phonological exponent*

As observed above, for most intents and purposes in the description of English endophoric referential expressions, we can identify the binding graph of α with the elementary graph that contains α: this is related to the fact that the distribution of pronominals and anaphors is complementary when the pronoun and its antecedent are co-arguments (thus, *John$_i$ admires *him$_i$/himself$_i$* but *John$_i$ counted four people in the room apart from himself$_i$/him$_i$*). Now, it is necessary to define *binding* formally. In MGG, *binding* is a relation between two distinct nodes α and β, where

> *α is X-bound by β if and only if α and β are coindexed, β c-commands α, and β is in an X-position* [where 'X' stands for either A(rgumental) or A' (non-argumental)] (Chomsky, 1981: 185; see also Safir, 2013)

Of the three conditions, we are interested in the first two: co-indexing and c-command. We have emphasised that if walks define sequences of nodes and edges such that a node may appear more than once in a sequence, then there is no need to have an independent indexing mechanism in the theory: if α and β nodes in a graph G share a category and a referential index, it means that they have the same semantic value. Thus, they are the *same node*, whose address is called multiple times in G: α has the same semantic value as β iff $⟦α⟧ = ⟦β⟧$ for α, β nodes in G[8] (we must point out that not all co-construal relations are covered

8 This point is related to the use of 'indices' in frameworks such as Montague grammar (Montague, 1973): in the PTQ framework, *co-indexing* (as in 'different expressions sharing a referen-

in this approach, we refer only to what Safir, 2013: 516 calls *bound variable anaphora*, leaving much aside).

This approach entails that in a walk though G (where G may be an elementary or derived graph) α will be 'bound' if α *dominates** (transitively dominates) α or, symbolically:

(237) $(\alpha, \alpha) \in \rho^*$

This is so because, as we said before, coindexation can be dispensed with given the addressing system. The requirement that there be a *strict ordering* between nodes in a sub-graph determines that there is a unique walk in which α is visited twice.[9] While this approach may seem at first glance adequate or at least intuitive, it cannot be the complete story. One of the most immediate problems that arise is that the cases of *reflexivity* we looked at in **Section 6.2.1** remain unaccounted for. To remind the reader of the relevant configuration, in reflexivity we are dealing with *parallel arcs* (distinct arcs that have the same head and tail):

(238) $\rho = \langle (v_j, v_i), (v_j, v_i) \rangle$

For example,

(239) a. John admires himself.
b. $\rho = \langle$(admire, John), (admire, John)$\rangle$

And since the tails of the 1 and 2 arcs of *admire* are the same node (there is only one uniquely identifying address that is called for twice in a walk defined in the

---

tial diacritic') has no meaning: since the set of variables of the language is indexed by the set of integers, a single indexed variable may be used several times in a formula (see also Chomsky, 2021: 16–17. In contrast to MGG, no additional stipulations are required in the present model to ensure what Chomsky calls a principle of 'Stability').

9 The definition of anaphors and pronominals above, while too general, seems compatible with the approach taken in Reinhart & Reuland (1993) in their definition of a general condition on A-chains:

A maximal A-chain $(\alpha_1, \ldots, \alpha_n)$ *contains exactly one link* -α_1- *that is both +R and Case-marked.* (Reinhart & Reuland, 1993: 696)

The feature [+R] means that an expression can be referentially independent (such as proper names or common NPs). Given the fact that chains are defined in phrase structure trees and links on a chain must be ordered in terms of c-command (such that α_m, $1 < m$, must c-command α_n if $n > m$), a chain is an ordered sequence of nodes. In our graphs the order is determined by the GF required by a predicate: the ρ-set of a graph is ordered following the GF hierarchy.

graph corresponding to the structural description of (239a), more specifically), we can account for reflexivity as defined in Reinhart & Reuland (1993) without the need to resort to indexing or to distinct nodes to occupy the subject and object positions: crucially, in Reinhart & Reuland (1993) reflexivity is a property of *predicates*, and since it is predicates that determine the number of arguments that co-occur and the grammatical functions that are assigned to those arguments, their proposal is a natural match to our system (an alternative will be mentioned in **Section 14.4**). This is important in that the ungrammaticality of a sentence like **Mario made himself* (where *make* is intended as a causative verb) is not the result of a configurational violation, but rather a lexical property of causative verbs (cf. Rosen, 1984: 183). The simplification of structural descriptions, avoiding indices and multiplication of terminals, is possible because in the present framework GFs are primitives instead of being read off phrase structure: in the derivation of a grammar in canonical form multidominance is banned (for reasons we saw in **Chapter 1**), for the specific case of 1–2 reflexivity, the same symbol cannot *follow from* both S and VP in the sense of Chomsky (1956, 1959). But there is no such restriction in the definition of dependencies in our graphs (or indeed in more derivationally oriented graph-theoretic approaches, like RG or the Phrase Linking system explored in Gärtner, 2002, 2014). Because the GF are read off the order between arcs in the ρ-set of a given expression, there is no need to assume any extra structure. It seems, then, that we need to incorporate into the notion of *binding* the fact that a single node can establish more than one grammatical relation with a predicate. Sticking to classical Binding Theory, the domain within which those relations are established determines the principle of BT that applies to the distribution of a certain expression; if within the target's governing category, then Principle A; if outside, then Principle B. This means that Principle B allows for two elementary graphs to be *linked* at the node to which corresponds to the semantic value of a 'pronominal' object, as in (240):

(240) Bruce$_i$ wishes that Selina would love him$_i$

In this case, we have two lexical predicates, *wish* and *love* (the latter, modified by an auxiliary *would*), which define two elementary graphs.[10] These are linked

10 Recall that whereas there are reasons to have modal auxiliaries anchor their own elementary graphs in Spanish, English does not allow for the same distributional freedom for modals since they have no non-finite forms; Frank's (2013) analysis thus seems to be correct in defining a modal to be part of the elementary tree anchored by the lexical verb in English.

at ⦃Bruce⦄, which is dominated by *wish* (of which it is the subject) and *love* (of which it is the object). In the Lees-Klima/Postal view, in order to pronominalise *Bruce* as *him*, it is necessary that *John* is higher up in the tree structure than *him*, and furthermore that they are mutually accessible. In our terms, these conditions can indeed be captured: because graphs are ordered sets of arcs, we can propose the following preliminary conditions as part of the definition of *binding*:

> ***Graph Binding***:
> *An expression A in a graph G* ***binds*** *an expression B in G iff*
> a) *The arc of which A is a tail is ordered before the arc of which B is a tail in the minimal graph G that contains A and B*
> b) ⦃*A*⦄ = ⦃*B*⦄

The 'minimal' graph that contains A and B may be an elementary graph or a derived graph: in the former case, then A and B are co-arguments, and conditions are met for an anaphoric interpretation. In the latter case, the relation between A and B is that between a pronominal (not an anaphor) and its antecedent. In both cases, we will have a multidominated node: if ⦃A⦄ = ⦃B⦄, then A and B are collapsed into a single node. The advantages of multidominance approaches for the analysis of coreference (and in particular when more than one grammatical relation is involved) have been highlighted since the 1960s: perhaps the first explicit endorsement of this treatment of coreferentiality is Morin & O'Malley (1969: 182) and, as we have emphasised throughout this monograph, multidominance (or 'multiattachment') analyses of coreference were a staple of RG and APG (Johnson & Postal, 1980: Chapter 11). The definition of binding graph, which in English we can identify with the elementary graph that contains the expression we are interested in (cf. MGG's *Complete Functional Complex*), takes care of the co-argumental restriction in the interpretation of bound pronouns (Reinhart & Reuland, 1993: 661 and references therein): pronominals cannot be bound with a co-argument, which in our terms entails the impossibility of re-visiting a node corresponding to a pronominal expression in a walk through an elementary graph. We are assuming that both (a) and (b) are evaluated in digraphs which comply with the *total ordering condition* in (231) above; in other words, if the nodes in a graph are not *strictly ordered*, it makes no sense to ask further questions: it is not a well-formed graph.

The idea that graphs may be *linked* at inner nodes (that is, at nodes other than root nodes), if not further constrained, can be used to argue that expressions like (227b, c), (232b), and (234b) above are well-formed expressions of the language, contrary to fact. Let us look again at the ungrammatical cases, repeated below, and see if they are adequately excluded given the assumptions above (again, traces and indices are added for purely expository purposes):

(241) a. *Who$_i$ did the President, whose son$_i$ is involved in a collusion scandal, betray t_i
b. *Who$_i$, now that President Trump$_i$ has been offered Mexico's help in the wake of Hurricane Harvey, t_i may be accepting help from a country full 'bad hombres'?

There are some interesting things that we would like to call the reader's attention to in these sentences. The first has to do with the relation between the appositive and the matrix sentence: the adjoined objects are not *self-contained*. This is because the appositive relative in (241a) contains a node that is also dominated by a node outside that sub-graph and so does the adverbial clause in (241b). Second, as we illustrate below, the ungrammaticality of these sentences is caused by the presence of the appositive clauses, *regardless of their linear position*. That is, if the adjoined domain is extraposed such that the NP they contain is not linearly between the *wh*-operator and the trace, the result is still ungrammatical, so long as the indexing is kept the same:

(242) a. *Who$_i$ did the President betray t_i, whose son$_i$ is involved in a collusion scandal
b. *Who$_i$ t_i may be accepting help from a country full 'bad hombres', now that President Trump$_i$ has been offered Mexico's help in the wake of Hurricane Harvey?

In order to account for these cases, we need to consider what exactly is happening in *crossover* cases, and then see how the cases with linked adjoined objects fall into the same category as these. Let us start from a simple case of *strong crossover*:

(243) *Who$_i$ does John$_i$ admire t_i?

We know that *who* and *John* are intended to have the same semantic value, since they are co-indexed. Therefore, they have to be the same node. We will come back to the syntax of *wh*-interrogatives in **Chapter 10**, but a preliminary analysis is not beyond us. The crucial thing to bear in mind is that in (243) *John* is the 1 of *admire*, and *who* is the 2 of the same predicate; furthermore, we know that *John* and *who* are coreferential, which means that they share semantic value. This gives us all the elements we need to characterise (243). A look at the ρ-set for (243) should clarify things:

(244) $\rho = \langle(\text{admire}, \text{John}), (\text{admire}, \text{John})\rangle$

There is no reflexivity in (243), despite the presence of parallel arcs in the ρ-set: there is at least a property of a declarative sentence (see (239) above) which ceases to hold once a new relation is created in the *wh*-interrogative, which defines *three* contexts for the node corresponding to the semantic value of *John* (we know it is the same node because *crossover* cases require co-referentiality): the operator, the subject, and the variable. But this means that this node must *dominate** itself and *be dominated** by itself in different contexts:[11] the *wh*-context dominates the Subject, but the Subject and the Object contexts are in parallel arcs (again, they are the 1 and 2 of the same predicate *admire*); thus this node cannot be ordered with respect to itself. This violates the condition of *antisymmetry* in the definition of *strict ordering*, and yields an ill-formed graph. The contrast with a *wh*-interrogative without *crossover* is, of course, stark:

(245) Who$_i$ does Mary$_j$ admire t_i?

In this case, there are no parallel arcs, and following the direction of the edges 'from the root down' we get a unique walk that visits *who* twice (the analysis of English *wh*-interrogatives will be the topic of **Chapter 10**):

(246) $\rho = \langle(\text{who}, \text{admire}), (\text{admire}, \text{Mary}), (\text{admire}, \text{who})\rangle$

In this preliminary representation, the first visit to the node with address ⦃who⦄ is as an operator that dominates the root of the elementary graph that it has scope over, the second as a direct object (more on the semantic value of *wh*-expressions in **Chapter 10**). The node with address ⦃Mary⦄ still fulfils the GF *subject* of the lexical predicate *admire*; note that the addresses for the expressions *who* and *Mary* are distinct (and thus so are their semantic values). No violation ensues, and therefore the sentence is correctly admitted as a well-formed expression of the language.

All in all, the take-home message of this section and the previous one is that parenthetical clauses are not always as opaque or syntactically independ-

11 Recall that in **Chapter 2** we defined *dominance** (notated ρ*) as the transitive closure of the *dominance* relation.

ent as a syntactically monotonic approach would have us believe. Part of the literature has focused on the syntactic heterogeneity of parentheticals (e.g., Dehé & Kahvalova, 2007; de Vries, 2009b; Kluck, 2011, 2015), specifically distinguishing between 'free' and 'anchored' parentheticals based on the presence or absence of an 'anchor' in their host (a constituent in the matrix clause that the parenthetical provides additional information about). The cases we have analysed, then, correspond to *anchored parentheticals*. Still, anchored or free, parentheticals have their fair share of idiosyncratic properties. In some cases, the syntactic uniqueness of parentheticals has been attributed to the operations that deliver the structural descriptions for parentheticals: de Vries (2009b, 2012) and Kluck (2015), for example, propose a variant of Merge called *par-Merge* such that the output of the operation does not dominate the input, by definition (unlike regular Merge, at least if interpreted somewhat graph-theoretically). If parentheticals are introduced via *par-Merge*, their opacity for c-command is guaranteed (since c-command can be expressed in terms of sisterhood+dominance), but, even then, anchored parentheticals remain anomalous.

Another problem is to specify exactly how Merge can deliver an undominated syntactic object. Suppose that Merge is strictly set-theoretic: the closest correlate of *dominance* is *membership*, such that a node A dominates a node B in a phrase structure tree iff B belongs to (a set that belongs to) a syntactic object labelled A (leaving aside the problem that A, as a label, may not be part of the syntactic representation *sensu stricto*; see Seely, 2006: 189). Then, *par-Merge* would deliver membership without membership; a clear contradiction. Graph-theoretically, the problem is not any easier: suppose that the output of Merge(X, Y) in graph-theoretic terms needs to be a graph G where the label of the syntactic object corresponding to the output of Merge is a node in the graph that immediately dominates X and Y (see e.g. Zyman, 2023). Again, having any form of Merge under these assumptions without domination (understood as 'memership', as the closest set-theoretic analog) is contradictory.

If parentheticals were uniformly derived (that is: if all parentheticals were created by means of the same set of operations, targeting the same structural positions, delivering identical phrase markers), as most analyses based on *phase theory* and *multiple spell-out* propose, then any element embedded in the parenthetical should never be visible outside the parenthetical's root, contrary to fact (in addition to the data considered here, see also Hoffmann, 1998). A configurational approach combined with an appropriate definition of 'opacity', here given in (205) in terms of *self-containment* and of 'transparency' in terms of *linking* seems to be preferable to purely configurational approaches in which the opacity of certain labelled domains is defined *a priori*.

The analysis presented in this chapter can be extended to other phenomena that have frequently been studied alongside parentheticals: syntactic amalgams. Of particular relevance to graph-theoretic syntax is van Riemsdijk's (2006) *graft* approach, which we discuss in the following section.

8.1 Grafts and Graphs

In a series of papers, van Riemsdijk (2000, 2006, 2010) has proposed an operation *graft* as part of the family of configurations that 'follow from Merge'. The idea is intuitive enough, and shares much with earlier work on multidominance and discontinuity. Let A and B be root nodes of disconnected trees (such that no node of A dominates a node in B and vice-versa):

(247)

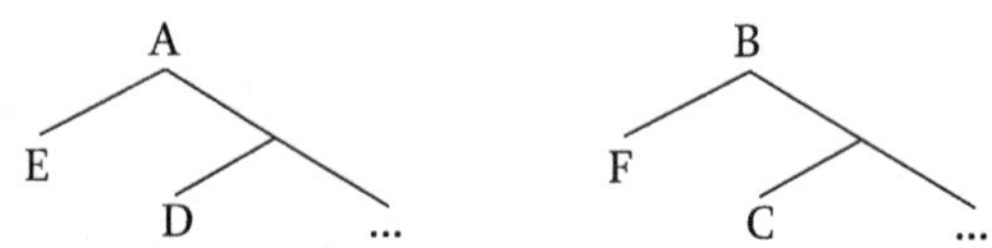

FIGURE 8.1 Input for Grafting

Only a stipulation, van Riemsdijk says, can prevent Merge from applying to the pair {B, D} and delivering (248), where B extends to β after Merge with D (van Riemsdijk, 2006: 19):

(248)

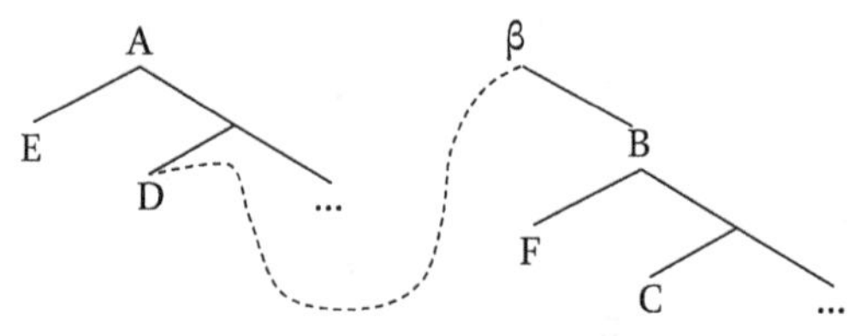

FIGURE 8.2 Grafting

Of the constructions that van Riemsdijk analyses using *grafts* (which include free relatives, transparent free relative clauses, and amalgams), here we will focus on one to show some technical problems with *graft* and how a sparser graph-theoretic grammar can handle it while keeping the aspects of the *graft* analysis that are empirically attractive.

Simple but crucial empirical cases are NPs such as *a far from simple matter* (Kajita, 1977). Van Riemsdijk poses the question of what the modifier actually is: if a projection of *simple* or of *far*. Intuitively, it is a projection of *simple*, since we have *a simple matter* but not **a far matter*. In this case, however, the analysis of *far from* is problematic, as *far from* is not a constituent: it thus cannot be analysed on par with pre-adjectival modifiers like as *very* or *quite*, which are simple adverbial phrases we can locate in the specifier of AP. Van Riemsdijk's analysis implies a radical departure from traditional phrase structure trees, in that it assumes two parallel structural analyses for part of the NP:

(249)

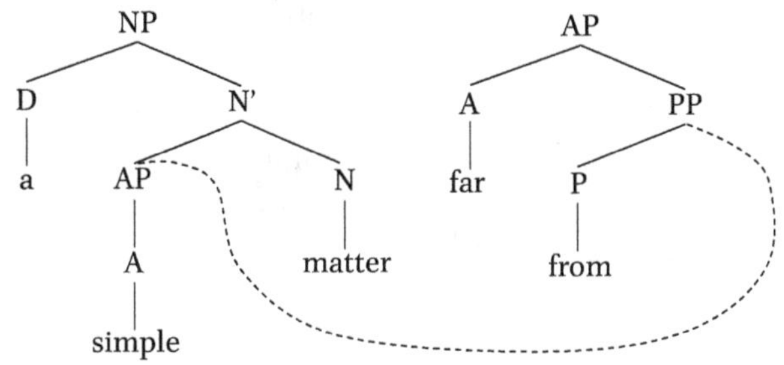

FIGURE 8.3 Grafting analysis of 'a far from simple matter'

In (249), minimally adapted from van Riemsdijk (2010: 291), *simple* is the head of the attributive AP modifying *matter*, 'both syntactically and semantically' (Op. Cit: 290). There are two pieces of structure: (i) the NP headed by *matter* and (ii) the AP headed by *far* which takes a PP as a complement and embedded in which we find the AP headed by *simple*. Because *simple* is also a prenominal modifier in the NP, this expression acts as node D in (248), connecting both local structures.

An immediate question that arises, within the boundaries of MGG, is whether grafts are indeed compatible with the current formulation of Merge, let alone follow from it. In contemporary generative grammar, as emphasised throughout this monograph, Merge is a strictly set-theoretic operation. Applied to any two syntactic objects, Merge outputs the set containing these syntactic objects and nothing else (Collins & Stabler, 2016; Collins, 2017; Epstein et al., 2015; Chomsky, 2020, 2021):

(250) Merge(X, Y) = {X, Y}

In this context, what *graft* requires is that the output of

(251) Merge({far, {from, {simple}}}, {a, {{simple}, matter}})

somehow unifies the instances of *simple* in both syntactic objects. But that is not possible: the diagram in (249) does not correspond to an object that can be defined under set-theoretic Merge. This is partly so because Merge-based grammars are concatenation-based, not unification-based (Jackendoff, 2011) due—among other things—to the so-called *No Tampering Condition* (Chomsky, 2008: 138): Merge of A and B leaves both A and B internally untouched. All Merge can do is deliver the set containing the two syntactic objects under consideration, namely,

(252) {{far, {from, {simple}}}, {a, {{simple}, matter}}}

A possible way to obtain the desired structural description under Merge would be to introduce the complex modifier before completing the DP: each object would be built in parallel (possibly along the lines of Uriagereka, 2002), and then Merge would need to apply as follows:

(253) **Workspace 1:**
Merge(from, {simple}) = {from, {simple}}
Merge(far, {from, {simple}}) = {far, {from, {simple}}}
Workspace 2:
Merge({far, {from, {simple}}}, matter) = {{far, {from, {simple}}}, matter}
Merge(a, {{far, {from, {simple}}}, matter}) = {a, {{far, {from, {simple}}}, matter}}

However, such a derivation is not very informative from a grammatical point of view: simply getting rid of intermediate nodes does not solve the problem of modification. The question now is how exactly *simple*, an object embedded in an atomised syntactic object, comes to determine the distribution and interpretation of that object (which presumably would be labelled by the highest head *far*). Note that while the specific example we are considering does not evidently require an operation to relate two trees (since at the derivational point where *far from simple* would be introduced in the NP, all we have is the terminal *matter*), it is easy to construct examples where relating complex objects is unavoidable:

(254) A far from simple matter of linguistics

In this case, the need for an operation that takes two complex structures as inputs and delivers one as output is more evident: a generalised transformation must be called upon. The output of this generalised transformation cannot, as de Vries' *par Merge*, eliminate dominance relations in its output, since we do need that relation to account for modification. The classical way of doing this (e.g., the original definition of Merge in Chomsky, 1995: 189–190) would require us to concatenate [matter of linguistics] with a placeholder, and replace the placeholder with the sequence [far from simple]. Recognising this, however, does not amount to a solution, because the specifics of the operation must be made explicit. Not just how the operation works within the limits of the so-called 'narrow syntax', but also how it allows for a compositional interpretation of complex structures. This is the second problem with grafts: there is no attempt so far as we know to provide a model of semantic interpretation for structures like the above. Finally, but by no means a minor issue, there is nothing restricting the size of grafts: the grafted tree *far from simple* could be of arbitrary complexity. Unless the size of elementary structures is somehow restricted, it is difficult to see how to avoid overgeneration if each structure in a graft is well-formed.

We contend that set-theoretic Merge is not the optimal framework in which to implement grafts, but van Riensdijk's structures immediately make more sense if viewed from the perspective of graph theory (see also McKinney-Bock & Vergnaud, 2014): from the viewpoint adopted in this monograph, his *grafts* are simply distinct elementary graphs linked at the common node. *Grafting*, then, is a name one could give to *graph composition under structure sharing* (i.e., graph union). Some minimal adjustment is necessary under the assumption that elementary graphs are anchored by predicates: following LTAG practice, lexicalisation is a way to restrict the size of elementary structures. In this context, a possible analysis for the case under consideration goes along the following lines: suppose that the structure *far from simple* is syntactically transparent (i.e., that *far*, *from*, and *simple* are all distinct categorematic basic expressions; for a different perspective see Brinton & Inoue, 2020 for an analysis under which *far from* is undergoing grammaticalisation, and thus may be a basic expression itself). The simplest graph for it would be (255):

(255) G = ⟨(far, from), (from, simple)⟩

At the same time, we have an arc for *simple matter*:

(256) G' = ⟨(simple, matter)⟩

Graph union applied to G and G' delivers precisely what van Riemsdijk intends, under the assumption that *simple* is not a lexical item token (as it is in Minimalism) but a uniquely addressed expression. Note that in this analysis the predicate *simple* dominates *matter*, as in van Riemsdijk's analysis (but in contrast to what would be the output of *par Merge*). The more complex case in (254), for completeness, requires minimal additions to G':

(257) G" = ⟨(simple, matter), (of, matter), (of, linguistics)⟩

In G", the preposition (which is not idiosyncratically selected, and thus corresponds to a node in the graph) is a 2-place predicate, the first of whose arguments is *matter* and the second of which is *linguistics*.

Grafting as proposed by van Riemsdijk has been applied to the analysis of constructions such as RNR (the analysis in de Vos & Vicente, 2005 bears resemblance to *grafting*, although they do not cite van Riemsdijk), Horn amalgams, and transparent free relative clauses (TFR). We can briefly examine the extent to which our framework can say something about the latter phenomenon now (leaving RNR for **Section 14.1**). The relevant cases are like (258) (258c, d taken from Grosu, 2003, exx. (2c, d), (258e) is taken from Wilder, 1998, ex. (3c)):

(258) a. She invited [what seems to be [a policeman]$_{NP}$]
b. Liz was [what you may call [unqualified]$_{AP}$]
c. He came out the next day, but I didn't get a chance to talk to him [what you might call [$_{AdvP}$ privately]].
d. He felt my mother was [what he called [$_{VP}$ poisoning my mind]].
e. A [what you may call [tricky]$_{AP}$] example (cf. A tricky example)

Transparent free relatives (TFR, given this name by Wilder, 1998: 191) are *wh*-clauses obligatorily introduced by *what* (and possibly *who*; see Schütze & Stockwell, 2019) which contain a predicative construction headed by a copula (*be*), a Raising verb (*seem*), or a propositional attitude verb (*consider, call, be described* (*as*), *regard* (*as*) ...) plus a small clause. The predicate in this small clause is what determines the distribution of the TFR: it must satisfy selectional requirements outside the TFR. This element, which seems to play a role both inside and outside the TFR, is called a *pivot*. An essential question is whether the pivot is interpreted only TFR-internally, only TFR-externally, or both, and what consequences this has for syntactic structure.

For example, consider (259):

(259) *What you may call stupid just walked in (Wilder, 1998: 193)

(259) is ungrammatical because the distribution of the TFR is determined by the adjectival predicate *stupid*: an adjective cannot occupy the subject posi-

tion of *walk in*. Furthermore, other properties such as definiteness (260b-c) and agreement (260d) may be inherited from this predicate:

(260) a. [What could best be described as pebbles] were strewn across the lawn (McCawley, 1998: 758)
b. There were [what could best be described as pebbles$_{\text{Indef}}$] strewn across the lawn
c. *There were [what could best be described as the pebbles$_{\text{Def}}$] strewn across the lawn
d. [What could best be described as pebbles$_{\text{Plural}}$] *was/were strewn across the lawn

These facts were initially used to argue that it is the pivot, and not the *wh*-element, that heads the TFR (Kajita, 1977). In Kajita's proposal, the TFR minus the pivot is essentially a parenthetical. Building on Kluck's (2015) typology, we may specify this and say it is an *anchored* parenthetical (to separate TFR from *free* parentheticals such as *Togusa is—I think, but I'm not really sure—a detective*).

Three main analyses of TFR have been proposed: deletion, relativisation, and multidominance (Kluck, 2011; Schelfhout et al., 2004). The deletion analysis, in the version in Wilder (1998) assumes that the matrix clause and the TFR are two distinct phrase markers. The TFR gets inserted like any other parenthetical, and the 'shared' constituent is (backwards) deleted inside the TFR under identity:

(261) a. *Syntax: independent phrase markers*
[John bought [$_{DP}$ a guitar]] *[what he took to be [$_{DP}$ a guitar]]*
b. *Phonology: parenthetical placement and deletion*
John bought ⟨what he took to be ~~a guitar~~⟩ a guitar
WILDER (1998: 195)

This approach has the advantage of satisfying selectional properties of the predicates in both anchor and parenthetical clauses: this allows Wilder to extend his analysis directly to amalgams:

(262) John invited ⟨you'll never guess what kind of ~~people~~⟩ people to his party

This is an important point, insofar as it emphasises the syntactic unity of amalgams and parentheticals.

The parenthetical placement approach is adopted and developed in Kluck (2011), who argues that these constructions are derived by means of *par Merge* + sluicing (where deletion is, as in Wilder, hypothesised to be a PF matter). The appeal of sluicing is clear in the cases of so-called Andrews amalgams (e.g., *Doug bought you'll never guess what*), but some adjustments are necessary for so-called Horn amalgams (e.g., *Doug bought—I think it was a guitar*). An alternative, proposed by Schelfhout et al. (2004) is to introduce the parenthetical *as is* into the matrix sentence, without containing an occurrence of the pivot. The incomplete predication structure that results from this approach is not problematic, the authors say, if we consider that other parentheticals have similarly incomplete argument structures (examples taken from Schelfhout et al., 2004: 88):

(263) a. "I don't think," Jones said, "that this would be a good idea."
b. That's not what your father meant, I think, but you could ask him

Next is the relativisation analysis. Grosu (2003) analyses TFR as relatives with a null antecedent. This antecedent shares category with the predicative TFR-internal XP:

(264) a. He made $[_{DP}\ e]$ $[_{CP}$ what$_i$ $[_{IP}$ $[_{SC}\ t_i$ may appear to be $[_{DP}$ a radically new proposal]]]].
b. He made an uninspired and $[_{AP}\ e]$ $[_{CP}$ what$_i$ $[_{IP}$ I'd describe $[_{SC}\ t_i$ as $[_{AP}$ catastrophic]]]]] decision.

However, Grosu (2014) proposes a unified representation for TFR and 'standard' free relatives, with a null D serving as a CP-external head. No mention is made of a null A (or a null supra-category) in this later work, and no examples of TFR with non-nominal predicates are considered. Unlike Wilder, Grosu takes the TFR to be *subordinated* to the main clause, being adjuncts to the null antecedent (building on the head-external analysis of relative clauses). TFRs are thus not analysed as parentheticals, but as (garden-variety) relative clauses. A potential problem for this analysis is that TFR do not display island sensitivity, as has been widely documented in the literature. If TFR receive the same syntactic analysis as standard free relatives, the obviation of islandhood effects in TFR remains unexplained.

Finally, the multidominance analysis has been explored, among others, in van Riemsdijk (2006) and Guimarães (2004). The basic insight in these proposals is that there is a single syntactic object, the pivot, that enters dependencies with a head within the parenthetical as well as in the matrix clause (or 'host'). Guimarães' approach is based on multidominance as re-Merge under the assumption that parentheticals and amalgams are *not subordinated* to their hosts. In a theoretical departure from MGG, he works within a top-down model of syntax. Parentheticals and amalgams are multi-rooted structures, where each single-rooted tree is built from a separate numeration: these numerations contain common elements, which end up being multidominated. This means that if multi-rooted representations are required (and they are, given the fact that parentheticals are not embedded in their hosts), multi-rootedness must derivationally precede multidominance (since the shared syntactic object will be transitively dominated by the root).

A crucial difference between the multidominance analysis and the relativisation analysis, in addition to the hypotaxis-parataxis debate, pertains to the occurrence of the pivot in the matrix clause. Note that in Grosu's analysis the pivot does not occur in the matrix clause, only in the embedded clause; selectional restrictions are presumably satisfied by the null antecedent. Do we need to assume that the pivot is a constituent of the host or only of the parenthetical? Examples such as (265) seems crucial to answer this question (see also van Riemsdijk, 2006: 35):

(265) They made [what could be charitably referred to as headway]

In their critique of the GB analysis of relative clauses (the *head-external* analysis), proponents of the *raising analysis* (according to which the antecedent of a relative clause begins its derivational life inside the relative and moves to its superficial position; see e.g. Schatcher, 1979; Kayne, 1994; also Bianchi, 2000 for a general overview) often appeal to the relativisation of idiomatic expressions as an argument that the antecedent must be a constituent of the relative. Compare in this respect (266a) and (266b):

(266) a. The headway they made was impressive
b. *The headway was impressive

The argument from the raising analysis was that the antecedent *headway* needs to occur inside the relative so that the idiomatic expression *make headway* forms a constituent at some level of representation or at some derivational step. The same argument could be invoked in the present context: we cannot propose that *make headway* is a multi-word basic expression insofar as relativisation and structures like (265) (in which *make* and *headway* are not adjacent linearly or structurally, and therefore cannot be a basic expression) are possible, but we can require that *make* immediately dominates *headway* within its elementary graph. This requirement can be straightforwardly satisfied under multidominance: the expression *headway* can be immediately dominated by *make* in the elementary graph anchored by *make*, which licenses the idiomatic interpretation, maintained under structure sharing.

There seems to be reason to think, then, that the pivot is a constituent of *both* the host and the parenthetical, in contrast to Grosu's analysis. Under present assumptions, this means that the host and the parenthetical must be independent elementary graphs: this also entails that selectional restrictions of each anchor must be satisfied *within* their respective elementary graphs: for a TFR such as

(267) Liz is [what you may call [insane]$_{AP}$]

We have:

(268) Elementary graph 1: [Liz is insane]
Elementary graph 2: [you may call what insane]

With Wilder, Guimarães, and van Riemsdijk, we propose that the TFR is not subordinated to its host: in the present framework, this means that no node in the elementary graph corresponding to the 'extended projection' of the host predicate immediately dominates the root/anchor node of the elementary graph corresponding to the TFR. We employ not deletion (in contrast to Wilder), but structure sharing as delivered via graph union: as highlighted previously, whereas deletion outputs always more than one syntactic object (only one of which receives a PF exponent), structure sharing outputs always only one. In contrast to Guimarães, we have no numerations or multiple Spell-Out, and whereas the specifics of how multidominance is obtained under Minimalist assumptions (given the fact that Minimalist numerations are sets of tokens) are left somewhat unclear in Guimarães' work, the addressing system proposed here delivers structure sharing as a result of the composition of local graphs with no further stipulations.

We close this section with a remark about bound pronominals inside TFR. Kluck (2011: 98) argues—convincingly in our opinion—that a multidominance approach along the lines of van Riemsdijk's grafts has problems to account for the bound reading of pronominals within TFR:

(269) a. Every professor$_i$ was kissing [what seemed to be his$_i$ mistress].
b. Every student$_i$ was kissing [what he$_i$ considered to be an attractive woman].

Under a graft analysis assuming phrase structure trees, according to which only *his mistress* and *an attractive woman* would be the shared constituents between the matrix clause and the TFR in (269 a, b) respectively, the bound reading for the pronouns is hard to explain: in (269a) the quantified NP *every professor* does c-command the pronoun *his* in the matrix clause, and thus the bound reading is allowed. But there is no c-command between *every student* and *he* in (269b), since no constituent of the matrix clause c-commands any constituent of the TFR other than the pivot.

A way to solve this problem is to assume, as we do, that bound pronominals arise 'transformationally': there are not two distinct nodes *every student* and *he* in (269b), but only one which is multiply dominated (the 'unbound' occurrence corresponds to the tail of the arc that is ordered before all others with the same tail in the derived graph). The TFR and the matrix clause would be two different elementary graphs, *linked* at ⦃woman⦄ and ⦃student⦄. This is distinct from saying that TFR are subordinated to their hosts (Kluck, 2011: 98), in that in our sketched analysis there is no node in the matrix clause that dominates the root of the TFR: the TFR is not a subgraph of the matrix clause but a distinct elementary graph.

In this chapter we developed the distinction between self-contained graphs and linked graphs introduced in **Chapter** 7 to analyse anchored parentheticals that pose interesting challenges to traditional approaches to the syntax of referentially bound expressions. In connection to these considerations, the following chapter will deal with aspects of complementation within the NP, more specifically, aspects of the syntax of non-restrictive (a.k.a. *appositive*) relative clauses and their similarities with what we have called *parentheticals*.

CHAPTER 9

Complementation within the NP

In this chapter we will deal with some aspects of the syntax of *non-restrictive*, or *appositive*, relative clauses and their relation with *restrictive* relatives. This builds on the discussion in the last couple of chapters, insofar as we are dealing with relations across elementary graphs in what derivational syntax would define as non-monotonically derived structures. Our goal is to refine the formulation of the conditions under which these relations are allowed in English. We derive these structural properties from conditions on linking across elementary graphs, and finally propose structural descriptions for restrictive and appositive relative clauses (having briefly mentioned transparent relatives in **Section 8.1** above).

The relation between appositive relatives and the NPs that contain them is a problematic one. The level of syntactic integration of appositives to their host has been object of controversy: some linguists propose that appositives are not syntactically integrated at all (e.g., Haegeman, 2009; also Peterson, 2004) and constitute separate syntactic objects perhaps only linked at the discourse level (see also Fabb, 1990: 75–76 for a related view), whereas others assume that the relation between an appositive and its antecedent is akin to inter-clausal relations (coordination, as in de Vries, 2006; see also Ross, 1967 and Emonds, 1979; or subordination, as in Kuroda, 1968 and Jackendoff, 1977). The effects of parenthetical placement on syntactic relations at the host, however, seems to be somewhat agreed upon. In this respect, we side with McCawley in that '*parentheticals are placed by that changes word order without changing constituent structure*' (1982: 95). Crucially, as emphasised in the previous chapter, this does not entail a 'radical orphanage' approach to appositive relatives (see also Arnold, 2007 for a critique of 'radical orphanage'): the extent to which appositives are syntactically integrated into their hosts cannot be underestimated, at the risk of missing important empirical generalisations. In this chapter we will build on what we've seen in **Chapter 8** and provide further evidence to the effect that there are exceptions to McCawley's claim that '*all grammatical phenomena to which the constituency of the* [target of parenthetical placement] *is relevant behave as if the parenthetical were not there*' (1982: 96), which reveal some interesting aspects of the conditions we can impose over relations among sub-graphs. It is interesting to note at this point—as an epigraph of sorts—that Emonds observes that Parenthetical Formation allows for violations of Subjacency (Chomsky, 1973), for example:

 | DOI:10.1163/9789004542310_010

(270) We introduced a diplomat who got, $\left\{ \begin{array}{l} \textit{it seemed to us} \\ \textit{in our opinion} \end{array} \right\}$, too much attention (Emonds, 1979: 212, fn. 1)

The McCawlean view, according to which the rule of Parenthetical Formation (which he refers to as Parenthetical Placement) is an *order changing rule* which does not affect constituency (i.e., it does not change grammatical relations) renders the apparent problem of the violation of Subjacency moot: if there is no syntactic displacement, then constraints on displacement of course do not apply.[1] We will return to this line of reasoning when briefly dealing with Right Node Raising, in **Section 14.1.**

This section begins with an uncontroversial observation, which is that English appositive relative clauses are non-recursive (this does not mean that they are not iterable, of course). The point is illustrated by Emonds and McCawley with examples like (271a), to be contrasted with (271b):

(271) a. *John, who_1 goes to MIT, who_2 likes math, will get a job (ungrammatical in a reading in which the antecedent of *who*$_2$ is *John, who goes to MIT*)
b. People who go to MIT who like math will get jobs (Emonds, 1979: 222)

Note that it is possible to save (271a) in a strictly paratactic reading, but in that case the structural description is different, not monotonically recursive. In Emonds' own terms,

> *the string* ***John, who goes to MIT*** *is not a constituent correferential with the pronoun following* (*that is, the string is not even a constituent*) *under the MCH* [Main Clause Hypothesis] (Emonds, 1979: 222)

This means that an appositive relative cannot take as an antecedent a structure of the kind [NP [Appositive]]. Put differently, *appositive relatives do not allow recursive modification.* An appositive relative, however, can take as an ante-

1 For a Minimalist analysis of Italian and (some) English data, according to which there is post-Spell-Out movement of the parenthetical, see Del Gobbo (2007). We will not discuss this analysis here, given the fact that it depends on several theory-internal assumptions (displacement-as-movement; overt vs. covert operations; a Y-model architecture with independent PF-LF levels of representation) none of which is independently justified in the present context.

cedent an NP modified by a restrictive relative clause (see e.g. (277a), below); this seems to indicate that the structural relation between restrictives and their antecedents is closer than that between appositives and theirs.

Restrictive relative clauses (both *wh*-relatives and *that*-relatives), on the other hand, *are* monotonically recursive, in the sense that a restrictive relative clause can take a structure of the form [NP [Restrictive RC]] as its antecedent. (272) illustrates this fact, with all possible combinations of *wh*- and C:

(272) a. [[Every psychologist [**that** Emily talked to]]$_i$ **who**$_i$ insisted on helping] ended up making a disaster
b. [[Every psychologist$_i$ [**who**$_i$ Emily talked to]] **that** insisted on helping] ended up making a disaster
c. [[Every psychologist$_i$ [**who**$_i$ Emily talked to]]$_j$ **who**$_j$ insisted on helping] ended up making a disaster
d. [[Every psychologist [**that** Emily talked to]] **that** insisted on helping] ended up making a disaster

McCawley (1998: 435) explicitly claims that because restrictive relative clauses (RRC) are sister-adjoined to N' (excluding determiners and quantifiers),

> [*t*]*he structure with the restrictive relative modifying an N' yields the correct prediction that restrictive relative clauses can be stacked, since there is nothing to prevent the N' of an* [$_{N'}$ *N' Comp'*] [read: N' S'] *from itself having the form N' Comp'*

The exclusion of the determiner in an NP from the scope of the relative is a crucial aspect of both the structure and interpretation of restrictive relatives. An early semantic argument in favour of an [N' Comp] analysis is to be found in Partee (1975), who argues that a compositional interpretation for restrictive relatives has the head noun of NP and the relative clause (S' / CP) form a derived category CN (common noun), which is then modified by Det / Quant. In this view, then, the correct interpretation of *the man that I saw* is 'the unique *x* such that *x* is a man and I saw *x*', which requires not that there is a unique man, but rather that there is a unique man that I saw (see also Bach & Cooper, 1978). If, on the other hand, restrictive relatives were adjoined to NP (where NPs are assumed to rewrite as Det, N') this would yield an inadequate compositional interpretation (bear in mind that in a Montagovian setting each syntactic rule has an associated semantic interpretation rule; see Partee, 1975: 213, 223 point 9; and for a very similar perspective in a different syntactic framework, the presentation of rule format in Gazdar, 1981: 156). This approach can be seen also in the head-external analysis in Demirdache (1991: 109), according to which restrictive RC are NP adjuncts, and appositive relatives are DP adjuncts (see Krivochen, 2022 for an overview of head-external, raising, and matching analyses and their competing structures).

Perhaps more important for our purposes, though, is the observation that [N' Comp] sequences behave, under certain tests, like *constituents*: transformations targeting N can (and in some cases *must*) equally target N' Comp. *Pronominalisation* is one such transformation:

(273) Tom has [[a violin]$_i$ which once belonged to Heifetz]$_j$, and Jane has one$_{?i/j}$ too (example taken from McCawley, 1998: 445; indexes and judgments are ours)

Stripping is another one, as we can see in the following examples ((274a) is taken from McCawley, 1982: 96; (274b) is taken from McCawley, 1998: 450):

(274) a. John sold Mary, who had offered him $600 an ounce, a pound of gold, but Arthur refused to Ø. (Ø = refused to sell Mary a pound of gold; ≠ refused to sell Mary, who had offered him $600 an ounce, a pound of gold; ≠ refused to sell Mary)
b. John sold Mary, who had offered him $600 an ounce, a pound of gold, and Arthur did Ø too (Ø = sold Mary a pound of gold; ≠ sold Mary, who had offered him $600 an ounce, a pound of gold; ≠ sold Mary)

Also relevant in the present context is the observation that an N' plus a restrictive relative can be conjoined with another N' plus a restrictive relative (as in (275a)); that is not possible with appositive relatives (as in (275b); examples taken from McCawley, 1998: 446):

(275) a. [Some violins [$_{\text{RRC}}$ that have been owned by famous performers]] and [flutes [$_{\text{RRC}}$ that were played by Frederick the Great]] are going to be sold at auction.
b. [*These violins, [$_{\text{Appos}}$ which were made by Cremonese masters]], and [pianos, [$_{\text{Appos}}$ which were made in nineteenth-century Paris]], are expected to fetch high prices.

Appositive relative clauses can co-appear with restrictive relative clauses, albeit under specific conditions. Emonds (1979) restricts the possibilities of co-appearance in the following descriptive statement:

a restrictive can follow an appositive if it is the only constituent that follows (Emonds, 1979: 222)

Examples of this structure (Appositive + Restrictive) are given in (276):

(276) a. We found that movie, which cost plenty, that you so highly recommended (Emonds' ex. (22))[2]
b. It was Fred, who you met at my party, that I was just talking to on the phone (McCawley, 1998: 449, ex. (19 a); the appositive is here followed by a free relative clause)

We may add that given the adjacency condition imposed over Bare Relatives (a.k.a *abridged* relative clauses, or S' subjected to a transformation *Whiz-deletion*, which deletes a *wh*-operator and auxiliary *be* in a restrictive relative clause; see Ross, 2012: 10) we will only deal with *Wh*-relatives and *that*-relatives in this section, because we are interested in the recursive properties and relative positioning of restrictives and appositives, and bare relatives are neither recursive nor can they be in any position that is not strictly adnominal (although, as usual, they are attested; see e.g., McCawley, 1998: 433, ex. (15b)).

In contrast to the distributional condition formulated by Emonds (and cited above) with respect to restrictives following appositives, an appositive *can* follow a restrictive relative clause even if it is not the final constituent in the string:

(277) a. The children that you brought, who were charming, got sick later (adapted from Emonds, 1979: 222)
b. *The children, who were charming, that you brought got sick later (Emonds' ex. (21))

Is there any way in which the syntactic approach advanced here can shed light on the distribution and properties of restrictive versus appositive relatives? We believe there is, and that the differences in distribution as formulated by Emonds (and also illustrated by McCawley) can be accounted for if we assume that (a) structural descriptions for English sentences can take the form of maximally connected graphs, and (b) the relevant conditions over graph

2 Not all our informants were happy with this sentence, but we keep Emonds' own judgment. If the antecedent NP appears in subject position, the acceptability of the sentence decreases drastically, as indeed noted by Emonds:
i) ??/*That movie, which cost plenty, that you so highly recommended ended up being a disappointment.

well-formedness from syntactic and semantic points of view are expressed in terms of possible and impossible relations between nodes and between sub-graphs. In this sense, recall that above we introduced the notion of graphs being *ordered*, and the condition that order be *strict* and *total* (see (231) and the discussion that follows) at the level of individual arbores. This order is *transitive*: if there is a relation R applying to nodes *a* and *b*, for *b* the root of a graph G, then R will hold between *a* and every node in G because G (if well-formed) will be totally ordered. This means that, given a string featuring a number of relative clauses (note that we have not qualified 'relative clauses'; in principle they could be either restrictive or appositive), the graph G corresponding to the structural description of that string will be well-formed if and only if there is a unique strict ordering $O(G)$ in the derived graph.

A notion of ordering is inherent to McCawley's observation that a restrictive relative clause can take a structure [N' Comp] as an antecedent (however the reader chooses to represent those nodes, we will just use the term Complex NP, CNP, henceforth). Let us consider (272c) again, repeated here as (278):

(278) Every psychologist who$_1$ Emily talked to who$_2$ insisted on helping ended up making a disaster

The antecedent of *who*$_1$ is the NP *Every psychologist*; the antecedent of *who*$_2$ is the CNP *Every psychologist who Emily talked to*. The 'size' of the antecedent grows monotonically with the introduction of new RRC, and the ordering between these is unambiguous. The point is that the size of the antecedent of a restrictive relative grows continuously and always at the same rate (one CNP per step), and so does the function corresponding to the interpretation of antecedents. Graphically, the monotonic growth in antecedent size can be represented as in (279):

(279) [[Every psychologist] who$_1$ Emily talked to] who$_2$ insisted on helping

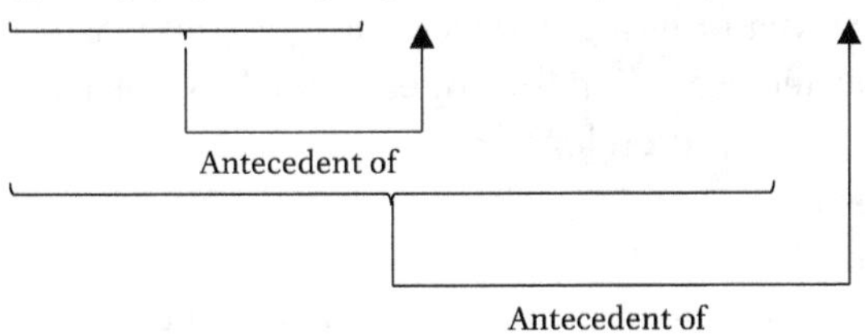

The monotonic growth in the structure gives us precisely the kind of strict ordering that we require as a well-formedness condition on graphs. In this case, the order depends on *n*-order functions corresponding to the interpretation of

the relative operators. This view is compatible with the observation in Bach & Cooper (1978) that a relative clause can denote properties of second order: *who insisted on helping* can denote (in these authors' terms) not only the property of insisting on helping asserted of x, but also the property of insisting on helping *and* having property P (in the particular case of (279), P is *Emily talked to* x). Because there is no principled limit to this process, this gives us the desired recursive structure, indefinitely (see also Bach & Cooper, 1978: 149).

The question now is, what happens in the case of appositives? The argument we put forth here is that appositive relatives are *not* monotonically recursive *because there cannot be a strict ordering for a set of recursive appositive relatives.* Crucially, the diagram that we used for (272), with its recursive semantic interpretation and monotonically growing antecedent size does *not* correspond to the interpretation of (271a), repeated here as (280):

(280) *John, who$_1$ goes to MIT, who$_2$ likes math, will get a job

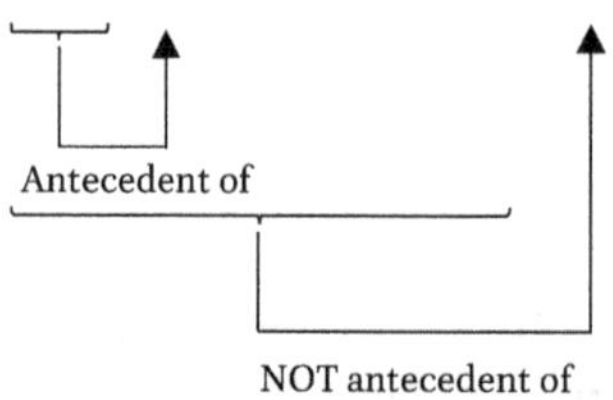

It is evident that in (280a) the antecedent of *who$_2$* is *not John, who goes to MIT*, but just *John*. But that is the same antecedent as *who$_1$*: whatever relation exists between *John* and *who$_1$* also exists between *John* and *who$_2$*, and neither of these takes the other as an argument; there are no first- and second-order functions (i.e., functions that take functions as their arguments) in the correct representation for (271), as opposed to the situation in (279). This means that there cannot be a strict ordering (total, antisymmetric, irreflexive, intransitive) between these two nodes *who$_1$* and *who$_2$* in the appositive case; in turn, this means that there is no strict ordering between the sub-graphs which contain these nodes, because if there was, then the nodes would be transitively ordered (recall that *transitivity* is a condition for *strict ordering*). That is: if G" is a graph properly containing G and G', and if (the root of) G which contains *who$_1$* was ordered with respect to (the root of) G' which contains *who$_2$*, then—as we said above, because the order is *total* and *transitive*—*who$_1$* would be ordered with respect to *who$_2$*, contrary to fact.

We said that there cannot be a strict ordering over a set of appositives such that we get a monotonically recursive interpretation (following Emonds and McCawley); it must be noted that the reading in which multiple appositives following an antecedent NP are strictly paratactic receives a different structural description. Recursive restrictive relative clauses can (and in fact must) take as antecedents ever-growing structures [N' Comp]: it is *not* possible to have a sequence of *restrictive* relatives in which all take the same N terminal as their antecedent, ignoring other relatives. To give a concrete example, the constituent segmentation and indexing in (281) below for a sequence of *restrictive* relatives is impossible to obtain, because restrictive relatives are monotonically recursive:

(281) [[Every psychologist]$_i$ [who$_i$ Emily talked to] [who$_i$ insisted on helping]] ended up making a disaster

But a non-(center) recursive, flat structure like that is precisely the kind of structural description that we have in the case of several *stacked* appositives, as we do in (282) (an example encountered in the wild):

(282) Donald Trump is a man who will spare no effort to get different parts of the country to hate and fear each other, who will do everything he can to damage the U.S. position in the world, who will set things up so his family members benefit financially from his presidency, …

The only possible structural description is one in which dependencies are strictly paratactic; computationally, the set of NRRCs define a *finite-state* sequence:

(282') Donald Trump is [a man]$_i$ [who$_i$ will spare no effort to get different parts of the country to hate and fear each other], [who$_i$ will do everything he can to damage the U.S. position in the world], [who$_i$ will set things up so his family members benefit financially from his presidency], …

We want to highlight that what we have here is *head-tail* recursion (Uriagereka, 2008: 228; see also his finite-state treatment of iterated small clauses on pp. 204, ff.). In the terms of Krivochen & Schmerling (2016a), the appositive relatives in (282) and examples of the sort are *que-coordinated*, each being a state in a Markov chain (see **Chapter 12**): the corresponding phrase marker would exhibit flat dependencies (Lasnik, 2011; Krivochen, 2021a, 2022). There is no embedding / hypotaxis between the appositive clauses (as is widely agreed upon in the literature), thus, no order imposed in terms of proper containment. The order among the appositives in cases like (282) is strictly linear, with parataxis being the only structural option to save the representation. All the relative clauses in (282) have the same antecedent, [a man]: we could shuffle these relatives around (change their relative linear position) and still get a well-formed sentence with no change in meaning. In other words, the subgraphs corresponding to the relative clauses all share the nodes corresponding to the expression *a man*: the derived ρ-set of (282), with some simplifications, is (283):

(283) ρ = 〈(spare, man), (spare, effort), (do, man), (do, everything), (set up, man), (set up, things) ...〉

If we follow the approach in Schmerling & Krivochen (2017), according to which the denotations of proper names in syntactic structures are sets of properties (based and expanding on the treatment of common NPs in Montague, 1973), then we can give some more precisions: all the linked relatives form a graph whose global interpretation constitutes the contextually relevant set of properties that define the proper name Donald Trump, with address ⦃Donald Trump⦄ and semantic value λPP{^*Donald Trump*} for purposes of interpreting (283). As in the theory of generalised quantifiers, the denotation of *Donald Trump* is the set of sets that have Donald Trump as a member. We assume, in line with Schmerling & Krivochen (2017) that sentences are interpreted relative to coordinates (Cartesian or other coordinate system) of sets of possible worlds, sets of times (which we take to be intervals), set of speakers, set of places, and so on. Combining these two assumptions (plus the definition of an operator which takes an NP-type extension as input and returns another NP type extension as output, namely, a contextually determined subset of the set of properties denoted by the input) allowed us to provide an account of sentences like *During the debate, Trump was just being Trump*, which are not tautologies: there is a contextually salient set of properties that is selected as being relevant in a particular set of coordinates. Importantly, the argument in Schmerling & Krivochen (2017) carries to the present context without modifications.

Let us now turn to 'mixed' cases, in which we have an appositive and a RRC to be linearly ordered with respect to each other. Relevant examples are like (276a) and (277a), repeated here as (284a) and (284b) respectively (with minor annotations added):

(284) a. We found that movie$_i$, which$_i$ cost plenty, that you so highly recommended e_i
b. The children$_i$ that you brought e_i, who$_j$ were charming, got sick later

As noted in fn. 2 in this chapter, (284a) elicited mixed responses from our informants. Emonds' condition for a restrictive to follow an appositive was that the restrictive clause was string-final (i.e., if there is no other constituent after that). The reason why (284a) is not fully acceptable may have to do with the fact that there is no growth in the size of the antecedent for the restrictive relative: both the appositive and the restrictive relative have exactly the same antecedent *that movie*. However, there are semantic differences between the two clauses which allow for a *partial* ordering (but not a *strict* ordering) to be imposed in interpretation; however, we don't have a contradictory situation as in *crossover* cases, which required a node to be ordered both *before* and *after* itself. To the extent that (284a) is grammatical and acceptable, that acceptability is accounted for outside of the grammar as it is conceived of here (see also Haider, 2019); at least if the grammar includes a condition on *strict* node ordering. It is important to distinguish the anomalies generated by a *partial* ordering imposed over a set of nodes (which can be solved *outside of the grammar*, by choosing an ordering as an interpretative hypothesis and seeing where things go from there) from the hard violation of *ordering* requirements that we have seen *crossover* cases generate.

Example (284a), with its somewhat dubious acceptability, forced us to make some additional considerations. (284b), however, receives a simpler analysis due to the fact that there is indeed a possible strict ordering between the restrictive and the appositive clauses. In this case, the antecedent of the restrictive clause is the NP *the children*, whereas the antecedent of the appositive clause is the complex [NP S'] (importantly, *not* [N' Comp], as in McCawley's quotation above). The condition imposed by Emonds seems to be relevant for these cases as well, for we *cannot* have (285b) below as an extraposed version of (284a)—which does *not* mean at all that (285b) is ungrammatical, it just means that it cannot receive an interpretation in which an operation of *Relative Clause Extraposition* has applied to (285a)—:

(285) a. The children that you brought which your sister loved, who were charming, got sick later
b. The children that you brought, who were charming, which your sister loved, got sick later

The relevant interpretation for (285a) is that there is a set of children, a subset of which were brought, and in turn a subset of this subset were also loved by someone's sister: we have a recursive restriction over the extension of the set of children (as Arnold, 2007: 272 puts it, restrictive relatives are interpreted intersectively). Call the set which results from this double restriction *s*. Then, a property is assigned to the members of *s*, namely, that of being charming. Syntactically, the antecedent of the first restrictive relative clause, *that you brought* is simply the NP *the children*; the antecedent of the second restrictive relative is (as we would expect), the CNP *the children that you brought*. So far, so monotonic. The antecedent of the appositive clause is the whole CNP *the children that you brought which your sister loved*; there is a straightforward strict ordering to be imposed among the sub-graphs (and transitively, to all the nodes in each of these sub-graphs) in (285a). But the same interpretative procedure is not available for (285b): the relative clause *which your sister loved* does not receive a restrictive interpretation; rather, it is interpreted as another appositive: the set of children who were charming and the set of children that the addressee's sister loved are strictly co-extensive. In this case, the finite-state syntax for stacked (only *tail* recursive, not *freely* recursive; see Uriagereka, 2008: Chapter 6 for some discussion about this point; also Krivochen, 2022) appositives that we proposed above in the analysis of (282) is called upon again: the sub-graphs which constitute the structural descriptions for the appositives are *linked* at the node that is the root of the CNP *the children that you brought*. We may add that the different *wh*-words used in the relative clauses (*who, which*) make no difference in terms of the configurational properties of the graph that proves (285b) to be a well-formed expression of the language since the addresses all point to the same semantic value. There is only one node, albeit one visited at different points of the walk defined through the graph. The *context* of each of those visits (recall that the context of a node *v* is the set of nodes that *v* is immediately connected to) may favour one or the other morphophonological exponents, but the semantic interpretation of that node remains the same. In MGG terms, *who* and *which* (and *the children that you brought*) are co-indexed.

Finally, we can summarise the structures proposed for restrictive and appositive relatives. We follow traditional practice in assuming that in restrictive relatives modifying a quantified NP the determiner is excluded from the relative antecedent: only the set term (in Barwise & Cooper's terms) is shared between the antecedent phrase and the relative. Appositive relatives can take antecedents of different categories. In addition to proper and definite NPs, we have:

(286) a. Vader told Luke that he was his father, which Luke refused to believe (sentential antecedent)
b. John looks like Richard Nixon, which my uncle does too (VP antecedent; taken from McCawley, 1998: 447)

c. Doug is very interested in neuroscience, which Paul is not (AP antecedent)

What this means, in the present context, is that any of these antecedents can be structure-shared between the graph corresponding to the host and the graph corresponding to the appositive.

The amount of shared structure varies between appositives and restrictive relatives. In **Chapter 4** we mentioned that quantified NPs such as *every N* would require some further consideration, since the determiner cannot be part of the relative, only the N can. We will expand on this idea in **Section 14.4**, but for now we can say that this property of restrictive relatives can be captured by having the N be shared between the host graph and the relative graph, but the determiner be only part of the host graph. We will refer to this as 'partial multidominance' insofar as only a part of the NP (or DP) is shared. The other option is to have 'total multidominance', whereby both the determiner and the N are shared nodes (see Krivochen, 2022 for extensive discussion). This distinction results in the following scenario:

– Partial multidominance with quantified antecedents corresponds to RRC:

(287) Every boy who Jane met arrived
$\rho_{\text{host}} = \langle(\text{every, boy}), (\text{arrived, boy})\rangle$
$\rho_{\text{relative}} = \langle(\text{met, Jane}), (\text{met, boy})\rangle$

For stacked restrictive relatives the amount of shared structure increases with each new relative (Krivochen, 2022).

– Total multidominance with quantified antecedents corresponds to NRRC:

(288) That boy, who Jane met, arrived
$\rho_{\text{host}} = \langle(\text{that, boy}), (\text{arrived, boy})\rangle$
$\rho_{\text{relative}}\ \langle(\text{met, Jane}), (\text{met, that}), (\text{that, boy})\rangle$

This analysis captures the traditional observations about the relative position of appositives and restrictives, the idea that appositives are somehow 'more external' than restrictives (Jackendoff, 1977; Demirdache, 1991; McCawley, 1998). Furthermore, if total multidominance delivers only NRRC, the fact that proper names do not take RRC follows: in *John, who goes to MIT, found a good job*, the anchors *found* and *goes* immediately dominate *John*, and there is nothing else to the antecedent of the appositive. When a proper name is preceded by a determiner partial multidominance is possible, and RRC—abridged and not abridged—become available:

(289) a. Michel Camilo is [the Oscar Peterson [of the Dominican Republic]$_{\text{AbridgedRRC}}$]
b. She was acting so weird. That was not [the Ellen [that I know]$_{\text{RRC}}$].

In this chapter we analysed aspects of complementation within the NP: we have derived empirical restrictions on the distribution and combinatorial properties of restrictive and appositive relative clauses (originally noted by Emonds, 1979 and McCawley, 1983, among others) from independently motivated requirements of strict ordering to be imposed over a graph in order to assign a compositional interpretation to that graph. Recursive restricted relatives require monotonic growth of the structure, with each relative operator taking the entirety of previous structure as its antecedent. In contrast, appositive clauses can only be *stacked* in finite-state sequences, displaying strictly paratactic dependencies between themselves (Krivochen, 2022 provides a TAG analysis that captures these differences). The structural description assigned to these is no different from other cases of non-scopal dependencies involving, e.g., multiple VP adjuncts (Uriagereka, 2008; Krivochen, 2015a). As exemplified in (282), appositive relatives may also establish paratactic dependencies, if they are all linked at a single node which corresponds to the antecedent. Because each arbor corresponding to a RC is strictly ordered, but the global structural description for multiple appositive relatives does not specify an order among them, permutation of linear order is permitted *salva veritate* and without changing meaning.

CHAPTER 10

Wh-Interrogatives: Aspects of Syntax and Semantics

Interrogative sentences have been at the core of structuralist, transformational, and non-transformational formal grammar for more than half a century. The generative tradition introduced a way to formalise the idea that interrogative sentences are derivative, by formulating transformations that allowed for reordering of syntactic objects: in this way, an interrogative sentence like *what did Mary buy?* could be conceived of as the result of reordering the direct object of *buy* to sentence-initial position (plus the insertion of the dummy auxiliary *do* to express tense and agreement; we will not deal with *do-support*). The need for reordering rules in transformational generative grammar was, in the beginning, a consequence of the idea that there is a set of canonical sentential forms (Harrisian *kernels*), and that superficially non-compliant sentences could be reordered into a kernel or set thereof (Harris, 1957: § 2.6.2): reordering rules address the 'mismatch' between underlying relations and superficial word order. Overviews of treatments of unbounded dependencies, including *wh*-movement, from a variety of theoretical standpoints can be found in Kaplan & Zaenen (1995), Borsley & Crysmann (2021), Postal (1998), Kroch (2001), Putnam & Chaves (2020), Müller (2011), Asudeh (2012), among others. From the perspective adopted in this monograph, it is an interesting question whether the reordering transformations that have been proposed in the literature to account for interrogative sentences (in particular, *wh*-interrogatives) modify existing syntactic relations, introduce new syntactic relations, or preserve syntactic relations (only changing linear order). In this chapter we will build on the notion of *linking* in order to provide an analysis of so-called *filler-gap dependencies*, with a particular focus on what is usually referred to in MGG as *wh*-movement (a term that we will use only descriptively, seeing as there is no 'movement' in our framework; see also Postal, 1998). We dealt briefly with *wh*-elements in previous sections (primarily *relative wh*-pronouns), but in those analyses we treated *wh*-phrases as atoms, without paying much attention to their internal structure or their semantics because it was not required for our purposes. Thus, *wh*-phrases like *what* and *which of the books* would have received the same treatment in our initial presentation of *wh*-dependencies, since it made no difference for the cases under consideration. While that 'uniformity' assumption (roughly, '*all wh- are created equal*') helped us simplify the

 | DOI:10.1163/9789004542310_011

exposition, there are interesting data whose analysis requires further refinement. The graph-theoretic proposal advanced in this work interacts in interesting ways with paradigms concerning *wh*-movement, islandhood, and lexical restriction. We speak of *lexical restriction* in *wh*-phrases, when we have structures of the type [*wh*- NP], in which the *wh*-word quantifies over an overt NP (*which book*, *whose bag*, etc.). Following usual practice, we will refer to the NP as the 'lexical restrictor'.[1]

We need to clarify some issues pertaining to the syntax and semantics of partial *wh*-interrogatives (we will not deal with total *wh*-interrogatives with *whether* here; see Karttunen, 1977 for a discussion of matrix and embedded polar interrogatives in a Montagovian framework that is germane to the ideas pursued here). In principle, the structure of a *wh*-interrogative is not different from that of a declarative sentence in terms of the assignment of GF: in (290a) and (290b), *who* and *what* are still Subject and Object of their respective predicates (a 'non-projective' analysis, using Hudson's 2007: 131 terms):

(290) a. Who bought a book?
b. What did John buy?

The *preliminary* ρ-sets of (290a, b) are thus (291a, b) respectively:

(291) a. ρ = ⟨(buy, who), (buy, book)⟩
b. ρ = ⟨(buy, John), (buy, what)⟩

In the ρ-sets (291a, b) the *wh*-words *who* and *what* are assigned GF according to the rule in (115): note that—as in Dependency Grammars, e.g., Osborne (2019: Chapter 8)—no 'overlay' functions such as Question (in Relational Grammar; see e.g. Perlmutter & Postal, 1983b: 86), FOCUS (Bresnan et al., 2016; Börjars et al., 2019), or DIS (Dalrymple et al., 2019: 37, 653, ff.) are invoked for *wh*-expressions, due to our stronger focus on configuration and semantic values of expressions than theories that require specific GFs for dislocated constituents. A question that arises now is whether there are aspects of the semantics of (290a, b) that are determined by syntactic configuration and which we have not represented adequately in (291). We need to look at the structure and interpretation of *wh*-interrogatives in some more detail to provide an answer to this question.

1 In some varieties of LFG, such as Bresnan et al. (2016), structures like [Which class] are referred to as 'operator complexes'. The *wh*-element is the operator, and the whole [*wh*- NP] structure containing the operator is the 'operator complex'.

Let us begin by considering an approach with a heavy focus on semantics. In the framework of Montague (1973), we can say that *wh*-words, as operators, bind an occurrence of an indexed pronoun over which they have scope; Karttunen (1977) builds on the PTQ analysis and proposes an interrogative operator symbolised with ? which takes scope over a proposition that contains an indexed dummy pronoun *him*$_0$. He then (1977: 24–25) defines a rule of *WH-QUANTIFICATION* (which works in similar way to Montague's *quantifying-in*; see quantification rules S14–S16 in Montague, 1973: 252), a 2-place rule which takes an indexed *wh*-word (*what*, *which*) and the proposition plus its ? operator, replaces ? with the indexed *wh*-word and deletes the pronoun in the preposition that has the same index (see (293) below for some examples). This is so because an interrogative sentence translates to an expression which denotes a (possibly unary) *set* of propositions; these propositions constitute an answer to the question. Therefore, the translation of *What did John buy?* translates as an expression which denotes the set of propositions for each thing that John buys, that he bought it (a set that may contain a number of entities, possibly contextually restricted). The way in which we have phrased this is not accidental: it mirrors the syntactic structure assigned by an analysis tree to a *wh*-interrogative in Karttunen's (1977) treatment: in Karttunen, the denotation of an interrogative sentence consists of the set of its *true* answers; others have weakened this requirement to *possible* answers. In all cases, however, questions denote sets of propositions (see also Engdahl, 1986).

This *variable binding + replacement* process is not exclusive to *wh*-interrogatives: relations of *scope* can be represented in this way more generally. For instance, let us consider Montague's (1973: 228) analysis tree for the *de re* reading of *John seeks a unicorn* (omitting the rules that have applied at each point):

(292)

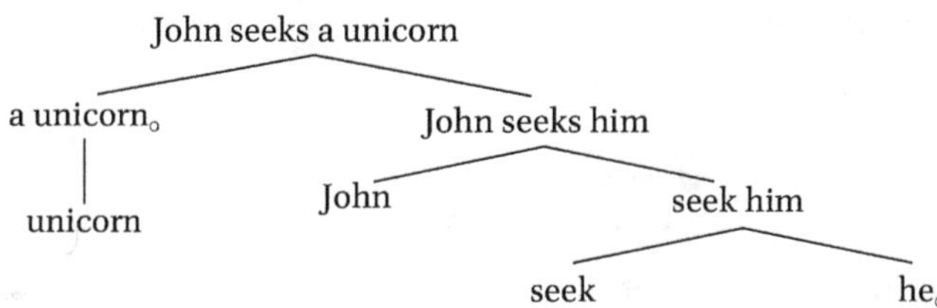

FIGURE 10.1 Simplified Montagovian analysis tree for 'John seeks a unicorn'

In Montague, the subindexes simply indicate that those expressions are to be regarded as arguments of a function, that function being specified in the rule that applies to yield a certain expression. Indices in Montague grammar are quite different from indices in transformational generative grammar: there is a set that indexes the set of variables of the language (usually, the set of integers). Thus, they are not assigned to structural positions (as they were in pre-Minimalist MGG): having a variable of the form $x_{(n)}$ is a way of uniquely identifying the *n*-th variable in the language. Thus, in any expression of the form x_n ... x_n we have not two coindexed variables (as in MGG; see e.g. Fiengo, 1977 and much subsequent work) but a single variable (the *n*-th) occurring twice (as in proof theory, what Chomsky 2020, 2021 calls 'general recursion'). If we interpreted the analysis tree as a PS tree in terms of the definition of scope, the *de re* NP would have scope over the intensional verb *seek*, but this is not essential to the analysis. For example, in Keshet's (2010) theory of *split intensionality* there may be a scope position *below* an intensional verb (but above an abstract intensional operator) where a NP can be interpreted as *de re*, in addition to a position *above* an intensional operator where a nominal can move to take scope. The take-home message that we want to focus on is that the *de re* reading involves an NP that appears structurally higher than its *de dicto* counterpart; this structural height has to do with the requirement that the NP has *scope* over the indexed pronoun that it binds (and this holds in both Montague's and Keshet's proposals). In this sense, Karttunen's (1977) *quantifying-in* proposal naturally extends the fragment in Montague (1973), which did not provide an account of the syntax or semantics of *wh*-elements.

The semantic treatment of questions that we will follow, and which suffices for our purposes is, as we said, Karttunen's (1977), in turn based on Hamblin (1973) (also Cooper, 1984; Engdahl, 1986; see Groenendijk & Stokhof, 2011 for a very complete comparative overview of formal theories of the semantics of questions). Similar approaches have been adopted in GPSG, see e.g. Gazdar (1981), who assumes Karttunen's semantic representation. In Karttunen (1977), *wh*-phrases are taken to be quantified terms, thus succeptible to be 'quantified in' as in Montague. Karttunen's analysis of *wh*-interrogatives is exemplified, in a simplified way, in (293):

(293) a. Who loves Mary?
b. *who*$_0$? *he*$_0$ *loves Mary*
c. Substitute **?** for *who*$_0$, adjust the case of *who*$_0$ to match the case of the first indexed pronoun in the proposition (here, *he*$_0$ is Nominative case), and delete the first occurrence of that pronoun. This outputs: *Who loves Mary?*

In this proposal, the treatment that *wh*-phrases receive is does not differ in essence from the treatment that garden-variety NPs receive: *wh*-interrogatives feature a quantified NP (in our example, *wh*+someone$_0$) and a free variable in the form of an indexed pronoun (in our example, *he*$_0$). The interpretation of *who* does not differ from the interpretation of *a man* in terms of how their semantic values are formalised. The interrogative in (293) translates as an expression which denotes a set that contains, for each individual that loves Mary, the proposition that he or she loves Mary. Identification of indexed categories allows for the construction of an appropriate semantic form, in exactly the same way in which *de re* and *de dicto* readings are constructed in Montague (1973); see Karttunen (1977: 24–25) for a precise formulation of the relevant quantification rule in PTQ-style. This much is not very different—superficially, and sticking strictly to subject *wh*-interrogatives, which do not feature *do*-support or subject-auxiliary inversion—from the MGG way of doing things. We can give an example of a simplified MGG derivation to illustrate some aspects of this similarity:[2]

(294) a. $C_{[Q], [+wh]}$ $NP_{[+wh]}$ loves Mary
b. $NP_{[+wh]i}$ $C_{[Q], [+wh]}$ t_i loves Mary
c. Who loves Mary?

This move is not without its semantic effects: we are saying that someone loves Mary, and we want to know the identity of that someone (see Culicover & Jackendoff, 2005: 308, ff. for a similar perspective which relies on lambda abstraction at the level of Conceptual Structure to deliver these effects). A *wh*-interrogative is, semantically, a request for the identification of an individual or set of individuals via *wh*-phrases which informationally constitute the *presupposition* such that it or they fulfil the requirements specified in the *focus*. Syntactically, this reading is represented by having the *wh*-element taking scope over the proposition via a *root transformation* (regardless of whether it is successive-cyclic or not, it always targets the *root* of the tree; see Postal, 1972; Abels & Bentzen, 2012 for further discussion): in the derivation in (294), the subject NP (which carries a [+wh] feature) moves from Spec-TP to Spec-CP, generating a configuration where there is Spec-Head agreement (since the interrogative C head also bears a [+wh] feature). In this configuration, the formal [+wh] feature gets discharged (or deleted, depending on the specific proposal) and the resulting representation is interpretable by the interface systems.

2 We must note that the idea that subject interrogatives involve movement of the subject from Spec-TP to Spec-CP (e.g., Pesetsky & Torrego, 2001; Adger, 2003) is not uncontroversial within MGG. Examples of approaches that consider movement from Spec-TP to Spec-CP to be 'too local' include Grohmann (2003) and Erlewine (2020).

Similarly, in Metagraph grammar (Postal, 2010), scope can be represented in terms akin to Montague's: in a sentence like *Mary saw no student*, whose underlying structure would be [⟨[no student]$_1$⟩ [Mary saw DP$_1$]], '*each member of the set of coindexed DPs would represent a single phrase, each X occurrence marking a separate arc sharing as head the single phrase represented by X*' (Collins & Postal, 2014: 14). The representation assumed in Collins & Postal's work is, for all present intents and purposes, equivalent to Montague's. In all the aforementioned frameworks (procedural as well as declarative approaches), the scope of a node is indicated by its height in a tree/graph, so the existence of certain common themes are not surprising given otherwise very different foundational assumptions.

A cautionary note is in order before continuing. It is almost otiose to say that the syntax of *wh*-dependencies is one of the core issues in grammatical analyses of English (and natural languages more generally) and has motivated an enormous body of literature from generative (transformational and non-transformational) and non-generative approaches; we will not attempt to provide a detailed treatment of the full range of English *wh*-dependencies in this work (see e.g., Engdahl, 1986; Cable, 2010; Fox, 2012 and references there; Dayal, 2016; Kotek, 2020 for detailed approaches to the syntax-semantics of *wh*-interrogatives). We will, however, propose a way to look at the problem which, to the best of our knowledge, has not been pursued before (although there are connections with traditional TAG analyses; see e.g. Kroch, 2001; Kroch & Joshi, 1985: § 6, 1987;[3] Simpler Syntax; and evidently with Montague-style intensional semantics, to the extent that it is necessary to provide an adequate syntactic representation).

The question we were dealing with is whether we need to amend the ρ-sets in (291) to have a better representation of the semantics as read off by the structural description, and this depends entirely on the relation between syntax and semantics that we assume. One way to revise these ρ-sets along the lines of transformational theories and scope theories is to have the *wh*-element dominate what would otherwise be the *root* node of the relevant graph, analogously to how questioned constituents are reordered to S' or Spec-CP (capturing the idea that *wh*-movement is a *root transformation*; Emonds, 1970; Postal, 1972). In this way, if scope is read off dominance relations (the closest analogous to MGG's c-command in our framework), the *wh*-word would have scope over the whole elementary graph where it occurs. We can adapt our dominance sets to

3 The reader may also want to compare the present approach with that of Frank (2006), where the LTAG analysis of long-distance extraction is directly compared to MGG phase-based locality.

be more like Karttunen's representation. Thus, we would have (295) instead of (291), where the *wh*-words and their subcategorising predicates form a *bicircuit*:

(295) a. ρ = ⟨(who, buy), (buy, who), (buy, book)⟩
b. ρ = ⟨(what, buy), (buy, John), (buy, what)⟩

This analysis is the closest thing to *wh*-movement that can be formulated in our terms, either overt or covert; whereas (291) is compatible with approaches to *wh* in-situ without LF movement, where most of the work is done by the semantics and/or feature percolation through the tree (see Kotek & Hackl, 2013 for a concise summary of the two main approaches to filler-gap dependencies in MGG). In (295), *who* and *what* are assigned a GF in the configurations specified in (115); those configurations remain identical in (295) and the preliminary ρ-sets in (291).[4] We have added another visit to the *wh*-nodes, at the very beginning of the walk: this position corresponds to the place of the operator ? in the translation of interrogatives into intensional logic (Montague, 1970, 1973) in Karttunen (1977). The second visit in the ρ-sets in (295) corresponds to the position of the bound indexed pronoun within the proposition, which is the structural context in which the *wh*-element receives a GF. Note that there is no dissociation of operator from variable in terms of their internal composition: all that differentiates them is their position in the walk through the graph-theoretic structural description of a *wh*-interrogative. This representation puts the *wh*-word and the predicate that subcategorises for it (in a simplex clause) in a *bicircuit* relation, whose irreducible expression is (296):

(296)

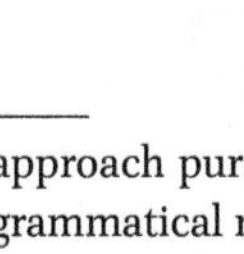

FIGURE 10.2 Bicircuit relation

4 The approach pursued here, in which *wh*-movement is an *order changing rule* (and therefore grammatical relations are preserved without the need to invoke additional elements in representations), satisfies Kroch's requirement below without further constraints required:

> *we never want to allow derivations under which thematic roles, once established, are altered by further adjunctions, and we will block such derivations by, in every tree, placing a particular local constraint on every node that is assigned a thematic role by a governor.* (Kroch, 2001: 11).

Part of the motivation for having two visits to the *wh-* node is semantic: Johnson (2020) points out, from the perspective of a Minimalist framework with multidominance (i.e., which allows the indegree of a node to be greater than 1), that a *wh*-element does not have at the same time the meaning of an operator and a variable. We will come back to this issue, but anticipating part of that discussion, Johnson revives the traditional analysis of *wh*-expressions whereby there is an abstract feature or morpheme (call it Q; the idea goes back to at least Katz & Postal, 1964: 79) which determines the interpretation of a DP as a *wh*-operator (see (294) above). However, instead of being a feature, Johnson's proposal involves an extra layer in the syntactic representation: an XP with head Q (the functional element we just mentioned) which takes DP as a complement (see also Baker, 1970b, who proposed a Q morpheme that bound *wh* in-situ). In this case, only the low layer DP is multidominated (by XP and VP): the functional layer XP is only dominated by CP. In this way, the operator interpretation and the variable interpretation are dissociated. The *wh*-word itself, however, is considered in Johnson (2020) to be a D head: the analysis of *which flower* he proposes is the following:

(297)

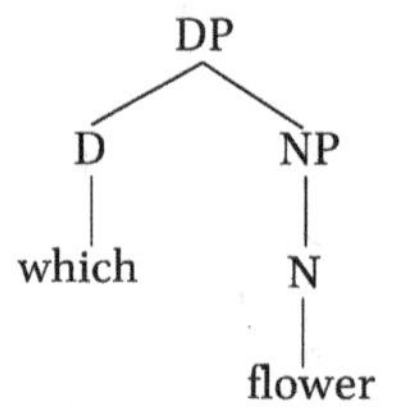

FIGURE 10.3 DP analysis of 'which flower'

In a sentence like *which flower should she bring?*, the DP would be immediately dominated by VP (being the complement of V) and by an XP with head Q (see **Section 14.4**).[5] The DP is interpreted in the lower position of a movement chain as a definite description, and it is the Q morpheme that licenses an operator interpretation when parsing the sentence. Baker's (1970b) approach, as summarised in Reinhart (1998), is somewhat different: Q unselectively binds all *wh*-variables that have not moved in a multiple-wh interrogative. Under standard Minimalist assumptions about bottom-up sequential structure building, it seems strange to dissociate Q from *which* since there would be a derivational point where *which flower* has been assembled by Merge but there is no Q

5 Johnson's (2009) proposal differs from his more recent (2020) analysis in having the *wh*-word be the Q head itself, and requiring *wh*-complexes to have a covert definite D head: the structure of *which flower* under Johnson's (2009) analysis would be: $[_{QP}$ which $[_{DP}$ ~~the~~ $[_{NP}$ flower$]]]$.

morpheme (in the Katz & Postal treatment, based on a top-down rewriting system, Q would be introduced in the derivation first): how exactly *what* comes to be and what its contribution is before Q is Merged is not clear (Johnson, 2016: 27 proposes that Q and D enter an Agree relation, and that *which* is the exponent of the Q morpheme, see also Cable, 2010; the issue of derivational timing is, however, still present). Presumably there could be another morpheme (call it R) for relative *wh*-words, and we could propose the same mechanism: as we will see in **Section 14.4**, there is a connection between this treatment of *wh*-phrases and the theory of generalised quantifiers, which allows to have an appropriate characterisation of the semantics of interrogatives under multidominance.

Johnson's analysis of multidominance in *wh*-interrogatives is focused on issues of linearisation under Kaynean antisymmetry, and aims at defining a sequence of terminals in terms of paths in multidominance trees. These *paths* are sets (of sets) of phrases (see also Citko & Gračanin-Yuksek, 2021): intermediate nodes are of crucial importance in the calculation of paths for linearisation purposes (even though only terminal nodes are assigned phonological exponents). It is important to point out that the kinds of representations proposed by Johnson and Citko & Gračanin-Yuksek are not trees in the graph-theoretic sense, since they contain closed walks (just like van Riemsdijk's). A certain ambiguity between sets and graphs has become frequent in contemporary generative grammar, given the strong set-theoretic commitments in the definition of Merge while the use of tree diagrams (and in some cases, the formulation of operations that require an order over nodes, such as search algorithms) is generalised.

There seems to be a strong semantic motivation behind dissociating operators from variables in multidominance approaches, in addition to syntactic arguments. In our terms, translating Johnson's concerns into graph-theoretic representations would indeed require a ρ-set along the lines of (295), defining a sequence $\Sigma = \langle \text{who, buy, who, book} \rangle$. If this was the input for a linearisation mechanism, some additional condition would have to ensure that only the first visit to the node with address ⦃who⦄ received a phonological exponent for purposes of the LCA (a property that Johnson calls 'terseness';[6] see also

6 Note that L- and R- extractions are constrained by terseness (with some exceptions: resumption being one of them), but not reflexivity (under our analysis): if the same mechanism underlies both, one may ask, why does terseness only apply to one? Put differently: why is (i) grammatical but (ii) ungrammatical?
(i) Who does Pat admire?
(ii) *Pat admires (intended: Pat admires himself)
Despite the formal mechanism being the same (multidominance), the configuration is differ-

Nunes' 2004 *chain reduction*): a Minimalist graph-theoretic approach along these lines can be found e.g. in Yasui (2002). It is unclear, however, whether the two visits to the *wh* nodes are necessary in the syntax: below we will deal with what the content of the addresses assigned to ⦃who⦄ and ⦃what⦄ could be, and suggest that we do not need the syntax to do anything other than (291), much discussion pending. This requires semantic interpretation to do significantly heavier lifting. In either case (having one or two visits to *wh* nodes), there is no need to have empty terminals in the graph: in this respect, our approach differs from the treatment that long-distance dependencies receive in TAG with links (Kroch, 2001; Frank, 2006) and gets somewhat closer to representations like LFG's *c*-structure for interrogatives (see e.g. Mycock, 2007 and Dalrymple et al., 2019: Chapter 17; trace-less analyses depart from the original proposal in Kaplan & Bresnan, 1982 which did include empty terminals).

Johnson's work points us towards a relevant question that we need to address: what is the nature of *wh*-elements? In other words: how do *wh*-elements come to be? The Standard Theory answer was that *wh*-elements emerged transformationally:

> T_{w2}*: Structural analysis: NP–X*
> *Structural change:* $X_1 - X_2 \rightarrow wh + X_1 - X_2$
> *where wh + animate noun → who*
> *wh* + [in]*animate noun → what* (Chomsky, 1957: 112)

That is, argumental *wh*-elements were (indefinite) NPs with a diacritic feature [+*wh*] (see also Katz & Postal, 1964). This diacritic has remained throughout the history of transformational generative grammar (note the similarities with Johnson's 2009 approach), recently in the form of an uninterpretable feature which has to be checked against a *wh*- head (this is the so-called *wh-criterion*; see Rizzi, 1996; also Epstein, 2015; Chomsky, 2015). The transformation above is supposed to account for two things, distinct although intimately related: first, that if an NP is questioned (endowed with a *wh* feature), it will surface as a *wh*-

ent: only (ii) could contain parallel arcs. If parallel arcs were involved in (i) (such that *who* and *Pat* were coindexed), the result would be a case of strong crossover. This shows that the configurations are different. As for why terseness applies to only one of these configurations (or, rather, why it *cannot* apply to (ii)), it is possible that the answer lies in what Grohmann (2003: 78) calls a *drastic effect on the output* (a PF exponent, specifically), triggered by the assignment of two different GF to the same expression. Only one GF is involved in L- and R- extractions, and conditions for these 'drastic effects' to arise do not hold (see Krivochen, 2023b).

word (either *who* or *what*); second, that these *wh*-words have the distribution of NPs. Current analyses take *wh*-words and operator complexes to be either DPs or QPs endowed with a specific feature [*wh*]; in either case, the *wh*-word is the head of a functional projection (a Q or D head).

In this monograph we are not concerned with the morpho-phonological form of these expressions, but their distribution is something we do care about. In our first approach to the analysis of relative clauses, the *wh-pronoun* was formalised as another visit to the node that corresponds to the antecedent of the relative clause: the relative clause and the NP it modifies were *linked* at that node:

(298) a. The man who came seemed drunk
b. Elementary graph 1: [The man seemed drunk]
Elementary graph 2: [the man came]
$\rho_1 = \langle(\text{seem, drunk}), (\text{drunk, man})\rangle$
$\rho_2 = \langle(\text{come, man})\rangle$

In (298), EG 1 and 2 are linked at the node with address ⦃man⦄, one visit to which—we assume—corresponds to what surfaces as *who* given specific structural conditions.[7] This analysis delivers the 'propositional value' of restrictive relatives as in Brucart (1999: 398). We have not gotten into exactly how the *man* → *who* change happens and, because we are concerned with structure and not with how that structure surfaces, and we will not get into it in detail either.

7 This analysis of relative clauses, as pointed out in **Section 4.1**, may superficially bear similarities with the so-called *raising* (or 'promotion') analysis, whereby the antecedent of a RC starts its derivational life inside the CP and moves to a position outside the CP where it takes the role of antecedent (Brame, 1968; Schatcher, 1973; Kayne, 1994; Bhatt, 2002). The similarities pertain to the fact that there is a single syntactic element that is a pivot between the NP and the relative clause, not an antecedent and an operator (as in the 'head-external' analysis that became standard in GB; see Chomsky, 1977; Demirdache, 1991). These similarities end when we consider the mechanism that underlies the connection between antecedent and relative clause: A'-movement (within the relative clause, to Spec-CP and one more, whereby the head noun adjoins to a projection outside the relative) in the *raising* analysis, *linking* (structure sharing) in the present approach (see Krivochen, 2022). In this context, our treatment may better be thought of as a version of the *matching* analysis: the antecedent and the relative operator contain corresponding heads, and after adjunction of the RC the embedded head gets deleted under identity (Sauerland, 1998; Citko, 2001; Salzmann, 2019). The main difference between the present approach and both the traditional raising and matching analyses is that, due to the addressing system, there need not be copies, traces, or deleted objects: the 'matching' relation may be subsumed to *linking* in the derived graph, such that the matrix clause and the relative are *linked* at the head NP.

The important point here is that the reasoning that we have applied to relative clauses (restrictive and appositive) does *not* hold for *wh*-interrogatives, because there is no *antecedent-operator* relation. It is not possible to say that in the examples in (290) above (*Who bought a book?* and *What did John buy?*) *who* and *what* link distinct elementary graphs or indeed that they have 'antecedents' at all. They need to be syntactic objects in their own right. This means that *interrogative wh*-words receive a different analysis from *relative wh*-words, on syntactic and semantic grounds. In Krivochen (2022: 312) we proposed that relative *wh*-operators always have nominal complements, and these complements are structure-shared with the antecedent of a relative. Interrogative *wh*-operators, on the other hand, do not necessarily structure-share their complements, which may be phonologically null (in which case they are not interpreted as bound variables).

Let us look at the Categorial Grammar perspective, which provides the semantic foundations for our own analysis. We agree with Karttunen (1977: 18) in that the assignment of *wh*-words to a category requires us to have a model for the semantics of the sentences in which they occur. With ordinary NPs that was not a problem, but *wh*-words—as we have seen—present different challenges. To flesh these challenges out (and see how they can be addressed), we need to introduce some basic aspects of Montague semantics (see Fox & Lappin, 2005 for a complete exposition of an intensional system. Here, we only take from intensional semantics what we need to provide some interpretations to our graphs; the focus of this monograph is syntax, not semantics). Montague (1973) defines two atomic categories *e* and *t*; *e* and *t* are then used to recursively define categories which are assigned to expressions more familiar to the linguist. In PTQ, proper NPs (or 'terms') are assigned to the category *t*/IV (they combine with intransitive verb phrases to yield *t*, which we can think of as analogous to the category *S*(entence) in PSG: PTQ has no category *S* distinct from *t*). The point of using the symbol *t* is its mnemonic value: *t* stands for 'truth (value)', and truth values are the kinds of objects that closed sentences (propositions) denote. Common NPs are *t*//*e*, where double slashes are used to indicate that a category differs from its single-slashed counterpart in its semantic role but plays an identical syntactic role (PTQ: 222). 'Terms' (proper Ns like *John* and variables such as he_n in (292)) are of category *t*/IV. In this context, Karttunen (1977: 19) defines interrogative *wh*-words as *t*//IV: modified expressions of category *t*/IV. This means that *wh*-words play the same syntactic role as 'terms' (recall also that in GB R-expressions and *wh*-traces are claimed to behave in the same way for purposes of Binding Theory; Chomsky, 1981: 193, ff.), but differ in their semantic role: *wh*-words are, in terms of their IL translation, equivalent to existentially quantified (indefinite) common NPs

(Karttunen, 1977; Reinhart, 1998: 44). Reinhart (1998) identifies some problems with Karttunen's original proposal, and builds on a modified version of it within the framework of the Minimalist Program. Interestingly for our purposes, she introduces a *choice function* (CH) for lexically restricted *wh*-phrases, such that a multiple *wh*-question like (299) is analysed as (300) (taken from Reinhart, 1998: 41; as in Montague, 1973, for any expression E, [^E] denotes the intension of E):

(299) Which lady read which book?

(300) a. for which $\langle x, f\rangle$(lady(x)) and (x read f(book))
b. {P|($\exists\langle x, f\rangle$) (CH($f$) & lady($x$) & P =^($x$ read f(book)) & true(P))}

where P is a set of true propositions such that the conditions that follow hold. Much detail aside, what matters to us is that the meaning of a bare *wh*-word like *what* or *who* can be assimilated to lexically restricted *wh*-phrases: *which lady* in (299) means 'choose an entity that belongs to the set of ladies (such that ...)'; *who*, in the same context, means 'choose an entity (such that ...)', with the added specification that the entity be animate and perhaps even also human (for inanimate or nonhuman entities, we have *what*). The 'such that ...' clause depends on the structure where the *wh*-word or phrase appears; in Reinhart's analysis of (299), that would be P =^(x read f(book)) $\wedge$ true(P)). As Reinhart observes, *which* does not correspond to any quantifier in logic, of first or higher orders. Karttunen (1977: 19) proposes that the IL translation of bare *wh*-words, e.g. *who*, is the same as the IL translation of an indefinite common NP, like *someone*: there is an existentially quantified term that corresponds to a member of a set of entities, the *wh*- operator introduced the choice function that selects an entity from that set based on whether it satisfies the conditions imposed by the predicative structure in which *wh* occurs. In Karttunen's approach, *who* and *what* have IL translations *who*' and *what*': $\hat{P} \bigvee x P\{x\}$;[8] again, whether we get *who* and *what* depends on whether x is animate/human. Crucially, x is the entity over which *wh* quantifies, it is semantically distinct from *wh*

8 The symbol $\bigvee$ must not be confused with $\vee$: the former, in Montague (1973: 229), is used to denote *existential quantification* (i.e., $\exists$), the latter is the logical disjunction *or*. In turn, *universal quantification* in Montague is denoted by $\bigwedge$, not to be confused with logical conjunction, $\wedge$. In turn, $\hat{P}$ abbreviates λP (see **Section 14.4**), and P{x} abbreviates $^{\vee}P(x)$ (the extension of P applied to the intension of x). Therefore, we can offer the reader a variant of $\hat{P} \bigvee x P\{x\}$ which uses lambda notation: $\lambda P\ \exists(x)\ [^{\vee}P(x)]$. Thanks are due to Susan F. Schmerling for her invaluable help in decoding Montagovian notation.

itself. This becomes clearer when we consider the IL translation that Karttunen assumes for a restricted *wh*-phrase like *which girl*: $\hat{P} \bigvee x[girl'(x) \wedge P\{x\}]$ (i.e., $\lambda P\ \exists(x)\ [\text{girl'}(x) \wedge {}^{\vee}P(x)]$). This has a direct impact on our framework, because we have the addresses point to semantic values: we said in **Section 2.2** above that semantic values could be intensions. If so, then the semantic value of a basic expression may be the IL translation of that basic expression, and we can have *who* and *what* as nodes in graphs without the need to assume that there is a phonologically empty N *in the syntax*. This view contrasts with Reinhart's, who proposes that bare *wh*-words take a null N complement, yielding a structure like (301):

(301) [who $[_N\ e_i]$] (Reinhart, 1998: 44; see also Panagiotidis, 2001: Chapter 5)

In the generative view, empty heads are nothing to worry about, and their proliferation is not customarily seen as something to be avoided. We, however, want to avoid them to the extent that we can: paraphrasing Postal (1972: 215), the problem is not that descriptive adequacy cannot be attained with empty heads (or formal features); rather, the problem is that the theory risks becoming unrestricted.

As an interim summary, the scenario we are left with is then one in which we could have the following IL translations for *wh*-words, which constitute their semantic value (based on Karttunen, 1977: 19; see also Hamblin, 1973: 49):

(302) a. $[\![\text{who / what}]\!] \equiv \lambda P\ \exists(x)\ [{}^{\vee}P(x)]$
b. $[\![\text{which N}]\!] \equiv \lambda P\ \exists(x)\ [N'(x) \wedge {}^{\vee}P(x)]$

Assuming in this context that semantic values are IL translations (a possibility we mentioned in **Section 2.3**), we can provide some additional details. An interrogative sentence that contains *who* or *what* is asking about all *x* such that P(*x*), P being 1-place predicate (an IV in categorial terms). As an example, the full translation of (293a) (*Who loves Mary?*) applying Karttunen's quantification rule would be (303) (from Karttunen, 1977: 20):

(303) Who loves Mary?
a. $[\![\text{who}]\!] \equiv \hat{P} \bigvee xP\{x\}$ [i.e., $\lambda P\ \exists(x)\ [{}^{\vee}P(x)]$]
b. ?-he$_1$-loves-*Mary'* $\equiv \hat{p}\ [{}^{\vee}p \wedge p = {}^{\wedge}love'_*\ (x_1, Mary')]$ (where $love'_*\ (x_1, Mary')$ is the IL translation of *there is an individual x with index 1 and x_1 love Mary*; *Mary'* is the IL translation of the NP [Mary]; this can be the set of sets that have Mary as a member)
c. who-loves-*Mary'* $\equiv \hat{p}\ [who'(\hat{x}_1\ ?\text{-}he_1\text{-}loves\text{-}Mary'\ (p))]$

In terms of how many nodes in a graph we need to represent the relations between expressions, note that the IL translation of *which N* includes the IL translation of N, notated *N'* (or, unabridged, $\lambda P[[{}^{\vee}P]({}^{\wedge}N)]$, the extension of P applied to the intension of N). The complement of *which*, of course, can be an object of arbitrary complexity:

(304) *which* $\left\{\begin{array}{c}\textit{boy}\\ \textit{student of mathematics}\\ \textit{picture of himself that you liked}\end{array}\right\}$ did John talk about?

It makes sense, in this context, to have *who* and *what* as single nodes and assign *which*-N a more complex structure, depending on its complements. We will return to this point in **Section 10.1** as well as **14.4**.

There is an aspect of the relation between the **?** operator (or Katz & Postal's Q morpheme) and the pronoun that is particularly interesting to the effect of formulating conditions over dependencies in our graphs. As anticipated above, we will briefly compare two possible analyses for the ρ-sets of *wh*-interrogatives; further research is needed in order to decide between them. Let us begin by assuming that *wh*-elements open scope and that said scope needs to be closed *syntactically*. This much is basically Johnson's approach. In principle, not very different from Koopman & Sportiche's (2000) Bijection Principle:

> *There is a bijective correspondence between variables and positions*. (Koopman & Sportiche, 2000: 23)

Plainly: every variable is bound by a single operator, and every operator binds only one variable. More specifically,

1) *A variable is locally bound by one and only one Ā-position*
2) *An Ā-position locally binds one and only one variable.*

One of the consequences of the bijection principle is that, if it is assumed as a principle of the grammar, both vacuous quantification and unbound variables (thus, open sentences) become ungrammatical. The importance of the bijection principle for the present discussion is that *open scope counts as vacuous quantification*, and as such needs to be *lexically* restricted (a scope-opening expression needs to bind an indexed lexical variable): going back to Karttunen's IL translations, **?** must quantify over an indexed *he*. This quantification requirement is met in the IL translation of *wh*-words. But syntactically we need to make sure that *wh*-arguments (i.e., Subjects and Objects which are questioned) receive GF within the proposition: we indicate this in the ordered ρ-set. In the Montagovian view (with some extrapolation), it is the pronoun that receives a

GF and not—*stricto sensu*—the operator (in LFG and RG, the operator receives an *overlay* function, and the variable a *primary* function); in our graphs the operator and the pronoun are the same node: a *wh*-node is both a *trigger* and a *variable* using Ladusaw's (1980: 112) terms.[9] There are two possible analyses for these cases: the first one we considered, which corresponds to the ρ-sets in (218), is that in which there is a single visit to the *wh*-node: what Karttunen captures by having two distinct elements (operator and pronoun) is taken care of by the semantic value of the node (cf. Engdahl, 1986; and Johnson, 2020 for an opposite view). Syntactically, what we care about is that the GF of the *wh* element does not change, regardless of how much structure is introduced. In other words, the interpretation of *what* is the same in (305a) and (305b):

(305) a. What did Mary read?
b. What does Susan think that Mary read?

In both cases, the node with address ⦃what⦄ is the Object of *read*: if all syntactic relations are established at the level of elementary graphs (Frank, 2002, 2013), we want to be able to say that the introduction of recursive structure does not change the relation between the expressions *what* and *read* (LTAG's *non-local dependency corollary*). In this case the resulting graph will be single-rooted: in (305a) the *wh* node is contained in an elementary graph with root and anchor *read*, and in (305b) it is the derived graph that is single-rooted (with root *think*). The single-visit analysis (which does not distinguish between operators and variables as different visits to a node in the graph) shifts some of the weight to the semantics and out of the ρ-sets. In the most extreme version of this view,

> *there is no reason to expect any instances of wh-movement that are caused by the semantic needs of the wh-words themselves* (Kotek & Hackl, 2013: 3)

Changes in word order, to the extent that they need to be represented in the syntax, must be independently motivated (in current MGG, for example, this is accomplished by an 'EPP' feature in C which requires its specifier position to be filled). The version that Kotek & Hackl themselves propose is somewhat

9 It is important to distinguish the notions of *semantic* and *syntactic* saturation of a predicate (Chung & Ladusaw, 2003: Chapter 1); here we are primarily concerned with how semantics can inform an adequate theory of syntax: how to appropriately construct and restrict the theory of syntax so that it assigns each expression of the language a structural description that adequately represents semantic dependencies between grammatical objects in that expression.

in the middle: *wh*-phrases move only to a propositional node (where they are interpreted at LF), but not to other intermediate landing sites.

The second possibility, which can be thought of as closer to the MGG view, is to have distinct visits to the *wh* node in a walk through the derived graph: one for the filler, another for the gap. In this case, the result is a non-rooted graph: this is so because the *wh* node dominates whatever the matrix predicate is and is dominated itself by its own governor. There is thus no undominated node in the graph: the result is a circuit. If we require all graphs to be (at least single-)rooted, this analysis is automatically excluded on configurational grounds. If not, we can define well-formedness conditions for derived graphs in this view as well. Specifically, let v_{wh} be a *wh*-node (whose semantic value is either (302a) or (302b); we will not deal with the interpretation of adverbial *wh*-words like *when, where* or *why* in this work). Then, we propose the following condition over *wh*-walks:

(306) There must be a *unique* walk w in G such that $(v_{wh}, v_{wh}) \in \rho^*$

The idea is simple enough: open scope must be closed, and what closes the scope opened by a *wh*-expression is the 'pronoun' bound by **?** (in terms of Karttunen, 1977; see also Cooper, 1984 for an approach to interrogatives that dispenses with *quantifying-in* but is very much in the spirit of PTQ). Furthermore, scope closing must be *unambiguous*, this is the significance of *unique walks* in the more general context of a *strict ordering* requirement on graphs. There cannot be more than one walk in G that, starting from the scope-opening position, reaches the address of the *wh*-expression in a context where it is assigned a GF: a *wh*-node must be *d-ordered* with respect to itself. The only thing is, for purposes of representing relations between expressions, the nodes that correspond to the operator **?** (the *trigger* in Ladusaw's terms) and the pronoun $\mathbf{he}_n$ (Ladusaw's *variable*) is *the same node, visited twice in a local walk*. Here, 'locality' is to be understood as a dependency that satisfies the conditions for *licensing* (note that there is no mention of specific nodes in the definition of *locality*: no barriers, phases, or the like). There is a further requirement, which is that there be a biunivocal relation between operators and indexed pronouns: in reality we are dealing with multiple visits to the same node, in distinct contexts (from different neighbouring nodes each time). This is important because the requirement in (306), as it stands, relates scope-opening contexts (i.e., root contexts) to restrictor contexts (i.e., non-root contexts) in terms of indexed categories, not lexical items. Because expressions with the same semantic value bear the same index in the algebra of the language, by referring to an index we are simply making reference to an expression of the language in isolation from its structural context. In a semantic framework like Cooper's (1984), walking through a graph and finding an interrogative NP requires us to store a binding operator and inserting in its place a variable as a placeholder (so-called *controlled quantification*). The second visit to the same NP would allow us to take the operator out of storage and replace it, pretty much along the lines of Karttunen's approach (see also Engdahl, 1986: 36, ff.). Interpretation takes place by means of substitution, as in PTQ. *Controlled quantification* can be related to specific lexical entries or to specific syntactic rules: in a non-transformational framework, only the former option is available. This means that if we wanted to implement a Cooper-storage approach to the semantics of *wh*-interrogatives (an enterprise that goes beyond the scope of the present monograph), we would need to specify controlled quantification as a property of the semantic value of *wh*-elements (as it cannot be encoded in terms of syntactic configuration).

As highlighted in the previous paragraph, we have introduced a requirement for well-formed *wh*-walks in graphs to be *strictly ordered*: under the view that there is only one visit to a *wh* node (and thus that interrogative sentences differ from declaratives not configurationally, but in terms of the semantic value of one or more of its argument nodes) this requirement can be met without further stipulations. Under the single-visit analysis, *dominance* (note: not *immediate dominance*) remains irreflexive, asymmetric, and transitive. However, under the two-visit view, according to which interrogatives differ from declaratives configurationally in terms of the dominance relations established between nodes in a graph (and not just in the content of the addresses involved), we need to make some further considerations. Dominance is no longer an asymmetric relation, due to the bicircuit configuration created via transitive dominance. In the relation between both contexts of occurrence of a *wh*-node (call it v_{wh}) the instance that is ordered first in a sequence opens scope, and the one that is ordered after that one closes scope. Furthermore, in *who* and *what*, there will be a visit to v_{wh} in the context of a lexical predicate (i.e., there will be an edge $e\langle \text{verb}, v_{wh} \rangle$), such that v_{wh} receives a GF. For example, under the two-visit approach, we would have the following analysis (see Krivochen, 2023a, b; also the commentary on Johnson's 2016 proposal in **Section 14.4**, below):

(307) a. What did John buy?
b. $\rho = \langle$(what, buy), (buy, John), (buy, what)$\rangle$
c. $\Sigma = \langle$what, buy, John, what$\rangle$

↓	↓
– Opens scope	– Closes scope
– Operator interpretation	– Receives GF

The *unique walk* condition (306) can thus be interpreted in relation to what Sternefeld (1998a) refers to the *Scope Condition* (based on much previous work, see Ladusaw, 1980; Barss, 1986; Lebeaux, 1994; Giannakidou, 2002, among many others), which he formulates as follows:

> NPIS [Negative Polarity Items] *as well as bound variables must be* [in] *the scope of* (*i.e., c-commanded by*) *the operators they depend on* (Sternefeld, 1998a: 151)

There are, however, two crucial differences between Sternefeld's condition (which we take to be representative of the GB-MP approach, but also shared by non-transformational frameworks like LFG; see Darlymple, 2001: Chapter 11, §2; Bresnan, 2001: 212, ff.). One is that *scope* in our framework is *not* defined in terms of c-command in L-trees, but rather in terms of the existence of a directed edge (or a sequence of directed edges for *transitive dominance*) between the relevant nodes (e.g., licensor and NPI / operator and variable). The other difference (and perhaps a more important one) is that in the case of operator-variable relations, the 'dependency' is not determined by *scope* in the traditional way (as a relation of c-command between two distinct nodes), since *operator* and *variable* are the same node: if we want to have the classical distinction between operators and variables be represented in ρ-sets, the problem must be formulated in a different way.

The relevant question, in terms of the present proposal, is whether there is a unique walk w in G containing v_{wh} such that

(a) There are as many visits to v_{wh} in G as there are predicates selecting v_{wh} if v_{wh} corresponds to an argument **plus** one which dominates the root, **or**

(b) There are as many visits to v_{wh} in G as there are predicates being modified by v_{wh} if v_{wh} corresponds to an adjunct (*why, how, where* in non-locative construals) **plus** one which dominates the root

The two proposals suggested above differ in terms of whether the requirement after **plus** must hold: if *interrogative formation* is a lexical rule (such that it depends on the semantic value of a node in the graph and not on configurational relations), then the conditions that mention the root node do not apply. In the case of argumental *wh*-elements, we additionally require that the structural context of v_{wh} in each case satisfies the semantic requirements imposed by neighbouring predicates. We ask, then: does v_{wh} in at least one structural context (where *context* is defined as the set of nodes immediately dominating v_{wh} or that v_{wh} immediately dominates) receive a GF from a predicate, thus partially satisfying that predicate's valency?

It is worth noting that having *wh*-movement be formalised graph-theoretically captures Postal's (1972) observation that there are reordering transformations that do not proceed in a strictly cyclic manner (which he calls *U-rules*, for *unbounded rules*): that is, whose application need not resort to intermediate reconstruction sites (a similar approach is taken in TAGs, where recursive structure is factored out in computing filler-gap dependencies). Postal argues that *wh*-movement is one of these rules (at least in English); we would like to restrict that claim by saying that *wh*-movement is unbounded when the structure crossed over by the *wh*-dependency grows monotonically (see also Bach & Horn, 1976: 273, ff.): for example, a sequence of *Equi* verbs, propositional attitude verbs, *Raising to Subject/Object* verbs, and the like (plus possible combinations of these). In this way, the ρ-set for a sentence like (308a), where the long-distance dependency relates a *wh* term with the object position of the most embedded predicate in a sequence of clausal complements, is (308b) (indices and gaps are purely illustrative):

(308) a. What$_i$ does Mary think that John believes that he should tell Peter that Susan read __$_i$?
b. ρ = ⟨(think, Mary), (think, believe), (believe, John), (believe, tell), (should, tell), (tell, John), (tell, Peter), (tell, read), (read, Susan), (read, what)⟩

Some comments are due: first, we have not assigned the complementiser a node in the graph, since it has no semantic value: it is syncategorematic. We can omit it with no consequence, in (308). In LFG, for example, *that* would appear at *f*-structure as the value of an attribute COMPFORM: complementisers have no PRED value (e.g., Dalrymple et al., 2019: 60). Also, we have followed the analysis whereby there is only one occurrence of the *wh* expression, in the position where it receives a GF; its status as an 'operator' status may be part of its semantic value. We have four lexical predicates in (308): *think, believe, tell,* and *read.* In each case, a verb that takes a causal complement directly dominates that clause's root (since everything that that root dominates is the direct object of the V). We can indicate Object relations as follows, in increasingly dark colours (such that the object of *think* is in italics, the object of *believe* is in bold, and the object of *tell* is in bold and italics):

(309) ρ = ⟨(think, Mary), (think, *believe*), (*believe, John*), (*believe,* **tell**), **(should, tell), (tell, John), (tell, Peter), (tell,** ***read*****),** ***(read, Susan), (read, what)***⟩

The node assigned address ⦃what⦄ never leaves the elementary graph where it is licensed, and the addition of other elementary graphs makes no difference in this respect: in this sense, all dependencies are locally defined, as in LTAGs (there is no real 'long-distance' dependency). Another important point is that in (308b) is that there is no need to assume additional occurrences of *what* in intermediate positions: *wh*-movement is *not* modelled as a successive-cyclic rule (*contra* Chomsky, 1973, 1986, 1995 and much related work). Using the terminology of Abels & Bentzen (2012: 433), the ρ-set in (309) is *uniform* in the sense that all intermediate objects in what we can descriptively refer to as the 'movement path' (the structure that mediates between both visits to *what* in the digraph defined in (308b)) are equally unaffected.

10.1 Simple Wh-Questions

In this section we will present sample analyses for a variety of *wh*-interrogatives involving a single *wh*-word or phrase.

Subject and Object *wh*-interrogatives have been analysed in the previous section, where we also defined the content of the addresses corresponding to ⦃who⦄ and ⦃what⦄. We have not, however, given an explicit analysis of Subject and Object *wh*-interrogatives with lexically restricted *wh*-phrases. Relevant examples are like (310a) and (310b) respectively:

(310) a. Which student read *Infinite Syntax*?
b. Which book did John read?

In the line of what we said above about the semantic structure of *which*-N, in which the *wh*- instantiates a *choice function* over the set of entities denoted by the N, we will analyse *which N* as ρ = ⟨(which, N)⟩: the *wh*-word dominates the N. In this relation, the *wh*-word quantifies over the N, and the N restricts the *wh*-word.[10] Transitively, anything that dominates *which* will dominate *N* also, but a node may dominate N and exclude *which*: this will become important in **Section 14.4**. The idea of *partial* multidominance was introduced already, in **Section 8.1**. The ρ-sets of (247a, b) are thus (248a, b) respectively:

10 In LFG, an *wh*-operator can inherit the syntactic rank of its 'operator complex': this is a phrase that gets obligatorily 'pied piped' with the operator (such as *whose brother's friend*). This is important because an operator complex can 'promote' an operator to a GF that outranks an expression bound by the operator complex. (Bresnan et al., 2016: 225).

(311) a. $\rho = \langle$(which, student), (read, student), (read, *Infinite Syntax*)$\rangle$
b. $\rho = \langle$(which, book), (read, John), (read, book)$\rangle$

In (311), the lexical restrictor is interpreted only in a local relation with its subcategorising predicate (cf. Sauerland, 2004; Saito, 1989). If we turn our attention to more complex examples, like (312a, b), the advantages of an approach that maximises local relations become more apparent:

(312) a. [Which picture of himself$_j$]$_i$ did John$_j$ say Mary likes __$_i$ (Uriagereka, 2011: 5, ex. 11)
b. John$_i$ wondered [which picture of himself$_{i/j}$]$_k$ Bill$_j$ saw __$_k$ (Chomsky, 1995: 205, ex. 36 a)

There are several things to focus on in these examples. One is that the *self*-forms that appear are *not* anaphors in the sense of **Section 6.2.1**, there are no *parallel* arcs in the graphs that define the structural descriptions for these expressions. These *self*-forms have the interpretation of Binding-theoretic *pronouns*, in that there is no *reflexivity*. Another important aspect of the sentences in (249), and perhaps the most relevant for our current purposes, is that they display a complex structure as complement of *which*; this structure contains a node that needs to be multidominated in order to yield the appropriate correference: *himself* is not a reflexive in the sense we take *reflexive* to work here, but it is referentially bound to an expression that is superficially outside the *wh*-complex where *himself* appears (which we have indicated with indices).

Let us focus on the first example for concreteness. The derivation of (312a) along classical MGG lines requires, at least, the following mechanisms:

- A set of operations (phrase structure rules, Merge, etc.) to generate the terminal string *John said Mary likes which picture of himself*
- A movement rule that displaces the syntactic term [which picture of himself] from its base position as the complement of *like* to the Spec-CP position / 'left periphery', call it *Wh-movement* (Chomsky & Lasnik, 1977: 434), *Move-α* (Lasnik & Saito, 1992), *Internal Merge* (Chomsky, 2000), etc.
- An indexing rule that keeps track of occurrences of syntactic terms. It needs to be able to assign the same index to *John* and *himself*, but also to *which picture of himself* and the gap in the complement position of *like* (e.g., Chomsky's 2021 *FormCopy*).
- A rule that inserts the auxiliary *do* to spell-out tense and agreement features

Interestingly, much of this complication emerged in the early days of the generative theory because interrogatives were assumed to derive via movement transformations that map trees onto trees. The declarative version is assumed to be generated by PSR or some equivalent mechanism (e.g., Merge); transformational rules then move things around extending the phrase marker and leaving derivational crumbs behind which allow interpretative systems to keep track of what has gone where (traces, copies, slash-features ...). In addition to the process of reordering, at some point there must be an indexing algorithm in play, which is capable of keeping track of the location of referential expressions in the phrase marker (or a search algorithm dedicated to tracking gaps when finding a filler, or an operation that, given two expressions, assigns the relation *Copy-of* to them).

Simplifying matters slightly, we can consider three stages in the implementation of displacement within MGG and still in use today. The first, formalised in Fiengo (1977), is the familiar *trace theory*: syntactic terms affected by reordering rules leave behind 'traces'; these are variables that are assigned identical indices to their antecedents and thus become constants. In Fiengo's terms, which we repeat from **Chapter 1**,

> *... movement of* NP_i *to position* NP_j *(where A and B are the contents of these nodes) in (30) yields (31) as a derived constituent structure.*
>
> (30) ... NP_j ... NP_i ...
> | |
>   *A*   *B*
>
> (31) ... NP_i ... NP_i ...
> | |
>   *B*   *e*
>
> *On this view,* NP_i *and its contents are copied at position* NP_j*, deleting* NP_j *and A, and the identity element e is inserted as the contents of (in this case the righthand)* NP_i*, deleting B under identity.*

This operation involves two steps: copy and deletion under identity, plus co-indexing (as a side-product of copying).

Early Minimalism redefined things, although in practical terms the approach to Move-α remained largely similar to the exposition of trace theory in Fiengo (1977) and related works. The second stage can be characterised as a move towards a strongly derivational approach to displacement (in contrast to the more representational approach that came before). Kitahara (1997: 24) provides a very clear definition of *Move-α* in this theoretical context:

Input: Σ (containing α)

Concatenate α and Σ, forming Σ'

Output: *Σ'*

In this case, the Extension Condition is respected. We take an object α of arbitrary complexity from a syntactic term with root Σ, and we concatenate α to Σ. This operation extends Σ to Σ' (which properly contains Σ and α), creates a trace t_α in the position where α was initially Merged, yielding a chain CH = $\{\alpha, t_\alpha\}$. The process is diagrammed as follows:

(313)

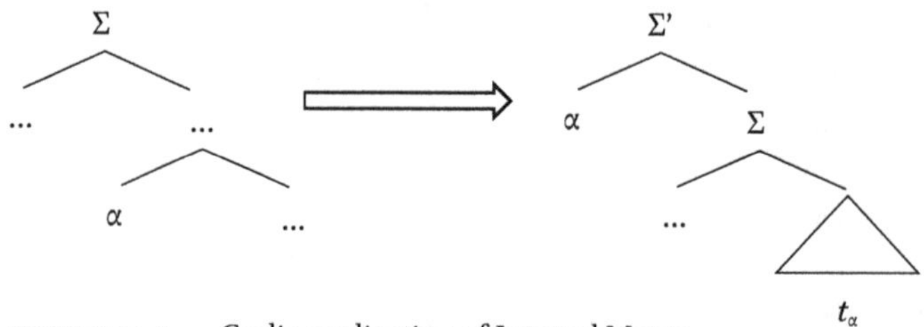

FIGURE 10.4 Cyclic application of Internal Merge

If $\Sigma = C_{[+wh]}$, we could say that $\alpha = NP_{[+wh]}$. The NP is displaced to Spec-C, creating a new node t_α so as to maintain the relation between the NP and its lexical governor. In the extension of C to CP via concatenation with α, another non-terminal node is created (namely, Σ' = CP).

But there is another option: the displaced NP does not extend the phrase marker, but lands in an intermediate position. This is the counter-cyclic version (which is required in order to implement uniformly successive-cyclic movement and operations like Richards' 1999 *tucking-in*). Kitahara's definition makes the workings of the operation explicit:

Input: Σ (containing *K* and α)

Concatenate α and K, forming L = $\{\gamma, \{\alpha, K\}\}$
Replace K in Σ by L, forming Σ'

Output: *Σ'*

The diagram is different this time, because the target of the operation is not the root of the tree, but an intermediate node (i.e., a node with indegree 1 and outdegree 2). This time, the operation looks like this:

(314)

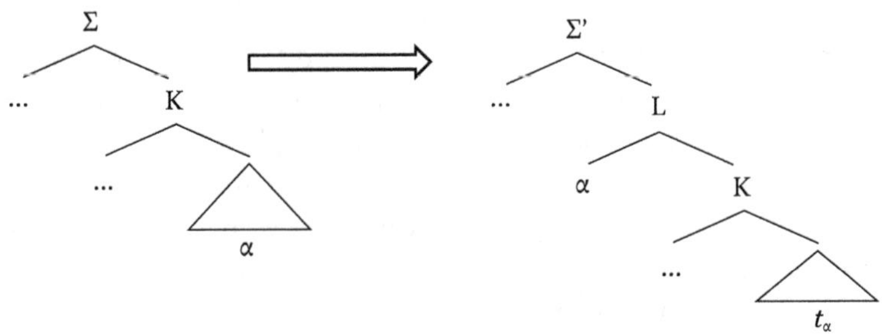

FIGURE 10.5 Counter-cyclic application of Internal Merge

The counter-cyclic version creates a new intermediate node: the term L which properly contains α, K, and t_α; this node is created in order to yield a configuration in which there is a syntactic term where α is excluded from K.

The third stage involves a further move towards a more general system. Chomsky (2000) argues for the elimination of traces in the theory of movement, since they imply a violation of the Inclusiveness Condition (traces are not specified in the Lexical Array from which derivations are built). The idea is that syntactic terms can be *copied* and Internally Merged, with no access to the Lexical Array or the Lexicon. The Copy+Merge theory of movement, explored among others in Uriagereka (2002) and Nunes (2004), has a very similar implementation to Kitahara's formulation:

a. *Copy one of two independently merged phrases*
b. *Spell out the lower copy as trace* [i.e., a phonologically null terminal]
c. *Merge the trace*
d. *Merge the higher copy (possibly in a separate derivation)* (Uriagereka, 2002: 61)

Independent operations of the Copy+Merge theory of movement

a. *Copy*
b. *Merge*
c. *Form Chain*
d. *Chain Reduction* (Nunes, 2004: 89)

These proposals are less clear with respect to the number of extra nodes created in the process, however. Further complications arise when we consider the fact that copied elements must be stored somewhere until they are Merged again (Stroik & Putnam, 2013; Krivochen, 2023b): the nature of this temporal storage has rarely been addressed in MGG. To address these problems, Frampton & Gutmann (1999), and more explicitly Stroik (2009) and Stroik & Putnam (2013) have proposed that elements in the Numeration used to derive a particular sentence remain active (in Stroik's terms, they *survive*) if they have unchecked/unvalued features: this allows terms to be re-introduced in the derivation without having a designated Copy operation, and at the same time dissolving the problem of how and where copies are stored before they are re-Merged (see also Gärtner, 2021 for discussion about the set-theoretic inconsistencies of the most recent developments of the Copy Theory of movement). Here we are not so much concerned with the details of temporary storage in derivations or the dynamics of 'derivational spaces' (see Krivochen, 2023b for detailed discussion), but with the consequences that these approaches to displacement have in terms of the multiplication of nodes in structural descriptions.

The third stage is related to multidominance theories, insofar as syntactic terms can get 're-Merged' in a derivation. Somewhat recently, there has been work emerging on what is called 're-Merge' theory of movement (e.g., Johnson, 2014, 2016, 2020; Larson, 2016), whose generative power is the same as the Copy+Merge theory (in particular, the version explored in Nunes, 2004), so far as we can tell (see Larson, 2016 for discussion). The differences arise mostly in the diagrams they allow for L-trees not so much in the L-trees themselves: in other words, the drawings change, but not the formal relations between elements in the tree as a mathematical construct. This is so because the mechanism involved in structure building is still Merge, which has not been redefined graph-theoretically (see also Citko & Gračanin-Yuksek, 2021; Krivochen 2023a, b proposes such a redefinition). The version of re-Merge theory in Johnson (2014, 2020) would assign a phrase marker like (315b) to (315a) (note that instead of having intermediate projections Johnson proposes multi-segment phrasal categories):

(315) a. Which script did Jane write?

b.

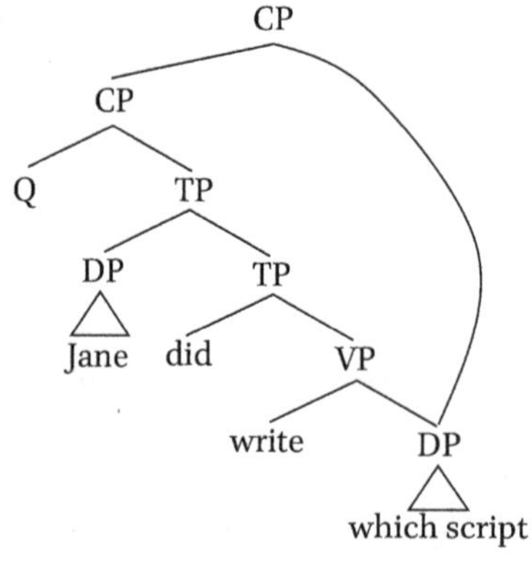

FIGURE 10.6 Multidominance analysis of wh-*interrogative in Johnson* (2014, 2020)

Prima facie, the representation in (315b) is rather close to our graphs, minus non-audible structure (all nonterminal nodes and labels corresponding to phrasal projections). There is no multiplication of DP nodes, but a single DP node *which script* that has more than a single mother node. However, the relation between these theories is not as close as it may seem. Leaving aside the evident use of intermediate nodes as part of the syntactic representation (which allows for asymmetric c-command relations required for LCA purposes to hold), which we dispense with, and the focus on defining sets of pairs of syntactic objects rather than proper graphs, there are differences that pertain to aspects of the interpretation of multidominated nodes. In this respect, Johnson (2014: 268) argues that

> *Note, then, that a phrase which resides in two positions, as* which dish *does in* [(315b)], *need be semantically interpreted in only one of those positions. The normal requirement that everything in a syntactic representation must be interpreted by the semantic component must be allowed to permit* (29).
>
> (29) *If a term is in more than one position in a phrase-marker, it need be semantically interpreted* ***in only one of them***. (highlighting ours)

Johnson argues that in a case like (*Mary wonders*) *which flower she should bring*, *which flower* is multidominated (by CP and VP), but that

> *The single phrase,* which flower, *would not seem to be able to simultaneously have the meaning of a variable and the meaning of the term that binds that variable*. (Johnson, 2020: 120).

That depends on exactly what the semantic value (or 'meaning') of *which flower* is defined to be, and to what extent that semantic value is represented syntactically. One way to solve the issue of assigning an operator and a variable reading to a single phrase (Larson, 2016) is to multiply the terminal nodes: dissociate the projection of *flower* (which would be the variable) from a Q-projection (which would be the operator) in *which flower* and then have only *flower* be multidominated: in that configuration, the Q-abstract morpheme can be dominated exclusively by CP, whereas the variable is multidominated. This *partial* multidominance approach (**Chapter 9**) will be capitalised on in **Section 14.4**. It is interesting to observe that one of the first multidominance approaches, Peters & Ritchie's (1981) Phrase Linking Grammar, assumes *total* multidominance: *which flower*, in their version, would be exhaustively dominated by S and VP (see Gärtner, 2014: 2, ff.).

Our model is based on the view that if a node occurs in a certain context (where, remember, the *context* of a node is the set of nodes it is *directly* connected to) then it is because that node is assigned an interpretation in that context: that is due to the connection between dominance relations and compositional interpretations. In this sense, Johnson's condition (29) plays a similar role to Nunes' *Chain Reduction*, and is at odds with our approach. For example, in the ρ-set that corresponds to (315a), which is (316),

(316) ρ = ⟨(which, dish), (eat, Sally), (eat, dish)⟩

which dish receives GF Object in the context *e*⟨⦃eat⦄, ⦃dish⦄⟩. Note that the dissociation between Q and N in Johnson's multidominance representation for operator complexes can be straightforwardly captured in ρ-sets under current assumptions. An advantage of *dependency*-based approaches, including ours (to a certain extent) is that semantically vacuous positions or occurrences are eliminated, because the motivation to have an element in a certain position in a structural description is *semantic* as well as *syntactic*. This contrasts with the transformational approach: historically, semantic interpretation was initially restricted to Deep Structure because transformations did not change meanings or grammatical function assignment (Katz & Postal, 1964); in later models Surface Structure played a role since it contains all the same information as Deep Structure *plus* traces and aspects of suprasegmental phonology determining *focus-presupposition* dynamics (Chomsky, 1970a; see also Schmerling, 2018a: xii). Finally, in classical Minimalism, semantic interpretation is handled by a component external to the generative computational system ($C_{(HL)}$), the Conceptual-Intensional system C–I (Chomsky, 1995 et seq.). Three assumptions are behind these treatments: (i) interrogatives are derived by the application of transformational rules (ii) structural descriptions assigned by the grammar are phrase structure trees, and (iii) in a separation between a syntactic and a semantic component in the grammar, only the syntactic component is generative, semantics only reads what syntax delivers.

As we have emphasised, our aim is to provide an exhaustive map of relations between expressions in a sentence. In this context, we can do away with the requirements that derive from PSGs, and begin by considering (a) how many elementary graphs we have and (b) at which nodes they are linked. In this sense, interrogatives receive the same treatment as declaratives, without there being a mapping from one to the other. This allows us to provide the following ρ-set for (312a) (*Which picture of himself did John say Mary like*?):

(317) Elementary graph 1: [John say [EG2]]
Elementary graph 2: [Mary likes which picture of John]

ρ_1 = ⟨(say, John), (say, like)⟩
ρ_2 = ⟨(like, Mary), (like, picture), (which, picture), (picture, John)⟩

We have two elementary graphs, with anchors *say* and *like*. These two structures are linked at ⦃John⦄; and the matrix verb dominates the anchor node of what is its complement, the clause [Mary likes which pictures of himself]. Because the anchor is also the root of EG2, it transitively dominates every node that root dominates.

It is important to bear in mind that we have left aside a number of questions pertaining to the semantics of *picture*-Ns; in particular, we treated *picture* as if it was an argument, but it would be also possible to take it as a function with domain *John*. In this case, and if we assume Montague's (1973) *translation rule* 1, given *John* in the domain of *picture of x*, then *John*' translates, informally, as *picture of* (*John*). As for the preposition, since it is a 'non-semantic' use (not locative or directional), we follow usual practice in LFG (Falk, 2001: 14; Dalrymple et al., 2019: 61, ff.) and CG (Schmerling, 2018a: 63) in considering them *case desinences* rather than *case assigners*: in this case, *of* is an exponent of Genitive case. This approach to prepositional case marking is not unheard of in MGG, see e.g., Uriagereka (2008: 146). Considering in particular Schmerling's view, it is worth highlighting that the mechanisms of Case in CG are quite different from Case as seen in MGG, where Case is the reflex of specific structural configurations in which arguments appear in the domain of specific functional heads (see Schmerling, 2018a: Chapter 5 for further discussion of Case in German from a pure CG perspective).

As for simple *wh*-interrogatives where the *wh*-element is not an argument, we will assume here that so-called VP adjuncts dominate the predicate that they modify. This goes along the lines of some versions of (neo-)Davidsonian semantics for VP adjuncts (Davidson, 1967, 1969; Parsons, 1990; this framework is assumed, e.g., in Uriagereka, 2008; also Higginbotham, 2000 and much related work; Maienborn, 2011 gives a general perspective of event semantics including Davidsonian and Neo-Davidsonian approaches) in a specific (and perhaps *sui generis*) sense. Consider the following sentence:

(318) Mary read a paper quickly in the park

A rough Davidsonian representation of (318) would go along the lines of (319):

(319) $\exists(e)$ [READ(Mary, a paper, e) $\wedge$ QUICKLY(e) $\wedge$ IN(e, the park)]

Because we have made no use of thematic roles in (319) to account for the relation of the event and its participants (such that *Mary* is the Agent of *e* and *a paper* is the Theme of *e*), that representation does not count as 'neo' Davidsonian (Maienborn, 2011: 811; Parsons, 1990), but this does not really matter in the present context. Here we are concerned with a different point: note that in (319) the modifiers *quickly* and *in* take the event as an argument (in the case of *in*, we are in the presence of a dyadic predicate, so the event is one of its arguments, *the park* being the other). Part of Davidson's crucial contribution to the theory of semantic representation for natural language sentences is that

> *Adverbial modification is* [...] *seen to be logically on a par with adjectival modification: what adverbial clauses modify is not verbs but the events that certain verbs introduce.* (Davidson, 1969: 298)

The relevant part of Davidson's quote for present purposes is *logically*. The semantic interpretation of something like *really fast* (see (78)–(79) above) is licensed by the same syntactic mechanisms that the interpretation of *black suit* (see (60) above): nodes corresponding to adjectives dominate nodes corresponding to the nouns they modify, just like nodes corresponding to adverbs dominate nodes corresponding to the verbs they modify.[11] In both cases, modification does not change the syntactic category of its input (Dowty, 2003: 33–34): an A modifying an NP delivers an NP, not an AP; an Adv modifying a VP delivers a VP, not an AdvP (*mutatis mutandis* for CG categories). In the present context, phrasal labels are simply shorthands for sets of nodes and edges, but it is important to consider the interplay between syntactic category, configurational relations, and semantic type.

In this context, the ρ-set that we propose for (318) is (320):

(320) ρ = ⟨(read, Mary), (read, paper), (quickly, read), (in, read), (in, park)⟩

The syntactic similarities between (320) and the Davidsonian representation (319) should be apparent; we indeed intend to capture Davidson's insight in our ρ-sets. Now, let us consider two interrogative sentences that can be formed from (255), questioning *quickly* and *in the park*:

11 It is important to bear in mind that the only parallel between adverbial and adjectival modification that is relevant for purposes of this work is *syntactic* in nature (defined in terms of *dominance*), and has nothing to do with the generative definition of 'lexical categories' using binary features [±N] and [±V] (Chomsky, 1970b). We also depart from Davidsonian assumptions in that modification does not express (syntactic) conjunction (cf. Higginbotham, 1985: 563).

(321) a. Where did Mary read a paper quickly?
b. How did Mary read a paper in the park?

The ρ-sets of (321a) and (321b) are (322a) and (322b), respectively:

(322) a. ρ = ⟨(how, read), (read, Mary), (read, paper), (in, read), (in, park)⟩
b. ρ = ⟨(where, read), (read, Mary), (read, paper), (quickly, read)⟩

The dominance sets (322a) and (322b) are, for all syntactic intents and purposes, exactly parallel to those assigned to *wh*-interrogatives in which an argument is questioned; the crucial difference is whether the *wh*-element receives a GF in the graph; if it does not, then there is no indexed pronoun dominated by a lexical predicate for the operator **?** to bind, in contrast to the structures considered above in which an argument (subject or object) was being questioned. Crucially, in (322a) the preposition is not a case marking (and thus syncategorematic, not corresponding to a node), but appears in the dominance set as a categorematic element with a locative meaning. In this respect, the difference between the interpretation of *of John* (in *a picture of John*) and *in the park* can be captured as long as we do not require that every orthographical word correspond to a syntactic terminal or a node in a graph.

To summarise the discussion in this chapter, we propose an analysis of *wh*-interrogatives in which there is no derivational relation with respect to declaratives: interrogatives and declaratives are equally 'basic' structures for purposes of a definition of the relations between basic expressions in a digraph. The main issue we faced was the choice between two different analyses: one in which operator-variable relations are syntactically represented by means of different visits to a node and one in which operator-variable relations are not syntactically represented, but rather formalised as lexical specifications. The two analyses can be exemplified as follows:

(323) What did Sue say?
a. ρ = ⟨(say, Sue), (say, what)⟩ (single visit analysis)
b. ρ = ⟨(what, say), (say, Sue), (say, what)⟩ (multiple visit analysis)

In (323a) we rely on the content of the address ⦃what⦄ to do much of the semantic work: operator-variable relations in *wh*-interrogatives do not affect the definition of the digraph that corresponds to the well-formed expression of English *What did Sue say?*. Under this analysis, the ρ-set of *Sue said something* and *what did Sue say?* do not vary in configurational terms, only in terms of the content of one (or more) of the addresses. One may say that in this approach *wh-interrogative* is a lexical process, not a syntactic rule. In (323b), in contrast, the definition of the digraph varies with respect to a declarative, since it features a *bicircuit* relation between the predicate that anchors the elementary graph and the expression that is questioned: ⦃what⦄ dominates ⦃say⦄ and ⦃say⦄ dominates ⦃what⦄. Configurationally, it seems important to distinguish between multidominance which yields bicircuits and multidominance which yields parallel arcs: only in the latter case is there an expression establishing two grammatical relations with a predicate; differences between interrogatives and reflexive sentences in terms of how these objects are assigned morphophonological exponents may be related to this configurational difference; see fn. 6 of this chapter. Operator-variable relations are represented in terms of the order in which each of the addresses are accessed in a sequence defined for the graph (see (307) above). *Ceteris paribus*, we would prefer an analysis like (323a) over (323b) as we can maintain the requirement that nodes in a graph be strictly ordered (see also Peters & Ritchie, 1981: 1; they explicitly reject the kind of closed walks in graphs that would be inevitable in the two-visit analysis), but as of writing these pages there seems to be no simple way to determine which of the two analyses is more empirically fruitful given the data under consideration.

CHAPTER 11

MIGs and Prizes

This chapter ties in with **Chapter 4** in terms of assessing the adequacy of the present model to provide adequate structural descriptions to sentences with limited crossing dependencies. Specifically, it offers an analysis of so-called Bach-Peters sentences, whose self-containing antecedent-anaphora relations have proven problematic for classical PSG-based theories. We also build on the discussion of binding phenomena in **Chapter 8**, and formulate graph-theoretic versions of what Lees & Klima (1963) referred to as the *pronoun rule* and *reflexive rule*.

In this chapter we will consider some English sentences which will lead us to explore the limits of the formal mechanisms explored here. A theory that maximises relations and allows for multidominance can provide new insights in the analysis of MIG-sentences (a.k.a Bach-Peters sentences; see Bach, 1970; Karttunen, 1971a; McCawley, 1967; Sampson, 1975). Relevant examples are like (324) and (325):

(324) The man_i who shows he_i deserves it_j will get the prize_j he_i desires

(325) Every pilot_i who shot at it_j hit the MIG_j that chased him_i

It has been noted in the literature that getting the indexes right on these sentences is problematic for a transformational approach to pronominalisation based on NP identity (Bach, 1970); this motivated the rejection of a Lees-Klima-style transformation of *pronominalisation* in much subsequent work.

We want to underline that an adequate structural description for Bach-Peters sentences should capture the fact that there are dependencies of varying formal complexity intertwined. From a derivational perspective, we have proposed (Krivochen, 2015a, 2016a, 2018, 2021a; Krivochen & Saddy, 2016) that there are several *local* structural layers at play, each displaying varying levels of computational complexity within the Chomsky Hierarchy, and that the assignment of a strongly adequate structural description[1] to these sentences

1 We understand this notion in the somewhat programmatic sense of Joshi (1985: 208):
*A grammar G is weakly adequate for a string language L if L(G) = L. G is **strongly adequate** for L if L(G) = L and for each w* [a string in L] *in L, G assigns an **'appropriate' structural description** to w* (Joshi, 1985: 208)

 DOI:10.1163/9789004542310_012

requires the system to be sensitive to local changes in computational complexity. That is, what we need is *mixed computation*. If we look at local structures, the dependencies that emerge are simpler than what we see if we try to analyse the global structure at once using a single kind of syntactic template. This procedure is a hallmark of computation in complex dynamical systems (Binder & Ellis, 2016). Let us illustrate this point using (324) as our example:

(326) a. The⁀man; the⁀prize; he⁀deserves⁀it ⟹ Regular
b. [The man *S'* [will get [the prize *S'*]]] ⟹ Context-Free
c. [*the man* [who shows [he deserves **it**]]] will get [**the prize** [*he* desires ***e***]]

⟹ (Mildly) Context-Sensitive

In generative terms, the simplest layer consists of units that can be generated with a grammar that does not go beyond rules of the kind A → *a*B (a regular grammar), (327a) for the substrings *the man* and *the prize* and (327b) for the substring *he deserves it* (see Uriagereka, 2002 for discussion about strings like the latter in the context of a strongly derivational framework):

(327) a. NP → D, N' (terminal, nonterminal)
N' → N (terminal)
b. S → Prn, VP (terminal, nonterminal)
VP → V, Prn (terminal, terminal)

The second layer (326b) involves relations between non-terminals in the form of rules of the kind A → BC, thus, context-free. As for context-sensitive depend-

---

In this context, the question has been raised to us whether *strong adequacy* is equivalent to *strong generative capacity* (as defined in Chomsky, 1965: 60). The difference is crucial: the strong generative capacity of a grammar is the set of structural descriptions it generates; the definition given by Chomsky says nothing about that set being formally heterogeneous. The grammar that generates structural descriptions in MGG is computationally uniform, and thus the set of structural descriptions is uniform as well. In contrast, Joshi's requirement for *strong adequacy* for structural descriptions can incorporate aspects of mixed computation if the grammar is made sensitive to semantic dependencies between syntactic objects. Chomsky's *strong generative capacity* makes sure there is a set of structural descriptions, Joshi's *strong adequacy*, in the way we interpret it, makes sure that set is (under present assumptions) minimally adequate (not assigning any more structure than strictly necessary to capture semantic dependencies in local evaluation domains), but not procrustean.

encies, the need for yet another increase in generative power is given by the crossing dependencies between referential indices: note how co-indexing paths do not embed, but rather cross. Because we are still within the limits established by two sets of symbols, it is not *full* context-sensitive power that we need, but merely *mild* context sensitivity (of the kind delivered by TAGs with links).

Note that the computational differences arise *within local derivational units* and pertain to relations of co-reference within or across graphs. From a procedural perspective, the grammar needs to be flexible enough to accommodate for oscillatory dependencies in assigning structural descriptions to sentences, going *up and down the Chomsky Hierarchy* in local domains. In Krivochen (2018) we defined syntactic cycles as *emergent properties* from a computational system that does not commit to a 'one-size-fits-all' theory of phrase structure, but to one that models linguistic computation as a non-linear, complex dynamical system (see also Saddy & Uriagereka, 2004; Saddy, 2018, and related work; Beim Graben et al. 2004, 2008 present a neurocognitive perspective centered more on processing than on grammar, but still very much related to this agenda): the proposal in Krivochen (2018, 2021a) was that the change in computational complexity is what delimits local syntactic domains, not the presence of designated nonterminal nodes. In such a 'mixed' system, computational uniformity (the norm in MGG: Chomsky, 1995; Fukui & Narita, 2013; Kayne, 2018; see Jackendoff, 2011: 277–279 for some additional discussion) would in fact have to be stipulated. To summarise the situation, in the structural description of (326a) we have uniformly binary branching minimal treelets; following Greibach (1965: 44) and Uriagereka (2012: 53), a finite-state grammar can capture all relevant dependencies within these substrings. The structural description of (326b) features phrasal 'constituents' related within structures containing placeholders for root nodes (the roots of the relative clauses); center embedding pushes us up to CF grammars. The appearance of crossing dependencies between two sets of categories in combination with embedding in (326c) (note that the linear distribution of indices is *i* ... *j* ... *j* ... *i* ... *j*) forces us up at least 'half' a level, to 'mild' CS (Joshi, 1985 and much related work; see **Section 4.3** above).

One of the reasons why Bach-Peters sentences bear a particular syntactic and semantic interest is that, under a transformational view of how pronouns come to be in Surface Structure, they must be assigned Deep Structures of infinite complexity (which is why they are sometimes referred to as Bach-Peters *paradoxes*). Under a Lees-Klima view of pronominalisation, Bach argues, the Deep Structure of (324) would have to be:

> *The man who shows that the man deserves the prize that the man who shows that the man deserves the prize that the man ... (ad infinitum) will get the prize that the man who shows that the man deserves the prize that the man who shows ... (ad infinitum)* (Bach, 1970: 121)

While it is possible to save the transformational view by restricting the applicability of *Pronominalisation* to specific contexts (as argued by e.g. Lakoff, 1976: 329, ff. and Postal, 1969, among others), that modification creates a potential overgeneration problem. One of the advantages of the framework we have sketched here is that should a 'transformation' *Pronominalisation* be formulated, it need not make reference to *two* nodes (as in the Lees-Klima version and its more modern incarnations), but only *one*, which is multiply dominated. In the theory proposed here, as we have highlighted a few times already, the surface elements *the man, who,* and *he* in (324) are *the same node* (with address ⦃man⦄) with indegree 3 in the derived graph. All visits to a single node in a walk, in distinct contexts, involve reading an address that points to the same semantic value. What we need to do, then, is specify how the digraph that is the structural description for (324) satisfies all the constraints over allowed walks that we have formulated and delivers the right dependencies.

In particular, we need to provide an account of how pronominalisation works in extreme cases like Bach-Peters sentences. As mentioned above, it is possible to recast the *Pronoun Rule* and the *Reflexive Rule*[2] of Lees & Klima (1963: 23) in terms of conditions over walks in well-formed graphs. Grouping together the conditions formulated in **Chapter 5** and **Section 6.2.1**, we can recast the Lees-Klima approach for pronominalisation in English (the idea that bound pronouns and anaphors are not lexically inserted as such but arise from specific syntactic relations, revamped within a Minimalist framework in, e.g., in Kayne, 2002; Hornstein, 2001; Hornstein & Idsardi, 2014: 14, ff.; Grohmann, 2003: Chapter 3; Gärtner, 2014, among others) as follows:

2 For reference, Lees & Klima's original formulations are as follows:
Reflexive Rule: X-Nom-Y-Nom'-Z → X-Nom-Y-Nom'+Self-Z where Nom = Nom' = a nominal, and where Nom and Nom' are within the same simplex sentence.
Pronoun Rule: X-Nom-Y-Nom'-Z → X-Nom-Y-Nom'+Pron-Z where Nom = Nom', and where Nom is in a matrix sentence while Nom' is in a constituent sentence embedded within that matrix sentence (Lees & Klima, 1963: 23).

For any v_i, v_j nodes in a graph G,

(328) **Pronoun rule:**
v_j can pronominalise v_i iff:
a. *v_i and v_j denote sortal entities*
b. *$\{\!|v_i|\!\} = \{\!|v_j|\!\}$, **and***
c. *$(v_i, v_j) \in \rho^*$ in a derived graph G*

(329) **Reflexive rule:**
Let v_j be a predicate and v_i an argument of v_j in an elementary graph G. Then, v_i is an anaphor iff
$\rho = \langle \dots (v_j, v_i), \dots (v_j, v_i) \rangle$

Or, alternatively, and under the same conditions,

... *iff v_i is the tail of parallel arcs*

In Bach-Peters sentences it is particularly interesting to see how many elementary graphs we have and how they are linked such that the appropriate correference relations hold. Let us now define the local domains we are dealing with in (324):

(330) Elementary graph 1: [man [EG 2] will get prize [EG 3]]
Elementary graph 2: [man show [EG 4]]
Elementary graph 3: [man desires prize]
Elementary graph 4: [man deserves prize]

Putting all local domains together by substituting each elementary graph in where it is supposed to go, we get (331):

(331) [man [man show [man deserves prize]] will get prize [man desires prize]]

And after applying graph union to our elementary graphs, we get the following ρ-set for the derived graph, where nodes corresponding to arguments are assigned GF:

(332) ρ_{derived} = ⟨(show, man), (show, deserve), (deserve, man), (deserve, prize), (get, man); (get, prize), (get, desire), (desire, man), (desire, prize)⟩

In (332) the object of *show* is identified with the root of the graph corresponding to the subordinate clause *he deserves it*, which is the node with address ⦃deserve⦄ by virtue of not being dominated by any other node within that subgraph; the analysis here follows the lines of **Chapter 6**.

It is interesting to see that some aspects of the present analysis are prefigured in McCawley (1970) and Sampson (1975), and it is possible that only minimal adjustments are needed in order to make it compatible with the variable-free semantic proposal of Jacobson (2000). McCawley (1970: 176–177) considers the following sentence, attributed to S. Kuno:

(333) A boy who saw her kissed a girl who knew him

(333) is a run-off-the-mill Bach-Peters sentence (to the extent that these creatures can be said to be 'run-off-the-mill' at all). McCawley notes—like Bach, even though neither cites the other- that an approach to *pronominalisation* like the one presented in Lees & Klima (1963) requires an *ad infinitum* proliferation of antecedents if applied to *all instances of pronominal reference*. Like Lakoff (1976), McCawley proposes to keep *Pronominalisation* as a transformation (that is, he accepts that at least some pronouns are derived transformationally in local relations between NP nodes or S nodes), but makes an important change to the way in which the transformation is conceptualised with respect to the original Lees-Klima version:

> *Pronominalization consists not of replacing repetitions of a noun phrase by pronouns, but rather of determining which occurrence of an index will have the corresponding noun phrase substituted for it. Those occurrences of indexes for which the substitution is not made are then filled by pronouns* (McCawley, 1970: 176)

As it is formulated there, McCawley's proposal rests on notational conventions, namely, indexes. But this need not be: an occurrence of an index in a structural description, under present assumptions, is simply the occurrence of a node in a walk. That is: giving up the SMC and adopting an address system makes it unnecessary to resort to additional indexes, for structural descriptions and structural changes in the transformational approach simply make reference to the context of a node (that is, the set of nodes that are *directly* connected to it). This said, it is interesting to look at the 'deeper structure' that underlies (333) in McCawley's conception (taken from McCawley, 1970: 177; see also the structure proposed in Altham & Tennant, 1975: 55):

(334)

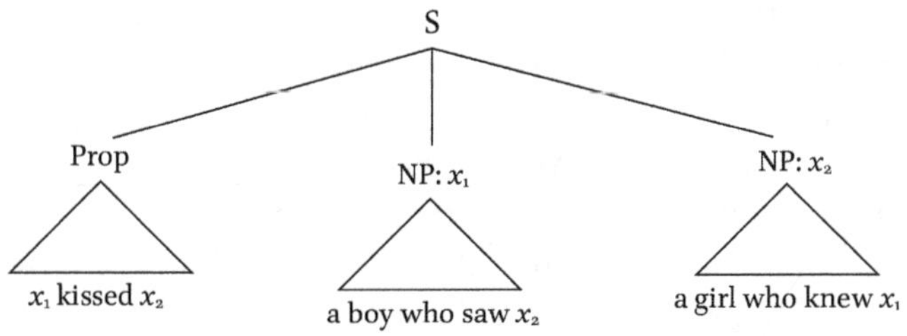

FIGURE 11.1 Analysis of a Bach-Peters sentence in McCawley (1970)

McCawley's proposal is to substitute x_1 and x_2 in Prop by the indexed NPs, leaving the indexed occurrences of variables *within* the NPs untouched; because these are not substituted, they surface as pronouns *her* and *him* respectively. Note that the local domains in (334) coincide with those identified in our analysis of (324) (with the caveat that (324) has a complement clause as the object of *show*); the only formal difference being that in McCawley's approach, substitution of each variable for the relevant NP takes place sequentially and—presumably—at *shallow* or *surface* structure. The reason for this is that in his proposal (as is customary in transformational generative grammar) the phrase marker must somehow determine the morpho-phonology. What is relevant for our purposes, however, is that (335a) is a possible variant of (333) but (335b) is *not*; this depends on structural conditions over the choice of nodes to substitute for NP:

(335) a. A boy who saw a girl who knew him kissed her (substitute x_2 in x_1 for NP: x_2; substitute x_2 in Prop for Prn)
b. *He kissed a girl who knew a boy who saw her (substitute x_1 in Prop for Prn; substitute x_1 in x_2 for NP: x_1)

The ill-formedness of (335b) can be accounted for as a violation of the so-called *Novelty Condition*:

> *An anaphorically dependent element cannot have more determinate reference than its antecedent.* (Wasow, 1979: 36)

The Novelty Condition can be interpreted as pertaining to the interpretation of visits to nodes in walks in digraphs: because there is a *strict order* requirement on graphs, in endophoric correference—where, recall, we have more than one visit to the same node—one visit will be ordered before the other. We have said that a node pronominalises another if they have the same address, they are connected, and there is a specific ordering between them; the Novelty Condition essentially determines the relative ordering between the visit to a specific node that is interpreted as the antecedent and the visit to that same node that is interpreted as being referentially dependant.

To summarise: we share with McCawley a strongly cyclic approach to the structural description of Bach-Peters sentences, but our proposal differs from his in terms of the topological properties of structural descriptions and the necessity to appeal to the substitution of indexed nodes. We also share aspects of Wasow's (1979) theory of anaphora, but dispensing with deletion operations (most notably, *Equi*, see Wasow, 1979: Chapter 7). Thus, the operation *substitution-by-NP*, which McCawley needs to convert the tree (334) into a *surface structure* without variables, can be eliminated if we allow nodes to be visited more than once in a walk (thus effectively giving up the requirement that structural descriptions be trees). Moreover, there is no need to stipulate that to each index in a structural description corresponds exactly one NP (McCawley, 1970: 178), because restriction this follows from the fact that addresses assigned to nodes in the graph are unique identifiers, without additional stipulations.

It does not go amiss to point out that we do not need to multiply the NPs by distinguishing NPs from indexes due to our use of addresses. At the same time, we can capture some aspects of the informal constraints on pronoun-antecedent pairs proposed by Wasow (1979: 61):

> *Given an NP and a definite pronoun in the same sentence* [note: 'sentence', not 'S node'], *the NP may serve as the antecedent for the pronoun, unless*:
> (a) *the pronoun and the NP disagree in person, gender, or number;*
> (b) *the pronoun is to the left of the NP and the pronoun is less deeply embedded than the NP*
> (c) *the pronoun is to the left of the NP, and the NP is indefinite*

What we see in (335b) is precisely a situation like (c). But we can think of a variant with a definite NP just as easily:

(335) c. *He kissed a girl who knew the boy who saw her

The argument that we have made here has some interesting consequences for the further study of English sentences displaying a combination of *crossing reference + embedding*. Karttunen (1971a: 157) attributes to Bach, McCawley, and Kuno the idea that the infinite regress paradox in Bach-Peters sentences arises because the following three assumptions are held simultaneously:

(a) *There is a rule of pronominalization that operates on two identical noun phrases.*
(b) *The rule requires that the noun phrases in question be (i) structurally, (ii) morphemically, and (iii) referentially identical.*
(c) *Pronominalization is an obligatory cyclic rule*

In our opinion, there is a *fourth* assumption involved which underlies the previous three, formulated here as (d) (see also Sampson, 1975: 7):

(d) *Pronominalisation (and, more generally, the establishment of referential dependencies) operates over* **distinct** *nodes* ***in SMC-complying*** *tree structures*

Conditions over licensing in the cases that interest us depend on (a-c) as much as they depend on (d), and we would go as far as saying that (d) is an even deeper and more fundamental assumption for it restricts the class of possible solutions (that is: the SMC is an admissibility condition which applies to whatever 'deep' or 'deeper' structural description we propose for a certain class of sentences). Karttunen objects to McCawley's structural description (in (334)) by saying that it is not capable of distinguishing between non-synonymous sentences like (325) above (*every pilot who shot at it hit the MIG that chased him*) and its variant in (336) below: the reason is that the same deeper structure would underlie both, and there is no distinct level of semantic representation nor is there a set of semantic interpretation rules. However, the objection is not quite fatal insofar as we can rescue the McCawlean intuition at least in its rejection of an infinite regression at deep structure (see also footnote *l* of the reprint of McCawley, 1970 in McCawley, 1973: 152–153). Moreover, since by definition our nodes are indexed by addresses which point to semantic values (and these, we hypothesised, are intensions), Karttunen's proposal, which incorporates a distinction between *individuals* and *descriptions*, can be incorporated into our theory with minimal adjustments at the level of the interplay between syntax and semantics: in Karttunen's view, the pronouns *him* and *it* in (325) do not refer to individuals (the pilot and the MIG respectively), but to the definite descriptions *every pilot who shot at x* and *the MIG that chased y* (see also Fox, 2002 for a 'trace conversion' process that may be invoked here as well, if pronouns are derived along the lines of Hornstein, 2001 or Grohmann, 2003). Giving up the SMC, and more fundamentally changing the way in which structural descriptions are conceived of, does not affect the aspects of Karttunen's argument that we are interested in, which pertain to the nature of referential expressions in Bach-Peters paradoxes. The error in McCawley's conception (as he himself acknowledges) is to treat NPs as 'referential' in the sense of Donnellan (1966); McCawley's revision of his (1970) paper in McCawley (1973) uses Karttunen's definite descriptions as the correct representation of what NPs stand for in 'deep structure' (= semantic structure, given the fact that McCawley worked within a Generative Semantics framework, where there was no independent level of Deep Structure generated by lexical insertion only as in interpretative semantics).

We will now consider some aspects of licensing in Bach-Peters sentences in more detail. When we consider a modified version of (325), in (336) below, we may ask how it is possible to get all relevant referential dependencies to hold (as usual, indices are purely illustrative):

(336) [Every MIG$_i$ [that chased a pilot$_j$ [who shot (at) it$_i$]]] was hit by him$_j$

Let us proceed carefully. First, we may point out that the referent for *him* in (336) is in the object position within a restrictive relative clause (in bold): *Every MIG [that chased* ***a pilot who shot at it****]*. Importantly, the reading notated in (336) with indexes should not be possible under a strict Subjacency-inspired view of locality: to begin with, *him* appears within a *by-phrase* adjunct; then, there are at least two bounding nodes on top of *a pilot*: S'/CP (*that* …) and NP (*Every MIG* …).

There is an additional problem which arises in strongly cyclic approaches to structural descriptions: *a pilot* must be accessible to *him* at the point of establishing a dependency, despite the presence of a potential governor *it* (corresponding to *every MIG*) in a local domain, in flagrant violation of Relativised Minimality (Rizzi, 1990) and similar principles. *Locality-as-intervenience* in a theory of syntax in which the syntactic component is an autonomous blind combinatoric engine does not have access to properties of the elements it manipulates like their denotation or other semantic properties. At most, it can have access to their category labels (NP, VP, etc.) or aspects of configuration (a given syntactic object can appear in an argumental or non argumental position—A vs. A'—). Neither seems to help in this particular case.[3]

3 Friedman et al. (2009), Rizzi (2013) and related work, propose a reformulation of Relativised Minimality which considers the morphosyntactic composition of syntactic objects alongside their structural position. In addition to the inherent problem posed by the theory of syntactic features in the absence of a meta-theory that restricts what possible features are (see Postal, 1972. Panagiotidis, 2021 is a recent attempt to formulate a meta-theory of features in Minimalism), this approach has its own weaknesses when it comes to predicting intervention effects (see Villata & Franck, 2016 for discussion), and still neglects semantics. Partly, this is a problem of granularity (i.e., at which level of 'syntactic organisation' is semantic interpretation determined?), and partly it is an architectural problem (syntax is still the only generative component, autonomous and severed from semantics, which is interpretative; see e.g. Lakoff & Ross, 1967 [1976]).

How can we derive the correct reading and provide an adequate structural description for (336)? To begin with, the fact that we do not require for each node to be visited only once allows for *every* MIG and *it* on the one hand, and *a pilot, who,* and *him* on the other to be superficial morpho-phonological realisations of just two nodes: ⦃MIG⦄ and ⦃pilot⦄. As pointed out above, this avoids the infinite regression problem noted by Bach (1970). But we still need to be able to define the composition of local graphs, with the concomitant identification of common components. A crucial point here is that *none* of the sub-graphs in the structural descriptions for (324)–(325) is *self-contained*, because all those contain nodes that are dominated directly by nodes in other graphs as well. We repeat the analysis of (324) in (337a) and that of (325) in (337b):

(337) a. Elementary graph 1: [man [EG 2] will get prize [EG 3]]
Elementary graph 2: [man show [EG 4]]
Elementary graph 3: [man desires prize]
Elementary graph 4: [man deserves prize]
b. Elementary graph 1: [pilot [EG 2] hit MIG [EG 3]]
Elementary graph 2: [pilot shot MIG]
Elementary graph 3: [MIG chased pilot]

Let us focus on (337a) first. We see that nodes corresponding to the expressions *man* and *prize* appear in all four elementary graphs, linking them. This means that none of them is *self-contained* because some of their internal nodes are also dependents in other elementary graphs, as can be seen in the ρ-set for (324) given in (332) above. The same happens in (337b): ⦃pilot⦄ and ⦃MIG⦄ are dominated by nodes in all elementary graphs. The corresponding ρ-set for (337b) is (338), where we have separated the ρ-sets which correspond to different elementary graphs, each containing a single lexical predicate and its arguments:

(338) ρ_1 = ⟨(hit, pilot), (hit, MIG)⟩
ρ_2 = ⟨(shot, pilot), (shot, MIG)⟩
ρ_3 = ⟨(chase, MIG), (chase, pilot)⟩
$\rho_{derived}$ = ⟨(shot, pilot), (shot, MIG), (hit, pilot), (hit, MIG), (chase, MIG), (chase, pilot)⟩

All three elementary graphs are linked at ⦃pilot⦄ and ⦃MIG⦄. Crucially, in this description, both pronouns *it* and *him* are bound (see also May, 1985: 36; in Jacobson's 2000 analysis, the first pronoun *it* does not behave like a garden-variety bound pronoun). The combination of sub-graphs via linking thus allows for all conditions required for licensing to hold, in both the structural descriptions of (336) and (337). We repeat the definition of *licensing* in (206) for the reader's convenience:

> ***Licensing***
> *Let G and G' be sub-graphs and v_i and v_j be nodes. Then, $v_i \in G$ may license $v_j \in G'$ iff*
> i. $(v_i, v_j) \in \rho^*$, *and* [alternatively, *there is a walk W that contains v_i and v_j and v_i is ordered before v_j*]
> ii. *G' is not adjoined to the root of G, and*
> iii. ***Neither G nor G' are self-contained***

And where *self-containment* is defined as follows (repeated from (206), above):

> *A graph G is a self-contained syntactic object iff $\nexists(v_i)$, $v_i \in G$ such that*
> i. $\langle v_j, v_i\rangle \in \rho^*$ *holds for $v_j \in G'$ and $G' \neq G$,* ***and***
> ii. *v_i receives a grammatical function GF in G*

The account we formulated is based on the idea that local graphs are lexicalised and respect a strong version of *economy of expression* (which heavily restricts the possibility of positing non-overt nodes and in general nodes that do not correspond to expressions in an input string), there is no such thing as 'bar levels' or 'projection' in our system. Furthermore, the classical Langacker-Ross-Reinhart definition of (c-)command (which is in turn based on phrase structure trees which introduce additional structure to terminal nodes in the form of branching intermediate nodes) has been replaced by a set of conditions on total order over walks, which are defined over possibly cyclic graphs. Neither of these two aspects are necessary conditions in a graph-theory based account. For example, Gärtner (2014) presents an approach to the derivation of Bach-Peters sentences inspired by Peters & Ritchie's (1981) Phrase Linking Grammars, which excludes closed walks in structural descriptions. In Phrase Linking Grammar, paradoxical constituency is banned:

> *in language it is impossible for one phrase to be a constituent of another phrase, and for the latter also to be a constituent of the former—except in the special case where the two are in fact the same phrase. This fact motivates a restriction to employing only acyclic graphs as structural descriptions, which restriction we adopt henceforth.* (Peters & Ritchie, 1981: 1, taken from Gärtner, 2014: 3–4)

Because of this restriction, Peters & Ritchie must include a formal device, *links*, which relate displaced constituents to their 'base-generation' position. Immediate *tree* domination and immediate *link* domination are defined as distinct relations. A node may be uniquely *tree* dominated and uniquely *link* dominated, but not multiply *tree* dominated. If the indegree of a node is greater than 1, then it must be both *tree* and *link* dominated.

In Gärtner's 2014 work (see also Gärtner, 2002), binary branching is respected throughout, although multidominance is allowed: every node has at most two daughters, but a node may have more that a single mother if (and only if) one of the motherhood relations is established by means of a *link*. As defined in Peters & Ritchie, these are similar to TAG-links (which relate an overt node with a c-commanded node dominating an empty terminal; see Joshi, 1985: 214, ff.); they relate a displaced constituent to a 'gap'. This is done in order not to violate the Kaynean condition that antecedents c-command pronouns: Phrase Linking Grammar-style loops can be licensed at LF, not in the 'overt syntax', since the Kaynean view on pronominalisation involves *movement*. The structural description assigned to a Bach-Peters paradox in Gärtner (2014: 9) is the following (minimally adapted):

(339)

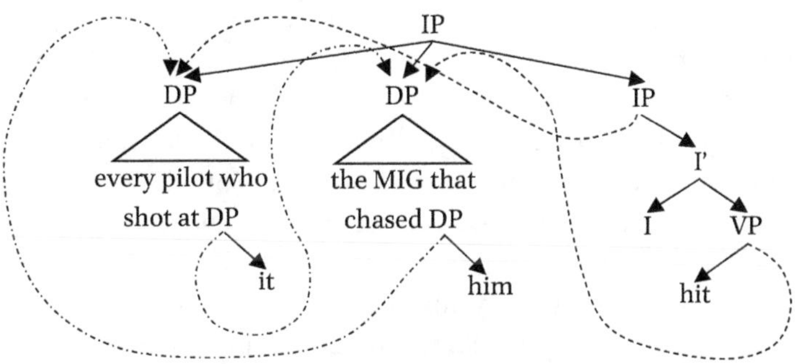

FIGURE 11.2 Phrase-linking analysis of a Bach-Peters sentence

There are three kinds of relations which we have represented with distinct graphical tools: solid arrows, in Gärtner's notation, correspond to *immediate (tree) dominance*. Dashed arrows correspond to *quantifier raising* (QR): the DPs have moved to adjoin to the two-segment category IP. Finally, dashed-dot arrows correspond to pronominalisation. Note that this view restricts multidominance to 'transformationally generated' relations, not to 'base-generated' relations. Gärtner, in his treatment of *Bach-Peters* sentences, allows for complex NPs to be raised, adjoined to the root (via QR at LF); in this way, it is possible to define links as Peters & Ritchie do.

Pronominalisation in Gärtner (2014) is treated not as base-generated multiple visits to a single node, but in terms of *movement* relating pronouns and their antecedents (see also Hornstein, 2001; Grohmann, 2003); these dependencies are subjected to the usual MGG constraints on movement. This analysis differs significantly from our treatment, in which there are no movement transformations, and pronominalisation is defined in a single-level syntax. Gärtner's is a graph-theoretic framework which, by virtue of adhering to some foundational assumptions of MGG (specifically, binary branching and the presence of intermediate phrasal nodes) constitutes an alternative to the present approach to Bach-Peters sentences. The ramifications explored by Gärtner are germane to some of the ideas explored in the present monograph, but at this point, more research is required in order to properly evaluate the relation between these proposals.

The graph-theoretic approach pursued here can provide an exhaustive description of relations between expressions in elementary and derived graphs while maintaining the advantages of a local approach to syntax and minimising the number of nodes (in particular, dispensing with nonterminals and projections; cf. the proposal in Gärtner, 2014: 9, which makes use of intermediate node labels like IP and VP). Specifically, we see that the elementary graphs that are composed to form the structural description of (324–325) can be linked at the relevant nodes giving us the desired reading, without violating any of the conditions for *licensing* or *linking*.

In connection to the description of conditions over possible dependencies between nodes within and across graphs, the following section will focus on some aspects of the structure of coordination, and the formulation of admissibility conditions for relations across conjuncts (building on Ross' groundbreaking 1967 dissertation). We will argue that a single structural template for coordinated structures is descriptively inadequate, and attempt to capture the advantages of a computationally mixed approach to coordination (particularly, the case made in Krivochen & Schmerling, 2016a) in terms of possible relations between elements belonging to distinct elementary graphs.

CHAPTER 12

The Structural Heterogeneity of Coordinations

Chapter 12 presents a comparison between the structural descriptions that Dependency Grammar, APG, and PSG assign to coordinated structures, and proposes a treatment of true coordinated structures that can accommodate syntactic and interpretative differences between coordination types.

Coordination was a topic that we very briefly touched upon in **Chapter 4**, but we did not go into a deep analysis of the structural descriptions assigned to coordinated structures. That is the topic of this chapter. We begin the discussion by recalling Fillmore's (1963) distinction between two kinds of generalised transformations: (a) *embedding transformations*, which inserts a sequence into another thus generating *hypotactic dependencies*, and (b) *conjoining transformations*, which take A and B and form C containing A and B, generating a *paratactic* dependency between them (where neither A nor B are embedded into one another). The general format of a *conjoining generalised transformation*, repeated from **Chapter 2**, is as follows (taken from Fillmore, 1963; see also Chomsky, 1955a, b):

$$\left.\begin{matrix} P \\ P' \end{matrix}\right\} \rightarrow P''$$

where P and P' are so-called *pre-sentences* (structures to which embedding transformations and preliminary singulary transformations have already applied).

The general format of an *embedding generalised transformation* may be described by a recursive rewriting rule:

> *Given P a pre-sentence, A a constant, and WAY a terminal string,*
> *A → P' in context W … Y* (Fillmore, 1963: 212)

The strings thereby generated involve different dependencies, and thus receive distinct structural analyses; we therefore must be careful to assign the appropriate structural description to natural language sentences obtained by conjoining and embedding. Crucially, from a descriptive point of view, there is no reason to assume *a priori* that all coordinations are structurally or semantically identical. Whereas coordination receives a unified analysis in logic, such that Boolean *and* has a unique interpretation and a unique truth table across all contexts

 | DOI:10.1163/9789004542310_013

of appearance (the so-called *Unique Readability* principle), natural language *and* behaves in a different way, as has been noted in the literature (Altshuler & Truswell, 2022 is a recent and very complete overview of coordination in syntax and discourse which highlights the semantic and syntactic heterogeneity of coordinated structures). However, the extra step of proposing that different interpretations of coordination in natural language may correspond to different structural descriptions has not usually been taken.

So far, we have been mostly analysing either simple sentences or complex sentences displaying different kinds of *embedding* (mostly, non-finite complementation); we have not, however, dealt with *conjoining* yet. We will do that now. In Krivochen & Schmerling (2016a, b) and Krivochen (2015a, 2016c, 2018, 2021a) we argued that *true coordination*[1] is not a unified phenomenon syntactically or semantically. Rather, structural descriptions assigned to strings of the general form [X and Y] need to take into consideration both syntactic and semantic features, which cluster coordinate structures in two classes. In order to illustrate our point, consider the following Latin examples:

(340) *Perdiderint cum me duo crimina, carmen et*
Ruin.3PL.AOR with me two crime.NOM.PL poem.NOM.SG and
error
error.NOM.SG
'Two crimes ruined me, a poem and an error' (Ovid *Tristia* II, 207)

---

1 'True' coordination is to be distinguished from what Krivochen & Schmerling (2016b) call 'mirage coordination' (see also De Vos, 2005; Biberauer & Vikner, 2017; Bravo, 2020, among others, who use the term 'pseudo-coordination'). Krivochen & Schmerling present arguments that the relevant expressions are not built in the syntax (against e.g. De Vos, 2005: 99), but are rather multi-word basic expressions, where *and* is syncategorematic. These are sequences of the kind [V ... and VP], involving finite Vs—most frequently two—, the last of which is a fully fledged VP. We argued that these structures ...

> ... *appear to enter into verb coordinations—but we will argue that on close examination these can be seen to involve something other than coordination. The fact that the structures we will consider appear at first to be coordinations but in fact are not gives us our name for them: mirage coordinations* (Krivochen & Schmerling, 2016b: 1)

Mirage coordination examples are, by virtue of not being real coordinations, exempt from the usual constraints on coordinate structures, including the CSC; they also display strong restrictions which do not apply to garden-variety true-coordinations (e.g., only two Vs can appear in mirage coordination, as opposed to the initially unbounded nature of true coordination; moreover, only a very limited number of verbs can appear as the first 'mirage conjunct' in these structures). Examples of mirage coordination include the English *go and, try and, up and, take and* in examples like:

(341) *effodiuntur* *opes,* *inritamenta*
Arise.3PL.PAST.PERF.IMPER wealth.ACC.PL incitement.ACC.PL
malorum. *iamque* *nocens* *ferrum* *ferro-que*
bad.GEN.PL and-now harmful iron.NOM iron.DAT-and
nocentius *aurum* *prodierat* [...]
harmful.COMPARATIVE gold.NOM come forth.3SG.AOR
'There arose wealth, incitement of bad things. And now came forth the harmful iron, and gold, (which is) more harmful than iron' (Ov. *Met.* I, 140–142)

Summarising much discussion in Krivochen & Schmerling (2016a), examples (340–341) are interesting because they showcase an important aspect of the heterogeneity of coordinated structures (certainly not the only one). In (340) there is a plural morphological mark in the V (*perdiderint* is a poetic form for the usual prosaic *perdiderunt*, such a change of vowel being common in verse), which agrees with the plural NP subject *crimina*. The following coordination *carmen et error* is a further clarification of the plural NP *crimina*: the meaning is roughly 'two crimes ruined me; namely, X and Y'. The presence of the coordinate conjunction *et* is revealing: we argue that each N, *carmen* and *error* is presented here as a separate entity, which has correlates in a plural N *crimina* and plural V agreement. It is our goal to capture that interpretation in the structural description that we assign to (340).

In contrast, (341) features a verb in singular form *prodierat* and a coordinated subject, *nocens ferrum ferroque nocentius aurum*. This is unexpected, *prima facie*, since a coordinated subject should agree with a plural verb. Crucially, (341) also differs from (340) in the choice of coordinating conjunction: *que* rather than *et*.

Semantically, it is interesting (and relevant to our point) that *nocens ferrum ferroque nocentius aurum* can be grouped under *inritamenta malorum*: the bad things that arose consist of iron, and gold (which is worse than iron). The claim is that here the coordinated N are presented, semantically, as an internally unanalysable whole, not as separate entities.

The crucial difference between (340) and (341) is, then, the way in which entities are presented (in the Fregean sense of 'presentation'): as multiple independent entities which are interpreted separately (coordinated with *et*, in

---

(i) She's *gone and ruined* her dress now. (Ross' 1967 (4.107a))
(ii) She's *up(ped) and ruined* her dress (note that there is no RNR interpretation available)
(iii) She *took and replaced* the hose (RNR interpretation irrelevant).

(340)) or as a single whole whose internal structure is opaque to interpretation and syntactic operations (coordinated with *que*, in (341)). At this point, tt is essential to point out that the morphological exponent of the conjunction (in the case of Latin, *et* or *que*) is *not univocally related* to the syntactic and semantic characteristics of its output: English or Spanish (the languages that are the focus of this monograph) only display a single morphological exponent for coordination (*and* or *y*, respectively), but, we will argue, *both modes of presentation.*

We will, then, refer to these classes as *et-coordination* vs. *que-coordination* (adopting the Latin terms due to their descriptive resemblance). The empirical specifics of this distinction are currently under research (see Krivochen and Schmerling, 2016a for extensive discussion and examples), but we can summarise the main characteristics of each:

***Que*-coordination:**
- No internal structure that can be probed by an external element: *que-coordinated* outputs are opaque for all syntactic intents and purposes (in particular, extraction)
- The arguments are interpreted as a single entity; thus no probing into a conjunct is allowed
- Triggers singular agreement when it is NPs being coordinated due to internal opacity

***Et*-coordination:**
- Allows for either hypotactic or paratactic dependencies between terms: the terms of the coordination are distinct syntactic elements which may be organised in different ways
- Each argument is a separate entity, allowing for syntactic operations to probe inside a conjunct
- Triggers plural agreement when it is NPs being coordinated due to its internal accessibility

Let us give a further example (see Krivochen & Schmerling, 2016a for more)

(342) a. The sudden rise and the equally sudden fall of the stock market have economists worried.
b. The sudden rise and equally sudden fall of the stock market has economists worried.

(343) a. *La abrupta subida y la igualmente abrupta bajada de la*
the sudden rise and the equally sudden fall of the
Bolsa preocupan al Gobierno
stock-market worry.3PL to-the Government
'The sudden rise and the equally sudden fall of the stock market worry the Government.'

b. *La abrupta subida e igualmente abrupta bajada de la*
The sudden rise and equally sudden fall of the
Bolsa preocupa al Gobierno.
stock-market worry.3SG to-the Government
'The sudden rise and equally sudden fall of the stock market worries the Government.'

The English example (342b) and the Spanish example (343b) are both instances of *que-coordination*. In each of these examples the conjoined NP's are understood as having a single referent, albeit a complex one: stock market fluctuation. Sentences (342a) and (343a), exhibiting *et-coordination*, are not semantically equivalent to their *que-coordinated* counterparts. Consider the *et-coordinated* example (342a): this sentence could describe a situation where particular economists were worried about the sudden rise of the stock market but not about its sudden fall, whereas others were concerned about its fall and not its rise. This interpretation is not possible for the *que-coordinated* example (342b), where economists must be worried about the combination of these phenomena.

It is important to observe that the distinction between *et-* and *que-coordination* does *not* overlap with that between *symmetric* and *asymmetric* coordination (see Schmerling, 1975 for a Gricean approach to the latter). Distinguishing between *symmetric* and *asymmetric* true coordination requires us to have access to the terms of the coordination; this therefore means that the distinction between symmetric and asymmetric conjunction falls entirely within the realm of *et-coordination*. *Que-coordination* is internally opaque, thus its terms cannot be either symmetric or asymmetric. A feature of *symmetric* coordination is the possibility of inverting the terms of the coordination *salva veritate*, as in (344) (344 c–d are taken from Schmerling, 1975):

(344) a. John had a beer and Mary had wine
b. Mary had wine and John had a beer
c. Rome is the capital of Italy and Paris is the capital of France
d. France is the capital of France and Rome is the capital of Italy

The conditions under which (344a) is true are the same as those under which (344b) is true (as is also the case with the logical conjunction ∧); furthermore, there is no implicature of a specific temporal order between the events of John having a beer and Mary having wine (the events could be simultaneous or not). The same happens in (344c-d), where the lack of order is more evident due to the fact that both terms of the coordination are states. We can have the terms of the coordination in any order and the compositional interpretation is exactly the same: there are two events, both accessible and independent of each other: the dependency between these terms (which remain distinct syntactic and semantically) is *paratactic*. Most generative approaches disagree with this point, in terms of the structural description assigned to coordinated structures: see for example, Camacho (2003) who argues in favour of a uniformly *asymmetric* phrase marker for coordination, ensuring generalised hypotaxis; similar approaches can be found in Munn (1993, 2000), Kayne (1994, 2018), Chomsky (2013), Progovac (1998), Neeleman et al. (2020), among many others. These are strictly binary-branching analyses of coordinated structures, where coordination is taken to be a syntactically—configurationally—uniform phenomenon: all coordinated structures are assigned the same phrase marker (see Borsley, 2005 for critical discussion). We can ask to what extent this approach is descriptively adequate.

Let us focus on one particular property of the interpretations assigned to the sentences in (344): the symmetric relation between conjuncts. At the core of our proposal is the idea that this symmetry needs to be represented in the syntactic structure (which feeds semantic interpretation and the computation of inferences; see e.g. Escandell & Leonetti, 2000, 2006). The paratactic reading is not possible with *asymmetric* coordination, since there is an ordering between the terms of the coordination (often implicating temporal or tempocausal ordering between events denoted by those terms):

(345) a. Mary-Jane went out with Harry and broke Peter's heart
b. Mary-Jane broke Peter's heart and went out with Harry

Note that changing the order of the terms changes the meaning of the sentence and also the conditions under which each is felicitous. For (345a) to be an adequate characterisation of a sequence of events, it must be the case that Peter's heart was broken *after* (and most likely *because*) Mary-Jane went out with Harry. But that is not at all the case in (345b), where Mary-Jane *first* broke Peter's heart and *then* went out with Harry (there seems to be no cause-consequence relation between the events in this latter reading). In both cases, the terms of the coordination are accessible, such that there are *two* events, not a complex one; however, these events are not independent of each other. Again, we defend the position that this asymmetry in interpretation must correspond to a structural asymmetry between conjuncts. Not because the nature of the interpretation is strictly syntactic, but because syntactic structure needs to provide enough information to license a particular semantic-pragmatic effect so that not any structure can license any interpretation.

We said that the distinction between *symmetric* and *asymmetric* coordination is not the same as the one made between *et*- and *que*-coordination: we can now examine why. *Et*-coordination preserves all terms in the conjunction as distinct syntactic elements. These may be related by parataxis or hypotaxis. In contrast, *que-coordination* is neither paratactic nor hypotactic, in the sense that the output of *que-coordination* is a single complex entity (sortal or eventive) which we cannot internally manipulate with syntactic rules or semantic interpretation principles: these must apply to the *whole* term as an atom. Neither *symmetric* nor *asymmetric* coordination can be outputs of *que-coordination*, in the light of the brief discussion about (344) and (345).

In order to formulate an empirically adequate theory of coordination that captures the cases we are interested in, it is useful to take a look at what other graph theory-based frameworks have proposed. APG, for instance, recognises a class of relational Arc dubbed *Con*, a Structural (not Label or Linear Precedence arc), Non-nominal R-sign (Johnson & Postal, 1980: 198). Further conditions specify that the heads and tails of *Con* arcs must be labelled with the same major category and *Con* arc heads must bear the same category label (essentially, a condition on the identity of coordinated terms, also known as the 'law of coordination of likes'). The relevant definitions, pertaining to heads and tails of *Con* arcs, are the following (Johnson & Postal, 1980: 209):

(346) a. Coordinate(a) $\leftrightarrow$ ($\exists$A) (Con arc(A) $\wedge$ Tail (a, A))
b. Conjunctive(a) $\leftrightarrow$ ($\exists$A) (Con arc(A) $\wedge$ Head (a, A))

For example, in the asymmetric example *The fiend shot and knifed its victim* (example taken from Johnson & Postal, 1980: 222), the coordinated Vs are the heads of Con arcs. If we consider a symmetric coordination, like *Ted sang and Melvin danced*, the arc pair description looks like (347) (from Johnson & Postal, 1980: 207):

(347)

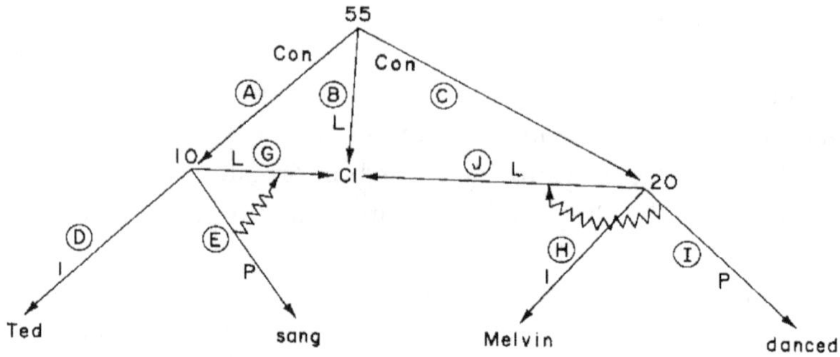

FIGURE 12.1 Arc-Pair Grammar analysis of coordination

In and of itself, so far as we can see, the APG treatment of coordination does not capture the difference between *et-* and *que-coordination* in its structural descriptions, which is empirically motivated at its very core. It is also unclear how (if in any way) APG could represent the difference between *symmetric* and *asymmetric* coordination; note that both predicate arcs E and I sponsor arcs G and J. Johnson & Postal offer no arc description of an asymmetric coordination that we can contrast (347) with. This does not mean that APG or MG could not be modified in order to capture these distinctions, but how to do that is not evident based on available analyses.[2]

Let us continue our brief survey, proceding to review some basic aspects of coordination in Dependency Grammars. Treatments of coordination within Dependency Grammar are not quite homogeneous (see e.g. Pickering & Barry, 1993; Popel et al. 2013 for surviews), and we will only consider a representative example of a specific approach: for example, Mel'čuk (1988: 26–28) (as a representative of Meaning-Text Theory MTT) argues that coordinated structures are headed (otherwise, there could be no dependency relation), based on the purported fact that '*In the majority of cases there is no reversibility in coordinated structures*' (Mel'čuk, 1988: 26); in these asymmetric structures, the left conjunct is the head, with the right conjunct *depending* on it (although in principle the head can be either the rightmost or leftmost conjunct). In Mel'čuk's approach, the coordinating conjunction is attached under the penultimate conjunct, and the last conjunct is attached under the conjunction. Some versions of MTT are closer to PSGs than others, see e.g. Mel'čuk & Pertsov (1987) for an account of coordination that admits some level of 'phrase structure'.

2 We must highlight that Johnson & Postal themselves acknowledge (1980: 207) that there are many questions raised by coordinate structures which they '*have not had the chance to study*', so it is possible that further research would have made it possible to distinguish between *symmetric* and *asymmetric* coordinations in terms of distinct arc pair descriptions.

The reversibility argument seems to us to only be (partially) valid for a specific kind of coordinate structures, specifically, *et-coordinations*. Furthermore, the lack of reversibility need not imply headedness, it only requires hypotaxis (i.e., an *asymmetry* between coordinated terms). Syntactically, there is no argument or reason to make all strings containing *and* belong to the same class; semantically and pragmatically, the heterogeneity of coordination only becomes more evident. Consider the following example, used in Krivochen (2018, 2021a) as part of an argument in favor of *mixed computation*:

(348) (talking about afternoon tea) John got milk and Bill brought some biscuits

We have two conjuncts, *John got the milk* and *Bill brought some biscuits*. The question is, what is the syntactic relation between them? In this context it is important to take note of the fact that (348) below, where the order of the conjuncts has been reversed, is perfectly acceptable (since there is no particular order in which the purchase of milk and biscuits should be presented; the events could very well take place simultaneously):

(349) Bill brought some biscuits and John got milk

The reversibility of conjuncts with no consecuence for truth value or meaning points towards a strictly *paratactic* structural description in which neither term has scope over the other; this is consistent with the lack of order in the interpretation of events, since there is no embedding. The coordination in (348) corresponds to logical conjunction in being commutative. Let us go one step further, and consider (349) embedded in a bigger structure:

(350) Bill brought some biscuits and John got milk, and we all had a wonderful afternoon tea

The terms of coordination are two once again (the first of which is itself a coordination), but in this case the relation seems to be asymmetric: we had a wonderful afternoon tea *after* (and probably also *because*) John and Bill contributed with milk and biscuits respectively. Besides the fact that symmetric coordination allows for the reversibility of conjuncts *salva veritate*, we need to point out that the grammar also allows for a *combination* of hypotactic and paratactic coordinations in the same sentence: if all instances of true coordination were assigned the same structural description (a headed one, in Mel'čuk's view and MGG), among other issues, we would be faced with the problem of getting the meaning right in each specific case since we would get no help from the syntactic structure. Thus, we cannot consider (350) as an instance of an ordered triple ⟨A, B, C⟩, which is an ordered pair the first of whose terms is itself an ordered pair: ⟨⟨A, B⟩, C⟩ (under the Wiener-Kuratowski definition of ordered pairs, see Dipert, 1982 for discussion about the formal problems brought about by different definitions of ordered pairs in axiomatic set theory). This is not possible because A and B are *not* ordered with respect to one another: any relation R between A and B (other than precedence) is symmetric (i.e., R(A, B) = R(B, A)). It is the relation between the pair A, B on the one hand and C on the other that is asymmetric. But nothing hinges on pairs: the first term of the asymmetric conjunction could be of arbitrary complexity, as could the second: each may contain in turn other coordinations (symmetric or asymmetric, *et* or *que*).

Ultimately, structurally uniform approaches to coordination as headed structures (either dependency based or phrase structure based) seem to be empirically inadequate, if the goal is to provide grammatical descriptions of languages: depending on how *head* is defined, we would be in the uncomfortable position of saying that *drank and danced* (a VP coordination) and *books and magazines* (an NP coordination) are both CoordP / &P despite the evident lack of intersubstitutability between these expressions. It must be said, however, that a good portion of the work in Dependency Grammars (in particular, Universal Dependencies) is oriented towards the goal of characterising coordinated structures for purposes of natural language parsing and grammar engineering, which is a different goal from those of grammatical description (Shieber, 1988; Pollard, 1997). There are, thus, different treebanks implementing distinct schemata for coordinated structures, with differences arising not only in linguistic terms (e.g., tests to determine when something is a coordination) but also in strictly implementational terms. These concerns are orthogonal to the aim of the present work, and thus we will only consider those aspects of the proposals that are grammatically relevant.

A view different from Mel'čuk's, but still within DG, is expressed by Tesnière (1959: 80, ff.). Tesnière assigns coordinating conjunctions to a category *junctives*, which are *functional elements* (or *empty words*, 'mots vides' in the original). Junctives are structurally between the terms they conjoin (called *nuclei*), but remain outside these terms. In brief, coordinated structures are not only not headed structures, but they are not even dependency-based. His diagrams, which he calls *stemmas*, are similar to ours insofar as they can violate the SMC if necessary: any term can have more than one *governor*. However, there is no requirement of *strict ordering* in his *stemmas*, unlike in our graphs. A simple coordinated structure like *Alfred et Bernard aiment leurs parents* ('Alfred and Bernard love their parents') is analysed thus:

(351)

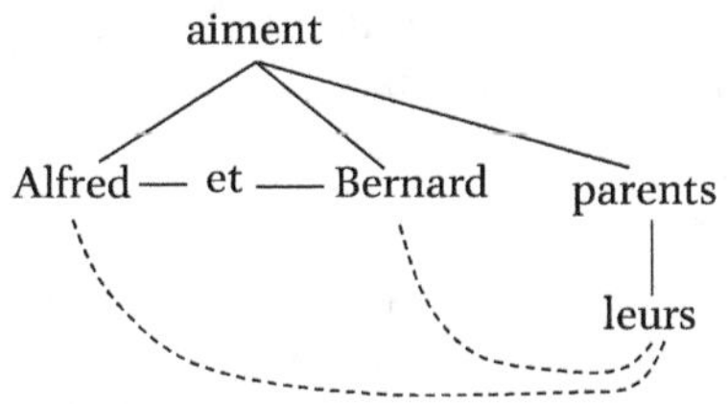

FIGURE 12.2 Dependency Grammar stemma for NP coordination

Observe that, because the sentence means that Alfred loves his parents and Bernard loves his parents (but neither Alfred nor Bernard love each other's parents), there is a dependency between the possessive *leurs* and both *Alfred* and *Bernard*; it is possible to dissociate the reference of the possessive NP without multiplying entities or indices as long as we assume that to each NP corresponds a different *leurs* token. We will not provide a full review of the kinds of coordinated structures considered by Tesnière, which constitutes a remarkably rich landscape, but we do want to point out that his view on coordination is particularly interesting insofar as he deals with discontinuities and 'crossings' in coordinated structure frees from the axioms of Phrase Structure Grammars and diagrams of L-trees. For instance, the *stemma* he assigns to a sentence featuring a coordinated subject and a coordinated predicate phrase, like *Alfred et Bernard jouent et rient* ('Alfred and Bernard play and laugh') is the following (Tesnière, 1959: *Stemma* 261):

(352)

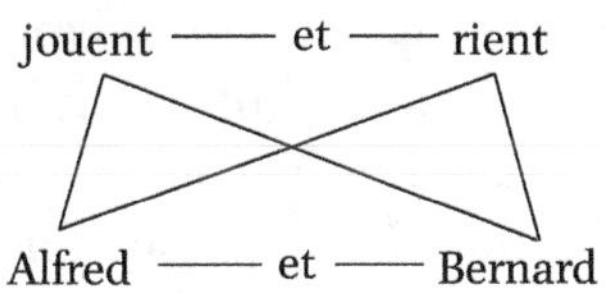

FIGURE 12.3 Dependency Grammar stemma for NP and VP coordination

It is important to remember that there is nothing in Tesnière's *stemmas* that specifies the linear order of the sequence for which the *stemma* provides a structural specification. In that sense, our perspective is close to his. However, there is no overt identification of grammatical functions *in the stemmas*; in a sentence like 'Alfred gives the book to Charles', *Alfred, the book*, and *Charles* are all equally *actants*, without distinguishing between grammatical functions (see Tesnière, 1959: Chapitre 48). Dependencies in these stemmas are not enough to fully represent argument structure and thematic relations (we cannot tell *who did what to whom* based on a stemma alonw: (351) could also be assigned to the sentence *Their parents love Alfred and Bernard*). The distinction between

Subjects and Objects is, according to Tesnière, a 'remnant' from the Port Royal grammarians. His argument is rather strongly worded:

> *§5 Indeed, all arguments that can be invoked against the concept of the verbal node and in favor of the opposition between subject and predicate come a priori from formal logic, which has nothing to do with linguistics.*
>
> *§6 Concerning strictly linguistic observations about the facts of language, the conclusions drawn a posteriori are of a much different nature. There is no purely linguistic fact in any language that suggests the existence of the subject-predicate opposition.* (Tesnière, 1959: Chapitre 49. Translation by Timothy Osborne & Sylvain Kahane)

It should go without saying that we disagree with this specific aspect of Tesnière's view, given the prominent role that grammatical functions play in our proposal. There are well-documented asymmetries between subjects and non-subjects in English and Spanish (the languages we have drawn almost all of our examples from), for example in terms of the availability of extraction: summarising much syntactic literature, filler-gap dependencies targeting objects are fine, whereas extraction from non-objects is a rather problematic issue.[3] Within MGG, several conditions have been proposed to account for the relevant facts: the *Condition on Extraction Domains* (Huang, 1982), the *Subject Condition* (Chomsky, 1973, 1986), the *Sentential Subject Constraint* (Ross, 1967), among many others (Bianchi & Chesi, 2014; Uriagereka, 2002; Freidin, 1992; Müller, 2011: 48–49; Dalrymple et al., 2019: 656), not all of which have been grammatical-function sensitive (but rather related to specific structural positions). While the universality of these conditions has not gone uncontested (e.g., Mayr, 2007 shows a lack of subject-object asymmetries for purposes of

---

3 Subject-object asymmetries have been object (no pun intended) of inquiry in psycholinguistic studies, dealing with acquisition of asymmetries and processing of extraction from both positions. We will not refer to these studies because the interest of the present work is the grammar as a formal system, without the requirement that it be implemented psychologically: a theory of the grammar need not be a theory of human knowledge of language or of the use of language. This statement may sound contentious to some, but it is actually a very common assumption in theories of language that do not identify themselves with the goals of the so-called 'biolinguistic' enterprise or with theories that see language only in relation to its users (human or not). Examples of the kind of approaches we identify with are, for instance, Ajdukiewicz- and Montague-style Categorial Grammar (e.g. Dowty, 1978: fns. 2 and 3), (some versions of) Dependency Grammar (e.g., Tesnière, 1959: Chapitre 20, §21) and Arc-Pair grammar (Postal, 2010: 3).

extraction in Bavarian; Falk, 2009 suggests that some speakers of Navajo can violate the Sentential Subject Constraint), they do capture (with varying degrees of success) English and Spanish phenomena. Extraction *from* subjects aside, extraction *of* subjects is also distinct from extraction of objects: a subject cannot be topicalised, but only left-dislocated (a resumptive pronoun is obligatory). Not all is extraction: as we will see in **Section 13.1**, an adequate formulation of *Passivisation* requires having a distinction between subjects and objects (Perlmutter & Postal, 1983a). Theta-marking is similarly sensitive to the subject/non-subject distinction: as shown in Marantz (1984: 25, ff.), is the whole VP (V + object) that theta-marks the subject, not just the verb. It seems to us that there is enough syntactic evidence for subject-object asymmetries to maintain the central role of grammatical functions in the theory of the grammar. It is, then, important for structural descriptions to represent this asymmetry somehow.

A particularly interesting approach within Dependency Grammars is that of Osborne (2006) (see also Osborne, 2019: Chapter 10 for an introduction to coordination in DG). Osborne distinguishes three kinds of theories about coordination based on the assumed size of conjuncts (Op. cit.: 41–44):

a. **Large conjuncts:** all conjuncts correspond to full sentences at some deep level of representation. This approach requires deletion operations to match conjunct size, and via deletion it can repair situations where non-constituents are coordinated.
b. **Small conjuncts:** conjuncts only contain the elements visible in the surface string. No deletion is required. Problems arise when considering non-constituent coordination.
c. **Eclectic approach:** assume small conjuncts when possible, and large conjuncts where necessary. The problem here is to adequately restrict the theory: how do we decide when non-audible structure is 'necessary'?

Osborne (2006, 2019) sides with the *small conjunct* approach, rejecting the large conjunct approach (due to the mismatch between syntactic and semantic representations that is repaired via deletion) and the eclectic approach. The reasons for rejecting this one are particularly interesting: according to Osborne (2006: 43), a requirement of 'theoretical stringency' disqualifies the eclectic approach from the start. We disagree. A similar requirement would support a uniform analysis of all adjective sequences as monotonically growing phrase markers, as in Cinque (2010) or Scott (2002): in this way, assigning a finite-state description to the intensive iteration of adjectives only, and embedding that FS unit in a CF structure would be impossible (see **Section 1.5**). While we agree with Osborne in that no more structure than needed should be assumed, and that PSGs multiply the number of 'constituents' by multiplying the intermediate nodes (Osborne, 2006: 54; Maxwell, 2013), we do not think that structural uniformity yields a descriptively adequate theory, or that 'theoretical stringency' is, *per se*, a good evaluation metric.

Osborne's approach to coordination, then, assumes that (a) coordination terms are only those that are seen in the surface string, and (b) that coordination is assigned a flat, potentially multi-rooted structure. In his coordination analysis, material that is external to the conjuncts must be shared: this requirement makes instances of RNR as well as VP coordination simple to analyse. A sentence like *Fred sent a letter to Sue and a package to Jane* receives the following analysis (adapted from Osborne, 2006: 61; compare with the TAG approach to *gapping* in Sarkar & Joshi, 1997, here in **Section 2.2**):

(353)

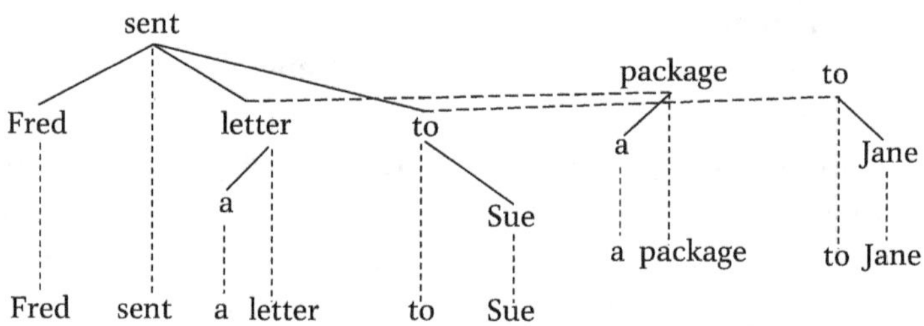

FIGURE 12.4 Three-dimensional Dependency Grammar analysis of non-constituent coordination

The dashed lines correspond to parallel roots between the conjuncts: note that the second conjunct contains two nodes that are not dominated by any other node, thus, it is more of a *vine* than a *tree*. In (353), the roots *package* and *to* (with dependent *Jane*) are in an equi-level to and behind *letter* and *to* (with dependent *Sue*): this 'three-dimensional' approach to coordination has relatives in MGG, including the work on parallel structures in Goodall (1987) and Moltmann (1992, 2017). This is important because there is a requirement that each root in a conjunct have a matching parallel root in all other conjuncts: again, this provides a good DG account of ATB rule application. Furthermore, the requirement that shared material be external to the conjuncts, in combination with the requirement that shared material may not follow the first root in the initial conjunct (the combination of which Osborne refers to as the *Contiguity Requirement*) is supposed to filter the same structures as Ross' (1967) *Coordinate Structure Constraint*. The analysis in Osborne (2019) elaborates on the structure of coordination, and in particular RNR, to which we will come back in **Section 14.1** below.

Osborne's account is focused on a subset of coordinated structures, for which it offers a uniform account. It is not clear to us, however, that the distinction between *symmetric* and *asymmetric coordination* can be captured in that framework, if all coordinations are equally flat (the same can be said of LFG's analysis of coordination, which treats all conjuncts as members of an unordered set at *f*-structure; see e.g. Kaplan & Maxwell, 1988; Dalrymple et al., 2019: Chapter 16). Furthermore, the distinction between *et-* and *-que* coordination seems orthogonal to the DG treatment. In other words: it does not seem possible to adapt the small conjunct dependency approach to provide an account of the syntactic and semantic variety that, we argue, coordinate structures display.

After APG and DG (and having looked at TAG's analysis briefly in **Section 2.2**, see also **Section 14.2**), we have MGG left to review. During the Standard Theory days, it was the norm to see *n*-ary branching representations of coordinated structures (e.g., Ross, 1967; Koutsoudas, 1971), and as a matter of fact, unbounded coordination was one of the phenomena usually invoked in the justification of transformations, *singulary* or *generalised* (Lees, 1976: 33, ff.; Chomsky, 1957: 37–38 respectively). However, as we briefly saw in **Chapter 4**, during the GB-MP days MGG attempted to make coordinated structures fit the X-bar schema: binary-branching, headed, projecting syntactic objects (as pointed out above, analyses of this form are defended in Kayne, 1994; Zoerner, 1995; Progovac, 1998; Chomsky, 2013; among many others). In one version of the binary-branching analysis, the coordinating conjunction projects a garden-variety phrase (CoordP / &P), with the coordinands as specifier and complement (if multiple specifiers are allowed, then coordinated structures of more than two terms may be accommodated). An alternative structural description proposed in the literature (Munn, 1993; Zhang, 2010; Truswell, 2019—the latter, including aspects of multidominance-) has the second conjunct adjoined to the first (a variant of the adjunction analysis has coordination as 'mutual adjunction'; see Neeleman et al., 2020) We can illustrate these approaches, like we did in **Chapter 4**, as follows:

(354) a. b.

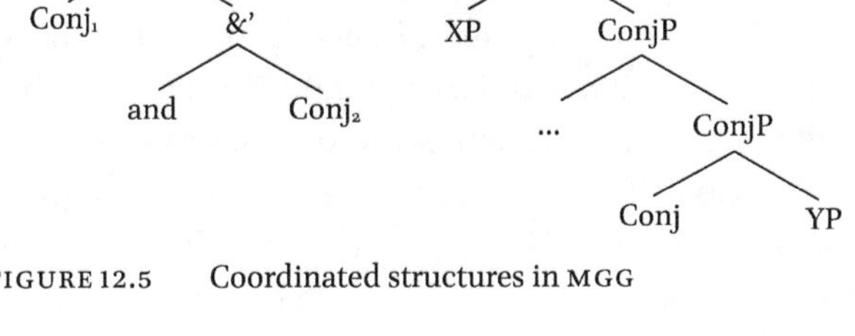

FIGURE 12.5 Coordinated structures in MGG

The result in either case is that coordinated structures *always* display hypotaxis, either (i) with one term of the coordination c-commanding the other as specifier and complement of a phrase whose head is the coordinating conjunction (e.g., an &P or CoordP) or (ii) all non-first conjuncts being adjoined to the first in, again, a strictly binary-branching fashion. Borsley (2005) provides an excellent overview of empirical arguments against structures like (354a, b), and observes that in fact such analyses is *not* assumed widely outside (present-day) MGG (see e.g., Pollard & Sag, 1994: 203; Sarkar & Joshi, 1997; Dalrymple, 2001: Chapter 13; Culicover & Jackendoff, 2005: 277). There are two main problematic aspects of a templatic approach like the ones illustrated in (354): (i) structural uniformity (binary-branching and obligatory hypotaxis) and (ii) projection of the coordinating conjunction to phrasal level (which means that if we coordinate NPs, the result will not be an NP, but a ConjP / &P). In turn, this means that the selectional properties of predicates, for example, would need to somehow include ConjPs / &Ps while at the same time having access to the internal categorial specification of the coordinated terms. For example, if we use an NP coordination as the direct object of a verb, as in *John read books and magazines*, and wanted to provide a specification for the lexical entry of *read* as a monotransitive verb, we would be forced to add something like *read* /__ {NP, &P, CP} to generate *John read* [*a book*]$_{NP}$, *John read* [*a book and a magazine*]$_{\&P}$, and *John read* [*that the President was impeached*]$_{CP}$. But this would predict that any other coordination could also be a suitable object, contrary to fact (e.g., **John read* [*sang and danced*]$_{\&P}$, or **John read* [*that the President was impeached and a book*]$_{\&P}$), since all coordinations would have the same categorial specification, &P/CoordP. Including &P to the subcategorisation frames of predicates would greatly overgenerate, in addition to not being particularly informative. We want to maintain the traditional generalisation that a coordination of NPs has the distribution of NPs, a coordination of VPs has the distribution of VPs, and so on (with exceptions being considered special cases and analysed as such).

Here we will not deal in detail with these objections in detail but rather propose three structural schemata for coordinated structures, which intend to highlight the structural (syntactic-semantic) heterogeneity of coordination, building on the TAG approach in Sarkar & Joshi (1997) and Banik (2004). We argued that coordination is not a unified phenomenon syntactically, and our structural schemata for coordination will highlight this fact. Furthermore, we align with the proposal that an adequate structural description for a coordination is *never* an &P, CoordP, or anything of the sort: we agree with Tesnière (1959), Osborne (2006), and Borsley (2005) in that there is no 'head' in coordinated structures (where 'head' is interpreted in X-bar theoretic terms:

no coordinating conjunction projects a phrasal label). Thus, there cannot be a *dependency* in all cases (in the DG sense), even though there is dominance. We are interested in capturing the contrast between *et-* and *que-coordination* in a way that maintains the core aspect of the analysis: both *paratactic* and *hypotactic* dependencies can be established in true *et-coordination* and *que-coordination* yields a perfective (internally opaque) entity; this simply cannot be captured with an *a priori* structurally uniform approach to coordination. That is: to a string *X and Y* featuring true coordination, where X, Y, and Z are variables over expressions (basic or derived), there can correspond three structural descriptions, roughly along the lines of (355):

(355) a. ⟨X, and⟩, ⟨and, Y⟩
b. ⟨and, X⟩, ⟨and, Y⟩
c. Z

(355a) corresponds to the *asymmetric et-coordinated* case, in which the root of the sub-graph X dominates the root of the sub-graph Y. Therein lies the essence of *hypotaxis* in the present view, which is the structural hallmark of *et-coordination*. (355b) corresponds to *symmetric et-coordination*, in which case the roots of the conjuncts are both dominated by *and*, which is thus the root of the whole graph. In this latter case, both conjuncts' roots are *sisters* (the structural relation between them is strictly paratactic); in the former case, the second conjunct's root is within the ρ-domain of the first conjunct's root, which means that the graph is not extended at the root, but at the frontier as one conjunct is embedded into the other. (355c), in contrast, features a *single object* corresponding to a string *X and Y*, distinct from both X and Y and without internal syntactic complexity: this corresponds to the *que-coordinated* case (e.g., *red beans and rice*, or *fish and chips*). This is a way (perhaps not the best, certainly not the only one) to represent the internal accessibility of *et-coordination* and the opacity of *que-coordinated* terms, as well as the different structure of symmetric and asymmetric coordination.

An important point to make here is that we will not attempt to provide an account of examples of coordination of unlikes such as *John is* [*a Republican*]$_{NP}$ *and* [*proud of it*]$_{AP}$ or *Sue is* [*healthy*]$_{AP}$ *and* [*in good shape*]$_{PP}$ (Pollard & Sag, 1994: 202, ff.; Dalrymple, 2017; Bruening & Al Khalaf, 2020; Przepiórkowski, 2022, among many others). There are reasons to believe that there are at least two kinds of coordination of unlikes: those in which the second conjunct contains an anaphoric element referring to a constituent containing the first conjunct (e.g., *John is a Republican and proud of* ***it***) and those in which both conjuncts are referentially independent (e.g., *Sue is healthy and in good shape*).

In Krivochen (forthcoming a) we apply Sarkar & Joshi's coordination schema applying at the S level plus pronominalisation (for ... *a Republican and proud of **it***) or at the VP level, delivering a flat, paratactic structure (for ... *healthy and in good shape*). Essentially, we follow McCawley (1998: 430) in analysing *it* as an anaphorically bound pronoun and assigning our sentence a reading like [[John is a Republican] and [John is proud of [John is a Republican]]], with the most embedded [John is a Republican] pronominalised under identity given an asymmetric order between both occurrences of the same elementary tree in the derived structure. Some evidence in favour of having (at least) two distinct structural descriptions for coordinations of unlike categories comes in the form of order permutation tests: only in the coordination of terms where the interpretation is intersective can the terms of coordination change order:

(356) a. Sue is in good shape and healthy
b. Sue is healthy and in good shape

The sentences is (356) are true iff Sue belongs to the intersection of the sets defined by *healthy* and *in good shape*. The same is true for cases such as *Pat is stupid and a liar* (AP and NP) or *they ate very quickly and without restraint* (AdvP and PP). The coordinating conjunction behaves like logical $\wedge$ in its commutativity. This interpretation is not available for cases like (357), in which the second term of the coordination contains a bound pronoun which refers to the first term of the coordination:

(357) a. John is a Republican and proud of it (it = John is a Republican)
b. *John is proud of it and a Republican

The derivational TAG analysis in Krivochen (forthcoming a) is based on structure sharing under graph union (such that common nodes between the elementary trees corresponding to the terms *John is a Republican* and *John is proud of* are unified) and S pronominalisation (such that the most embedded occurrence of *John is a Republican* is pronominalised along the lines explored in this work). This said, we leave a fuller discussion of coordination of unlikes under present assumptions to future work.

Let us now consider the ρ-sets of the two conjuncts in (358) below, an instance of *asymmetric et-coordination*:

(358) John had a beer and fell asleep

(359) ρ_1 = ⟨(have, John), (have, beer)⟩
ρ_2 = ⟨(fall, John), (fall, asleep), (asleep, John)⟩
$\rho_{\text{coordination}}$ = ⟨(have, and), (and, fall)⟩

The root of the first conjunct is the verb *have*, because there is no node that dominates it. In the asymmetric case, the first conjunct is strictly ordered with respect to the second such that the first conjunct always precedes the second conjunct. The interpretation of a temporal order, which is eminently pragmatic (an inference that can be cancelled) is, in the present view, guided by properties of the syntactic structure (as in some versions of Relevance Theory, the presence of *procedural* categories, which guide the interpretation of conceptual content; e.g. Leonetti & Escandell, 2015).

The *symmetric et-coordinated* case is different, as we anticipated. We would like to propose that in this case both conjuncts share the same root, which is the coordinating conjunction. It is crucial to observe that this does *not* mean that 'and' is the label, or head, of the coordinated structure, *because our structures are neither labelled nor endocentric*. However, because there is no dominance relation between the terms of the coordination, we predict that any linear ordering of these terms should be not only grammatical, but also preserve all grammatical relations and semantic representation (at least those aspects of semantics which can be accounted for configurationally; see Schmerling, 2018b for discussion). Recall that one of the features of symmetric coordination was the reversibility of conjuncts, which we saw in (348) and (349), repeated here as (360a–b):

(360) a. Bill got biscuits and John bought milk
b. John bought milk and Bill got biscuits

Our argument is that (360a) and (360b) are derived expressions whose structures display the same dependencies between basic expressions; therefore, there is a single graph that describes them. The only difference is word order, not syntactic relations. What we have in (360) is, we argue, something along the lines of (361) for the derived graph:

(361) ρ_{derived} = ⟨(and, get), (and, buy), (get, Bill), (get, biscuits), (buy, John), (buy, milk)⟩

We have two lexical verbs, *buy* and *get*, each of which is the anchor of its own elementary graphs, and also the root of it: these are conjoined as in schema (355a). They contain each verb's nominal dependants: *Bill* and *biscuits* for *get*, and *John* and *milk* for *buy* (plus the temporal modifiers of these verbs, which we omit for convenience). Note that, in contrast to (358), here there is no asymmetric relation between the conjuncts: they are *paratactically related* and thus the order between them is commutative.

The kinds of coordination we have identified are, in principle, not limited to specific categories: we can *et-* or *que-coordinate* any category that can be coordinated. For example, let us see what happens when we *que-coordinate* Ns or NPs:

(362) a. John had red beans and rice for dinner, which was/*were quite good
b. The rise and fall of the stock market has/*have economists worried

The *que-coordinated* reading of (362a) depends on *red beans and rice* being a single dish (there is, thus, some potential for variation depending on the reader's culinary background; this particular example was suggested to us by Susan Schmerling, p.c., as a traditional dish in Southern US. British readers may replace it with *fish and chips*). Likewise, the *que-coordinated* reading of (362b) depends on *the rise and fall of the stock market* referring to the oscillation of the stock market, where *the rise and fall* denote a single process.

In addition of *et*-coordination of VPs, we can also have *et-coordinated* Ns. The following are examples of *symmetric et-coordination* of Ns; symmetry is probed for by changing the order of the terms of the coordination:

(363) a. John drinks beer and whisky
b. John drinks whisky and beer

Obviously, in both (363a–b) John is drinking *two* distinct things, not one; a *que-coordinated* reading is not an option in this case (see also Moltmann, 2017). The value of the conjunction is simple logical conjunction as in the propositional calculus: both sentences in (363) are true if and only if John drinks beer and John drinks whisky. However, some coordinations are ambiguous, in the sense that both an *et*-reading and a *-que* reading are available. Is there any test that we can apply to dissociate *et-* from *-que* readings in coordinations that are otherwise ambiguous? At this point, we need to give some more detail about the conditions under which each of these readings are available. In Krivochen & Schmerling (2016a) we argued that the possibility of having structures of the kind *both X and Y* or *either X or Y* is a test for *et-coordination*, since *both* requires access to the terms of the coordination. If *que-coordination* yields internally opaque units, then it is impossible to access the terms separately. In this respect, let us take a look at the following paradigm, where *both* and *either* necessarily require accessibility of two distinct terms in a true coordinated structure, and thus serve as a reliable test for *et* vs *que* coordination:

(364) a. #John had both red beans and rice for dinner (# in the interpretation in which *red beans and rice* is a single dish)
b. #John had either red beans or rice for dinner (same as above)
c. John drinks both beer and whisky
d. John drinks either beer or whisky

In this chapter we introduced some evidence against a structurally uniform approach to coordination, as assumed in MGG. We argued that there are two kinds of coordination: *et*-coordination and *que*-coordination, which differ both semantically and configurationally. Considering aspects of the treatment of coordination from a variety of theories, we formulated a framework to capture some of the aspects of *et*- and *que-coordination* with the tools available to us in the theory of syntax sketched in the present work. Even if the specifics of the analysis turn out to be only partially on the right track, we think that there are aspects of the study of coordinated structures on which a graph-theoretic approach can shed light, combining insights from APG, PSGs and DGs.

CHAPTER 13

A Small Collection of Transformations

In this chapter the distinction between *relation-changing* and *relation-preserving transformations* comes to the forefront again, and the processes that have been discussed in all previous chapters are classified in terms of whether they involve new nodes or only new edges between existing nodes in a graph. This chapter argues that *Passivisation* is one of the very few *relation-changing transformations* in English. This point is crucial for the argument at the core of this book: most syntactic processes do not in fact change grammatical relations (erasing old ones and creating new ones), but simply add new relations on top of existing ones or modify only word order.

We will now classify some well-known empirically motivated transformations from the *aetas aurea* of generative grammar in terms of whether they change existing grammatical relations or they just create new relations on top of what was already there. The following examples must be understood as referring *only to English*, and encoding traditional insights on what reordering transformations do (understood as descriptive devices). There is, in this collection of transformations and their classification, *no* aspiration of aprioristic universality. All in all, what is important here is to see how many 'transformations' actually change grammatical relations, which directly impacts on the descriptive adequacy of our theory. In this sense, Epstein et al. (1998: 3) say that

> *From Syntactic Structures through the Extended Standard Theory framework, transformations were both language-specific and construction-specific: the linguistic expressions of a given language were construed as a set of structures generable by the phrase structure and transformational rules particular to that language. Transformational rules thus played a crucial role in the description of particular languages.*

They go on to make the contrast with P&P-based theories (GB/MP), which focus on universal restrictive principles (as highlighted in Lasnik & Uriagereka, 1988: 7, in GB all transformations—which can be reduced to *Affect α*—were optional, with apparent obligatoriness being a consequence of output filters that would be violated if a rule didn't apply). We think that the descriptive power of phrase structure rules, which was due to the close attention paid to the analysis of constructions and the explicit formulation of inputs and outputs of transformational rules, is something to *recover* rather than to abandon

 | DOI:10.1163/9789004542310_014

in procedural syntax (see Postal, 2004: §§ 9, 12, 13 for detailed discussion; also Falk, 2001: 28). In the current framework, we will stick to transformations as powerful descriptive devices, which in this context correspond to different organisations of nodes in a graph.

What follows is a preliminary classification of well-known and time-tested transformations in terms of whether the processes create new relations while maintaining existing relations or whether no new relations are created, only linear order; we refer again to McCawley's (1982) distinction between RCT vs. RPT. It is important to highlight that RPTs come in two variants, if we adopt a transformational standpoint (and see transformations as mappings from trees to trees): a process may create new relations in addition to 'base-generated' relations or it may only change linear order while leaving all previously established relations intact.

Next to each transformation we include a cross-reference to the place in this monograph where we have dealt with the relevant process and sometimes an important bibliographical reference that we find particularly close to our own view.

I. **Relation-preserving transformations:**

 I.1 **Create *new* grammatical relations (leaving old relations intact):**
 - a) Raising to Subject (**Section 6.1**)
 - b) Raising to Object (**Section 6.2**)
 - c) Topicalisation (**Section 6.5**)
 - d) Focalisation (**Section 6.5**)
 - e) NPI fronting (**Section 6.5**)
 - f) *Wh*-interrogative formation (**Chapter 10**)

 I.2 **Maintain all existing relations *only* changing linear order:**
 In these cases there are no grammatical relations added or deleted: only the linear order between constituents or expressions is changed.
 - a) Right Node Raising (**Section 14.1**; Levine, 1985; McCawley, 1988; Citko, 2011; Citko & Gračanin-Yuksek, 2021; Sabbagh, 2014)
 - b) Wrap (as in Bach, 1979; **Chapter 4**)
 - c) Rightwards extraction, including Heavy NP Shift and Relative Clause Extraposition (**Section 14.1**)
 - d) Location fronting (a.k.a. Locative inversion: *An old woman lives in the forest* → *In the forest lives an old woman*; Ross, 2012: 12)
 - e) Though- / as- preposing (*Although Bill is rich, he eats a mean waffle.* → *Rich though/as Bill is, he eats a mean waffle*; Ross, 2012: 6)
 - f) Parenthetical insertion (see McCawley, 1982 and **Chapters 7** and **8**)

g) Clitic climbing (**Section 7.1**)
h) Pseudocleft formation (*I ate an eel → what I ate was an eel*; *as per* Ross, 2011)
i) Classical NEG raising (see Collins & Postal, 2014 for extensive discussion and Krivochen, 2020, 2021b for an alternative syntactic analysis based on *lowering* rather than *raising*)

II. **Relation-changing transformations:**

13.1 Passivisation

Being one of the very few relation-changing transformation that we have identified in the grammar of English (and, historically, one of the first to be formulated in transformational grammar, making an appearance already in Chomsky, 1955a: Chapter 9, 1955b; and Harris, 1957), *Passivisation* merits some detailed discussion. We follow Perlmutter & Postal's (1983a) characterisation of the *passive* in terms of grammatical relation changing, regardless of how that change is attained: via transformations, as in Williams (1982), Collins (2005), among others; via lexical rules, as in Bresnan (1982b), Dowty (1978) and much subsequent work; or in terms of conditions over relational networks, as in Johnson & Postal (1980), Postal (1986: 7). It is not evident that the same arguments for or against a transformational analysis hold universally: for instance, Müller (2000) provides arguments against an Object-to-Subject raising analysis of the German passive, but it is not clear whether those arguments can be extrapolated directly to English. For purposes of this work, we will restrict ourselves to a claim about the grammar of English and Spanish.

We have classified *Passivisation* as a relation-changing transformation (in the descriptive sense in which we have been thinking about transformations all throughout) because, in RG terms, it requires not only the advancement of a 2 to 1 (*1-advancement*), but also that the relevant NP is *not a 2 anymore* (and what was a 1 becomes a *Chômeur*). An important caveat is due: this does not mean that the NP under consideration is no longer an argument of the V, just that its grammatical function changes: this is fully compatible with RG's view that Passivisation (in English) is an entirely relational process (Perlmutter & Postal, 1983a; Postal, 1986). This means, importantly, that accounts of Passivisation purely in terms of *word order*, *verbal morphology*, or *case* cannot form the basis of an adequate universal characterisation of the process (see Perlmutter & Postal, 1983a: § 3 for insightful discussion about these points). The view adopted here is also similar to that of LFG and some versions of Montague-style Categorial Grammar (e.g., Dowty, 1978, 1979), to an extent, since there is no transformational relation between actives and passives: neither derived from the other. This departure from the Harrisian tradition (based on 'kernelism') allows us to study passives in their specificity, without the need to invoke specific *ad hoc* mechanisms (such as *smuggling*, as in Collins, 2005 or *grafting*, as in Geraci, 2020) which are only motivated within specific versions of the transformational tradition. It is important to note that even in the generative transformational approach, the analysis of passives is not homogeneous. For example, Chomsky (1981: 120–121) distinguishes between syntactic and lexical passives (following the line inaugurated in Chomsky, 1970a of combining lexical rules with transformational rules), and even within syntactic passives he suggests that the specific rule structure may vary cross-linguistically. From these considerations stems our cross-linguistic caution and the restriction of our empirical basis.

In English, there are at least two interesting examples of constructions that do not fit the mould of traditional accounts of the passive in terms of verb typology ('only transitive verbs can be passivised') or morphology ('subjects of passive clauses cannot be non-nominative'[1]): namely, (i) passives of intransitive verbs (so-called 'pseudopassives'; Postal, 1986: 30–32; Chapter 6; Findlay, 2016 for a perspective from LFG; Padovan, 2016, 2021a for a transformational view closer to MGG and decompositional approaches to prepositional phrases) and (ii) non-nominative subjects in passive clauses. In RG terms, passives of intransitive verbs (also known as 'prepositional passives') arise when the advancement of the 2-arc leaves a preposition stranding, a factor that we will not analyse here (but see Postal, 1986: Chapter 6 for discussion). The second group, non-nominative subjects in passives, present a different quirk, since they are dialectally restricted (we exemplify them in (366), below). As for the first group, prepositional passives, relevant examples are like (365):

(365) a. No one has ever sat on this chair
b. This chair has never been sat on (by anyone)
c. His parents never talked to him
d. He was never talked to (by his parents)

The subjects in (365b, d) are morphologically nominative, as are all subjects in what Postal (1986) names 'pseudo-passives'; we can probe this by pronominalising them: *They/*them have never been sat on* (note also that the prepositions are both categorematic). Several ways of analysis become possible, depending

1 Importantly, here we are referring only to Nominative-Accusative languages. See Dixon (1972) and Schmerling (1979) for an analysis of Dyirbal (an Ergative-Absolutive language) which suggests that a characterisation of the Dyirbal passive strictly in RG terms may be inappropriate.

on theoretical assumptions: if the sequence V + P is 'reanalysed' as a monotransitive 'complex predicate' (e.g., Hornstein and Weinberg, 1981), then prepositional passives are derived in the same way as garden-variety passives. A problem with this approach is that not all prepositional regimes allow for the corresponding prepositional passive (Padovan, 2016, 2021a):

(366) a. Many people have slept under this bed
b. *This bed has been slept under

The RG analysis of the passive, which is seen as a modification of the relational network of a lexical predicate (involving object promotion and subject demotion), is particularly attractive to us. But not without issues. A potential problem with the RG approach, which is noted in Hübler (2016), if there may be only one advancement to 1 in a clause (Perlmutter's 1978 *1-advancement law*) it is not obviously compatible with the fact that not only (some) *unergative* predicates (like *sleep* or *talk*) but also some *unaccusative* predicates (like *sit* in our (365), which would require *two* advancements to 1 and are thus excluded by the *1-advancement law*) allow for the prepositional passive (Hübler does not consider English examples, she cites Blake, 1990: 65 and an example from Turkish that native speakers we consulted categorically reject in the intended interpretation as a passive unaccusative[2]). The cross-linguistic issues that arise are not the center of our attention for the time being, since we are focused on English. Still, if indeed there were passive unaccusatives in English, it would be problematic. The facts need to be considered very carefully, however: stative unaccusatives like *exist* or *stand* do not allow for the prepositional passive, nor do dynamic unaccusatives like *appear*:

2 The example in question, cited by Hübler, is:
(i) *Burada dü-ül-ür* [sic. The root should be *düş-* instead of *dü-;* the verb is *düşmek*: to fall]
Here fall-PASS-AOR
'Here it is fallen.'
The other example provided in Blake (1990: 65) also, according to some of our native informants, 'does not make any sense', and when accepted, it was as an impersonal construction (whose syntactic structure is arguably not the same as a passive)
(ii) *Bu yetimhane-de çabuk büyü-n-ür.*
this orphanage-LOC quick grow-PASS-AOR
'In this orphanage it is grown quickly.'
Given the fact that the native speakers we have consulted do not unanimously accept the sentences (in particular, (ii) was regarded as either impossible or very marginal), using them as an argument to make a claim as strong as that passive unaccusatives are possible in Turkish in any sense other than morphological marking seems to be a stretch (either in RG or any other framework).

(367) a. Gorillas still exist in Africa (taken from Postal, 1986: 31)
b. *Africa is still existed in by gorillas
c. A statue stands in the middle of the square
d. *The middle of the square is stood in by a statue
e. A pirate ship appeared on the horizon (taken from Levin, 1994: 86)
f. *The horizon was appeared on by a pirate ship

The availability of the prepositional passive seems to be lexically restricted at a level of granularity that is finer than just *unaccusative* vs. *unergative*. Much discussion pending, and as always centering out attention in English, we will assume that a version of the RG characterisation of the passive is correct: passive voice affects the relational network of a lexical predicate.

Transformational generative grammar, from its very early days (e.g., Harris, 1957: rule 4.11; Chomsky, 1955a: 504f.-g), derives a passive sentence via the application of a mapping rule to an active sentence. In the transformational view, and using slightly anachronistic terminology, an NP is base-generated in the object position (in phrase structure terms, as a sister to V, daughter to VP) at the base component (Deep Structure later on), before the application of all transformational rules. This object NP is then moved to the position of subject (which can be Spec-Infl / Spec-T / Spec-AgrS, depending on the framework of choice) leaving a trace or copy behind in order to satisfy the requirement that clauses have subjects (the so-called *Extended Projection Principle*; Chomsky, 1982: 9–10); the NP that played the role of subject at the pre-transformational level becomes an adjunct, and an auxiliary verb *be* is inserted (counter-cyclically), which in turn selects for a participial form in its complement. Interestingly, the transformational analysis of the passive encounters many problems when mappings trees to trees as in an LSLT / SS framework is replaced with stepwise recursive combinatorics. As a reminder, the classical transformational formulation of *Passivisation* is due to Chomsky (1957: 112) (see also Harris, 1957: 288, 309):

Passive—optional:
Structural analysis: NP-Aux-V-NP
Structural change: $X_1–X_2–X_3–X_4 \rightarrow X_4–X_2 + \text{be} + \text{en}–X_3–\text{by} + X_1$

In Minimalism, the situation is quite different. Because operations do not take entire structural descriptions as their input, and because there is no 'passive' as a rule, the only way to proceed is to derive passives by a combination of Agree and Merge (External / Internal). But since the VP/*v*P is constructed before the T head is Merged, when auxiliary *be* is inserted in the derivation, structure needs to be reconfigured (in particular, whatever operations have taken place for Case assignment within the VP are now invalid), in violation of the No Tampering Condition (Chomsky, 2008), which bans modification of structure previously constructed. The passive requires drastic reanalysis of the structure below TP, once auxiliary *be* has been inserted. Some approaches based on the notion of 'grafting' (van Riemsdijk, 2006; Geraci, 2020; see also **Section 8.1**) allow for Merge to the non-root, probing into already derived structure and inserting a piece of tree counter-cyclically (somewhat similarly to TAG-*adjunction* in its counter-cyclicity). Geraci's analysis is somewhat different from van Riemsdijk's, however, in that grafting counter-cyclically removes the structure above the grafting point. In Geraci's (2020) analysis, the internal DP argument is assigned Accusative case in Compl-V. After the *v*P is completed, auxiliary *be* as a T head is grafted at the VP level, and Nominative is assigned to the same DP. This grafting removes the *v*P layer, and replaces it with a TP projection of auxiliary *be*. The *by*-phrase is analysed as a PP, which gets adjoined perhaps to TP.

Closer to traditional movement-based MGG there are analyses like Collins' (2005) which, in a very condensed summary, (i) assume that there is a Part(iciple)P on top of VP, where V moves, itself dominated by a *v*P and a VoiceP and (ii) the whole PartP moves from the Complement of *v* to the Specifier of VoiceP; this movement allows the direct object to escape the *v*P phase and move again to Spec-TP (a process that Collins refers to as 'smuggling', since the moved PartP 'smuggles' the direct object so that it can move again). A *smuggling* analysis would look like (368) (see Collins, 2005: 90):

(368)

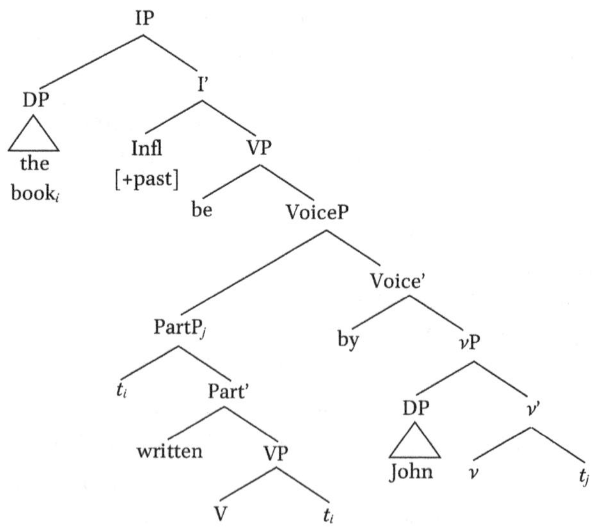

FIGURE 13.1 'Smuggling' analysis of Passivisation in Collins (2005)

The 'external argument', Collins says, is Merged in the same position as in the active (Spec-*v*P) and indeed as a DP, not a PP: *by* is the head of Voice, and does not form a constituent with the DP (which makes the analysis of sentences such as *By whom was Bill killed?* unnecessarily difficult). Collins argues that *by* has no interpretable features, but does not acknowledge the existence of syncategorematic expressions (which could take care of *by John* without adding a projection whose head bears no interpretable features). Theta-marking in this structure depends on the active participle (the one that appears as a complement of perfective *have*, and which can take accusative complements) and the passive participle (the one that appears as a complement of passive *be*, and which cannot take accusative complements) being identical in all syntactic respects (Collins, 2005: 85), certainly a controversial position (see Fabb, 1983 for a summary of differences between active and passive participles in English; Bravo & García Fernández, 2016; García Fernández & Krivochen, 2020 for arguments to distinguish between active and passive participles building on Spanish data). Importantly, even in this Minimalist transformational view the passive and the active do not share a numeration or a reference set (in the sense that the active and the passive do not use the same lexical resources): before movement of PartP to Spec-VoiceP there is a PartP instead of a VP immediately contained by *v*P and a PP Merged at Spec-*v*P, which means that even before VoiceP is introduced, the system must know it is dealing with a passive. Finally, the passive auxiliary *be* is analysed as a V head, which is somewhat strange considering it is not a lexical verb.

We find both the grafting and the smuggling approach problematic due to their strongly intra-theoretical nature: they seem to be consequences of particular Minimalist assumptions and address issues that only emerge under very specific theoretical conditions. The arguments from RG pertaining to the relevance of GF for the characterisation of the passive (and diathesis being a property of relational networks) remain unaddressed, for the most part, in MGG. The empirical and theoretical advantages of a transformational approach to the passive are, to us, unclear.

Non-transformational approaches to the passive generate passives as structures that are as 'derivationally basic' as actives. In these cases, *Passivisation* is not a *syntactic* rule, but a *lexical* rule. For example, in LFG actives and passives differ on how the parallel levels of functional structure and argument structure map to constituent structure; but there is no mapping from constituent structure to a derived constituent structure (Bresnan, 1982b, 2001: 25, ff; Dalrymple, 2001: 207, ff.). In LFG, *lingking theory* takes the role of relating argument structure to functional structure: the former encodes semantic roles, whereas the latter encodes grammatical function. Grammatical functions in LFG are classified according to two binary features: [± object] and [± (thematically) restricted]. This gives us the following characterisation of GF:

> **Subject:** [- objective] [- restricted] (since there are no thematic restrictions on the subject position)
> **(Primary) Object:** [+ objective] [- restricted]
> **Secondary Object** (or Obj_{θ}): [+ objective] [+ restricted] (since secondary objects are restricted to bearing a theme thematic role)
> **Oblique argument** (Obl_{θ}): [- objective] [+ restricted] (locative/benefactive/malefactive thematic roles)

The LFG typology of objects is not as rich as, say, Postal's (2010), but some correspondences can be drawn (so, for instance, Postal's *subobject* would cover similar ground to LFG's thematically restricted *object*$_{\theta}$). Theta-roles are also assigned features, such that:

i. Patient-like roles get assigned [-r]
ii. Secondary patient-like roles get assigned [+o]
iii. The elsewhere cases get assigned [-o]

In this context, linking theory provides two *mapping principles*, which we can summarise as follows (see Butt, 2006: 129; Dalrymple et al., 2019: 337):

- Highest-ranked theta-role marked [-o] is mapped to the grammatical function *subject* when initial in *a-structure*
- Else, the highest-ranked theta-role [-r] is mapped to subject

In the classical LFG approach (Bresnan, 1982b), Passivisation suppresses the highest argument, taking the highest-ranked theta-role and mapping it to Ø. In a more recent *demotion* account, the highest argument is not mapped to Ø, but rather assigned an additional [+r] feature. In either case, the result is the same:

(369) a. **Active:** *hit* ⟨Agent, Patient⟩
[-o] [-r]
↓ ↓
Subj Obj

b. **Passive:** *hit* ⟨Agent, Patient⟩
[-o] [-r]
↓ ↓
Ø Subj

The suppressed argument is unavailable for linking (in the LFG sense), and becomes a thematically restricted oblique. The lexical entries for hit_{active} and hit_{passive} would then be

(370) a. **Active:** PRED 'hit⟨SUBJ, OBJ⟩'
b. **Passive:** PRED 'hit⟨SUBJ, OBL_θ⟩'

It may be noted that LFG's analysis (which has *Passivisation* as a lexical rule) was heavily inspired by Emonds' (1970) *Structure Preservation* hypothesis, and that *Passivisation* was indeed identified as a *structure-preserving transformation*: the D-Structure NP object can move to subject position because there is an independently motivated phrase structure rule that generates the configuration $[_S$ NP VP].

What do passives look like under graph-theoretic assumptions? We want to emphasise that, in principle, the framework that we have been describing applies straightforwardly to passive clauses, with no additional stipulations (rules, levels of representation, indices) needed. Lexicalised elementary graphs lend themselves to representing relational networks rather naturally. Consider, for example, the following pair:

(371) a. Bill killed John
b. John was killed

Focusing on expressions and relations, the preliminary ρ-sets for (371a) and (371b) are given as (372a) and (372b) below:

(372) a. ρ = ⟨(kill, Bill), (kill, John)⟩
b. ρ = ⟨(be killed, John)⟩

(372b) features a 'detransitivised' expression *be killed*, the GF *object* cannot be assigned in this structure. Here, the lexical anchor is *be killed*, with a nominal dependent *John*; participial morphology prevents it from taking an object, and since there is only one NP, the highest ranked GF is assigned to it. And this, following RG, is the fundamental property of the *passive*: the change in the distribution of grammatical relations, which are atomic and primitive in this theory and its successors (within which we can count ourselves, to a certain degree). But, what happens with the *by*-phrase that may express the agent optionally? We argue that *by Bill* modifies the event of *John being killed* (along Neo-Davidsonian lines), with *Bill* the agent of the event. Thus, provisionally, we propose that *by Bill* (which, given that *by* is syncategorematic, would be just a node *Bill*) dominates the root node of the elementary graph whose lexical anchor is the passivised verb. Crucial to our point is the fact that in the present view both (372a) and (372b) are equally 'basic', in the sense that there is no mapping from one to the other. Because in classical transformational grammar GFs are defined configurationally, a transformational account of *Passivisation* says nothing about what we think is the most important property of this process: it is a *relation-changing transformation* (as has been noted in non-transformational frameworks, particularly RG and later on LFG).

There is, however, a potential problem with (372b): Schmerling (1983b; 2018a: Chapter 6), in her analysis of the English auxiliary system, provides evidence that English finite auxiliaries are modifiers of Nominative subjects instead of VP heads (or first expressions of IVs; in either case the auxiliary would be grouped with the lexical verb). Her arguments involve a combination of prosodic and syntactic evidence, and account for the restriction of certain classes of auxiliaries, in particular, modals. The fact that modals do not show an alternation of tense (with the exception of *can-could* and *will-would*) would follow from the fact that they are not expressions of category IV (the category of intransitive verb phrases, of type $\langle e, t \rangle$), but rather modifiers of nominative subjects. The reasoning extends to all auxiliaries. This view, although controversial, provides basis to account for cases of VP ellipsis such as *Mary will go to the party, but John will not* without deletion. In this case, Schmerling analyses *John will not* as an expression of category FC//IV: the category of *modified* expressions that must concatenate with an IV to form a finite clause. Note also that this allows us to have expressions like *John has* (in a context like *John has arrived*) to be modified FC/IV, and *Has John* (in the interrogative *Has John arrived?*) be IFC/IV, with IFC being the category of interrogative finite clauses: there is subject-auxiliary inversion without movement (see also Schmerling, 2018a: Appendix B; in this work she uses IIC—Inverted Indicative Clause—instead of IFC). Similarly, finite auxiliaries form a unit with the subject in instances of VP topicalisation: *They said that John would be murdered, and murdered he was*. In addition to the elegant account that Schmerling's categorial system provides for patterns of VP ellipsis and topicalisation in auxiliary verb constructions, it also allows us to tackle the issue of cross-linguistic variation in an interesting way.

In light of these arguments, we assume two fundamental properties of passive *be* in English: (i) it has no argument structure, with all selectional properties being determined by the participle; and (ii) it forms a derived expression with the subject of the lexical verb (see Schmerling, 1983b: 26–27. Note also that this second property distinguishes passive *be* from Raising verbs). Like Schmerling's categorial approach, this marks a departure from classical phrase structure assumptions, where passive *be* would form a constituent with the participial VP. This is not surprising if we bear in mind that pure Categorial Grammar is an IP system, not an IA one, and constituency *per se* does not exist. Under Schmerlingian assumptions, the ρ-set of (371b) would be (373),

(373) ⟨(killed, John be)⟩

where the subject is a modified expression.

The properties of the passive auxiliary and constructions where it appears need not be universal. If we consider passive *be* alongside its Spanish counterpart, ⟨*ser* + participle⟩, there are some differences in terms of the syntactic processes that can affect passives in both languages. We follow Schmerling (2018a) in assuming that English passive *be* is syncategorematic (see also Johnson & Postal, 1980: 153 for an analysis that does not assign passive *be* or its Spanish counterpart *ser* a node in R-graphs), therefore, it does not correspond to an independent node in the graph. This much seems to be common with Spanish passive *ser*. However, this does not mean that passive constructions in Spanish and English behave the same. We can observe that only property (i) holds for Spanish *ser* (see also Bravo et al., 2015; García Fernández et al., 2017 for discussion about the functional nature of Spanish passive *ser*); as pointed out above, Spanish does not have English-style VP ellipsis:

(374) a. Mary wasn't helped, John was ~~VP~~ (= murdered)

b. **María no fue ayudada, Juan*
M. NEG be.AUX.PASS.PAST help.PART.FEM, J.
fue ~~*VP*~~ (= *ayudado*)
was.AUX.PASS.PAST ~~VP~~ (= helped)

We propose, following Schmerling, that it is the existence of a relation between *be* and *John* that licenses examples like (374a), and presumably that relation does *not* hold in Spanish: *ser* is much more of an inflectional element than *be* is in English (it is more grammaticalised). Furthermore, if Schmerling's (1983b) analysis of English auxiliaries is in the right path, a source of cross-linguistic variation would be whether auxiliaries form a derived expression with the subject or with the (saturated) verb phrase: English auxiliaries do (and thus an expression like *John was* is a well-formed expression assigned to the category of expressions that combine with a passivised IV to form a finite clause FC; see Schmerling, 1983b: 27), but Spanish auxiliaries do not. This would translate into the claim that passive *ser* forms an expression with the participle that it selects and not with the subject of the clause: the ρ-set for a sentence like

(375) *María fue ayudada*
M. be.AUX.PASS.PAST help.PART.FEM
'Mary was helped'

would be simply

(376) a. ρ = ⟨(ser ayudada, María)⟩

and *not* (377a) (a pseudo-Rossian representation) or (377b) (the Schmerlingian one for English):

(377) a. ρ = ⟨(ser, María), (ser, ayudada), (ayudada, María)⟩.
b. ρ = ⟨(ayudada, ser María)⟩

There is no relation between *ser* and *María* (nor is *María ser* a modified expression), which readily blocks the possibility of having an expression *María fue* in so-called VP ellipsis contexts (cf. (373b), *mutatis mutandis*): passive *ser* is *not* a basic expression in Spanish, nor does it form a derived or modified expression with the subject. Rather, in Spanish as opposed to English, it is to be grouped with the predicative elements, what in IC analyses would be called the VP: passive *ser* is a modifier of the lexical predicate, not of the Nominative subject.

This proposal finds some additional support in the analysis of examples like the following:

(378) *María no fue ayudada, pero Juan lo= fue*
M. NEG be.AUX.PASS.PAST help.PART.FEM, but J. CL.ACC= be.AUX.PASS.PAST
'Mary wasn't murdered, but John was'

In (378), the pro-form *lo* stands for the IV *ayudado*, deleted under sloppy identity (because the agreement features in the participle must change, since *María* is a feminine N and *Juan* is a masculine N). It is also a phonological clitic, in the sense that it is not a well-formed phonological word on its own. Now, assume that *ser ayudado* is not a two-word basic expression, but rather a derived expression: CG can help us make this clearer. Let *ayudado* be assigned to the category of expressions that must concatenate with an NP to form a finite clause (i.e., FC/NP). Then, passive *ser* (should it be a basic expression of its own), could be assigned to the category of expressions that must combine with an expression of category FC/NP to form an expression of category FC/NP: that makes *ser* an expression of category (FC/NP)/(FC/NP) (see Bach, 1983: 111 for such an analysis in the framework of Generalised CG). But that would predict that a rule of functional application (or a 'transformation') can affect the participle (or the pro-form) independently, for instance, clitic climbing. This is a testable prediction. A model of Spanish passives in which *ser* does not form an expression with the participle it selects but with the subject could generate the ungrammatical (379):

(379) **María no fue ayudada, pero Juan lo=*
M. NEG be.AUX.PASS.PAST help.PART.FEM, but J. CL.ACC=
tuvo que ser
had to be.AUX.PASS.INF
'Mary wasn't murdered, but John had to be'

If, on the other hand, there is a basic expression (of category FC/NP, in our example, but not necessarily) *ser asesinado*, (379) is adequately filtered out: the pro-form *lo* cannot be reordered on its own, and of course the (complex) unit *ser+lo* cannot be targeted by clitic climbing.

By adopting the CG distinction between basic and derived expressions and implementing it in the ρ-set of English and Spanish passives, we can capture the empirical differences between English and Spanish in terms of the availability of VP ellipsis (building on Schmerling's original insight). As highlighted in Krivochen & Schmerling (2022), these differences cannot be straightforwardly accounted for in structurally uniform, templatic theory based on a fixed hierarchy of functional projections where terminal nodes are most often than not equated to orthographic 'words' (or, in some recent developments, smaller units like morphemes or features—e.g., Embick & Noyer, 2007; Baunaz & Lander, 2018 respectively-). As we have emphasised throughout the monograph, the distinction between 'words' and 'basic expressions' is at the core of our approach and, we think, of a descriptively adequate theory of natural language grammar.

13.2 Dative Shift

Following Dowty (1978), we analyse Dative Shift (also known as 'dative alternation', but see below) as an RCT. Dative Shift, in the transformational tradition, is the name of the transformation that applies to a *prepositional indirect object construction* (PIOC) to produce a *double object construction* (DOC). If, as proposed in Larson (1988), DOCs derive from PIOCs, we can have a rule with (380a) as its input and (380b) as its output:

(380) a. Sue sent a letter to Mary
b. Sue sent Mary a letter

The Larsonian treatment of DOC is based on the fundamental requirement that syntactic representations be exhaustively binary branching trees; similar requirements are to be found in the treatment of the PIOC-DOC alternation in Hale & Keyser (2002: § 5), Adger (2003), Runner (2001), and many others within the generative tradition.

As Harley (2003, 2011) observes, the transformational route is not the only one taken in generative grammar. Some proposals, like Pesetsky's (1995) and Harley's own take on the issue, analyse DOC and PIOC as distinct base-generated structures, not derivationally related. The distinction, then, is captured lexically lexical rather than transformationally. We can exemplify the alternative analyses as follows (see Harley, 2003; Harley & Miyagawa, 2017 for extensive discussion):

(381) a. PIOC, transformational (Larson, 1988)

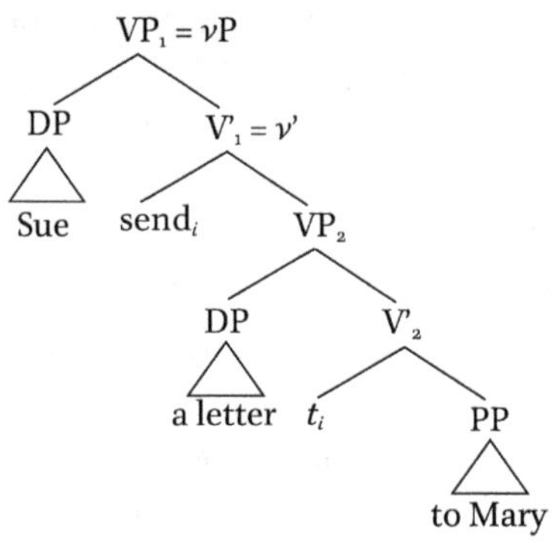

FIGURE 13.2 Analysis of PIOC in Larson (1988)

b. DOC, transformational

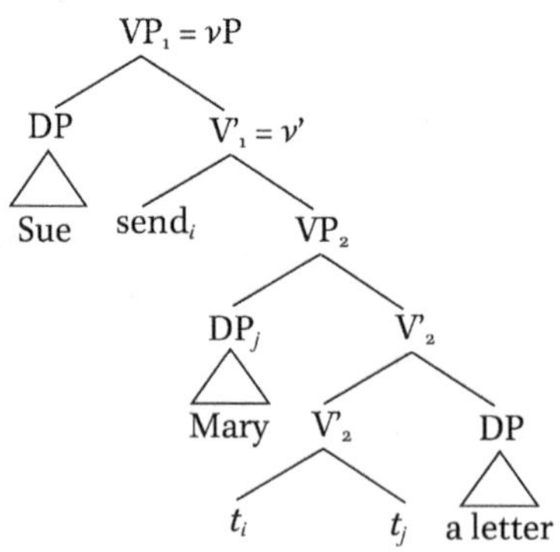

FIGURE 13.3 Analysis of DOC in Larson (1988)

Larson's analysis assumes a 'clause-like' structure for the VP: in the PIOC, 'a letter' is analogous to a 'subject' in the lower VP, and 'to Mary', analogous to an 'object' (this 'clausal' structure then gets predicated of the external argument, introduced by a higher functional projection). Note that in (381a) the verb and the oblique form a constituent that excludes the object, in the same way that a verb and its direct object form a constituent that excludes the clausal subject. Under this analysis, the DOC alternation is a sort of 'Passivisation' within the lower VP (Larson, 1988: 335–336): the oblique is 'promoted' (that is, it comes to occupy the position that was reserved for the highest available GF; here the Spec-VP), and the constituent generated in Spec-VP gets demoted to an adjunct position.

The lexical approach presents some variation in terms of the category of the relational predicate: whereas Pesetsky and Harley (among others) assume a P head with two flavours (central and terminal coincidence), Hale & Keyser (2002) change the categorial specification of the predicate that relates theme and location in DOC constructions from P to V. The alternative analyses are given below:

(382) a. PIOC, lexical (Hale & Keyser, 2002)

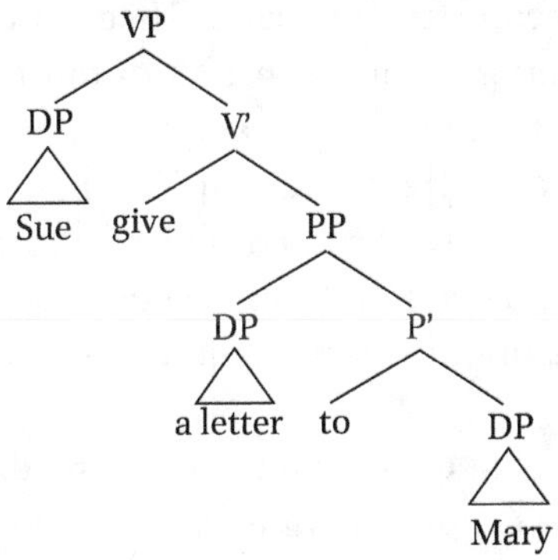

FIGURE 13.4 Lexical analysis of PIOC in Hale & Keyser (2002)

b. DOC, lexical

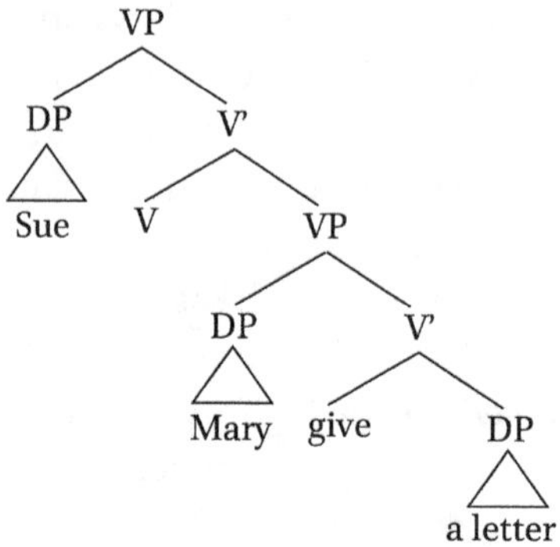

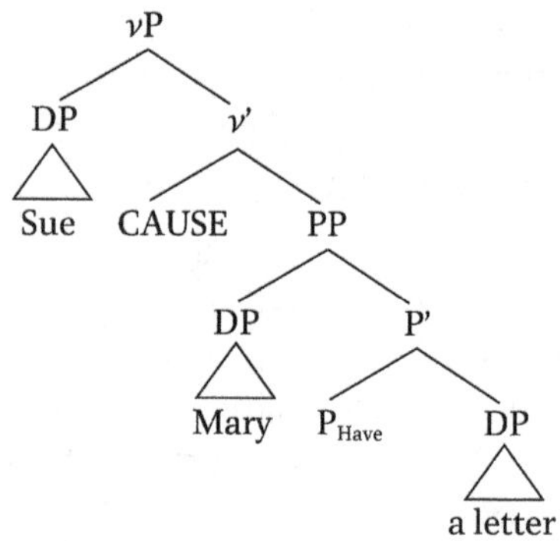

See Hale & Keyser (2002: 162) See Harley (2003: 32)

FIGURE 13.5 Lexical analyses of DOC in Hale & Keyser (2002) and Harley (2003)

For present intents and purposes, it is important to bear in mind that the arguments used in generative grammar for or against a specific configurational approach to the English PIOC-DOC alternation are based on binding paradigms (Barss & Lasnik, 1986), the distribution of NPIs, possession entailments, the position and scope of secondary predicates (Hale & Keyser, 2002), etc. In the MGG literature, no mention is made to GFs in most analyses of Dative Shift, in that whatever syntactic operations apply do not specify GFs as part of their input, only configurational information. This means that, despite the insights provided in the MGG works cited above (and considering Larson's important analogy between Dative Shift and Passivisation), we need to look elsewhere if we want to justify the classification of Dative Shift as a RCT.

That 'elsewhere', in this context, is LFG. We can build on the discussion of *Passivisation*, above. Recall that the canonical argument structure of an active transitive sentence includes a SUBJ and an OBJ (in RG terms, a 1 and a 2); in the passive, the OBJ becomes a SUBJ (2 to 1 promotion). In a ditransitive sentence, there is a subject, a primary object, and a secondary object (or 'indirect object'): a 1, a 2, and a 3. In this context, we can characterise the English PIOC-DOC alternation in terms of GF change (see e.g. Bresnan, 2001; Falk, 2001):

(383) a. PIOC

Send: PRED 'send ⟨SUBJ, OBJ, OBL_θ⟩'

↓ ↓ ↓

Sue a letter to Mary

a. DOC
Send: PRED 'send ⟨SUBJ, OBJ, OBJ_{θ}⟩'
↓ ↓ ↓
Sue Mary a letter

The LFG approach allows us to characterise the PIOC-DOC alternation in terms of GF change, which is exactly what we want to do. In the PIOC, *a letter* is a direct object, and *to Mary* is an oblique argument or OBL_{θ}, for θ = goal / beneficiary. In the DOC, *Mary* is the object, and *a letter* has been 'demoted' from 2 to 3 (an OBJ_{θ}, for θ = theme). Thus, we have the same number of participants and thematic roles remain the same, but their GFs change. We can then define the ρ-sets of a PIOC and a DOC sentences as follows:

(384) a. Sue sent a letter to Mary
ρ = ⟨(send, Sue), (send, letter), (send, Mary)⟩
b. Sue sent Mary a letter
ρ = ⟨(send, Sue), (send, Mary), (send, letter)⟩

we see that in the DOC construction, the Goal outranks the Theme in terms of GFs: the Goal becomes the 2, and the Theme is a 3 (the specific arc identifiers may change should one adopt Postal's 2010 rich array of objects). Evidence in favour of this analysis comes from Passivisation: if in passives it is the OBJ (and not the OBJ_{θ} or OBL_{θ}) that becomes the SUBJ, we can make sense of the paradigm in (385):

(385) a. I gave her a book (DOC)
b. She was given a book
c. *A book was given her
d. I gave a book to her (PIOC)
e. A book was given to her
f. *To her was given a book

Under the assumptions about the distribution of GFs in (383), the contrast between (385c) and (385e) shows that only when *a book* is a 2 can it become a 1 via Passivisation (see Postal, 2010 for extensive discussion). In (385a) the goal *her* is a 2 and the theme *a book* is a 3: the ungrammaticality of (385c) follows from (a) the lexical approach to the passive that we saw above, and (b) the distribution of GFs in (383). In contrast, when the theme *a book* is a 2 (in the PIOC), it can become a 1; the prepositional object *to her* cannot become a subject. Because ρ-sets are ordered sets of arcs, the asymmetries required to account for the binding facts that played a prominent role in Larson's (1988) analysis still hold, even without phrase structure.

As a conclusion to this section, we have seen that there are indeed grounds to classify Dative Shift as an RCT, as argued in Dowty (1978) and much work in lexicalist, non-transformational theories. Dative Shift changes the distribution of GFs among internal argument NPs: for example, *give* is always a 3-place predicate and always takes a 1, a 2, and a 3. However, which syntactic objects fulfil those functions varies between PIOC and DOC. This does not mean, crucially, that there is a derivational relation between these constructions (and thus that one is more basic than the other); at most, that the lexical entry of a ditransitive verb that allows for the PIOC-DOC alternation should be underspecified in terms of the linking between thematic roles and GFs (see Kibort, 2008 for critical discussion; Kibort finds shortcomings in the traditional LFG approach to Dative Shift in terms Lexical Mapping Theory for its applicability beyond English, but in the present context those objections have little force).

13.3 Transformations vs. Alternations

Before closing this chapter, we will make a brief note on the distinction between argumental alternations and transformations. In the early days of generative grammar, in particular under Generative Semantics assumptions (see e.g. Lakoff, 1965), the so-called *inchoative alternation* was handled syntactically. This means that there was a syntactic rule, or set thereof, that transformed e.g. *the sauce became thick* into *the sauce thickened*. This rule, in Lakoff (1965), was a replacement rule, which replaced an abstract pro-verb with a feature [+ inchoative] by a result adjective. This process is not unlike the rule of *conflation* in Hale & Keyser (1993, 2002) and related works (cf. also Pullum, 1996). We may also consider alternations such as the following (see Mateu Fontanals, 2014; Levin, 1994, 2014):

(386) a. John broke the vase
b. The vase broke (Causative alternation)
c. Bill dreamt
d. Bill dreamt the dream of the just (Unergative-transitive alternation)
e. John wiped the table
f. John wiped at the table (Conative alternation)

The question is whether these alternations are indeed transformations (thus, syntactically derived) or not (thus, lexical specifications). This is an ongoing debate, and there are numerous issues that we are leaving aside here (e.g., whether argument structure and event structure are distinct systems and, if so, how to define the mapping between them). Here we remain agnostic about this mainly because our system is constraint-based, not procedural. This entails that it is not the aim to formulate a mapping between, say, (386a) and (386b) (see Harley, 2011), but rather a constraint or set thereof that ensure that the structural description assigned to an expression is well-formed. Furthermore, we want those structural descriptions to represent GFs, and dependencies between basic expressions. In this context, the ρ-sets for the sentences in (386) would be as in (387) (simplifying matters a bit):

(387) a. $\rho = \langle$(break, John), (break, vase)$\rangle$
b. $\rho = \langle$(break, vase)$\rangle$
c. $\rho = \langle$(dream, Bill)$\rangle$
d. $\rho = \langle$(dream, Bill), (dream, dream)$\rangle$
e. $\rho = \langle$(wipe, John), (wipe, table)$\rangle$
f. $\rho = \langle$(wipe, John), (wipe, at), (at, table)$\rangle$

As long as these representations do not violate any constraint on well-formed graphs, they are allowed. The important thing for purposes of this book is that these alternations can be dealt with in the lexical entry of each predicate (which may include a number of options) rather than transformations in the classical sense; in any case, argumental alternations do not seem to be problematic for our approach.

The idea that we tried to illustrate in this chapter is that most 'transformations' identified in the generative tradition (see Ross, 2012 for a very complete overview of more than 200 transformations, including examples of each) either create new relations *while preserving existing ones* or only change linear order, thus not impacting on the dominance relations between nodes in graph-theoretic structural descriptions at all. We think that pursuing this path could lead to a radical simplification of the apparatus required in the study of the syntactic phenomena in natural languages, attending to the definition of their expressions as lexical predicates, lexical arguments, or functional modifiers (within the set of categorematic expressions). This allows us to explore the relations between expressions displayed by constructions cross-linguistically without an *a priori* template: all we have is the definition of *elementary graph* and, since it depends on the definition of *lexical predicate*, what counts as *lexical* is allowed to vary cross-linguistically (depending, for example, on grammaticalisation paths). The framework proposed here allows us to tackle the issue of variation beyond word order in a theoretically and empirically interesting way.

CHAPTER 14

Some Open Problems and Questions

This chapter closes the book by presenting some open issues to be addressed in this framework beyond the foundations of grammatical analysis presented in **Chapters 4–13**. Specifically, we will focus on several kinds of phenomena: (a) left and right extraction asymmetries, (b) deletion, (c) long distance dependencies and resumptive pronouns, (d) generalised quantification, and (e) implicit arguments. We will consider English and Spanish data and discuss possible treatments of these questions from the point of view of graph-theoretic based grammar.

The graph-theoretic approach to grammatical description that we have presented in this work is still in its infancy. Thus, it is only to be expected that there are many problems and questions still to be addressed in order to make our theory a competitive one. In this section, we present a sketch of an analysis for some of those as a roadmap for future research, together with some provisional answers and speculative explorations.

14.1 A Note on Leftward and Rightward Extractions

Interrogative formation and right-dislocation are different kinds of 'transformations' in the framework presented in this paper: their structural descriptions are *not* isomorphic. That might shed some light on why leftwards movement is unbounded (creates new relations, cumulatively) whereas rightwards movement—Extraposition, Heavy NP Shift, Right Node Raising—is differently constrained (only changing linear order, but not constituency; RNR especially being immune to island effects, see Wexler & Culicover, 1980 and Sabbagh, 2007 for critical discussion). Now, what is not clear is how to encode the specific filters (e.g., the Right Roof Constraint) as graph admissibility conditions. A possibility is that those are not constraints at all, if dependencies are arbor-bounded in self-contained objects *and* if elementary graphs are single-rooted sub-trees structured around a single lexical predicate. This view shares some aspects with TAG analyses of Extraposition (e.g. Kroch & Joshi, 1987) and Right Node Raising (e.g., Sarkar & Joshi, 1997; Han et al., 2010), notably the way in which rightwards movement is bounded.

We will now illustrate some of these points. Consider now the distinction between *et-coordination* and *que-coordination* which we introduced in

 | DOI:10.1163/9789004542310_015

Chapter 12 here, and in Krivochen (2015a, 2016a, 2018) and Krivochen & Schmerling (2016a, b), according to which coordinated structures can be internally complex (*et-coordination*) or internally opaque (*que-coordination*). Because of its relevance in the arguments in favour of discontinuous structure and multidominance, we will focus on RNR among the processes mentioned above. As observed in Postal (1998: 97) RNR is a *phenomenon*, not a rule: indeed, under present assumptions, all the 'transformations' in **Chapter 13** are. In RNR structures, there is a constituent that is shared between coordinated terms and which surfaces in a position to the right of all these terms in an ATB manner: the rightwards constituent needs to receive an interpretation in all coordinated terms, and all gaps must be coindexed, as we can see in the examples in (388) (from Postal, 1998: 97; De Vos and Vicente, 2005: 98):

(388) a. Ernest suspected e_i, Louise believed e_i, and Michael proved e_i [that she was guilty]$_i$.
b. I know a man who loves e_i and a woman who hates e_i London$_i$

What is the mechanism underlying this phenomenon? Grosz (2015) distinguishes three kinds of syntactic approaches to RNR (see also Sabbagh, 2014): (i) backwards deletion, (ii) rightwards ATB movement, and (iii) multidominance. Baltin (2006) includes a fourth alternative, originally suggested in Kayne (1994: 78, ff.), according to which there is leftwards movement of the object modified by the extraposed term (so as to have the phrase marker comply with the apriorisms of antisymmetry; more generally, Kayne formulates a ban on right-adjunction which has as a consequence that relative clauses must be complements to D; see Borsley, 1997, 2001 for critical discussion). There are also mixed proposals (e.g., Barros & Vicente, 2012). Deciding between those is not a straightforward matter, and we will not spend much time arguing *against* some account or another, but rather *in favour* of the kind of approach we defend.

First and foremost, we need to characterise the configurations that license RNR. Something that all accounts agree on is that in order to have RNR the terms of coordination need to be probed into, be it by a deletion rule, a movement rule, or the establishment of a direct dependency. In the light of the discussion in **Chapter 13**, this means that RNR arises in *et-coordinated* structures, symmetric or asymmetric (the examples in (388), for instance, are both *symmetric*), since *que-coordinated* structures are internally opaque (which is the main factor in triggering singular agreement). The set of elementary graphs and ρ-set for a simple case like (389) (arguably, an *asymmetric et-coordination*) are given, minimally simplified, in (390):

(389) Bill bought and John washed those china dishes

(390) Elementary graph 1: [Bill bought dish]
Elementary graph 2: [John washed dish]
ρ_{derived} = ⟨(and, buy), (and, wash), (buy, Bill), (buy, dish), (wash, John), (wash, dish)⟩

The composition of elementary graphs 1 and 2 reduces the number of nodes in the derived graph: nodes with identical addresses (here, ⦃dish⦄ occurs in the elementary graph of *buy* and of *wash*) are collapsed into one (see also Sarkar & Joshi, 1997: 614). In this respect, the mechanisms that give us RNR do not differ from the general case of graph composition, under our assumptions: all we needed to do is identify in the structural description of (389) the elementary graphs. Structure sharing takes care of most of the work. Because we know that an elementary graph contains a (possibly modified) saturated predicate, then ⦃dish⦄ must occur in the elementary graph anchored by both lexical predicates. The present account shares much with the multidominance approach (which is not a unified front, however), and specifies that multidominance is a consequence of looking at derived graphs as the result of graph union, where each elementary graph is a unit of argument structure. In the terms of Sabbagh (2014), ours is a *pivot-internal* analysis, in that the RNraised term belongs to one of the terms of the coordination (which is required by the definition of elementary graph) as opposed to appearing outside of the coordination (as in movement analyses); other analyses in this vein include Wexler & Culicover (1980), McCawley (1982) Levine (1985), Blevins (1990), among others.[1]

---

1 An interesting piece of evidence in favour of a pivot-internal analysis comes from apparent P-stranding in Spanish: unlike English, Spanish does not allow for P-stranding in *wh*-interrogatives, relatives, Heavy NP Shift, etc.:
(i) Who are you talking to?
(ii) *¿Quién estás hablando con?
McCloskey (1986) notes that Irish (a language that, like Spanish, does not allow for P-stranding in wh-interrogatives or relatives) allows for transitive prepositions to surface complementless in RNR contexts. Spanish behaves in the same way:
(iii) Le prestaste dinero a, pero no lo recibiste de, tu mejor amigo
'You lent money to, but didn't receive it from, your best friend'
Under a movement analysis of RNR, one would be forced to conclude that only some instances of rightwards movement allow for P-stranding, while no L-extractions or other R-extractions do (Heavy NP Shift, Extraposition, etc.). If the RNraised term never leaves either coordinand (i.e., if RNR is not syntactic movement), these facts follow.

There are interesting differences between our analysis and the DG treatment of RNR in Osborne (2019: 324, ff.). In his view, a RNraised object is only a dependent of the second conjunct. Thus, he argues against a dependency tree like (391), where the RNraised term is shared:

(391)

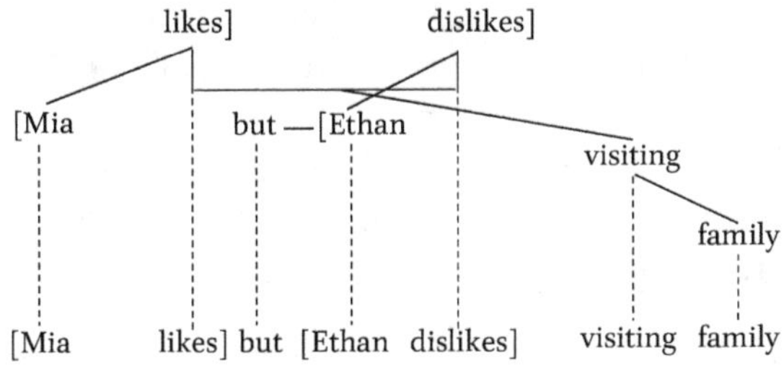

FIGURE 14.1 Dependency Grammar analysis of RNR, rejected in Osborne (2019)

Against (391), Osborne says

> *the analysis would have difficulty accommodating certain coordinate structures in which the matched words would be distinct in syntactic category,* e.g. [Bill **supports**], and [Susan has been speaking in favor **of**], the bill (Op. Cit., p. 325)

That is a fair point, although it depends on the height at which the dependency is established: both the transitive V and the preposition subcategorise for a nominal; thus, the local dominance relations ⟨of, bill⟩ and ⟨support, bill⟩ are legitimate and part of well-formed elementary graphs. But Osborne presents a second argument in favour of a second-conjunct-only dependency analysis: in a sentence like

(392) Bill thinks about his chances and that he has to try (Osborne, 2019: 326)

the clause headed by the complementiser cannot be a complement of the preposition, since *about* does not allow for finite clausal complements (cf. **Bill thinks about that he has to try*). However, this is only a problem if the *that*-clause is analysed as a dependent of *about*, which is an unwarranted assumption. There would be two elementary graphs, anchored by *think*. One of these, however, would take an oblique argument (*about his chances*), and the other a finite complement clause (*think that he has to try*). There seems to us to be no reason to have the coordination *under* the preposition.

A question that arises at this point is how RNR relates to other grammatical phenomena in English: are the mechanisms underlying RNR common to other structures? In this respect, Postal (1998: Chapter 4 and Appendix B) argues that RNR is analogous to L-extractions (leftwards extractions; this label includes *wh*-movement, topicalisation, etc.) and thus should be accounted for using the same mechanism. This approach predicts that RNR should not be clause-bound, because L-extractions are not (as in the case of long-distance *wh-movement*, which can span multiple clauses). In this sense, as long as we keep *et-coordinating*, we can RNraise without boundaries: *et-coordination* is indeed recursive and monotonically so. Thus, the availability of 'successive cyclic' RNR is only to be expected. For example:

(393) Bill bought e_i, John washed e_i, Mary dried e_i, and Susy carefully put away e_i, those delicate china dishes$_i$

In contrast to L-extractions, which in non-ATB contexts are restricted to single elementary graphs (Kroch, 2001; Frank, 2006), RNraised objects always link several elementary graphs: this is a crucial difference in configurational terms that our approach captures straightforwardly. This can be refined, in fact: the crucial distinction seems to be ATB dependencies (which are the result of structure sharing between several elementary graphs) vs. dependencies within a single elementary graph (as in *wh*-movement, focalisation, etc.), and which behave as unbounded rules (Postal, 1972).

There are indeed some common constraints between L-extractions and RNR, to get a full perspective on which we refer the reader to Postal (1998). An important difference between these two phenomena, however, is their sensitivity to island constraints (Wexler & Culicover, 1980). De Vos and Vicente (2005) point out that RNR, unlike L-extractions, do not display CNPC islandhood effects: compare (393) above with (394):

(394) *Which city$_i$ do you know a man who loves e_i and a woman who hates e_i?

However, it seems to be incorrect, or at least rushed, to say that RNR is 'island insensitive' without further qualifications (cf. De Vos & Vicente, 2005: 98, who point to the 'island insensitivity' of RNR under coordinated structures): McCawley (1982: 101) and Postal (1998: 121, ff.) observe that the CSC applies to RNR cases just as much as it applies to L-extraction ((395a) and (395b) are taken from McCawley, 1982: 101):

(395) a. Tom is writing an article on Aristotle and Freud, and Elaine has just published a monograph on Mesmer and Freud
b. *Tom is writing an article on Aristotle e_i, and Elaine has just published a monograph on Mesmer e_i [and Freud]$_i$
c. *Who$_i$ is Tom writing an article on Aristotle (and) e_i and Elaine has just published a monograph on Mesmer (and) e_i?

This fact must be taken carefully, for in and of itself it does not provide unambiguous support for a movement, deletion, or multidominance analysis. As far as we are concerned, Postal is right in saying that

> *the CSC is, of course, a fundamental condition on L-extractions* (Postal, 1998: 121)

However, this does not solve the problem of *what* the mechanism behind extraction actually is: if our approach to coordinated structures in **Chapter 12** is on the right track, then CSC effects can be captured in a no-movement constraint-based model just as much as in a transformationally-enhanced derivational model or even a declarative approach with sponsor-deletion arcs, etc.

In the context of this monograph, the distinction between operations that change relations vs. operations that change linear order was crucial; RNR was indeed one of the original order-changing processes in McCawley (1982). McCawley's 'order-changing' approach to RNR can be traced back to Wexler & Culicover (1980: 301) (also cited in Postal, 1998: 102)

> [A] *raised node* [by RNR] *always behaves, vis-à-vis all constraints on analyzability, just as it would if it were in its original underlying position. Hence, whereas it is apparently possible to apply* RNR *to a constituent of a relative clause, if we then try to analyze this raised node, we find that it acts as though it were still within the relative clause*

In our terms, this approach translates as the raised node being dominated by a node in each of the coordinated terms: grammatical relations assigned at the level of elementary graphs are, in this way, preserved under RNR. We must highlight at this point that RNR does not seem to make the targeted object internally opaque (which, in the present context, indicates that the RNR'd object is not *self-contained*): we are referring here to what Postal calls *interwoven dependencies* and De Vos & Vicente refer to as *CoRNR* ('Coordinated structures under RNR'). Let us illustrate the relevant structures:

(396) a. I didn't say that John had talked and that Peter had replied in a loud voice and with a whisper (De Vos & Vicente, 2005: 98)
b. John loves and Mary hates oysters and clams, respectively (Postal, 1998: 134)

Postal correctly observes that L-extractions allow for interwoven dependencies (if no other constraint is violated[2]), as in (397):

(397) [[Which nurse]$_1$ and [which hostess]$_2$]$_3$ did Fred date t_1 and Bob marry t_2, respectively? (taken from Postal, 1998: 134, ex. (109b). Indices and traces are his)

In principle, the availability of interwoven dependencies means that the relevant configurations do not involve *self-contained* objects in either case: if the L-extracted object (3) was self-contained, it should be internally opaque and thus not allow any of its sub-graphs (1 and 2) to establish a dependency outside of it; however, (397) proves that conjuncts 1 and 2 are independently accessible. The CSC is *not* violated in (397) because whatever rule we assume yields *wh*-fronting does apply Across the Board (to all terms of the coordination): as soon as one of the object NPs stays *in situ*, the structure becomes ungrammatical for all the usual reasons (e.g., **Which nurse did Fred date and Bob marry which hostess?*). Crucially, the S coordination in (397) is a case of *symmetric et-coordination*; neither term *Fred date NP* and *Bob marry NP* is embedded into the other. It may be useful for the reader if we used a *stemma*-like diagram (inspired by Tesnière, 1959) to see what is going on in (397) in terms of dependencies (see also McKinney-Bock & Vergnaud, 2013: 229, ff.):

2 This is an important caveat: some examples presented by De Vos & Vicente (2005) are ungrammatical not because of some problem with coordination itself, but because of an unrelated violation. For example, their example (5c) (reproduced below as (i)) can be taken to illustrate a *negative island* (given the fact that the extraction sites are c-commanded by NEG; see Ross, 1984; Rooryck, 1992; Abrusán, 2014 for a variety of perspectives. In Abrusán's approach, which we adhere to, the condition that there be a single exhsuative, most informative answer is not met in so-called negative islands, which results in their marginality) rather than L-extraction being incompatible with interwoven dependencies:
(i) *[How] didn't you say [[that John had talked *e*] and [that Peter had replied *e*]]?

(398) 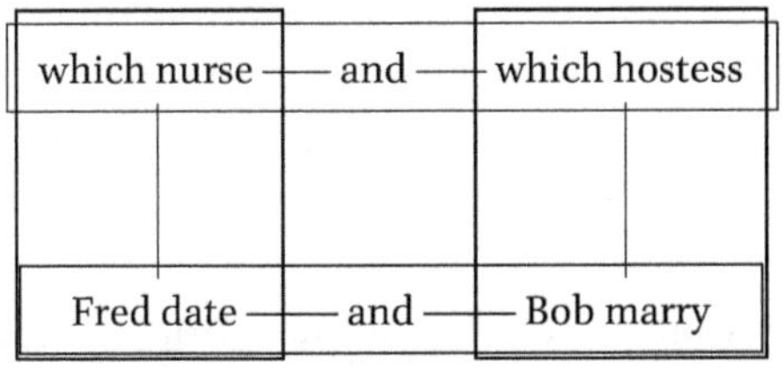

FIGURE 14.2 Stemma-like analysis of interwoven coordination

(398) visually illustrates that there is *no relation* between *Bob marry* and *which nurse*, or between *Fred date* and *which hostess*. We have two elementary graphs, because there are two lexical verbs with their respective nominal dependants and in the derived graph there are no crossing dependencies from one arbor probing into the other: locality is respected throughout. All syntactic dependencies can be defined within the limits of a single elementary graph, as required by the *Fundamental* TAG *Hypothesis*, which we borrowed. In addition to these, we have two symmetric coordinations, *Fred date and Bob marry* and *Which nurse and which hostess*. That means that, in total, we can identify four single-rooted syntactic objects, to make things simpler (which we have boxed in (398) above): these are *arbores*, but not elementary graphs. Let us provide the ρ-sets for them (we will use numerical subindices to distinguish between *which* in *which nurse* and *which hostess* for expository purposes only):

(399) Arbor 1 = ⟨(date, Fred); (date, which_1); (which_1, nurse)⟩
Arbor 2 = ⟨(marry, Bob); (marry, which_2); (which_2, hostess)⟩
Arbor 3 = ⟨(and, which_1); (and, which_2)⟩
Arbor 4 = ⟨(and, date); (and, marry)⟩

The sets of dominance relations in (399) capture what in our opinion are the crucial features of (395): first, that there are no *crossing* relations at the level of syntactic dependencies (cf. the examples analysed in **Section 4.3**); second, that we are dealing with *symmetric* coordination. Indeed, as Postal claims, RNR behaves in the same way, *mutatis mutandis*:

(400) Fred dated and Bob married a lovely nurse and a gentle hostess, respectively

Both RNR and L-extractions are in principle unbounded processes in English. This means, in this context, that as long as the structural conditions specified above hold (ATB rule application in *et-coordinated* structures), we can make the structure grow monotonically without violating any syntactic condition (although things get harder for humans to parse; this is orthogonal to the discussion in this monograph). For example, we can have the following rather extreme example of RNR which is by no means reader-friendly, but perfectly grammatical:

(401) Frank didn't admit to t_1 that he could t_2 nor deny to t_1 that he should t_2 [hire (t_1), train (t_1), and deal with t_1 as an equal]$_2$ nor did Glen reach any agreement with t_1 [that angry middle-aged person]$_1$. (taken from Postal, 1998: 199. Traces between brackets are ours, added for expository purposes only)

However, as we have already said, these considerations do not provide unambiguous evidence in favour of any particular treatment of RNR: it may still be multidominance or movement, as long as it is the same operation involved in L-extractions. Postal (1998: 178) claims that '*the base-generation/in-situ view is of course entirely impotent for those RNR structures that involve what are called interwoven dependencies*', but this claim needs to be appropriately restricted: Postal makes this rather strong claim in the context of a discussion of Bošković (2004) (which was then a manuscript not widely available). We must emphasise that the limitations recognised by Postal hold only for base-generation hypotheses which *also* assume SMC-complying phrase structure trees (see, for instance, Kayne, 1994: 67–68, who supports Wexler & Culicover's 1980 deletion analysis and rejects the multidominance analyses in McCawley, 1982 because the latter is not compatible with the basic assumptions of the antisymmetric enterprise about the format of structural descriptions).

To summarise our position, it seems to be the case that L-extractions and RNR behave in the same way under specific structural conditions: namely, *et-coordinated* structures to which a rule applies *Across the Board*. In that respect, we agree with Postal in that L-extractions and RNR are derived *by the same means* (see also Sarkar & Joshi, 1997), but in the present view those means are *multidominated nodes*, not relation-changing movement. It is not clear whether Postal's arguments indeed show that L-extractions and RNR change *constituency*, particularly considering that in APG or MG there are no constituents as such (and recall that RNR is, in McCawley's view, a rule that only changes linear order; just like *parenthetical insertion*). What seems clear, and is crucial for our approach, is that *neither RNR nor L-extractions modify existing grammatical relations*.

In this context, it would be a mistake to group RNR with other rightward displacements (in the sense that there is no unified category of 'R-extractions'; this is however also the case for L-extractions in the detailed analysis of Postal,

1998). Specifically, the rules of Relative Clause Extraposition (RCE) and Heavy NP Shift do not work in the same way that whatever process underliying RNR does (examples from McCawley, 1998: 529. Annotations are ours):

(402) a. That someone exists [who can beat you up to a pulp] is a foregone conclusion
b. *That someone exists t_i is a foregone conclusion [who can beat you up to a pulp]$_i$ (via Relative Clause Extraposition)

(403) a. That John sent to his mother [the money that you wanted him to give us] is understandable
b. *That John sent to his mother t_i is understandable [the money that you wanted him to give us]$_i$ (via Heavy NP Shift)

What do the structural descriptions of these sentences look like, and how do they differ from RNR?

In the McCawley-Levine view, to which we mostly adhere, RNR is a process which does not change grammatical relations (objects before RNR are still objects after being RNraised; only linear order is disrupted); in this sense it does side with L-extractions. In contrast to the unboundedness of RNR, we get heavily constrained displacement in the rightward displacement operations RCE and Heavy NP Shift, which are both cyclically bounded operations: specifically, S is a bounding node for Extraposition from VP and VP is a bounding node for Extraposition from NP (Kroch & Joshi, 1987: 132).[3] RNR preserves grammatical relations *across* elementary graphs in derived structures, satisfying the requirement that an RNraised object be present in all elementary graphs (which results in RNR being an unbounded operation, as pointed out above), whereas RCE and Heavy NP Shift change order *within* an elementary graph: we believe this to capture the core observation behind the Right Roof Constraint.[4] That is, not just that RCE and Heavy NP Shift are *cyclic* rules in the classical MGG sense, but that they *cannot be successive-cyclic*. In procedural terms, the problem seems to be one of derivational timing (pre- vs. post- vs. last-cyclic rules; see e.g. Ross, 1967:

3 A successive-cyclic approach to L-extractions, incidentally, runs into the problem that the set of bounding nodes for Extraposition is larger than the set of bounding nodes for L-extractions and RNR-yielding rules if all these are assumed to be derived by the same means, whichever these are (Kroch & Joshi, 1987: 132; Baltin, 1981).

4 In Ross' (1967: 307) terms,

> *Any rule whose structural index is of the form ... A Y, and whose structural change specifies that A is to be adjoined to the right of Y, is upward bounded.*

285, ff.) and the interaction between strictly locally cyclic rules and what presumably are successive cyclic rules, and it is not clear how to formulate those under the present assumptions: because there are no derivations in there is no 'timing' or rule ordering. The relations of bleeding and feeding must be formulated, if needed at all, in terms of material implication (as pointed out in Postal, 2010: 7). A different approach must be pursued, which reformulates the relevant conditions in terms of node admissibility conditions (see also McCawley, 1968). There is one possibility: because we have kept *root nodes* (as nodes that are not part of the ρ-domain of any other node; there being no reason to ban these to begin with), it is possible to specify that certain 'transformations' apply within the ρ-domain of a root. In other words, certain admissibility conditions make reference to a designated node and possible relations that can be established with said node in graphs that belong to the grammar. It is a good opportunity to re-evaluate Emonds' (1970) typology of transformations, given the importance of root phenomena. In this sense, we could informally propose the following descriptive classification:

(404) a. Processes targeting the closest root (i.e., the root node of the local graph containing the target of the process)
b. Processes targeting the matrix root (i.e., the root of the derived graph, after all linking operations have applied)
c. Structure-preserving processes

Recall that in a derivational framework

> [...] *A phrase node X in a tree T can be moved, copied, or inserted into a new position in T, according to the structural change of a transformation whose structural description T satisfies, only if at least one of two conditions is satisfied: (i) In its new position in T, X is immediately dominated by the highest S or by any S in turn immediately dominated by the highest S. (A transformation having such an effect is a* ***root transformation.***) *(ii) The new position of X is a position in which a phrase structure rule, motivated independently of the transformation in question, can generate the category X. (A transformation having such an effect is a* ***structure-preserving transformation***) (Emonds, 1970: ii. Highlighted in the original)

Let us insist on the fact that there is no movement, copy, or insertion in the theory developed in this monograph; however, a translation of Emonds' insights into the present approach seems possible. We can take a first, informal stab at it:

> *A process P* [read: a construction, we maintain Emonds' terminology] *may link a non-root node v_i in a single-rooted graph G to a node v_j in a single-rooted graph G' in* ***either*** *of the following ways:*
>
> I. *P creates an edge between v_i and v_j where*
>   a. *$⦃v_i⦄ \neq ⦃v_j⦄$ and $⦃v_j⦄$ is the root of G',* ***and***
>   b. *G' is the smallest arbor that properly contains G,* ***or***
>
> II. *P creates an edge between v_i and v_j where*
>   a. *$⦃v_i⦄ \neq ⦃v_j⦄$ and $⦃v_j⦄$ is the root of G',* ***and***
>   b. *There is at least one arbor G" such that $G \subsetneq G'' \subsetneq G'$,* ***or***
>
> III. *P creates an edge between v_i and v_j where*
>   a. *$⦃v_i⦄ = ⦃v_j⦄$*

Note that '=' is defined in terms of having identical addresses, which entails also having the same semantic value. It should be apparent that *condition I* refers to processes of the kind (404a), in which *strong cyclicity* comes into play (each arbor counts); *condition II* refers to processes of the kind (404b), in which only the *last cycle* is relevant; and *condition III* refers to processes of the kind (404c), in which graphs are connected by means of identifying common expressions (the most general version of structure sharing). In the context of the present discussion, we are interested in I and II. Of these, only I instantiates the classic *cyclic principle* in its strongest version: a root phenomenon cannot 'jump' across a root. II, in contrast, pertains to a different class of transformations, *last-cyclic* or *higher-trigger cyclic* (Postal, 1972: 212). To say that Relative Clause Extraposition and Heavy NP Shift are transformations of the kind (404a) amounts to establishing I as an admissibility condition over graphs corresponding to the structural descriptions of sentences displaying RCE and Heavy NP Shift. On the other hand, we have *wh*-movement and RNR as examples of constructions whose relevant admissibility condition is II; that gives them their unbounded character (without requiring *successive cyclicity* as an additional assumption). In a sense, I–III refine Emonds' original distinction, because it seems to be significant whether a process can only target the immediate root (the root of an elementary graph or arbor) or whether it can link a node with a 'remote' root (by applying after all linking operations have applied, and therefore operating on *global* rather than *local* structural terms).

We are aware that, as it is formulated, the set of conditions above is an *ad hoc* stipulation that captures the behaviour of some English rightwards extractions, but what it follows from, if anything, is still unknown and left for further research.

14.2 Deletion without Deletion

Bach (1964: 70) lists the possible things that PSGs and transformations can do, in a very general (variable-free) format:

(405) a. Delete: $a + b \rightarrow b$ (or $a \rightarrow \varepsilon$)
b. Replace: $a \rightarrow b$ [PSG]
c. Expand: $a \rightarrow b + c$ [PSG]
d. Reduce: $a + b \rightarrow c$
e. Add: $a \rightarrow a + b$ [PSG]
f. Permute: $a + b \rightarrow b + a$

Note that the cases where there is a single symbol in the left-hand side are within PSG power, but operating over more than one symbol entails rewriting a sub-tree: in turn, to do this we need to make reference to the derivational history of a sequence, and this requires the additional power of transformations (Chomsky, 1957). So far, we have dealt with constructions that have been traditionally described as involving additions, permutations, replacements, reductions, and expansions (in one way or another). However, we have said nothing explicitly about *deletion* (but see **Section 6.3–6.4** for an deletion-less account of what classically would have been *Equi-NP deletion*). So, the question is: what do we do with *deletion* operations? That is, how do we capture the phenomena that in transformational grammar are accounted for by means of a rule that replaces a non-null string by ε (either at the syntax or at PF)? First of all, let us give the reader an idea of the kinds of transformations we have in mind:

(406) a. **Gapping** (V ellipsis)
Luke drank juice, and Ellen ~~drank~~ beer
b. **Sluicing** (TP ellipsis)
Jane drank something, but I don't know what ~~Jane drank~~
c. **Bare Argument Ellipsis** (see Culicover & Jackendoff, 2005: Chapter 7)
A: What has Mike been drinking?
B: Beer, with Andy
…

It is crucial to bear in mind that, just like *movement* or *extraction*, we are using *deletion* in a purely descriptive sense, attending to the long history that the term has in generative grammar and its descriptive value. In a framework like the

Standard Theory (in which each operation needed to specify input and output as structural description and structural change, respectively), we can define a *deletion* rule as any rule that replaces at least one non-zero index in a structural description by zero. That is, roughly:

> *If the structural index of a transformation has n terms $a_1, a_2, a_3, \ldots a_n$ it is a* ***deletion*** *rule iff*
> i. *its structural change has any a_i replaced by o, and*
> ii. *a_i is neither adjoined to some a_j nor is it substituted by some a_k in the structural index*

In this sense, for instance, Ross' (1969: 267) formulation of *sluicing* (see (412) below for examples) is a deletion transformation: note how the indices 7 and 9 are replaced by o

(407)

W–	[X–	[-Def]$_{NP}$ – Y]$_S$–	Z–	[NP–	[$_S$ X–	(P)–	Y]]$_S$–	R–	
1	2	3	4	5	6	7	8	9	10
								→ OPT	
1	2	3	4	5	6	o	8	o	10

Condition: 2 = 7
4 = 9
6⌒7⌒8⌒9 is an embedded question

Note that the conditions for *sluicing* to apply include identity constraints: deletion operations apply to sets of objects (usually in pairs), and deliver structures where some of those objects receive a null exponent or are replaced by an internally unanalysable pro-form. Subsuming at least some deletion phenomena under structure sharing requires a different perspective: identity conditions must be explicitly formulated, as in transformational approaches, but the output of structure sharing applied to any set of objects which count as identical for all purposes of the grammar is always a single object (in our case, a single node as per graph union). The simplest and most restrictive hypothesis under current assumptions is that only structure sharing is available for 'deletion', since it is independently required in the grammar, and would have to be somehow blocked (possibly, by refining the definition of 'identity' that is relevant for structure sharing).

A specific characterisation of each of the rules in (406) is (although necessary in the long run) beyond the breadth of the present work, but a general characterisation of the class to which those rules belong is not. This general characterisation will necessarily overlook details and quirks of each rule, but we are confident it will provide some general guidelines for the interested reader to pursue a thorough study of any of the aforementioned processes.

A general requirement on deletion operations is that '*all syntactic deletion be recoverable*' (Hankamer, 1979: 2), which means simply that we *cannot* 'delete' syntactic objects such that afterwards (a) there is no indication that a rule has applied at all, and (b) we have no way to know what has been deleted (see also Katz & Postal, 1964: 80). This requirement will be important, as a reality check on the structural descriptions we propose: are relations preserved after 'deletion' (including structure sharing) such that a complete propositional form can be built?[5] The 'recoverability' approach (if taken perhaps too literally) entails, however, that at some point there is a fully overt terminal string, some parts of which are deleted in specific licensing configurations (e.g., by bearing a designated feature in a specific relation with a licensing head; as in Merchant, 2001; see Johnson, 2014 for an overview of *recoverability* conditions; Lasnik & Funakoshi, 2019 for an overview of MGG approaches to ellipsis). Put differently, syntactic deletion approaches posit that cases of ellipsis (gapping, stripping, and other deletion operations) are derived from underlying strings where no deletion has taken place (Lasnik & Fukanoshi, 2019). As emphasised before, deletion operations are mappings from pairs to pairs: the input to a deletion operation is a pair of syntactic objects, and its output is a pair of syntactic objects one of which does not receive a phonological exponent (but maintains internal structure, as in the case of VP ellipsis), or is replaced by an internally opaque pro-form (as in Null Complement Anaphora). The syntactic approach has not been uncontested: in some semantic-pragmatic approaches there is no silent structure at all in ellipsis (Culicover & Jackendoff, 2005; Jacobson, 2012; see van Craenenbroeck & Temmerman, 2019: for an overview), with recover-

5 It is possible that the requirement of recoverability needs to be relativised to specific languages. For example, Dasgupta (2021: 13) provides the following examples from Bangla that defy traditional conceptions of recoverability:

(i) *Kumbhokarnero* (***jā***) *ghum bhāngiye debe* ***èmon*** *āwāj* (Kumbhakarna's.even **which** wake up will **such** noise), '**such** a noise **which** will rouse even Kumbhakarna from his slumber' (the *wh*-operator ***jā*** may be deleted, but the demonstrative ***èmon*** remains, allowing—through mechanisms that need to be formalised even in MGG—for recoverability)

(ii) (***jā diye***) *bastā kāṭā jāy èmon kẫci* (which with sack cut can.be such scissors), 'such scissors which we can cut sacks with'

(iii) (***jā dekhe***) *cokh dhẫdhiye jāy èmon ujjal ālo* (which seeing eye dazzled are such bright light), 'such a bright light on seeing which one's eyes are dazzled'

Under more or less widely assumed approaches to recoverability, in (ii) and (iii), where both the *wh*-operator and its modifying demonstrative may be deleted there is not enough information in the string to recover an underlying structure.

ability being a semantic and/or discourse process (see fn. 6 of this chapter for some discussion of Jacobson's proposal).

The standard MGG syntactic view commits us to a derivational approach: at derivational step *n* we have a terminal string with no null symbols, and at step *n+1* some terms are 'erased' (replaced by ε) by a rule that must have access to the derivational history of a sequence (given the restrictions on rewriting imposed in Chomsky, 1959: 143). Alternatively, a node may be marked by an arbitrary feature that prevents the assignment of phonological exponents to everything that node dominates, or deletes these exponents at PF (depending on whether late insertion is assumed or not) (see Merchant, 2008). Is it possible to eliminate the assumption that sentences displaying *deletion* (now used as a descriptive term only) are transformationally derived, and provide a declarative account of the relevant effects? A comprehensive answer to this question goes far beyond this monograph. However, we can sketch a deletion-free analysis within our graph-theoretic approach, as evidence that, in principle, it is possible to have a deletion-less analysis of deletion in our framework.

Above, in **Section 2.2** we briefly mentioned Sarkar & Joshi's (1997) TAG approach to *gapping*: if elementary trees are graphs whose nodes are assigned uniquely identifying addresses, an operation they call *contraction* allows for the identification of common addresses in the composition of elementary trees like *John gave Mary a book* and *John gave Susan a flower*: the common addresses here are *John* and *gave*. Composition of these local graphs, under Sarkar & Joshi's assumptions, and our own, yield *John gave Mary a book and Susan a flower* (in which position in a walk through the derived tree the nodes to which common addresses correspond are materialised is, of course, a problem: Sarkar & Joshi follow Ross' treatment of *gapping* and assume that the lexical anchor *gave*, and its subject, are realised in the first conjunct). The elementary tree proposed in Sarkar & Joshi (1998: 614) for a ditransitive construction is the following:

(408)

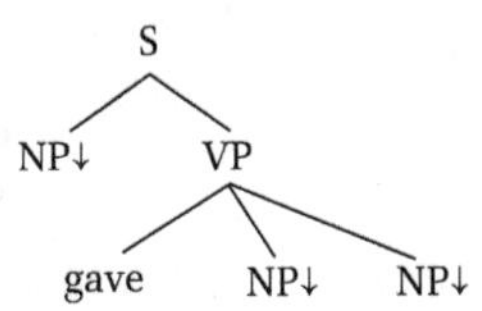

FIGURE 14.3 Anchored elementary tree for a ditransitive predicate

The nodes marked with downward arrows indicate *substitution* sites: because Sarkar & Joshi allow lexical NPs to be independent elementary trees (as Frank,

2002, 2013 and most work in LTAGs do), the subject, direct object and indirect object in the DOC must be introduced in the initial tree (408) via substitution. In turn, the elementary tree (408) is inserted twice in the symmetric coordination schema (409) below, once per object-oblique combination. In (409), X stands for any category label:

(409)

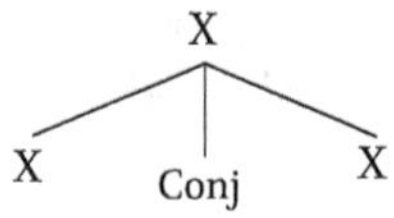

FIGURE 14.4 TAG coordination schema (Sarkar & Joshi, 1997)

After substitution and identification of common addresses, the derived tree for *John gave Mary a book and Susan a flower* is (410) (repeated from **Chapter 2**), where the effects of deletion are accomplished via structure sharing:

(410)

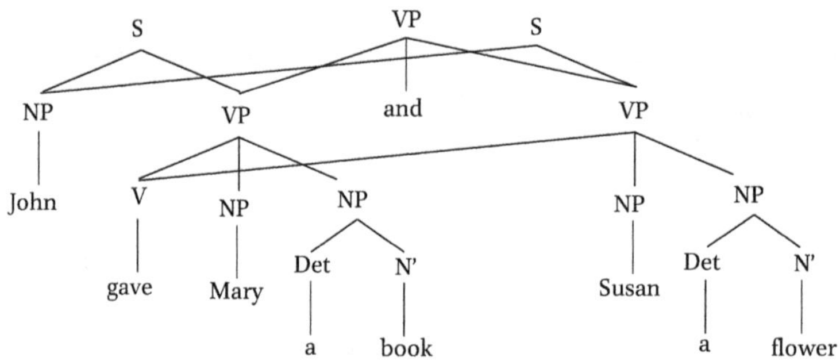

FIGURE 14.5 Derived tree with structure sharing for non-constituent coordination

We can define a minimally simplified ρ-set for *John gave Mary a book and Susan a flower* as follows:

(411) Elementary graph 1: [John gave book Mary]
Elementary graph 2: [John gave flower Susan]
ρ_1 = ⟨(give, John), (give, book), (give, Mary)⟩
ρ_2 = ⟨(give, John), (give, flower), (give, Susan)⟩
ρ_{coord} = ⟨(and, give), (and, give)⟩
$\rho_{derived}$ = ⟨(and, give), (give, John), (give, flower), (give, Susan), (give, Susan)⟩

We assume, with Sarkar & Joshi, that there are two elementary structures, each of which corresponds to a conjunct: the anchor of each is the lexical predicate *give*. As part of graph union, nodes with identical addresses in distinct elementary graphs are identified. The simplest cases of *gapping* and *stripping* can be receive an analysis along these lines: they are accounted for in terms of graph composition and node contraction (i.e., the identification of common addresses in distinct elementary graphs). In these cases there is neither deletion nor recoverability: no node is deleted from a graph, nor is there an operation to 'undo' the effects of deletion.

This analysis of deletion is compatible with a number of other analyses, in terms of the generalisations that are being captured. The structure-sharing approach seems to work best for what Hankamer & Sag (1976) termed 'surface anaphora': morphosyntactically dependent deletion (including gapping, stripping, and sluicing). Hankamer & Sag's 'deep anaphora', insofar as it requires reference to a shared discourse model, is not evidently configurational (crucial to the Hankamer & Sag analysis is the idea that not all 'deletion' is generated equal; see also Kempson et al., 2019 for a mixed syntax/discourse/processing approach). Among the approaches that our analysis seems to pair well with is the analysis in Osborne et al. (2012), Osborne (2019) in terms of *catenae*. This analysis predicts that only material that is continuous with respect to dominance can be deleted: thus, in a case like Ross' classical (1970) example *I want to try to begin to write a novel and Mary a play*, the possible deleted expressions need to be *catenae*: *wants to try to begin to write*, *want to try to begin*, *want to try*, *want*. Following the analysis of complementation in **Chapter 6**, the verbs in Ross' example would indeed be nodes visited in a walk through the graph: there is a walk from *want* to *write* that passes through *try* and *begin* (these expressions thus constitute a *catena*). It is important to note that the conditions under which ellipsis is licensed vary cross-linguistically: as observed above, Spanish, unlike English, does not have generalised VP ellipsis. We will see in what follows that a *d-order* between expressions cannot be both necessary and sufficient.

In the remainder of this section we will focus on those structures where reconstruction results in an *illicit* structural description: the reason is that a transformational approach to these runs into problems more obviously than in those cases where recoverability of silent structure is a viable option. Another way to look at these is to say that deletion has applied to *repair* a syntactic violation. Consider, for instance, cases in which the remnant of *sluicing* does not correspond to any plausibly recoverable syntactic object:

(412) a. They want to hire someone who speaks a Balkan language, but I don't know
{ *which* / **which they do* / **which Balkan language they want to hire someone who speakst* }
(*wh*-island repair under sluicing; taken from Merchant, 2008: 138)

b. Harriet drinks scotch that comes from a very special part of Scotland, but I don't know where ~~Harriet drinks scotch [that comes from *t*]~~. (ungrammatical if reconstruction takes place; Complex NP Constraint violation)

c. Harriet either drinks scotch or smokes cigars, but I can't remember which ~~of drinks scotch or smokes cigars Harriet does~~. (Taken from Culicover & Jackendoff, 2005: 268)

Examples like these have been analysed from a variety of perspectives, a review of which is outside the aim this monograph. Some remarks are in order, though. It is interesting to observe that, should an approach to ellipsis along the lines of Jacobson (2012, 2019) be combined with our approach (see fn. 6 below), no silent structure should be assumed: in that case, the circumvention of island effects follows directly. The problem that arises, however, is to determine the syntactic configuration that licenses the semantic operations proposed by Jacobson. The simplest compatible analysis would assume no silent structure or structure sharing; rather, partial multidominance as in **Chapter 10** when applicable (e.g., (412a): *which* [*Balkan language*]). The problem is that such an analysis is not always available (e.g., (412b) requires us to consider *where* an independent syntactic object, possibly dominated by *from*). The structure sharing analysis, which also circumvents island violations, has different difficulties: it essentially amounts to a version of MaxElide (Merchant, 2008: 141), in that there's no choice but to structure share when possible. In the most radical interpretation, if something does not get structure shared, there must be a different address involved, and thus a different semantic value. The problem is that MaxElide has been criticised for both being too restrictive and too lax (Takahashi & Fox, 2005; Messick & Thoms, 2016, among others). Structure sharing provides a straightforward definition of structural and semantic *parallelism*, at least in terms of configuration. However, post-syntactic operations (e.g., focus assignment) may also influence the availability of deletion and recoverability. Furthermore, A-movement has been claimed to deliver structures where more than one ellipsis possibility is available (e.g., *John is likely to attend the party, and Mary* {*is* / *is likely to*} *as well*), which suggests that just structure sharing is too restrictive.

The aforementioned problems are only part of the reason why we have relegated processes of deletion to 'open problems and questions'. Deletion must be handled with care: Berwick (1984) makes a strong case for blaming the computational problems in the early Standard Theory on unbounded deletion, which in turn leads to the necessity to reconstruct a pre-transformational phrase marker (a Deep Structure) in order to build a semantic representation (see also Peters & Ritchie, 1973 for an early argument). Unbounded deletion, in essence, can make a surface string arbitrarily shorter than its corresponding Deep Structure, which makes structure reconstruction for purposes of semantic interpretation potentially intractable. This point had already been made—with an empirical rather than purely computational motivation—in Ross (1970a: 249):

> *Deep structures can contain elements or even whole clauses which do not appear in surface structure, and the order in deep structure of elements which appear in both levels of representation may be far different from the surface structure order of the same elements. Furthermore, it seems to be the case that eyen in apparently simple sentences, the transformational mapping between deep and surface structure is extremely complex—far more so, in fact, than has previously been thought. These facts make it extremely difficult to ascertain the nature of deep structure, and necessitate the use of long chains of inference to this end*

If we follow this line of reasoning, we can plausibly think that the problem is not *recoverability* per se, but rather *recoverability of deletion*. Under current assumptions, if deletion is an effect of structure sharing, identity of semantic value is a necessary condition. This is a strong claim—as noted before, almost certainly too strong—, but if taken at face value, recoverability follows automatically: node contraction in structure sharing entails identity of addresses across nodes in distinct elementary graphs. Identity of addresses in turn entails identity of semantic values. In this context, if structure sharing is *possible* then the optimal scenario is that in which it *must apply*. Additional conditions must be formulated to *prevent* structure sharing in the relevant cases; these conditions are currently being looked into. Much research pending, let us briefly explore what this perspective can give us. Consider to this end the following example of *sluicing*:

(413) They want to invite someone who knows $\left\{ \begin{array}{c} \textit{a} \\ {}^{*}\textit{every} \end{array} \right\}$ version of Dependency Grammar, but I don't remember which (= which version of DG; ≠ which person)

In (413) the *wh*-word has an antecedent in the structure: the simplest analysis would assign a structure such as *they want to invite someone who knows a version of* DG *but I don't remember which version of* DG. Now the problem can be stated in relatively familiar terms: syntactically, we have partial multidominance again, with *version* being dominated by the D and the *wh*-phrase.

Having an overt antecedent in the structure is not a *sine qua non* condition. For instance, let us look at (414):

(414) He's writing (something), but you can't imagine $\left\{\begin{array}{c} \textit{what} \\ \textit{where} \\ \textit{why} \\ \textit{how (fast)} \\ \left\{\begin{array}{c} \textit{to} \\ \textit{with} \\ \textit{for} \end{array}\right\} \textit{whom} \end{array}\right\}$
(Ross, 1969: 252)

In cases like (414) the *wh*-word has no overt antecedent, which simply means that there is no constituent in the sentence with which it would be 'coindexed'. For our present purposes, both (413) and (414) present the same property: as noted in Chung et al. (1995: 241) and Reinhart (2006: 67): the *wh*-word in sluicing can only 'correlate' with an existentially quantified phrase (Chung et al. 1995: 241 specify '*an indefinite or other weak DP*'). The modifiers in (414) *where, why, how fast*, etc. would be analysed here as dominating the anchor of the elementary graph *write*, just like any other modifier. Part of the appeal of a framework that incorporates GF at the core of the formalism is that some asymmetries between arguments and adjuncts with respect to deletion can be characterised in these terms, but even the data is controversial.

In order to address the issues raised by (413–414), we will briefly focus on some aspects of semantics and configuration in *wh*-interrogatives with resumption. Recall that in **Chapter** 10 we followed Reinhart (1998) in saying that *wh*-words grammatically instantiate a *choice function* CH: *wh*-interrogatives require the selection of a member of a set (which is also related to the semantic accounts of negative islands that are based on the non-existence of an exhaustive, maximally informative answer, as in Abrusán, 2014). And, following Montague (1973) and Karttunen (1977), **?** (analogous to Katz & Postal's 1964 Q) has scope over an indexed NP. We are interested in the structural description assigned to (413), and the reason why the choice of quantifier matters. In syntactic terms, it is not particularly difficult to account for the interpretation of *which*: if there is no deletion, we can have a multidominated object *a version of* *dependency grammar*, dominated immediately by ? and V. But the difference in grammaticality given the desired interpretation remains unexplained: we need to be able to say something more.

In this context, it seems rather natural to assume that an existential quantifier can appear in the scope of CH, but a universal quantifier cannot: indefinites can be viewed as 'restricted free variables' (Reinhart, 2006: 68; also Heim, 1982), which means in this context that they can unselectively be bound by an operator—including ?. We will not get too deep into the semantic consequences of this assumption, but rather analyse the impact that it has on the structure. With Chung et al. (1995), we assume that there is no deletion, but unlike them we do not assume a separate level of Logical Form with its own 'covert' syntactic operations. This assumption is only required, we argue, if structural descriptions must deliver *paths* between operators/antecedents and variables: it is relevant to point out that one of the basic assumptions that Chung et al. make is that phrase markers have the format of X-bar trees, and sluicing is a process whereby there is *wh*-movement to Spec-CP and C^0 and IP are phonologically null ((415) is taken from Chung et al., 1995: 242):

(415)

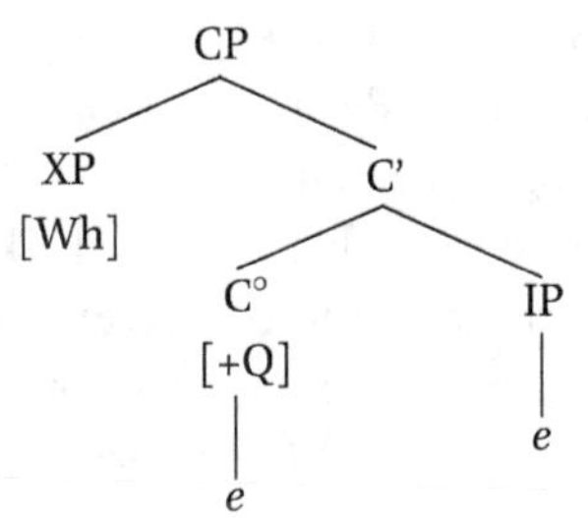

FIGURE 14.6 IP-deletion analysis of sluicing

If X-bar theoretic requirements on the format of phrase markers are dropped, the problem needs to be rephrased: it is not about constructing a Logical Form for the 'defective' structure in (415) from which the question type meaning can be determined; rather, we need to make sure that there is a well-formed graph that serves as the structural description to sluiced expressions of the language. In our view, there is no need to construct an LF representation from the surface structure of a sluiced sentence: if there is a legitimate graph that corresponds to a sluiced expression of the language, we will have everything we need at the level of syntactic description (along Jacobsonian lines). We may note that the approach to *wh*-movement developed here satisfies the specifications in Chung et al. (1995: 244):

> *On the syntactic side, the displaced constituent must syntactically bind a position within the IP complement of C^0. On the semantic side, the displaced constituent must contain a Wh-indefinite that is interpreted as a variable semantically bound by an interrogative operator. Third and finally, the displaced constituent must contribute to semantic interpretation just as if it were sitting in the syntactically bound position.*

The 'binding' condition is satisfied by the ?-replacement operation. The semantic condition is not a condition over structural descriptions, but rather over the expression itself (if indefiniteness is at least partly a matter of form). The final condition is satisfied by the specification of dominance relations in the graph: because what could amount to 'interpretation' is walking a *trail* and not a *path*, the 'displaced constituent' *is indeed sitting in the syntactically bound position*. The difference between *binder* and *bound* is simply one of structural context.

If we are going to propose that sluicing involves multidominance,[6] it is crucial to determine exactly *what* is being multidominated. More concretely: if we have *some version of Dependency Grammar* as the local antecedent for *which*, what is the syntactic object that appears in the contexts *e*⟨know, __⟩ and *e*⟨remember, __⟩? Specifically, is it the quantified NP *some version of dependency grammar* or only the lexical NP *version of dependency grammar*?

Let us assume that what gets multidominated is only the lexical NP, that is, that the representation 'bifurcates' (as it were) leaving the Q/D layer untouched (an idea we have hinted at throughout the monograph, specifically in **Chapter 10**, and which will be explored in more detail in **Section 14.4**). This

6 What follows is perhaps a conservative analysis, in that we assume that functions need to be saturated in a local structural domain (a natural assumption that follows from our definition of elementary graphs). Jacobson (2012, 2019), however, presents an interesting alternative (specifically focused on Antecedent Contained Deletion, the analysis of which has traditionally required LF movement to prevent an infinite regress; see **Section 14.7**): the idea that functions need to be saturated can be abandoned. In this context, there is no deleted structure to recover: the typed meanings of expressions are defined in such a way that the meaning of the remnant of deletion can be computed directly. This approach (see also Culicover & Jackendoff, 2005, whose analysis of sluicing proposes that the *wh*-phrase is a syntactic orphan, whose interpretation depends on contextual 'indirect licensing') gives a much more important role to the semantics, and says very little about syntactic configuration: expressions have more richly articulated meanings, with relations playing a more peripheral role. An implementation of Jacobson's analysis seems to us to require a revised definition of our elementary graphs (since by definition predicates' selectional requirements must be saturated within elementary graphs), which while possible, is not the road we will take.

assumption would have undesirable consequences: if the multidominated object excluded the quantifier, then both the existential quantifier and the universal quantifier should be grammatical since the *wh*-word would only have scope over the lexical N. In this context, it is useful to reproduce the translation rules for quantified NPs *every N* and *a/an N* in Montague (1973):

> *T2. If* $\zeta \in P_{CN}$ [a phrase of category Common Noun] *and* ζ *translates into* ζ', *then* **every** ζ *translates into* $\hat{P} \bigwedge x[\zeta'(x) \rightarrow P\{x\}]$, **the** ζ *translates into* $\hat{P} \bigvee y [\bigwedge x [\zeta'(x) \leftrightarrow x = y] \wedge P\{y\}]$, $F_2(\zeta)$ [i.e., **a**/**an** ζ] *translates into* $\hat{P} \bigvee x[\zeta'(x) \wedge P\{x\}]$. (Montague, 1973: 233)

We propose that a Montague-Karttunen representation of (413) could go along the (informal) lines of (416):

(416) A version of dependency grammar$_0$? they want to invite someone who knows it$_0$ but I don't know it$_0$

The indexed pronoun *it*$_0$, as in Montague (1973), stands for the generalised quantifier, not just for its lexical content. This is consistent with the translation of *wh*-words that we provided in **Chapter 10**:

(417) *which* N ≡ $\lambda P\ \exists(x)\ [N'(x) \wedge {}^{\vee}P(x)]$

This approach is also consistent with the requirements in Chung et al. (1995) in both syntactic and semantic aspects (see the quotation above). A potential advantage of the present system is that the graph-theoretic approach would eliminate the need to specify 'chains of inference', as a side-effect of eliminating *deletion* altogether as a class of operations over structural descriptions.

Needless to say, the topic of *deletion* is way too vast and complex for us to claim that the present approach works for all cases (particularly, phenomena like Bare Argument Ellipsis, as analysed in Culicover & Jackendoff, 2005: Chapter 7; see also Osborne, 2019: Chapters 12 and 13 for a DG view on ellipsis phenomena, which in his view involve deletion and null material). However, we expect to have provided enough arguments to consider a graph-theoretic approach as an empirically viable and theoretically consistent alternative to transformational accounts of sentence-bound deletion.

14.3 Long Distance Dependencies and Resumptive Pronouns

The model presented in this monograph requires some revision of the formulation of mechanisms and constraints involved in filler-gap constructions in general, and long-distance dependencies more in particular. We have given a programmatic introduction to the dynamics of *wh*-movement in **Chapter 10**, which is of course far from comprehensive. The most general condition for filler-gap dependencies is the satisfaction of the conditions imposed in the definition of *licensing* (see (206) above): informally, *licensing* aims at ensuring mutual *accessibility* between the grammatical objects involved in a referential dependency. In this context, consider a sentence like (418):

(418) Who$_i$ did John want to tell Mary that he had talked to $\__i$?

The structure of (418) is—essentially—monotonically recursive, with subordinate clauses being arguments of *want* and *tell*:

(419) [Who$_i$ did John want [to tell Mary [that he had talked to $\__i$]]]

In the single-visit analysis, licensing conditions are met trivially at the syntactic level, so let us consider the two-visit approach in order to say something more interesting. Given the tail-recursive nature of (419), there is no post-cyclic adjunction or anything of the sort that we must be on the lookout for: conditions *i* (the existence of a walk in which the relevant node transitively dominates itself) and *ii* (which pertains to root adjunction) in the definition of *licensing* are thus met. Now, what about condition *iii*, which filters out dependencies into self-contained graphs? If the structural description indeed features no adjuncts, then every node in the ρ-set of (419) is in the ρ-domain of *who* (including itself, transitively). This means that there is no self-contained domain in (419), and therefore licensing between 'filler' and 'gap' can take place without problems. So far, it seems, so good.

But let us now consider a different kind of example, slightly more complex:

(420) Which picture$_i$ did they all blush when John saw *(it$_i$)? (adapted from Kaplan & Zaenen, 1995: 139)

There is a fundamental difference between (418) and (420): only (420) features an adjunct, the *when* clause, which contains a bound expression. This contrasts with the monotonic nature of the structural description assigned to (418), in which—in phrase structural terms—*want* takes as a complement the clause headed by *tell* and *tell* takes as a complement the clause headed by *talk*. The * in (420) indicates that if the pronoun *it* is omitted, the sentence is ungrammatical. We can frame this more technically: *chopping* from the adjunct is not possible, but *copying* yields an acceptable output (we use these terms in the sense that they have in Ross, 1967: 427).

Following Ross, we assume that *chopping* and *copying* reordering transformations must be distinguished formally: here, we want to defend the idea that only if reordering of *which picture* is the product of *copying* (thus, if it leaves a 'copy' behind, in a sense that we will refine shortly) will we obtain a grammatical output. This is relevant because, in Ross' own words, '*chopping rules are subject to the constraints in Chapter 4* [i.e., the Complex NP Constraint, the Coordinate Structure Constraint, the Sentential Subject Constraint, the Left Branch Condition, etc.], *copying rules are not*' (1967: 428). In the case of (420), it is usual to invoke a constraint along the lines of an *Adjunct Condition* which bans extraction from a syntactic object SO if there is no subcategorisation involved in SO (e.g., Huang, 1982: 503; Johnson, 2003; see also Dalrymple et al., 2019: 656, ff.; Pollard & Sag, 1994: 384); the Empty Category Principle (ECP the requirement that a trace be properly governed; see Chomsky, 1981: 250; Lasnik & Saito, 1984) goes along the same lines. Transformational approaches define restrictions in terms of the relations between operators and variables, and specify the conditions under which these relations can hold. In this sense, they are in fact establishing conditions over the well-formedness of phrase markers, such that a phrase marker will be well-formed if and only if it violates no principle of the grammar. Nothing is said, however, about expressions of the language.

The theory of locality as conceived of in this work is based on two main concepts: *accessibility* and *locality*. *Accessibility*, as per the definition of *licensing* and *self-containment* aims at specifying the conditions under which bound occurrences of a syntactic object are legitimate. Conditions on *locality* pertain to those structures in which we have several lexical predicates with their corresponding functional modifiers and nominal dependants: more specifically, we are interested in defining the configurations in which we can *link* arbores (which in turn depends on whether the linking nodes are *accessible*: the notion of *self-containment* thus becomes relevant). In this context, the question is: what is the role of resumptive pronouns in expressions like (420)? First, we should define what we understand by *resumptive pronoun*. Descriptively, resumptive pronouns are pronouns that '*appear in positions where one would, in a certain sense, have expected to find a gap*' (McCloskey, 2006: 95): they spell out the tail of an A' chain or an A chain (the latter, only in Copy-Raising; see **Section 6.1.1**). This is important, theoretically, because it implies that the dynamics of resumptive pronouns are linked to whatever mechanism is assumed to account for displacement. Resumptive pronouns are obligatorily bound, unlike garden-variety pronouns (which can have exophoric reference). Resumption is a wider phenomenon than just resumptive pronouns, including for example epithet NPs (like *that poor idiot* in a sentence like *He$_i$'s a guy who we couldn't understand how that poor idiot$_i$ could get a job*); here we are only concerned with pronouns since we have given a programmatic account of *pronominalisation* in graph-theoretic terms: we can therefore ask whether resumptive pronouns constitute an exceptional phenomenon or not from the perspective of how these pronouns come to be. The null hypothesis is that resumptive pronouns are the result of *pronominalisation*, like any other *bound* pronoun in the grammar (see Postal, 1969; Hornstein, 2001: Chapter 5; Grohmann, 2003: Chapter 5; Kayne, 2002; Gärtner, 2014). However, explaining *why* resumptive pronouns appear in specific contexts is far from straightforward.

It is a usual claim that resumptive pronouns are somewhat of a *last resort* to rescue an otherwise ungrammatical structure: for some reason, an operator-variable relation that would have violated a constraint is saved from ungrammaticality if the variable is spelled out as a pronoun (e.g., Ross, 1967; Kroch, 1981; Shlonsky, 1992: 443; Boeckx, 2003; also Alexopoulou, 2010; Ackerman et al., 2018 from a processing perspective). Frank & Kotek (2022: 8) put it simply: '*the availability of a resumptive pronoun correlates with the presence of an island*'. We need to make some additional considerations, beginning with the fact that the various perspectives that have arisen since Ross (1967) about resumption as a last resort (transformational vs. base-generation origin of the resumptive pronoun, mainly) present similar problems when viewed from our perspective: in both cases it is necessary to specify the conditions under which a relation is legitimate that would otherwise violate a constraint; this relation may be operator-variable (if resumption is a form of *stranding*, as in Boeckx, 2003: 25, ff.) or co-indexing (if there is no movement rule involved in resumption, as assumed in Chomsky, 1977: 80–81). In either case there is a relation that needs to be appropriately characterised, including a specification of the structural descriptions where that relation can hold without violating a constraint of the grammar. Kroch (1981)—correctly, in our opinion—identifies some problematic issues with both movement and base-generation analyses of resumption; essentially, the rules that can be invoked in the grammar are either too restrictive or not restrictive enough. Furthermore, if pronoun insertion rules are ways to rescue sentences that would otherwise be violations of conditions like the ECP (as proposed in Kayne, 1981b), then they must be constrained to apply only when the ECP would be violated at a later derivational step; otherwise, pronoun insertion would overgenerate and replace traces with pronouns in cases where resumption is not necessary (e.g., *She's the only woman here that I know* (**her*) *very well*; see Kroch, 1981: 126). The derivational timing of resumption is also analysed in Frank & Kotek (2022).

It is important to note that these problems pertain to the *generative power* of the grammar, its capacity to recursively enumerate structural descriptions (Chomsky, 1965): things look very different if we consider constraints over *expressions* of the language instead of *generative functions*. In one case, there is an ordered set of rules that apply to derive a terminal string after a finite number of steps; in the other, it is necessary to determine whether the graph that captures dependencies between expressions in a sentence locally and globally satisfies all applicable constraints. As a consequence of adopting a constraint-based approach, there is no 'overgeneration' to be worried about, as the goal of grammatical theory under declarative assumptions is quite different from procedural proof-theoretic syntax (Chomsky has recently—conferences in 2017—expressed skepticism about problems raised by overgeneration, but has not so far as we know elaborated on the matter): we aim at proving (by means of defining fully explicit dominance sets which exhaustively characterise graphs) that specific sentences are or are not well-formed expressions of the language; in this sense our goals are close to those of Item-and-Process grammars (Hockett, 1954; Schmerling, 1983a, 2018a). For example, in a 'pure' (Ajdukiewicz-style) CG, a proof that an expression is (or isn't) a well-formed expression of category C (in the form of an analysis tree) defines the relations that basic and derived expressions establish within the algebra of the language. The theory developed here does not make use of indexed categories in the specification of relations (although it could, as observed in **Chapter 2**), but other than that, we share pure CG's goals. This is important, we want to highlight, because by doing this some problems pertaining to the adequacy of rules and constraints simply do not arise.

After that brief methodological parenthesis, let us go back to the analysis. Above, we asked 'what is the role of resumptive pronouns in structures like (420)?' Now, we have the elements to formulate an answer to that question: the need for a resumptive pronoun in (420); more specifically, the need for an occurrence of *picture*' in the *when*-adjunct in the expression of the language, arises so that the adjunct clause is *not self-contained*, and therefore the conditions for *licensing* (see (206)) hold. If the adjunct clause was *self-contained*, then the embedded verb *saw* would be left with no object, and *who* would have no thematic role. In the view presented here, resumptive pronouns are mechanisms to *join* elementary graphs, in the example that interests us: *it* corresponds to the same node as *which pictures*, getting pronominalised as it is visited for the second time in a walk through the derived graph. This analysis may provide some evidence in favour of the 'double occurrence' analysis of *wh*-movement which we suggested in **Chapter 10**, although inconclusive. Let us provide the ρ-set for (420) in (421):

(421) $\rho_1 = \langle(\text{which, picture}), (\text{blush, they}), (\text{blush, when})\rangle$
$\rho_2 = \langle(\text{when, saw}), (\text{saw, John}), (\text{saw, picture})\rangle$

We see that under the double-occurrence analysis, ⦃picture⦄ belongs to both elementary graphs.

The problem of resumption is much more complex than we can cover in a simple note, but it seems that we can capture the insight in Zaenen et al. (1983: 679) that

> *the binding relation between a wh-element and a "resumptive" pronoun is, at least in some languages, of the same nature as the binding relation between a wh-element and a trace*

Our framework lends itself particularly well to this because in both cases (resumption and filler-gap dependencies) we are dealing with the same kind of phenomenon: a walk in a derived graph visiting a node more than once. Crucially, if the *when* clause was *self-contained*, it would be opaque for purposes of operations at the matrix clause, because the output of any operation purporting to involve elements in the adjunct and the matrix clause would violate the licensing conditions. We also capture the resistance of monotonic structures to resumptive pronouns: they are simply not needed to comply with the licensing conditions in monotonically growing structures regardless of structural distance if no constraint is violated. The present view thus aligns itself with *resumption-as-last-resort* approaches: resumption is a mechanism that the grammar appeals to when there is no other option. No other option for what? Here we proposed that resumption is a mechanism by means of which the structural conditions for *licensing* can be satisfied and therefore arbores can be appropriately *linked*. Why, though? Because the sentences we are considering, with resumptive pronouns, are well-formed expressions of the language. We agree with Alexopoulou (2010: 487) in that '*it is not the case that grammars license* [intrusive] *resumption*'; rather, grammars license a range of structural configurations and resumption appears as a last resort in a limited number of those configurations in the place of an 'illicit' gap (in English, relevant configurations promimently feature relative clauses and *wh*-interrogatives). It is consistent to say that giving a node with at least two occurrences (one in a scope position, one in a bound position) a pronominal exponent if otherwise the relation between *operator* and *variable* occurrences of a node (or *scope* and *restrictor* occurrences) could not be unambiguously determined. As McCloskey (2006: 19) notes,

> *It is known that resumptive elements may serve the purpose of marking variable positions in unbounded dependency constructions. It is known that resumptive elements may occur in positions from which movement is impossible (hence apparently allowing greater expressive power than is permitted by movement alone). It is also known that resumption imposes a considerably lighter burden on the human sentence processor than does the production and resolution of syntactic movement configurations. Why, then, is movement used at all in the creation of these structures?*

McCloskey's paradox emerges only if resumption is, as he puts it, *used in the creation* of structural descriptions under a *displacement-as-movement* view (which would lead to overgeneration under an economy-based approach to language like early Minimalism; Chomsky, 1995); it does not arise, however, if resumption is viewed as a *repair* strategy related to the conditions of *licensing* rather than a *generative* one related to the dynamics of *Move-α* (cf. McCloskey, 2002). In other words, if resumption is an output strategy, rather than a derivational one.

The analysis proposed here seems to have some advantages when looked at in interaction with other syntactic phenomena. Here we will simply sketch an instance of such interaction. Going back to our discussion about *topicalisation* and *focalisation*, if adjunction creates the kind of structural configuration in which resumptive pronouns can be called upon to link arbores, and if *topicalisation* (but not *focalisation*) is indeed an instance of *adjunction* (as we proposed in **Section 6.5**), then we have a possible explanation for the following contrast:

(422) a. Syntax, Mary loves (it)
b. It is syntax that Mary loves (*it)

In (422a) we are in presence of topicalisation of the constituent *syntax*; if that topicalisation is indeed adjunction of *syntax* to the root (and *syntax* is its own arbor), it is to be expected that a resumptive pronoun *can* be called upon to link the main clause and the adjoined element, regardless of the structural distance between gap and filler. In contrast to (422a), in (422b) the resumptive pronoun is banned because it is not required to save an otherwise ungrammatical sentence. This is another way of expressing the *last resort* character of resumption as a grammatical procedure: if it is not *needed*, it is strongly dispreferred.

Furthermore, and building on the previous contrast, if we combine *topicalisation* with extraction from an adjunct (in this case, the source of extraction is a relative clause, which in phrase structural terms is adjoined to NP; Demirdache, 1991; Krivochen, 2022), things get much worse really quickly, as we can see in (423b):

(423) a. Syntax, John thinks that Mary said that she loves (it)
b. Syntax, John knows a girl who likes *(it)

(423a) keeps things structurally monotonic: we have a sequence of verbs taking saturated, finite clauses as complements. We have three elementary graphs, structured around *thinks*, *said*, and *loves* but none of those is *self-contained* and the elementary graphs of *said* and *loves* are the 2 of their respective selecting predicates. In this context, the relation between *scope* and *restrictor* occurrences in (423b) can be unambiguously determined, and since the *filler-gap* relation would violate the Complex NP Constraint, it is possible to resort to *resumption* as a way to generate a licit configuration in terms of locality without falling into the problems that derivational approaches have in terms of determining the derivational point at which the choice to materialise the tail of a chain.

A final observation is in order. For our hypothesis about what is causing the contrast between (422a) and (423a), and (422b) and (423b) to work, we would need to commit to the claim that clefting essentially occurs *within a single-rooted graph*. That is to say, that it is structurally akin to *focus*: recall that in **Section 6.5** we argued that *focalisation* involved structural monotonicity and thus NPIs could be licensed in focus constructions (but not in cases of *topicalisation*). This approach has correlates in procedurally-based syntax. If, as argued e.g. in Uriagereka (2002, 2012) monotonicity in phrase structure (what he calls 'command units' are syntactic terms created by continuous applications of Merge which introduce a single terminal node at a time, defining a finite-state unit) correlates with syntactic accessibility (and thus, with locality), the resistance of focus constructions to displaying resumption would be at least partially accounted for: there is no need to resort to resumption in sentences like (422b) because clefting, being parallel to focus, involves local enough relations (and the possibility of defining a total order over a derived graph where the licensed precedes the licensee). 'Local enough', in the present context, is defined with respect to elementary graphs. This is a working hypothesis, however: in the present framework, monotonicity (local structures that grow throughout the derivation, and always at the same rate) is not enough to define locality, there must also be a single lexical anchor for there to be an elementary graph and relations between elementary graphs are regulated differently than simply assuming that whatever does not belong in the same elementary graph as a given predicate is entirely inaccessible to it. Whether clefting being parallel to focus in the relevant aspects is actually the case is far from clear, although the semantic / pragmatic relation between focus and clefting has been noted as early as Jespersen (1985 [1937]), and even MGG-inspired research within the so-called *cartographic* approach has proposed that clefting in fact makes use of a functional Focus projection (FocP; see e.g., Rizzi, 1997; Kiss, 1999; Belletti, 2008). The data presented here provides further support to the idea, although more research is needed to determine to what extent the hypothesis is tenable.

14.4 Identity Issues in Local Reflexive Anaphora[7]

In this section we explore the much-anticipated partial dominance analysis of quantified NPs. The empirical domain of our interest now is the interpretation of local lexical reflexive anaphora, which presents additional problems with respect to the simplified analysis in **Section 6.2.1**. Recall that in **Section 6.4** we mentioned the case of

(424) Every candidate wants to win

as one of the arguments raised in the 1970's against a transformation of Equi NP Deletion as formulated, e.g., in Rosenbaum (1965): (424) does not mean *every candidate wants every candidate to win*, but rather *for every x, x a candidate, x wants x to win* (thus, Candidate 1 wants Candidate 1 to win, Candidate 2 wants Candidate 2 to win, etc.). Let us call this the 'identity paradox' (since the NPs involved in the transformation cannot be identical if we want to get an adequate semantic representation). The same problem arises for cases like (425) below under a Lees-Klima approach to *pronominalisation* (see also Ross, 1967: Chapter 5; Langacker, 1969; recent Lees-Klima-style treatments of reflexives are presented in Hornstein, 2001: Chapter 5; Grohmann, 2003: Chapter 3; Gärtner, 2014; Hornstein & Isardi, 2014, among many others):

---

7 This section owes much to discussions with Susan F. Schmerling, who we cannot thank enough for her guidance through the literature on generalised quantification and Montague semantics. Responsibility for any mistakes is exclusively ours.

(425) Every man shaves himself

Like (424), (425) does not mean 'every man shaves every man' but rather 'for every x, x a man, x shaves x' (and for any man y, if $x \neq y$, x does not shave y. Thus, John shaves John, Bill shaves Bill, etc., but John does not shave Bill). This is a problem if pronominalisation applies under identity, as in a Lees-Klima context.[8] In both cases, strict identity presents a problem when the subject position is occupied by a quantified NP. This problem is related to the interpretation of cases like *every republican loves his mother*: here, '*the value for the pronoun varies with that assigned to the subject bound by the universal quantifier, such that each loving Republican is matched with his own mother*' (Safir, 2013: 516). As Safir points out, while there is some consensus about semantic representation, how that relates to the syntax is controversial. In this section we present a sketch of what a treatment of the identity paradox could look like under graph-theoretic assumptions.

We need to introduce some additional technical details about the analysis of quantified NPs in order to address the identity paradox. The standard analysis of quantification in contemporary formal semantics, inspired by work in formal logic (Lindström, 1966) and developed in the analysis of natural language in Montague (1973), Barwise & Cooper (1981), Higginbotham & May (1981), and others, generalises the traditional inventory of quantifiers and accordingly is known as the theory of *generalised quantifiers* (Barwise & Cooper, 1981, see Westerhål, 2015 for an overview). The theory of generalised quantifiers was developed due to an inability of the first-order quantifiers ∀ and ∃ to characterise quantified expressions in natural language: just to give an example, it is not possible to express natural language quantified NPs like *most voters* or *less than half of the students* in terms of ∀ and ∃ (see especially Barwise

8 It is also a difficulty for approaches such as Chomsky's (2021), which we briefly mentioned in **Section 6.4**. Chomsky's 'neo-Equi' analysis of Obligatory Control is based on identifying that in [every candidate want [every candidate to win]] the relation between what he calls *occurrences* of [every candidate] cannot have been generated via Internal Merge, given the requirement that only External Merge, and not Internal Merge, correlates to theta-role assignment. Having two theta-assigners, as observed in **Section 6.4**, licenses deletion of the lowest occurrence. Chomsky recognises that what we called the 'identity paradox' is indeed a problem, but suggests that it is overcome under phase theory (the theory of locality in Minimalism), such that interpretation takes place 'at the phase level' (vP / CP). Equi verbs take phasal complements, unlike Raising verbs (Chomsky, 2000). It is unclear, however, how phase-level interpretation can 'delete' the determiner in a generalised quantifier to deliver an interpretation such as (i)

(i) for every candidate x, x wants [x to win] (adapted from Chomsky, 2021: 23).

& Cooper, 1981 for discussion). Furthermore, first order logic representations of natural language sentences provide no insight into the syntactic structrure of quantified expressions: translating *every man sneezes* as $\forall(x)[\text{man}(x) \rightarrow \text{sneeze}(x)]$ does not reflect the fact that there is a constituent *every man* in the syntactic representation of the sentence (Barwise & Cooper, 1981: 165) and introduces logical connectives (material implication in the case of ∀ or logical conjunction in the case of ∃) which do not correspond to anything in the natural-language sentences whose meanings and structure we want to capture.

In order to provide adequate representations for natural language quantified NPs, it must be possible to quantify over sets of entities, define functions from subsets to subsets, etc.: natural language quantifiers are *relations between sets*, such that *every politician lies* makes reference to (i) the set of politicians and (ii) the set of liars. *Every politician lies* is true iff the set of liars is a member of the set of sets of which every politician is a member. The theory of generalised quantifiers treats NPs as denoting sets of sets, and the analysis of predication in sentences containing generalised quantifiers is based on determining set membership. In Barwise & Cooper's (1981: 164) terms,

> *The truth of a sentence* $Qx[\varphi(x)]$ *is then determined by whether or not the set* $\hat{x}[\varphi(x)]$ [read: x such that $\varphi(x)$; $\hat{x}$ is equivalent to the familiar λx notation] *is a member of the quantifier denotation*

In this approach, generalised quantifiers (which Barwise & Cooper, 1981 define as the combination of a determiner and a set expression, as in $most_{Determiner}$ $students_{Set\ expression}$[9]) are functions from sets of entities to Boolean truth values ({0, 1}, or {false, true}). Different quantifiers partition the set of entities differently: if we have a set of entities E, for all subsets x, y of E, *every x* maps each subset y of E to {true} iff x is a subset of y (i.e., if each x is y; see Keenan, 2006). In contrast, *some x P* partitions the set E differently, such that only some sets will output {true}. Formally, we can define the semantic value of the determiner *every* as follows:

9 Formally,

> *If D is a determiner and* η *is a set-term, then* $D(\eta)$ *is a quantifier* (Barwise & Cooper, 1981: 168).

It is the whole NP that is the quantifier, in contrast to familiar usage whereby 'quantifier' refers only to the determiner.

(426) $[\![\text{every}]\!] = \lambda S[\lambda T[S \subseteq T]]$[10]

Here, S and T are variables of type $\langle e, t \rangle$, functions from entities to truth values (the type of the characteristic functions of sets): this captures the idea that generalised quantifiers are relations between sets. Von Fintel & Heim (2011) introduce first-order quantifiers in their formalisation of the semantic value of *every*, as does the treatment of *every* in Montague (1973).[11] But others, such as Barwise & Cooper (1981) or Keenan (1997), instead of using the first-order logical quantifiers ∀ and ∃ in their formalisation of the interpretation of determiners, directly express the semantic values of quantifiers in terms of the second-order formalisation—quantifiers as sets of sets—that we have been discussing. The approach to anaphoric binding in **Chapter 8** defined reflexivity graph-theoretically as an instance of parallel arcs: building on Reinhart & Reuland (1993), we defined a predicate as reflexive if and only if at least two of its arguments (i.e., its 1 and 2, or 1 and 3, or 2 and 3) were coindexed. Here we will focus our attention on what they call SELF anaphors: referentially dependent, reflexive expressions.

In his treatment of anaphora, Keenan (2006) proposes a reflexive function (SELF) where, for any relation R, the following equation holds

(427) $\text{SELF}(R) = \{a \mid (a, a) \in R\}$

In Keenan's analysis, SELF is a function from binary to unary predicates: when two co-arguments of a predicate are identical. On this approach, *reflexivisation* is a valency-changing transformation: a binary predicate becomes a unary predicate. A similar perspective is taken in Bach & Partee (1980), Reuland (2005), and Chierchia (2004) in terms of reflexivity involving a mapping between a 2-place predicate and a 1-place one (see also Safir, 2013: 553). In the context of the present monograph, reflexivisation would not, however, be an RCT because in these cases an NP does not stop being a 2 to become a 1 (as it does in *Passivisation*) or vice-versa: relations are created on top of already existing ones. If the 1 of a reflexive predicate is a GQ, as in (428) below, the compositional interpretation of a sentence where R is instantiated by a transitive verb will involve a function from a set to itself.

10 Or, in Montagovian terms, $\hat{S}[\hat{T}[S \subseteq T]]$ (read: S such that T such that S⊆T).

11 Montague (1973) formalises the semantics of expressions of the form *every N* in his Intensional Logic as in (i):

Keenan's SELF function would enable us to resolve the apparent identity paradox arising when the antecedent of a reflexive is a quantified NP. If *every N* is a generalised quantifier, then the interpretation of (428a) would be as indicated in (428b) (minimally simplified from Keenan, 2006: 9):

(428) a. Every poet admires himself
b. EVERY(POET) (SELF(ADMIRES)) = true iff POET $\subseteq \{a \mid (a, a) \in$ ADMIRE$\}$

We would thus derive that for each entity in the set of poets, that entity is in an admiring relation with itself (see also Chierchia's 2004 REFL operation).

We want to arrive at a structural description that resolves the identity paradox while avoiding a multiplication of nodes for reflexives (in MGG we need at least two nodes: one for the antecedent, another for the anaphor). In the framework developed in this monograph, the basic representation of (429a) is (429b), with parallel arcs (as the same expression establishes two distinct grammatical relations with a predicate):

(429) Dob shaves (himself)
$\rho = \langle$(shave, Dob), (shave, Dob)$\rangle$

However, for quantified NPs in subject position, it is not evident that a representation along the lines of (429) works: we want to exclude *every* from the interpretation of the object.

How can we make the syntax work? As seen in **Chapter** 10, some multidominance approaches to *wh*-interrogatives, like Johnson's (2016, 2020), allow only part of a nominal phrase (in Johnson's paper, a DP with head D) to be multidominated: for a case like the embedded interrogative (*I wonder*) *which flower she should bring*, Johnson proposes the following representation:

(i) *If* $\zeta \in P_{CN}$ [$\mathrm{P_{CN}}$ is the category of Common Nouns] *and* ζ *translates as* ζ', *then every* ζ *translates into* $\hat{P} \bigwedge x[\zeta'(x) \rightarrow P(x)]$.

Recall that, as noted, Montague uses $\bigwedge$ instead of $\forall$ in his notation for universal quantification. $\hat{\mathrm{P}}$ is an abbreviation for λP, where P is a variable of the type of intensions of set expressions like *students*; $\lambda P[P({}^{\vee}x)]$ is of the type of intensions of characteristic functions of sets of intensions of individuals (entities). The symbols ^ and ˅ are operators that. roughly, derive expressions denoting intensions and extensions from expressions respectively denoting extensions and intensions.

(430)

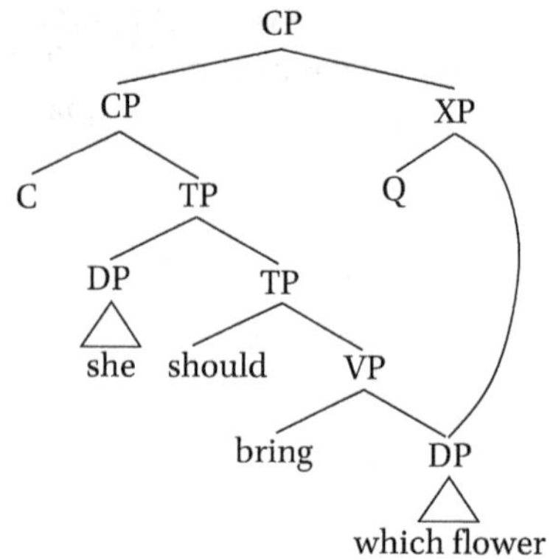

FIGURE 14.7 Multidominance analysis of wh-*interrogative*

Here, Q is an abstract morpheme that provides the operator interpretation to the DP (which, as mentioned above, can be traced back to Katz & Postal, 1964). As far as its semantics are concerned, Johnson (2020: 120) says, the denotation of Q is '*responsible for creating a binder out of the higher phrase*': a *wh*-interrogative involves an NP/DP introducing a variable and a Q morpheme that binds that variable; in some Minimalist analyses, Q enters an Agree relation with the *wh*-word in D (e.g., Cable, 2010). In a more traditional generative approach, moved *wh*-phrases are operators, and the traces they leave behind are variables bound by those operators (see Chomsky, 1981: 102, ff.; also Chomsky, 1986 for constraints on operator-variable relations). The important aspect of Johnson's proposal, for our purposes, is that the *wh*-operator and the variable (or the 'binder' and the 'bound variable', respectively) interpretations are dissociated not only semantically, but also syntactically: there is a specific head in charge of making a *wh*-operator out of a phrase in a specific syntactic context (dominated by CP).

The crucial aspect of Johnson's proposal, for our own purposes, is this dissociation. Analogously, we may propose that, in the graph-theoretic system laid out here, only *man* in *every man shaves himself* is the tail of parallel arcs: identity holds not between the meaning of an entire generalised quantifier and a reflexive but only between what Barwise & Cooper call *set terms* (which would correspond to the NP in a DP framework; in the present work, however, since we do not have phrasal categories or projection, the distinction is inconsequential). This, combined with the principle that semantic composition takes place 'bottom up' (Dowty, 1982), provides us with a potential way of addressing the identity paradox. The ρ-set of *every poet admires himself* would be (431), with a graphical representation (432) where the *parallel* relation is evident:

(431) $\rho = \langle(\text{every, poet}), (\text{admire, poet}), (\text{admire, poet})\rangle$

(432)

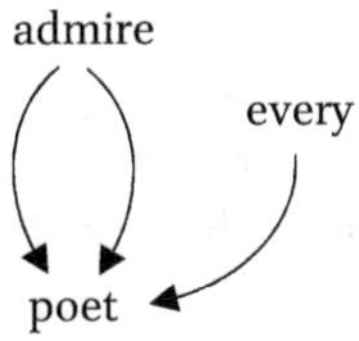

FIGURE 14.8 Graph-theoretic analysis for reflexivity with generalised quantifiers

It is crucial to remember that the set term of the generalised quantifier and the 2 of the lexical predicate are the same node; this is the core of the present proposal. *Every poet P*, then, means that for every *x*, if *x* is a poet, then P belongs to the set of sets that contain those who admire themselves. In our treatment of *wh*-interrogatives we suggested that, since (as observed by Johnson, 2020: 120) an expression cannot at the same time be a variable and an operator (i.e., a variable and a binder for that variable, in this case a binder in what traditionally would be called an A' position), an operator complex like *which flower* may only have *flower* multidominated, with *which* playing the role that in (430) would be assigned to the abstract morpheme Q: in our analysis, *which* would dominate *flower* but not be dominated by *bring* (see the analysis of *which dish should Sally eat?* in (316) above).

The analysis of reflexive anaphora that we need in order to avoid the identity paradox would be analogous: as with *which* (in (299) above *which lady read which book?*, taken from Reinhart, 1998), we can make use of the analysis of *every* given in (426) above. In Keenan's example *every poet admires himself*, under the analysis in (431), the expression *poet* receives as GF the 1 and 2 of *admire*: ⟨poet, poet⟩ belongs to the set of relations denoted by *admire*. Combining Montague's and Keenan's approaches to *every* and reflexivity, respectively, a representation like (421) can be seen to correspond to a semantic representation *for every x, x a poet, x is an admiring relation with itself*. Note that the separation between determiner and set expression that Barwise & Cooper (1981) emphasise in the definition of generalised quantifiers has become relevant for the syntax: only the set expression is multidominated. Without generalised quantifiers, the syntactic solution would not be available (or would be available only under additional stipulations).

What remains to be fully worked out is how to properly integrate the semantics in the definition of a walk through the graph for these and more complex cases.

14.5 Ghost in the Graph

A guiding principle in our approach has been that only overt expressions are represented in structural descriptions. Any deviation from that principle (which, as we have observed, is shared by some non-transformational approaches such as LFG) must derive from necessity: phonologically null expressions are not something we start with or which come 'for free'. In this section we will consider data that seems to point towards the need to include what we will call *ghost nodes* in structural descriptions. The term is heavily inspired by Arc Pair Grammar's *ghost arc* (Johnson & Postal, 1980: Chapter 10), which they use for arcs that are not sponsored (the corresponding term in RG is simply *dummy nominals*; these are non-referential nodes that head nuclear term arcs. See Perlmutter & Postal, 1983b: 101, ff.). APG's ghost arcs correspond to dummy nominals/expletives, and due to their semantic inertness, are never members of L-graphs (which are—*very roughly*—APG's version of Logical Form, not to be confused with our use of 'L-graph'). We depart from the APG usage of *ghost* insofar as expletives in our system are not assigned an address and thus have no semantic value: as argued in **Section 6.1**, they point nowhere (and that is their unique property). Furthermore, as we have emphasised throughout, whereas APG's principles and laws quantify over relations between arcs, we are focused on nodes. Our *ghosts*, then, are not arcs, but nodes, and do not correspond to dummy nominals (which are the 'elsewhere case' of the indexing system: if a node is not indexed by any member of the set of addresses, it is an expletive). They are syntactically active objects which participate in syntactic relations in the same manner that overt expressions do. At this stage of the development of the graph-theoretic approach, it seems that we need ghost nodes to correspond to unbound nominals whose interpretation does not depend on otherwise expressed verbal morphology *when needed*. Let us give an example: periphrastic passives need not have a null agent, unless there is some syntactic dependency that forces us to posit the existence of a null agent. These dependencies include control relations and predication, as illustrated in (433a-c) (enriched with empty categories for expository purposes only):

(433) a. The game was played [PRO wearing no shoes] (control)

b. The room was left (*angry) (secondary predication)

c. *pro llegaron* *pro cansados*
arrive3Pl.PST.PERF tired.MASC.Pl

d. John ate (*the meat) raw

In his detailed analysis of implicit arguments, Landau (2010) distinguishes between weak and strong implicit arguments: strong implicit arguments may licence secondary predication, weak implicit arguments can license control (i.e., Equi), but not secondary predication. If the reader recalls the treatment of Equi in **Chapter 6**, we did not appeal to implicit arguments (rather, we leveraged structure sharing; see also Sampson, 1975; Pollard & Sag, 1994; Börjars et al., 2019). It is possible that the framework proposed here allows us to dispense with weak implicit arguments. Landau proposes that strong implicit arguments are DPs, but weak implicit arguments are not: this is so because secondary predication must be predicated of a DP. We agree with Landau that it is not the case that weak implicit arguments are absent in the syntax, being perhaps discharged in the lexicon: the difference between our analyses and Landau's is that because all syntactic relations are defined at the level of elementary graphs there are no 'non-local' relations, strictly speaking. All apparent non-locality is the result of graph composition (as in a TAG, following the *Non-local dependency corollary*). So, for instance, whereas in Landau's view the relation between *the detective* and *himself* in (434) is '*only syntactic (via* PRO*)*' (Landau, 2010: 361), in our analysis there is no PRO: the relation is indeed syntactic, but involves structure sharing between elementary graphs:

(434) **The detective** promised Bill to disguise **himself** (Landau, 2010, ex. (15b).)

Exhaustive Obligatory Control (Equi), in this perspective, is neither coindexing nor predication (nor, for evident reasons, movement or Agree; see Landau, 2003: Chapter 2 for an overview of theories of lexical and syntactic control). The approach defended in this monograph respects the principle of locality of lexical relations (Landau, 2010: 362), but because the addressing system allows for structure sharing, there is no need to invoke null arguments in cases such as (434): the anaphor *himself* appears in an elementary graph anchored by *disguise*, whose Subject and Object is the node corresponding to *the detective*. All we have is an elementary graph with parallel arcs. The elementary graph anchored by *disguise* and the elementary graph anchored by *promise* are linked at the node with address ⦃detective⦄; no PRO is needed to bind the anaphor.

Strong implicit arguments are a different beast, insofar as they cannot—so far as we can see—be reduced to structure sharing. A secondary predicate needs an argument regardless of whether it is phonologically overt or not. The basic insight that leads us to allow for ghost nodes is that they are syntactically active: as Bhatt & Pancheva (2006) observe, the existence of syntactically active but not syntactically projected arguments is conceptually problematic; it is not any less problematic—we add—from an empirical viewpoint. Particularly so for an approach that explicitly endorses a version of direct compositionality (albeit a relatively weak one, where the units subjected to compositional interpretation are elementary graphs; as noted in **Section 2.2**, Jacobson, 2012 refers to such an approach as *Type 3 direct compositionality*).

Once we have motivated the existence of ghost nodes—partially, at least—on the grounds of the distinction between strong and weak implicit arguments, let us sketch the fundamental properties of ghost nodes, much research pending.

– Ghost nodes are categorematic

This must be so because they participate in predication relations. Ghost nodes, unlike APG's *dummy nodes* (which head *ghost arcs*) are not expletives (which are semantically inert), but both syntactically and semantically active elements in syntactic representations.

– Ghost nodes are nominal arguments

This follows rather closely the argument in Landau (2010). Ghost nodes need to satisfy some selectional requirement that would not be met otherwise within an elementary graph. We do not, however, make a distinction between DP and more impoverished phrases as Landau does, for obvious reasons. Given the treatment of binding in this monograph, it follows that ghost nodes must furthermore correspond to unbound arguments.

– Ghost nodes are syntactically present only when syntactically active

We want to avoid a multiplication of nodes in syntactic representations. There is no reason to posit the existence of a ghost agent in a passive like *Saito was shot*, because the agent is not syntactically active: it does not participate in binding, control, or predication relations. The analysis of passives in **Section 13.1** remains untouched in its fundamental aspects, but may be augmented by ghost nodes if and only if those nodes are syntactically active.

– Ghost nodes, by virtue of being categorematic, are uniquely indexed just like any other categorematic node

As such, ghost nodes can take part in garden-variety dependencies, including predication. This property in particular is important for the analysis of two constructions that we have not mentioned previously: partial control and arbitrary control. In the former, under classical assumptions, the interpretation of PRO corresponds to a 'semantic plurality' properly including the controller (Landau, 2010: 364). In the latter, PRO has no argumental controller, and is unbound. The relevant cases are like the following:

(435) a. The chair$_i$ found it frustrating PRO$_{i+}$ to gather without a concrete agenda (Landau, 2010, ex. (41b). Indices are ours)

b. Sue$_i$ said that [PRO$_{arb/*i}$ to buy her$_i$ nothing in Rome] would be unacceptable. (Landau, 2000, ex. (6b))

In Landau's notation, *i*+ means that the interpretation of PRO is only partially determined by the index of its controller. The availability of partial control is a matter of lexical semantics: some predicates license partial control, some do not (a list is provided in Landau, 2013: 158; an amended one in Pearson, 2016: 693). Landau's (2000, 2010) argument points to the fact that partial control may involve implicit arguments. Because control requires syntactically active arguments to act as controllers, there must be an implicit argument in (435a) that introduces the semantic plurality of which *the chair* is part. This implicit argument, we propose, is syntactically represented by a ghost node, which we will notate as *e* due to it being phonologically empty. In this context, the derived ρ-set for the Spanish sentence (433c) would be:

(436) $\rho = \langle$(llegar, *e*), (cansados, llegar), (cansados, *e*)$\rangle$

The behaviour of the ghost node is not distinct from the behaviour of phonologically overt nodes with respect to graph union. If the ghost node was syncategorematic, for example, it would not be able to license secondary predication (let alone agreement) or control.

It is important to bear in mind that, in our theory, Equi structures *always* have a syntactically active subject: the argument structure of each predicate will be satisfied within its elementary graph. As a consequence, lexical relations are established by construal at elementary graphs (including all requirements of co-argumenthood) and syntactic relations are a consequence of graph union, where the indexing mechanism is leveraged (following the typology of Landau, 2010). This is not a predication analysis of control, since each predicate (however many there are in a multi-clausal structure) will have its valency saturated locally: graph union establishes relations between arguments of different predicates. The availability of ghost nodes for partial control avoids one of the problems that Landau identifies with Hornstein's (2003) Movement Theory of Control: partial control does not arise with Raising predicates (i.e., there is no such thing as 'partial raising'). This is unexpected, Landau argues, if the syntactic mechanisms underlying Raising and control are the same (in Hornstein's view, movement), but expected if the mechanisms are different. In the present view, however, it is not the mechanisms that differ (we only have graph union), but the composition of elementary graphs: only in control constructions can there be ghost nodes. This restriction amounts to recognising the lexically governed nature of partial control: some verbs license a ghost node argument, some do not.

In sum, we have very briefly examined the possibility that the approach presented in this monograph may be augmented with syntactically active but phonologically null nodes for a very restricted set of constructions: as far as we can tell, only implicit arguments (and only when required to satisfy a syntactic relation such as control or predication) qualify for *ghost node*-hood. In what pertains to partial control, we agree with Landau that it is a syntactic phenomenon, that the predication analysis of control suffers from shortcomings, and that partial control provides evidence for the existence of implicit arguments. There are, certainly, pragmatic aspects involved in the interpretation of sentences containing partial control (see Lawler, 1972 for examples and discussion[12]), but the argument is different: the underlying mechanism whereby partial control comes to be in language is syntax, with semantics and pragmatics building on configurations licensed by syntactic conditions. Our treatment of Equi (exhaustive control, in Landau's terms) may be assimilated to existing analyses, to an extent: for example (as observed above), there is a resemblance to some versions of the LFG approach whereby Equi is an instance of structure sharing (e.g. Börjars et al., 2019). These approaches have clauses selected by Equi verbs be subjectless in their syntax (in the case of LFG, *c*-structure), with embedded subjects being however, present semantically (see also Dowty, 1985; Chierchia, 1989; Culicover & Jackendoff, 2005, 2006; Haug, 2014; Pearson, 2016 for variants of the *control-as-semantics* approach. Underlying many of these is the idea that infinitival complements of control verbs are properties, not propositions. See also fn. 7 in **Section 6.3**). However, there are two aspects that clearly distinguish ours from existing alternatives: first, unlike LFG and Simpler Syntax, our analysis is formulated exclusively in terms of syntactic configuration (specifically, structure sharing delivers multidominant graphs). Second, we have allowed for a phonologically empty element in the structural description of specific constructions (as there is no parallel semantic or functional level that can carry the burden of control, leaving syntactic structure unharmed). We restrict implicit arguments to what seems to be empirically

---

12 Lawler considers, among many others, pairs like (i) and (ii):

(i) *Mary decided to recognise Red China (Lawler uses *, we would prefer # for a pragmatic anomaly since no syntactic well-formedness conditions appear to be violated)

(ii) Nixon decided to recognise Red China

The metonymy licensed by *Nixon* (i.e., the US government during Richard Nixon's presidency) is not available for *Mary*; this metonymy is part of what underlies the *interpretation* of the implicit argument. However, the fact that the implicit argument needs to be syntactically active remains undisputed by examples like these (and, as far as we can tell, by all of Lawler's carefully discussed data).

irreducible considering the empirical basis of this monograph: unbound, syntactically active, categorematic arguments.

14.6 A Derivational Alternative?

The early stages of our inquiry were motivated by the possibility (suggested in McCawley, 1982) that most grammatical processes that are customarily modelled using transformations do not modify grammatical relations between elements, only linear order. In that respect, our system straightforwardly satisfies what Pollard (1997: 12) called *nondestructiveness*: grammars should not make reference to operations that destroy existing linguistic structure (see also Lasnik & Uriagereka's 2005: 53, 112 *Conservation Laws*). The fact that our theory satisfies these conditions is partly a result of lexicalising the grammar (such that the lexical anchors of elementary graphs determine what relations are needed to saturate their lexical properties), and partly a result of adopting a declarative approach: no grammatical relations can be destroyed if there are no operations to disrupt these relations. In formulating a theory of syntax, the use of graphs for structural descriptions seemed to suggest constraint-based syntax as the way to go (since we can define constraints over representations which provide a definition of graphs that correspond to well-formed derived expressions of the language). The constraint-based approach allowed us to formulate conditions on relations between nodes within an elementary graph and across elementary graphs without requiring a huge theoretical apparatus.

Does this mean that graph-theoretic syntax is *necessarily* declarative? No, so far as we can see. At least two roads are open to the derivationalist. The first is straightforward. As suggested above, it is possible to formulate a derivational alternative to our declarative formalism, perhaps closer to traditional Tree Adjoining Grammars: if Andrews (2021: 20) is correct in his evaluation of the distinction between procedural and declarative approaches as '*matters of convenience rather than fundamental importance*' (a controversial position, certainly, but an interesting one[13]), we may admit generalised graph union as an

13 If part of the explanatory burden of a theory rests in its *derivational generative capacity* (Schuler et al., 2000; also the comparison between CFG and PDA in Frank & Hunter, 2021), as it is the case in TAGs, it becomes harder to accept that structure building and filtering do not define fundamentally distinct classes of explanations. The notion of *derivational generative capacity* refers to the mechanisms by means of which structures are built, such that a grammar G has a greater derivational capacity than G' iff G and G' gen-

operation that delivers derived graphs from elementary graphs, maintaining a declarative approach to the description of elementary graphs for convenience. Derivational applications of a graph-theoretic framework with an addressing axiom under derivational assumptions are pursued in Krivochen (2022, 2023a, b), where Merge is taken not to deliver unordered sets, but digraphs (i.e., Merge(X, Y) = $e\langle X, Y\rangle$). McKinney-Bock (2013) and McKinney-Bock & Vergnaud (2014) provide a graph-theoretic analysis of sentence structure more closely couched on traditional Minimalist phrase markers (see the **Appendix**).

A second possibility, should the reader aim for a fully derivational system within the broad guidelines on grammatical analysis provided in this monograph, is to adopt a formalisation of graph theory closer to *graph dynamical systems* (GDS; see e.g. Mortveit, 2008; Macauley & Mortveit, 2009). In this case, instead of building graphs stepwise, derivations define sequences of graphs (not unlike singulary transformations, as originally defined). Formally,

> *A Graph Dynamical System is a triple consisting of:*
> *A graph Y with* [finite] *vertex set* $v[Y] = \{1, 2, \ldots n\}$.
> *For each vertex i a state* $x_i \in K$ *(e.g.* $K = \{0, 1\}$*) and a Y-local function*
> $F_i : K^n \to K^n$
> $F_i\,(x = (x_1; x_2; \ldots ; x_n)) = (x_1; \ldots ; x_{i-1}; f_i\,(x_{[i]}); x_{i+1}; \ldots ; x_n)$:
> *An update scheme that governs how the maps* F_i *are assembled to a map*
> $F : K^n \to K^n$ (Mortveit, 2008)

GDS allow us to have update schemas for graphs by defining functions that specify how v_t goes to v_{t+1}. Each vertex is assigned a state from a set K, and an update scheme determines whether updates to the set of vertices apply all at once (in which case we are dealing with a generalised cellular automaton) or sequentially. we can define update functions that operate over the set of ordered relations. We can diagram a GDS as a graph, each of whose nodes is itself a specification of the state of each node in a graph. This is particularly useful if one wanted to redefine the present system in derivational terms: like RG

erate the same string and tree languages, but G derives those in more ways. Consider, e.g., the case of a TAG and a CFG generating a strictly right-branching [Head-Complement] structure: TAGs have at their disposal both substitution and adjunction, whereas CFGs cannot operate counter-cyclically (rewriting internal nodes). Whatever explanatory burden rests on rule ordering, however, should not pose a problem, insofar as declarative theories can appeal to material implication as an ordering between filters (Postal, 2010: 7).

strata, we can define sets of sets of expressions and relations, linked by update functions. Graphs would still need to satisfy well-formedness conditions, but these conditions may apply at different stages. For example, consider the following PIOC:

(437) a. John gave a book to Mary
b. ρ_{PIOC} = ⟨(give, John), (give, book), (give, Mary)⟩ (recall that this *to* is syncategorematic; Schmerling, 2018a: 51–52)

As before, this structural description tells us that *John* is the 1 of *give*, *book* is the 2, and *Mary* is the 3. Dative Shift modifies these relations, making *Mary* the 2 and demoting *book*. Under GDS, we may define the relation between the ρ-set in (437b) and the one in (438)

(438) ρ_{DOC} = ⟨(give, John), (give, Mary), (give, book)⟩

in terms of an update to the graph. In other words: it is possible to formalise relation-changing transformations as graph update functions. Here too we can maintain the idea that RPTs are fundamentally different from RCTs: RPTs just create new edges in elementary graphs, but do not change the grammatical relations in either elementary or derived graphs. The result is likely to be a theory with a number of language-specific mappings each of which specifies input, updating schema, and output, not unlike transformations in the generative Standard Theory (which operated over sequences of variables, not graphs, however. See **Chapter 1** for some examples). As such, it is entirely possible that the same—or very similar—problems arise for graph dynamical systems as well. This point, of course, requires further research, and we leave it here as a promissory note of sorts.

14.7 Future Prospects

There are, of course, many crucial empirical phenomena left to be analysed; we can mention some here to give the reader an idea of the research agenda that we intend to pursue in future work. We hinted at a treatment of some deletion transformations, but a fuller account of ellipsis is still to be developed. The theory we have presented is flexible enough to allow for the variation in the structural descriptions of strings that seems to be needed in order to account for the fact that English and Spanish differ in the availability of VP ellipsis (**Section 13.1**), but phenomena like Bare Argument Ellipsis, Antecedent Contained

Deletion (ACD), and ellipsis under sloppy identity still remain unaccounted for. For ACD, it seems likely that the structural configuration that we have proposed for MIG-sentences can help: one of the main issues that arises with a transformational treatment of ACD (which makes use of operations like Quantifier Raising at the level of Logical Form), and which has been taken as an argument in favour of a Merge-based analysis (at least for authors like Hornstein, 1994 and Van Wyngaerd & Zwart, 1999) is that the reconstruction of the antecedent VP in the place of the deleted VP leads to an infinite regress (example (439) is taken from Van Wyngaerd & Zwart, 1999: 203, in turn based on an example from Lasnik, 1993):

(439) Dulles [$_{\text{VP1}}$ suspects everyone Angleton did [$_{\text{VP2}}$ *e*]]

If the antecedent of *e* is VP_1 (such that Dulles suspects everyone that Angleton suspected), then the antecedent properly contains itself (since VP_2 is a constituent of VP_1). This is a problem that we have faced before, when analysing Bach's (1970) objections to a transformation *pronominalisation*: Bach's case prominently featured an infinite regress of complex NPs. It is to be seen whether an analogous treatment would work for the canonical cases of ACD. Also a difficult phenomenon, sloppy identity (as in *Jimmy$_i$ is losing his$_i$ hair, but Robert$_j$ isn't ⟨losing his$_{*i/j}$ hair⟩*) is a challenge for the structure sharing agenda, in that only strict identity seems to be predicted if 'deletion' is always a surface effect of structure sharing.

Another set of phenomena that we only briefly commented on, and which deserves a much more detailed treatment, is *islands* (in the sense of Ross, 1967). We presented an approach to coordination that allows us to account for the canonical CSC cases (**Chapter 12**), our treatment of relative clauses also allows us to provide a description of CNPC phenomena (**Chapter 9**), and adjunct islands seem to be covered by the requirement that an arbor cannot be *self-contained* if one of its non-root nodes enters a relation R with a node outside that arbor (which, we argued, makes the correct predictions when it comes to parentheticals); however, not all islands seem to be reducible to configurational issues. To put it differently: not all locality phenomena depend on syntactic configuration, like 'coordinated structure', 'complex NP', or 'adjunct'. Within the set of islands that do not seem to be amenable to a purely syntactic (i.e., configurational) account we find things like *negative islands* (or 'inner islands'; Ross, 1984) and *factive islands* (e.g., Rooryck, 1992), which we exemplify in (440) (from Ross, 1984, his (2b)) and (441) respectively:

(440) *What did no imitation pearls cost__? (-£5.000)

(441) *How did he regret that his son had fixed the car__?

Here, semantic issues and possibly lexical specifications seem to play a bigger role than phrase structure configuration. There is an argument-adjunct asymmetry in factive islands (*What did he regret that his son had fixed?* is reported to be better than (441)), but still we need to consider whether the complement of a factive verb defines a closed domain for reason independent of the size of elementary graphs. Furthermore, as Erteschik-Shir (1973: 40) observes, not all factive predicates induce islands effects (thus, *This is the man that he is aware that you like __* is grammatical). As is well-known (Karttunen, 1971b), factive verbs can take non-propositional complements (by virtue of containing variables), which means that the opacity of factive complements cannot be blamed on their 'propositionality':

(442) a. Some senators regret that they voted to acquit Trump
b. *For some x, x a senator, x regrets that x voted to acquit Trump*

The variable *x* is free in the clause *x voted to acquit Trump*, and as such it cannot be a part of a proposition or a 'presupposed' sentence. It has also been proposed that complements of factives are *definite*, as a property selected by the main predicate (Melvod, 1989), that factives do not select CPs like most transitive verbs, but rather sentential complements dominated by N (and thus these complements lack an intermediate landing site for successive-cyclic movement, which would account for their opacity for purposes of long-distance dependencies; Kiparsky & Kiparsky, 1971; Rizzi, 1990), that factives are best described in terms of meaning postulates which state entailment relations (Karttunen, 1971b) ... however, it is not clear that any of these accounts actually involves a modification of the set of relations between expressions in a sentence. In other words: the structural descriptions for factive complements seems to follow the format for other transitive constructions. A unified account of factive islands is not obviously syntactic (for a semantic-pragmatic approach, see e.g. Oshima, 2006).

As for *inner islands*, Ross (1984) and McCawley (1998), among others, point out that these are islands only for the application of certain rules (so-called *inner rules*: e.g., parenthetical insertion, some instances of *wh*-movement): the presence of negation restricts the domain of application for these rules, yielding island effects where the island is smaller than a purely configurational approach would predict. Again, we cannot just say that the scope of negation becomes opaque. Configuration alone may have difficulties in filtering out (440) and allowing (443)—below, from Ross (1984), his (2a)—, since there is no evident way in which the structural description assigned to a transitive construction with *cost* in an IA framework differs from the one assigned to a transitive construction with *touch* (i.e., in absence of independent evidence to the contrary, in both cases we would expect a structural description like [$_{VP}$ cost / touch [$_{NP}$ £5000 / the hands of the Countess]], with the NP in the complement position of the VP in both cases):

(443) What did no imitation pearls touch? (-the hands of the Countess)

It is always possible to 'syntactify' phenomena if the meta-theory is lax enough. For example, we could simply encode arbitrary features that block specific operations assigned to specific lexical items (see Haider, 2020 for a similar critique of arbitrary 'syntactification'; also Postal, 1972 for a warning about the perils of unrestricted feature systems). We would like to avoid going that way. Particularly, given the fact that non-configurational proposals are available and well worked out (e.g., the semantic analysis of weak islands in Abrusán, 2014). What is there in syntax to be captured, we do attempt to capture. However, not every grammatical phenomenon needs to be a syntactic property (and not every syntactic property needs to be universal).

CHAPTER 15

Concluding Remarks

We began this exploration by asking a simple question: what if the grammar attempted to minimise the number of nodes and maximise relations between those nodes, instead of maximising the number of nodes and establishing 'unambiguous' paths between them? One possible answer to that question was sketched in this book. The grammar is a set of well-formedness conditions over relations between nodes in (elementary and derived) directed graphs. The choice of graph theory as the formalism within which to formulate our theory is a natural choice if we want to have a way to refer to a set of elements and the possible dependencies that can be established between them.

A graph, mathematically, is a set of nodes and edges. The rest is given by the syntactic theory. The nodes on what we have called *L-graphs* are addresses that can be called upon in different contexts and whose content is the semantic values of basic expressions of the language. Our theory is *lexicalist* (and 'endo-skeletal', in the sense of Borer, 2003) in that syntactic structure 'projects' from lexical requirements: the building blocks of syntax are irreducible graphs within which the subcategorisation properties of lexical predicates are satisfied, and thematic roles assigned by these predicates are discharged. Syntax, in this view, is not 'templatic': as the data from Spanish auxiliary chains eloquently showed, it is not possible to determine *a priori* what an EG will contain (i.e., how much structure will be projected).

Edges in these graphs are directed, going *from* predicates *to* their arguments (as in DG, and in contrast to RG and APG), defining the binary relation *immediately-dominates*. Unlike MGG, where the relation between a predicate and its argument is always mediated by an intermediate node (in X-bar theory) or a label (under set-theoretic Merge), our theory defines direct formal relations between expressions. Directed edges impose an order over the set of nodes: $e\langle a, b\rangle \neq e\langle b, a\rangle$. Furthermore, the set of edges is itself ordered: this allows us to map the order between edges to the Grammatical Function Hierarchy, as GFs are primitives of the theory (as in RG, APG, LFG, and HPSG).

The combination of graph theory as a formalism with a set of constraints over allowed relations in local graphs, and the idea that grammatical relations are almost always preserved in 'transformations' constitutes the core of the present approach to syntax. Consider as an illustration of this our treatment of long-distance dependencies as relation-preserving rules: instead of removing an expression from a syntactic position (as in *trace* theory) or duplicating

 | DOI:10.1163/9789004542310_016

it and thus creating a new object that enters into new relations (as in the *copy* theory), we implemented an account that the theory makes possible whereby a single expression enters into multiple relations within a single elementary or derived structure. Derivational systems whose constituent structures abide by the Single Mother Condition have no option but to disrupt relations and/or to multiply entities. For cases in which grammatical relations are not preserved, we recognised a class of lexically governed alternative orders of arcs (in *Dative Shift*) or that involve different derived expressions (in *Passivisation*).

In addition to the maximisation of connections between categorematic expressions, we entertained the additional requirement that only overt categorematic expressions correspond to nodes in graphs. A great deal of empirical ground can be covered with such a strong restriction: building syntax from the ground up, with a toolset that is very limited despite the 'expressive power' of the formalism, proved fruitful. We saw in **Section 14.5**, however, that such a requirement is too strong—that it is necessary to enhance our inventory with phonologically null categorematic argument expressions in restricted contexts.

A consequence of our work building on Lexicalised Tree Adjoining Grammars is that domains for semantic interpretation and for the definition of what counts as a 'local' syntactic relation are not delimited *a priori* by the presence of designated functional categories (as is the case in *phase theory*, for instance, where *v*P and CP are bounding nodes): an *elementary graph* is the irreducible graph where all selectional requirements of a single lexical predicate are saturated. The condition that the anchor of an elementary structure be a lexical predicate entails a restriction over garden variety LTAGs (e.g., Frank, 2002), where every lexical head (predicative or not) is allowed to anchor an elementary graph. This restriction proved very useful for the analysis of subject Equi and Raising (**Sections 6.1** and **6.4**) and auxiliary verb constructions (**Section 7.1**), where functional modifiers belong to the elementary graph anchored by the lexical predicate they modify.

Lexicalised grammars permit the definition of an upper boundary for the size of their local structural units without the need to introduce endmarkers in the alphabet or add stipulations about where syntactic domains end: locality is a consequence of lexicalisation. We also borrow from TAG s the possibility of relating syntactic domains at non-root nodes; however, we depart from TAG s in defining a set of conditions over *linking*, such that distinct graphs may be *linked* at common nodes. This is equivalent to having the same variable appear several times throughout a formula (cf. Chomsky, 2021), which is why we have followed Sarkar & Joshi (1997) (in turn based in Gorn, 1967) in assigning each node a uniquely identifying address, which in our theory points to the semantic value of the expression that corresponds to the node. Technically, the set of nodes is indexed by the set of addresses. The importance of the addressing axiom for our theory cannot be overstated, but two illustrations of its power as an analytical device can be seen in the treatment of reflexivity in terms of parallel arcs which capture the two distinct relations that a single expression establishes with a predicate, and of Bach-Peters sentences, where we simplified the Phrase Linking Grammar analysis by pruning the structure and reducing the number of relations.

The directed graph format also gives us a direct connection between syntactic structure and semantic interpretation, along the lines of *direct compositionality* (Jacobson, 2012). Interpretation is a function that has as its domain a set of nodes and as its range a set of semantic values; the compositional interpretation of graphs is determined by the directed nature of edges. In **Section 2.2** we defined a **semantic interpretation rule** according to which a directed edge from a predicate to an argument is interpreted as the application of the semantic value of that predicate to that argument (see also Bach, 1979: 516). We then expanded the system to 2- and 3-place predicates, building on Heim & Kratzer (1998), until reaching a level where interpretation can have access to whole arbores (as in the case of intensive iteration): Jacobson (2012: 113) calls this *Type 3 direct compositionality*, which she traces back to Partee (1974).

Where the architecture of the grammar is concerned, we emphasised the fact that in our theory there is a single level of syntactic representation. Unlike MGG, where syntactic operations generate sets of sets (under Merge), LFG, where syntactic information is distributed between at least a pair of levels (minimally, *c*- and *f*-structures), or RG/APG, where syntactic representations involve a sequence of 'strata' (with well-formedness conditions being defined over initial/final arcs), our approach aligns with DG and Schmerling's NSG in that there is only one level of syntactic representation, where all relevant relations are defined.

A final (methodological) consideration. It seems to us that, very broadly speaking, there are in the field three main conceptions about what a theory of 'grammar' is and what we expect of it:

1) The theory of grammar is a theory of knowledge of language (this is, in essence, the MGG perspective, especially under 'biolinguistic' assumptions)
   - How it is acquired by humans
   - How it is used by humans
   - What its physiological underpinnings are
   - How it evolved (or how humans evolved such that they now have that knowledge)

2) The theory of grammar is a theory of string recognition/parsing (the formal-language-theory perspective, historically important especially in computer science)
   - Recursively enumerate *all* and *only* grammatical strings of the language
   - Grammars are families of acceptors or producers of strings
3) The theory of grammar is a theory about expressions and relations
   - Identify basic expressions of the language
   - Specify the allowed relations between basic and derived expressions
   - Provide an account of variation between grammatical systems in terms of permissible expressions and relations

It is crucial that we recognise that these three perspectives are internally consistent alternatives, and that they ask different questions about objects that are not obviously the same. Some theories aim at unifying two or more of these views, or reducing one to another, and some remain firmly within one of these. In any case, what counts as a question, what counts as an answer, and what counts as an 'explanation' varies enormously between theories that align with any of these conceptions. In the present work we have explicitly followed the line of Tesnière (1959), for whom the laws of grammar, as an empirical discipline, are not to be looked for or formulated in terms of biology, psychology, or cognitive science, but on terms that are its own:

> *la syntaxe est tout-à-fait indépendente de la logique et de la psychologie* (Tesnière, 1959: Chapitre 20 §18)

APPENDIX

Some Notes on (Other) Graph-Based Approaches

Graphs have been used in linguistics for some time now, but not always with the same purpose in mind. From Peirce's work on *existential graphs* (see, e.g., manuscript MS 514; it appears with commentary in Sowa, 2017) and taken up in works such as He (2015) to graphs used in Frame Semantics (e.g., Baker & Ellsworth, 2017) and relational networks, the use of concepts from graph theory in linguistics has been widespread, and fruitful. Different theories make different use of graphs, sometimes as notational variants for other formal objects, sometimes as formal objects themselves.

Early generative grammar made use of graphs in structural descriptions (we have already stressed our debt to Morin & O'Malley, 1968, as Postal, 2010 also does), and all descendants of RG, in particular APG and MG, use graphs as their format of preference for structural descriptions. Within MGG, Minimalism signified an explicit departure from graph-theory and the adoption of set-theory as the formalism within which to define operations and relations. However, there are some non-standard Minimalist approaches that do work with graphs. For instance, McKinney-Bock (2013) and McKinney-Bock & Vergnaud (2014) implement a graph-theoretic approach in which arcs encode two basic relations: *selection* (represented by an *S* annotation on an edge) and (feature) *checking* (represented by a *C*). Crucially, in this view, Merge is a fundamentally *asymmetric* operation. Part of the motivation for their graphs comes from van Riemsdijk's *grafts* and the analysis of relative clauses (restrictives with split antecedents and transparent relatives) and coordinated structures. McKinney-Bock & Vergnaud (2014) assume that:

> *classical Phrase-markers seem to be the right objects to describe interpretive properties of expressions at the interface levels.* (McKinney-Bock & Vergnaud, 2014: 218)

This entails a commitment to binarity (combinatorics applies to 'pairs of formatives'), as well as non-terminal symbols in structural representations. In their view, '*Narrow syntax will be formalized as a graph in the general sense* [read: connected and directed, plus labelled edges indicating the kind of syntactic relation holding between nodes]. *Phrase-markers will be read from that graph, subject to various conditions*' (McKinney-Bock & Vergnaud, 2014: 218): this means that graphs and classical phrase markers should be related by a homomorph-

 | DOI:10.1163/9789004542310_017

ism, preserving relations within selected structure. Their alternative to set-theoretic Merge is the following:

(444) 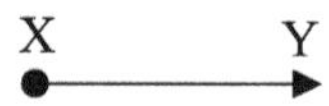

FIGURE 0.1 Graph-theoretic Merge and Projection (McKinney-Bock & Vergnaud, 2014)

Whereas in current generative theory Merge is set-theoretic and delivers only unordered sets (in particular if labels are eliminated, as in Epstein et al., 2015; Collins, 2017; see however Chomsky, 1994, 1995; Stabler, 2011), the digraph in (444) reads 'X Merges with Y, and Y projects'. In this context, the Minimalist clausal architecture for an intransitive sentence such as *the man laughed* (following the hierarchy C > T > V) is represented as follows:

(445) 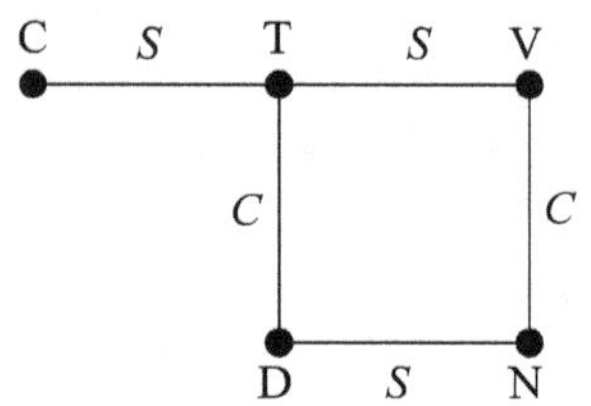

FIGURE 0.2 Graph-theoretic analysis of intransitive construal (McKinney-Bock & Vergnaud, 2014)

This format allows them to have multidominated nodes and structure sharing (which they capitalise on for the analysis of relative clauses, such that the antecedent is shared between the matrix and subordinate clauses, as in the matching analysis). A more recent work, Zyman (2023), argues that Merge as set-formation is problematic, and also proposes a graph-theoretic definition of Merge while explicitly attempting to stay closer to MGG than McKinney-Bock & Vergnaud (and Krivochen, 2023a, b). Zyman's approach is heavily based on selectional features (like Minimalist Grammars), and maintains the idea of projection: Merge(X, Y), where Y must match a feature in the selectional list of X, delivers a binary-branching tree whose root is Z = the 'reflection' of X, where X's reflection is 'identical to X at every stage of the derivation at which they both exist' (essentially, the BPS idea of multi-segment categories). Zyman's Merge multiplies the nodes in structural descriptions, just like set-theoretic Merge with projection does. Graph-theory is used, so far as we can see, to obtain more traditional phrase structure trees than are available under set-theoretic Merge, but not leveraged to minimise the number of nodes or deliver structure sharing (as opposed to McKinney-Bock & Vergnaud's or our works).

Dependency Grammars also use digraphs as their preferred format: in these, predicates always dominate their arguments (such that *dependency* is determined by *motherhood*). Dependency trees make use of the two-dimensional plane in which diagrams are drawn: the x-axis represents precedence, and the y-axis represents dominance (Osborne, 2019: 35). Dependency graphs do not use phrasal nodes (there are no VPs or NPs in a DG), since dependency contrasts with constituency: in a DG the number of nodes in a tree matches the number of words in the sentence whose structure is being represented (note: not *expressions*, but *words*; Osborne, 2019: 37):

(446)

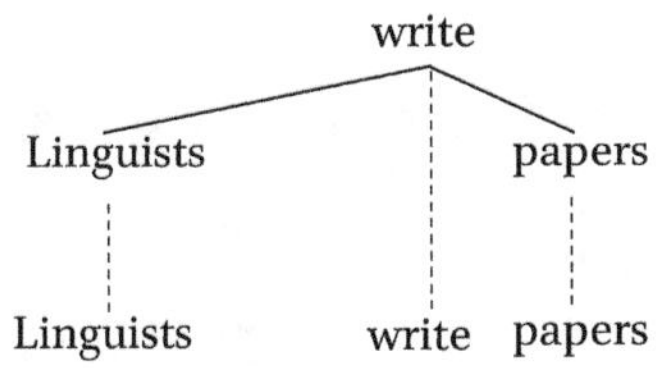

FIGURE 0.3 Dependency Grammar analysis of transitive construct

This style of dependency tree is particularly prevalent in linguistically-oriented Dependency Grammars, rather than computationally oriented versions of the theory. Osborne emphasises the 'economy' of dependency trees with respect to immediate constituent structures in terms of the relation between the number of nodes in a tree and the number of words in a sentence (see also Maxwell, 2013); in a way, that is related to the objections to phrase structure treatments of iteration and coordination based on assigning the relevant sequences 'too much structure' (**Section 1.5**). It is important to note that the number of nodes in a representation is not an automatic indication of economy or of empirical adequacy: in phrase structure grammars, intermediate nodes introduce structure that is leveraged in terms of explanatory adequacy (think, for example, about the importance of intermediate nodes for the definitions of asymmetric c-command, and thus of scope). It is possible for a structure to be 'too simple', in that it does not adequately represent hierarchy and/or asymmetries in relations between expressions. For instance, the dependency analysis offered by Maxwell (2013: 198) for a sentence like *John walks slowly* makes the subject and the adverb both daughter nodes of the verb, with no structural asymmetry between the argument and the adjunct:

(447)

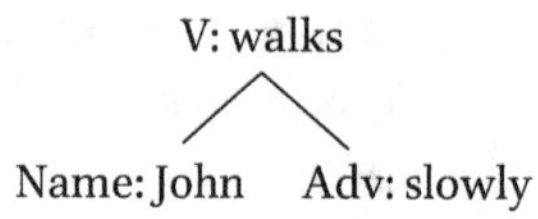

FIGURE 0.4 Dependency Grammar analysis of intransitive sentence with adverbial adjunct in Maxwell (2013)

Additional annotations are needed to capture that which configuration does not: for instance, having arrows instead of lines to indicate that a dependent is an adjunct. Far from being just a notational convention, that annotation encodes a structural notion (or perhaps semantic, depending on whether *adjunct* is a statement about the derivational history of a syntactic object or about its relation to propositional content) that is as part of the analysis as a phrase structure node. The diagram *looks* simpler than a PSG tree, but the formal object it represents (i.e., the system of expressions and relations in a sentence) is not.

A more developed perspective about the relation between contemporary MGG trees and DGs is presented in Osborne et al. (2011). These authors observe that, in order to properly represent the way language is processed (left-to-right), Merge cannot apply over constituents (since the rigidly binary-branching structures generated by classical Merge, if interpreted as trees, do not correspond to the result of most constituency tests). Rather, it must deliver units which are grammatically significant and continuous with respect to dominance: these are *catenae*. Osborne et al. identify Chomsky's (1994) 'bare phrase structure' and Collins' (2002) label-less trees as steps towards a grammar based on dominance alone, and include in this line also approaches that aimed at eliminating the notion of *specifier* (e.g., Starke, 2004; Jayaseelan, 2008). Their argument is that these developments in MGG have taken the theory of structure building (at least to the extent that trees survive in the theory) closer to a Dependency Grammar than to a traditional PSG (an independent argument that Minimalist structures do not represent IC relations can be found in Krivochen, forthcoming b). A crucial point to make the BPS-DG translation work, however, is that some graph-theoretic aspects of Merge-derived structures need to be maintained (such that mere membership / co-membership in unordered sets, as delivered by 'simplest Merge' is not enough to construct DG trees from).

Osborne et al.'s (2011) point can be summarised (but not really made justice to) in the following comparison:

(448)

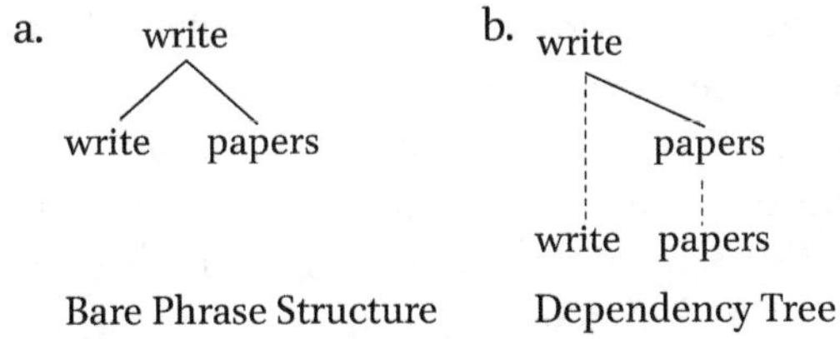

FIGURE 0.5 Comparison between Bare Phrase Structure and Dependency Tree

Dependency grammar can simplify the structure, by eliminating the redundancy in the BPS diagram (note that *write* is akin to a two-segment category). The cost of eliminating structure is eliminating formal relations that BPS and other constituent-structure based representations can express, and which are unavailable in Dependency Grammars (such as c-command and thus the c-command definition of scope).

Interestingly, BPS trees can be mapped to our irreducible graphs (without pre-terminal symbols) via edge contraction: in graph theory, edge contraction removes an edge from a graph, and merges its head and tail as a single node. Then, given (449a), we can apply edge contraction and obtain (449b), introducing the additional requirement that an edge may be contracted only under categorial identity of the nodes it joins (contracted edges are dashed):

(449)

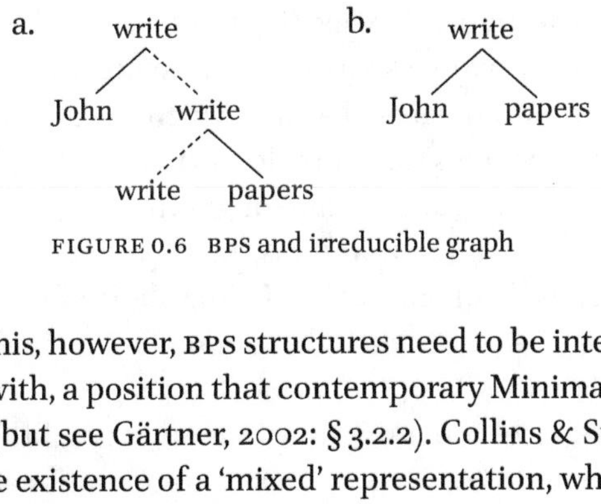

FIGURE 0.6 BPS and irreducible graph

To do this, however, BPS structures need to be interpreted as graph-theoretic to begin with, a position that contemporary Minimalism does not seem willing to adopt (but see Gärtner, 2002: § 3.2.2). Collins & Stabler (2016: 48) seem to suggest the existence of a 'mixed' representation, where Merge generates sets, but graph-theoretic diagrams are allowed in which non-terminal symbols are '*sets, syntactic objects, with arcs pointing to their elements*' (Op. Cit.). The edge contraction method would work in such representations, but these have not been widely adopted so far as we know. Note, incidentally, that the conversion is unidirectional: BPS can be mapped to irreducible graphs but not the other way around, since vertex splitting (Eades & de Mendoça, 1995), applied recursively, could in principle produce an unlimited number of intermediate nodes (unless restricted by conditions that would ultimately result in an analogue to X-bar theory, or LFG's *off-line parsability constraint*—Dalrymple et al., 2019: 245, ff.-).

Other frameworks have used graphs to formalise relations in feature structures rather than dependencies between lexical/phrasal expressions: for example, HPSG feature descriptions are digraphs (e.g., Pollard & Sag, 1994: 17–18) which represent a hierarchy between attributes (such that an attribute of order *n* will immediately dominate an attribute of order *n-1* until reaching a value, which will have outdegree 0). For example, features SYNSEM, LOCAL, CATEGORY, HEAD in the description of a form such as pronoun *she* would be related by immediate dominance (Pollard & Sag, 1994: 17). Syntactic configuration in GPSG, however, is represented by phrase structure trees, with annotated nodes; HPSG has moved away from PS rules (Müller, 2015: § 2.2). LFG uses graphs to represent templates which encode relations between functional descriptions of structures (not structures themselves, as in HPSG type hierarchies): these functional descriptions are called *templates*, and are used to capture generalisations about grammatical categories (Dalrymple et al., 2019: 231, ff.). Some versions of LFG deal with *c*-structures as if they were digraphs (e.g., Wescoat, 2005), but this is not generally assumed (see e.g. the reconceptualisation of *c*-structure in Lowe & Lovestrand, 2020).

Currently, perhaps computational and quantitative approaches are the ones using graphs to the greatest extent, very frequently couched on Dependency Grammar (e.g., Čech et al., 2011; Liu et al., 2017). Dependency parsing is based on defining a walk through an annotated dependency graph, which is optimised to minimise distances. Universal Dependencies (Nivre et al., 2020) is a significant development in this area, providing an inventory of syntactic dependencies, part-of-speech tags, and morphological features which are used to devise multilingual treebanks (UD 2.0 provides treebanks for 90 languages). However, these models make no predictions about structure or grammaticality: they are not grammatical frameworks, but annotation tools. As such, they can generate annotated trees for ill-formed sequences, thus falling short of the goals of linguistic theory but satisfying the requirements of grammar engineering (see Shieber, 1989 for insightful discussion).

The properties of graphs that become relevant in grammatical theory are not always the ones that are important for purposes of computational modelling. Dependency-based quantitative approaches are often heavily influenced by corpora analysis and aim at clustering languages based on quantitative properties: in some of these approaches, with a strong typological base, there is a measure of *distance* which is used to define some notion of 'language family' or equivalent and produce reliable phylogenetic linguistic trees (Marcolli, 2014, 2016; Siva et al., 2017; Shu et al., 2018). Graphs are used as models to represent these language clusters, with edges encoding the relevant distance measure. In some cases, these quantitative approaches use dependencies (in the technical sense) and valency analysis as a way to measure linguistic complexity in a way that is cognitively relevant (Fang & Liu, 2018; Liang et al., 2017): in these works, graphs are used in the framework of theories that aim for psychological plausibility as part of theories of parsing, but whose interests do not include grammatical description. Despite the differences between the theories handled in the references that we somewhat unfairly have grouped here, it is worth pointing out that none of these approaches adopt forms of PSGs or can be said to involve *derivations* in any meaningful sense, although because their focus is not set on providing definitions or characterisations of *well-formedness* for either expressions of the language or formulating second-order conditions over rules, they are orthogonal to the IP/IA distinction. Strictly speaking, they are neither *procedural* nor *declarative*: their focus is not grammar *per se*, but rather *grammar modelling*[1] (frequently, this implies models of *language processing*) and what Shieber (1988) and Pollard (1997) call 'linguistic engineering', whose aims and methodology are very different from those of theoretical and applied syntax (i.e., grammatical analysis).

---

1 This is evident in papers like Ferrer-i-Cancho's (2014) comment to a paper by Liu. The centrality of statistical methods in quantitative approaches contrasts with branches of linguistics focused on grammatical description, which rarely if ever rely heavily on statistical analyses of big datasets (which does not mean, of course, that syntactic work does not resort to corpora, but its goal is to characterise and make predictions about structures, and define a notion of well-formedness).

References

Abels, Klaus & Kristine Bentzen (2012) Are movement paths punctuated or uniform? In Artemis Alexiadou, Tibor Kiss & Gereon Müller (eds.) *Local modelling of non-local dependencies in syntax*. Berlin: de Gruyter. 431–452.

Abney, Steven Paul (1987) *The English noun phrase in its sentential aspect*. PhD dissertation, MIT.

Abrusán, Marta (2014) *Weak island semantics*. Oxford: OUP.

Ackerman, Lauren, Michael Frazier & Masaya Yoshida (2018) Resumptive pronouns can ameliorate illicit island extractions. *Linguistic inquiry* 49(4). 847–859.

Adger, David (2003) *Core syntax: a Minimalist approach*. Oxford: OUP.

Adger, David & Peter Svenonius (2011) Features in Minimalist syntax. In Cedric Boeckx (ed.) *The Oxford handbook of linguistic Minimalism*. Oxford: OUP. 27–51.

Aikhenvald, Alexandra (2004) *Evidentiality*. Oxford: OUP.

Aissen, Judith & David Perlmutter (1983) Clause reduction in Spanish. In David Perlmutter (ed.) *Studies in Relational Grammar 1*. Chicago: University of Chicago Press. 360–403.

Alarcos Llorach, Emilio (1994) *Gramática de la lengua española*. Madrid: Espasa.

Alexopoulou, Theodora (2010) Truly intrusive: resumptive pronominals in questions and relative clauses. *Lingua* 120. 485–505.

Alsina, Alex (2008) A theory of structure-sharing: focusing on long-distance dependencies and parasitic gaps. In Miriam Butt & Tracy Holloway King (eds.) *Proceedings of the LFG08 conference*. Stanford: CSLI. Available online at: http://web.stanford.edu/group/cslipublications/cslipublications/LFG/13/papers/lfg08alsina.pdf [accessed 15/02/2021].

Altham, J.E.J. & Neil Tennant (1975) Sortal quantification. In Edward Keenan (ed.) *Formal semantics of natural language*. Cambridge: CUP. 46–58.

Altshuler, Daniel & Robert Truswell (2022) *Coordination and the syntax-discourse interface*. Oxford: OUP.

Anderson, John (2011) *The substance of language*, Vol. I: *The domain of syntax*. Oxford: OUP.

Andrews, Avery (2018) Sets, heads and spreading in LFG. *Journal of language modelling* 6(1). 131–174.

Andrews, Avery (2021) A speculation about what linguistic structures might be. In I Wayan Arka, Ash Asudeh & Tracy Holloway King (eds.) *Modular design of grammar*. Oxford: OUP. 9–21. https://doi.org/10.1093/oso/9780192844842.003.0002.

Antonenko, Andrei (2012) *Feature-based binding and phase theory*. PhD dissertation, Stony Brook University. Available online at: https://www.researchgate.net/publication/299658117_Feature-Based_Binding_and_Phase_Theory [accessed 16/10/2022].

Arnold, Doug (2007) Non-restrictive relatives are not orphans. *Journal of linguistics* 43(2). 271–309.

Asudeh, Ash (2005) Control and semantic resource sensitivity. *Journal of linguistics* 41(3). 465–511.

Asudeh, Ash (2012) Unbounded dependencies in LFG. *SE-LFG* 8. Available online at: http://www.sas.rochester.edu/lin/sites/asudeh/handouts/asudeh-se-lfg8.pdf [accessed 20/10/2022].

Asudeh, Ash & Ida Toivonen (2012) Copy raising and perception. *Natural language and linguistic theory* 30(2). 321–380.

Asudeh, Ash & Ida Toivonen (2017) A modular approach to evidentiality. In Miriam Butt & Tracy Holloway King (eds.) *Proceedings of the LFG'17 conference*. Stanford: CSLI. 45–65.

Åfarli, Tor Anders (2017) Predication in syntax: toward a semantic explanation of the subject requirement. In Piotr Stalmaszczyk (ed.) *Understanding predication*. Berlin: Peter Lang. 73–96.

Bach, Emmon (1964) *An introduction to transformational grammars*. New York: Holt, Rinehart & Winston.

Bach, Emmon (1970) Problominalization. *Linguistic inquiry* 1(1). 121–122.

Bach, Emmon (1979) Control in Montague grammar. *Linguistic inquiry* 10(4). 515–531.

Bach, Emmon (1980) In defense of passive. *Linguistics and philosophy* 3(3). 297–341.

Bach, Emmon (1982) Purpose clauses and control. In Pauline Jacobson & Geoffrey Pullum (eds.) *The nature of syntactic representation*. Dordrecht: Reidel. 35–57.

Bach, Emmon (1983) Generalized categorial grammars and the English auxiliary. In Frank Heny & Barry Richards (eds.) *Linguistic categories: auxiliaries and related puzzles*, Vol. 2. Dordrecht: Reidel. 101–120.

Bach, Emmon & Robin Cooper (1987) The NP-S analysis of relative clauses and compositional semantics. *Linguistics and Philosophy* 2(1). 145–150.

Bach, Emmon & George Horn (1976) Remarks on 'Conditions on Transformations'. *Linguistic inquiry* 7(2). 265–299.

Bach, Emmon & Barbara Hall Partee (1980) Anaphora and semantic structure. In Jody Kreiman & Almerindo Ojeda (eds.) *Papers from the parasession on pronouns and anaphora*. Chicago: CLS. 1–28. [Reprinted in Partee, Barbara Hall (2004) *Compositionality in formal semantics*. Oxford: Blackwell. 122–152]

Baker, Carl Lee (1970a) Double negatives. *Linguistic inquiry* 1(2). 169–186.

Baker, Carl Lee (1970b) Notes on the description of English questions: the role of an abstract Question morpheme. *Foundations of Language* 6(2). 197–219.

Baker, Carl Lee (1995) Contrast, discourse prominence, and intensification, with special reference to locally free reflexives in British English. *Language* 71(1). 63–101.

Baker, Chris & Pauline Jacobson (2007) Introduction: direct compositionality. In Chris Baker & Pauline Jacobson (eds.) *Direct compositionality*. Oxford: OUP. 1–19.

Baker, Colin & Michael Ellsworth (2017) Graph methods for multilingual FrameNets. *Proceedings of TextGraphs-11: the workshop on graph-based methods for natural language processing*, ACL 2017. 45–50.

Baltin, Mark (1981) Strict bounding. In Carl Baker & John McCarthy (eds.) *The logical problem of language acquisition*. Cambridge, Mass.: MIT Press. 257 295.

Baltin, Mark (1982) A landing site theory of movement rules. *Linguistic inquiry* 13(1). 1–38.

Baltin, Mark (2006) Extraposition. In Martin Everaert & Henk van Riemsdijk (eds.) *The Blackwell companion to syntax*. Oxford: Blackwell. 237–271.

Banik, Eva (2004) Semantics of VP coordination in LTAG. In *Proceedings of the 7th International workshop on Tree Adjoining Grammars and related formalisms* (TAG+7). 118–125.

Barros, Matthew & Luis Vicente (2011) Right Node Raising requires both ellipsis and multidomination. In Lauren Friedman (ed.) *University of Pennsylvania working papers in linguistics* 17(1). 1–9. http://repository.upenn.edu/pwpl/vol17/iss1/2.

Barss, Andrew (1986) *Chains and anaphoric dependence*. PhD dissertation, MIT.

Barss, Andrew (2001) Syntactic reconstruction effects. In Mark Baltin & Chris Collins (eds.) *The handbook of contemporary syntactic theory*. Oxford: Blackwell. 670–696.

Barss, Andrew & Howard Lasnik (1986) A note on anaphora and double objects. *Linguistic inquiry* 17(2). 347–354.

Barwise, Jon & Robin Cooper (1981) Generalized quantifiers and natural language. *Linguistics and philosophy* 4(2). 159–219.

Baunaz, Lena & Eric Lander (2018) Nanosyntax: the basics. In Lena Baunaz, Liliane Haegeman, Karen De Clercq & Eric Lander (eds.) *Exploring Nanosyntax*. Oxford: OUP. 3–56.

Beim Graben, Peter & Sabrina Gerth (2012) Geometric representations for minimalist grammars. *Journal of logic, language, and information* 21(4). 393–432.

Beim Graben, Peter, Bryan Jurish, Douglas Saddy & Stefan Frisch (2004) Language processing by dynamical systems. *International journal of bifurcation and chaos* 14(2). 599–621.

Beim Graben, Peter, Dimitris Pinotsis, Douglas Saddy & Roland Pothast (2008) Language processing with dynamic fields. *Cognitive neurodynamics* 2(2). 79–88.

Bell, Sarah (1983) Advancements and ascensions in Cebuano. In David Perlmutter (ed.) *Studies in Relational Grammar 1*. Chicago: University of Chicago Press. 143–218.

Belletti, Adriana (1988) The Case of unaccusatives. *Linguistic inquiry* 19(1). 1–35.

Belletti, Adriana (2008) The CP of clefts. *Rivista di Grammatica Generativa* 33. 191–204.

van Benthem, Johan (1988) The Lambek calculus. In Richard Oehrle, Emmon Bach & Deirdre Wheeler (eds.) *Categorial Grammars and natural language structures*. Dordrecht: Reidel. 35–68.

Berinstein, Ava (1984) Absolutive extractions: evidence for clause-internal multiattachment in K'ekchi. In Carol Rosen & Laurie Zaring (eds.) *Cornell University working papers in linguistics* 5. 1–65.

Berwick, Robert (1984) Strong generative capacity, weak generative capacity, and modern linguistic theories. *Computational linguistics* 10(3–4). 189–202.

Bhatt, Rajesh (2002) The raising analysis of relative clauses: evidence from adjectival modification. *Natural language semantics* 10(1). 43–90.

Bianchi, Valentina (1999) *Consequences of antisymmetry: headed relative clauses*. Berlin: de Gruyter.

Bianchi, Valentina (2000) The raising analysis of relative clauses: a reply to Borsley. *Linguistic inquiry* 31(1). 123–140.

Bianchi, Valentina & Mara Frascarelli (2010) Is topic a root phenomenon? *Iberia* 2(1). 43–48.

Bianchi, Valentina & Cristiano Chesi (2014) Subject islands, reconstruction, and the flow of the computation. *Linguistic inquiry* 45(4). 525–569.

Binder, Philipe & George Ellis (2016) Nature, computation and complexity. *Phys. Scr.* 91. 064004. doi: 10.1088/0031–8949/91/6/064004.

Bjorkman, Bronwyn (2011) *BE-ing Default: The morphosyntax of auxiliaries*. PhD dissertation, MIT.

Blake, Barry (1990) *Relational Grammar*. London: Routledge.

Blevins, James (1990) *Syntactic complexity: evidence for discontinuity and multidomination*. PhD dissertation, University of Massachusetts, Amherst.

Blevins, James & Ivan Sag (2013) Phrase structure grammar. In Marcel den Dikken (ed.) *The Cambridge handbook of Generative Grammar*. Cambridge: CUP. 202–225.

Boeckx, Cedric (2003) *Islands and chains: resumption as derivational residue*. Amsterdam: John Benjamins.

Boeckx, Cedric (2012) Phases beyond explanatory adequacy. In Ángel Gallego (ed.) *Phases: developing the framework*. Berlin: de Gruyter. 45–66.

Boeckx, Cedric & Kleanthes Grohmann (2004) Barriers and phases: forward to the past? Presented in *Tools in Linguistic Theory* 2004 (TiLT) Budapest (May 16–18, 2004). https://people.umass.edu/roeper/711-05/Boexce%20barriers+phases%20tilt_bg_ho.pdf [Accessed on 21/06/2023]

Bondy, J.A. & U.S.R. Murty (2008) *Graph theory*. New York: Springer.

Borer, Hagit (2003) Exo-skeletal vs. endo-skeletal explanations: syntactic projections and the lexicon. In John Moore & Maria Polinsky (eds.) *The nature of explanation in linguistic theory*. Stanford: CSLI. 31–67.

Borsley, Robert (1997) Relative clauses and the theory of phrase structure. *Linguistic inquiry* 28(4). 629–647.

Borsley, Robert (2001) More on the raising analysis of relative clauses. Ms., University of Essex. https://www.researchgate.net/publication/254848315_MORE_ON_THE_RAISING_ANALYSIS_OF_RELATIVE_CLAUSES. [Accessed on 05/07/2023]

Borsley, Robert (2005) Against CoordP. *Lingua* 115. 461–482.

Borsley, Robert & Berthold Crysmann (2021) Unbounded dependencies. In Stefan Müller, Anne Abeillé, Robert Borsley & Jean-Pierre Koenig (eds.), *Head-Driven Phrase Structure Grammar: the handbook*. Berlin: Language Science Press. 537–594.

Bosque, Ignacio (2000) ¿Qué sabe el que sabe hacer algo? Saber entre los verbos modales. In Fernando García Murga & Kepa Korta Carrión (eds.). *Palabras. Víctor Sánchez de Zavala in memoriam*. Vitoria: University of the Basque Country. 303–323.

Bošković, Željko (2014) Now I'm a phase, now I'm not a phase: on the variability of phases with extraction and ellipsis. *Linguistic inquiry* 45(1). 27–89.

Bouchard, Denis (1982) *On the content of empty categories*. PhD dissertation, MIT.

Bowers, John (2001) Predication. In Mark Baltin & Chris Collins (eds.) *The handbook of contemporary syntactic theory*. Oxford: Blackwell. 299–333.

Börjars, Kersti, Rachel Nordlinger & Louise Sadler (2019) *Lexical Functional Grammar*. Cambridge: CUP.

Brame, Michael (1968) A new analysis of the relative clause: evidence for an interpretive theory. Ms., MIT.

Brame, Michael (1978) *Base generated syntax*. Seattle: Noit Amrofer.

Branan, Kenyon & Michael Yoshitaka Erlewine (2022) Locality and (minimal) search. To appear in Kleanthes Grohmann & Evelina Leivada (eds.) *Cambridge handbook of Minimalism*. https://ling.auf.net/lingbuzz/005791. [Accessed on 22/04/2022]

Bravo, Ana (2016a) Verbos auxiliares. In Javier Gutiérrez-Rexach (ed.) *Enciclopedia de lingüística hispánica* Vol. 2. London: Routledge. 152–162.

Bravo, Ana (2016b) Verbos modales. In Javier Gutiérrez Rexach (ed.) *Enciclopedia de lingüística hispánica*. Volumen 2. Londres: Routledge. 163–173.

Bravo, Ana (2017) *Modalidad y verbos modales*. Madrid: Arco.

Bravo, Ana (2020) On pseudo-coordination in Spanish. *Borealis* 9(1). 125–180.

Bravo, Ana & Luis García Fernández (2016) Perífrasis verbales. In Javier Gutiérrez-Rexach (ed.) *Enciclopedia de lingüística hispánica* Vol. 1. London: Routledge. 785–796.

Bravo, Ana, Luis García Fernández & Diego Gabriel Krivochen (2015) On auxiliary chains: auxiliaries at the syntax-semantics interface. *Borealis* 4(2). 71–101. http://dx.doi.org/10.7557/1.4.2.3612.

Brennan, Virginia (1993) *Root and epistemic modal auxiliary verbs*. PhD dissertation, University of Massachusetts, Amherst

Bresnan, Joan (1982a) Control and complementation. *Linguistic inquiry* 13(3). 343–434.

Bresnan, Joan (1982b) The passive in lexical theory. In Joan Bresnan (ed.) *The mental representation of grammatical relations*. Cambridge, Mass.: MIT Press. 3–83.

Bresnan, Joan (2001) *Lexical Functional Syntax*. [1st Edition]. Oxford: Wiley Blackwell.

Bresnan, Joan, Ash Asudeh, Ida Toivonen & Stephen Wechsler (2016) *Lexical Functional Syntax* [2nd Edition]. Oxford: Wiley Blackwell.

Bresnan, Joan & Jane Grimshaw (1978) The syntax of free relatives in english. *Linguistic inquiry* 9(3). 331–391.

Bresnan, Joan, Ronald Kaplan, Stanley Peters & Annie Zaenen (1982) Cross-serial dependencies in Dutch. *Linguistic inquiry* 13(4). 613–635.

Broekhuis, Hans & Ellen Woodford (2013) Minimalism and optimality theory. In Marcel den Dikken (ed.) *The Cambridge handbook of Generative syntax*. Cambridge: CUP. 122–161.

Brucart, José María (1999) La estructura del sintagma nominal: las oraciones de relativo. In Ignacio Bosque & Violeta Demonte (eds.) *Gramática descriptiva de la lengua española*. Vol. 1. Madrid: Espasa Calpe. 395–522.

Bruening, Bejnamin (2020) The head of the nominal is N, not D: N-to-D movement, hybrid agreement, and conventionalized expressions. *Glossa* 5(1). doi: https://doi.org/10.5334/gjgl.1031.

Bruening, Benjamin & Eman Al Khalaf (2020) Category mismatches in coordination revisited. *Linguistic inquiry* 51(1). 1–36.

Buszkowski, Wojciech (1988) Generative power of Categorial Grammars. In Richard Oehrle, Emmon Bach & Deirdre Wheeler (eds.) *Categorial Grammars and natural language structures*. Dordrecht: Reidel. 69–94.

Butt, Miriam (2006) *Theories of case*. Cambridge: CUP.

Cable, Seth (2010) *The grammar of Q: Q-particles, wh-movement and pied-piping*. Oxford: OUP.

Camacho, José (2003) *The structure of coordination*. Dordrecht: Kluwer.

Cardinaletti, Anna (2004) Toward a cartography of subject positions. In Luigi Rizzi (ed.) *The structure of CP and IP: the cartography of syntactic structures Vol. 2*. Oxford: OUP. 115–165.

Carnie, Andrew (2010) *Constituent structure*. Oxford: OUP.

Carrasco Gutiérrez, Angeles & Luis García Fernández (1994) Sequence of Tenses in Spanish. *University of Venice working papers in linguistics* 4(1). 45–70.

Cerrudo Aguilar, Alba (2015) Cyclic transfer in the derivation of complete parenthetical clauses. *Borealis* 5(1). 59–85.

Cháves, Rui & Michael Putnam (2022) Islands, expressiveness, and the theory/formalism confusion. *Theoretical linguistics* 48(3–4). 219–231.

Chen-Main, Joan & Aravind Joshi (2010) A dependency perspective on the adequacy of tree local multi-component tree adjoining grammar. *Journal of logic and computation* 24(5). 989–1022.

Chierchia, Gennaro (1989) Structured meanings, thematic roles and control. In Gennaro Chierchia, Barbara Partee & Raymond Turner (eds.) *Properties, types and meanings II*. Dordrecht: Kluwer. 131–166.

Chierchia, Gennaro (2004) A semantics for unaccusatives and its syntactic consequences. In Artemis Alexiadou, Elena Anagnostopulou & Martin Everaert (eds.) *The unaccusativity puzzle*. Cambridge: CUP. 22–59.

Chomsky, Noam (1955a) *The logical structure of linguistic theory*. Mimeographed ms., MIT. Available online at http://alpha-leonis.lids.mit.edu/wordpress/wp-content/uploads/2014/07/chomsky_LSLT55.pdf [Accessed on 15/02/2019].

Chomsky, Noam (1955b) *Transformational analysis*. PhD dissertation, University of Pennsylvania.

Chomsky, Noam (1956) Three models for the description of language. *IRE transactions on information theory* 2. 113–124.

Chomsky, Noam (1957) *Syntactic structures*. The Hague: Mouton.

Chomsky, Noam (1959) On certain formal properties of grammars. *Information and control* 2. 137–167.

Chomsky, Noam (1964) A transformational approach to syntax. In Jerry Fodor & Jerrold Katz (eds.) *The structure of language: readings in the philosophy of language*. New Jersey: Prentice Hall. 211–245.

Chomsky, Noam (1965) *Aspects of the theory of syntax*. Cambridge, Mass.: MIT Press.

Chomsky, Noam (1970a) Deep structure, surface structure, and semantic interpretation. In Roman Jakobson & Shigeo Kawamoto (eds.) *Studies in general and oriental linguistics presented to Shirô Hattori on the occasion of his sixtieth birthday*. Tokyo: TEC Corporation for Language and Education Research. 52–91.

Chomsky, Noam (1970b) Remarks on nominalization. In Roderick Jacobs & Peter Rosenbaum (eds.) *Readings in English transformational grammar*. Waltham: Ginn & Co. 184–221.

Chomsky, Noam (1973) Conditions on transformations. In Stephen Anderson & Paul Kiparsky (eds.) *A festschrift for Morris Halle*. New York: Holt, Rinehart, and Winston. 232–286. [Reprinted in Noam Chomsky, *Essays on form and interpretation*. New York: North Holland. 81–160.]

Chomsky, Noam (1977) On wh-movement. In Peter Culicover, Thomas Wasow & Adrian Akmajian (eds.) *Formal syntax*. New York: Academic Press. 71–132.

Chomsky, Noam (1980) On binding. *Linguistic inquiry* 11(1). 1–46.

Chomsky, Noam (1981) *Lectures on Government and Binding*. Dordrecht: Foris.

Chomsky, Noam (1982) *Some concepts and consequences of the theory of Government and Binding*. Cambridge, Mass.: MIT Press.

Chomsky, Noam (1986) *Barriers*. Cambridge, Mass.: MIT Press.

Chomsky, Noam (1994) Bare phrase structure. *MIT occasional papers in linguistics* 5. [reprinted in Gert Webelhuth (ed.) (1995) *Government and binding theory and the minimalist program*. Oxford: Blackwell. 383–439.]

Chomsky, Noam (1995) *The Minimalist Program*. Cambridge, Mass.: MIT Press.

Chomsky, Noam (2000) Minimalist inquiries: the framework. In Roger Martin, David Michaels & Juan Uriagereka (eds.) *Step by Step—Essays in Minimalist syntax in honor of Howard Lasnik*. Cambridge, Mass.: MIT Press. 89–155.

Chomsky, Noam (2001) Derivation by phase. In Michael Kenstowicz (ed.) *Ken Hale: a life in language*. Cambridge, Mass.: MIT Press. 1–52.

Chomsky, Noam (2004) Beyond explanatory adequacy. In Adriana Belletti (ed.) *Structures and beyond. The cartography of syntactic structures*, Vol. 3. Oxford: OUP. 104–131.

Chomsky, Noam (2007) Approaching UG from below. In Uli Sauerland & Hans-Martin Gärtner (eds.) *Interfaces + recursion = language? Chomsky's minimalism and the view from syntax-semantics*. Berlin: de Gruyter. 1–29.

Chomsky, Noam (2008) On phases. In Robert Freidin, Carlos Otero & Maria Luisa Zubizarreta (eds.) *Foundational issues in linguistic theory*. Cambridge, Mass.: MIT Press. 133–166.

Chomsky, Noam (2009) Opening semarks. In Massimo Piatelli-Palmarini, Juan Uriagereka & Pello Salaburu (eds.) *Of minds and language*. Oxford: OUP. 13–43.

Chomsky, Noam (2013) Problems of projection. *Lingua* 130. 33–49.

Chomsky, Noam (2015) Problems of projection: extensions. In Elisa Di Domenico, Cornelia Hamann & Simona Matteini (eds.) *Structures, strategies and beyond: studies in honour of Adriana Belletti*. Amsterdam: John Benjamins. 1–16.

Chomsky, Noam (2020) The UCLA lectures. Ms. https://ling.auf.net/lingbuzz/005485. [Accessed on 04/11/2020]

Chomsky, Noam (2021) Minimalism: where we are now, and where we can hope to go. *Gengo Kenkyu* 160. 1–41.

Chomsky, Noam & Morris Halle (1968) *The sound pattern of English*. New York: Harper & Row.

Chomsky, Noam & Howard Lasnik (1993) The theory of principles and parameters. In Joachim Jacobs, Arnim von Stechow, Wolfgang Sternefeld & Theo Vennemann (eds.) *Syntax: an international handbook of contemporary research*, vol. 1. Berlin: de Gruyter. 506–569.

Chomsky, Noam & George Miller (1963) Introduction to the formal analysis of natural languages. In Duncan R. Luce, Robert R. Bush & Eugene Galanter (eds.) *Handbook of mathematical psychology* 2. New York: John Wiley & Sons. 269–321.

Chung, Sandra & William Ladusaw (2003) *Restriction and saturation*. Cambridge, Mass.: MIT Press.

Chung, Sandra, William Ladusaw & James McCloskey (1995) Sluicing and Logical Form. *Natural language semantics* 3(3). 239–282.

Church, Alonso (1936) An unsolvable problem of elementary number theory. *American journal of mathematics* 58, 354–363.

Cinque, Guglielmo (1999) *Adverbs and functional heads: a cross-linguistic perspective*. Oxford: OUP.

Cinque, Guglielmo (2010) *The syntax of adjectives. a comparative study*. Cambridge, Mass.: MIT Press.

Citko, Barbara (2001) Deletion under identity in relative clauses. *Proceedings of NELS* 31. 131–145.

Citko, Barbara (2005) On the nature of Merge: external Merge, internal Merge, and parallel Merge. *Linguistic inquiry* 36(4). 475–496.

Citko, Barbara & Martina Gračanin-Yuksek (2021) *Merge. Binarity in (multidominant) syntax*. Cambridge, Mass.: MIT Press.

Collins, Chris (2002) Eliminating labels. In Samuel Epstein & T. Daniel Seely (eds.) *Derivation and explanation in the Minimalist Program*. Oxford: Blackwell. 42–64.

Collins, Chris (2005) A smuggling approach to the passive in English. *Syntax* 8(2). 81–120.

Collins, Chris (2017) Merge(X, Y) = {X, Y}. In Leah Bauke & Andreas Blümel (eds.) *Labels and roots*. Berlin: de Gruyter. 47–68.

Collins, Chris (2022) The complexity of trees, Universal Grammar and economy conditions. *Biolinguistics* 16. https://doi.org/10.5964/bioling.9573.

Collins, Chris & Paul Postal (2014) *Classical NEG raising*. Cambridge, Mass.: MIT Press.

Collins, Chris & Edward Stabler (2016) A formalization of Minimalist syntax. *Syntax* 19(1). 43–78.

Collins, Chris & Erich Groat (2018) Distinguishing copies and repetitions. http://ling.auf.net/lingbuzz/003809.

Comrie, Bernard (1986) Tense in indirect speech. *Folia linguistica* 20. 265–296.

Corbett, Greville (2012) *Features*. Cambridge: CUP.

van Craenenbroeck, Jeroen & Tanja Temmerman (2019) Ellipsis in natural language: theoretical and empirical perspectives. In Jeroen van Craenenbroeck & Tanja Temmerman (eds.) *The Handbook of Ellipsis*. Oxford: OUP. 1–16.

Culicover, Peter, Ray Jackendoff & Jenny Audring (2017) Multi-word constructions in the grammar. *Topics in cognitive science* 9(3). 552–568.

Culicover, Peter & Ray Jackendoff (2005) *Simpler Syntax*. Oxford: OUP.

Čech, Radek, Ján Mačutek & Zdeněk Žabokrtský (2011) The role of syntax in complex networks: local and global importance of verbs in a syntactic dependency network. *Physica A*, 390. 3614–3623.

Dalrymple, Mary (2001) *Lexical Functional Grammar*. New York: Academic Press.

Dalrymple, Mary (2007) Introduction to LFG. *LSA linguistic institute* 2007, Day 3. Slides available online at https://www1.essex.ac.uk/linguistics/external/LFG/www-lfg.stanford.edu/pubs/lfg-presentations/institute07/lecture3.pdf. [Accessed on 05/07/2023]

Dalrymple, Mary (2017) Unlike phrase structure category coordination. In Victoria Rosén & Koenraad De Smedt (eds.) *The very model of a modern linguist*. Bergen Language and Linguistic Studies 8. 33–55.

Dalrymple, Mary, Ronald Kaplan & Tracy Holloway King (2015) Economy of expression as a principle of syntax. *Journal of language modelling* 3(2). 377–412. http://dx.doi.org/10.15398/jlm.v3i2.82.

Dalrymple, Mary, John Lowe & Louise Mycock (2019) *The Oxford reference guide to Lexical Functional Grammar*. Oxford: OUP.

Dasgupta, Probal (2021) On certain consequences of the objectification of languages. *Linguistic frontiers* 4(2). https://doi.org/10.2478/lf-2021--0013.

Davidson, Donald (1967) The logical form of action sentences. In Nicholas Resher (ed.) *The logic of decision and action*. Pittsburgh: University of Pittsburgh Press. 81–95.

Davidson, Donald (1969) The individuation of events. In Nicholas Resher (ed.) *Essays in honor of Carl G. Hempel*. Dordrecht: Reidel. 216–234.

Davies, William & Stanley Dubinsky (2004) *The grammar of raising and control*. Oxford: Blackwell.

Davies, William & Stanley Dubinsky (2007) *New horizons in the analysis of control and raising*. Dordrecht: Springer.

Dayal, Veneeta (2016) *Questions*. Oxford: OUP.

Dehé, Nicole & Yordanka Kavalova (2007) Parentheticals: an introduction. In Nicole Dehé & Yordanka Kavalova (eds.) *Parentheticals*. Amsterdam: John Benjamins. 1–22.

Del Gobbo, Francesca (2007) On the syntax and semantics of appositive relative clauses. In Nicole Dehé & Yordanka Kavalova (eds.) *Parentheticals*. Amsterdam: John Benjamins. 173–201.

Demirdache, Hamida (1991) *Resumptive chains in restrictive relatives, appositives and dislocation structures*. PhD dissertation, MIT.

Dipert, Randall (1982) Set-theoretical representations of ordered pairs and their adequacy for the logic of relations. *Canadian journal of philosophy* 12(2). 353–374.

Dixon, Robert M. (1972) *The Dyirbal language of North Queensland*. Cambridge: CUP.

Doliana, Aaron & Sandhya Sundaresan (2022) Proxy control. *Natural language and linguistic theory* 40. 43–101.

Donnellan, Keith (1966) Reference and definite descriptions. *The philosophical review* 75(3). 281–304.

Dowty, David (1978) Governed transformations as lexical rules in a Montague grammar. *Linguistic inquiry* 9(3). 393–426.

Dowty, David (1979) *Word meaning in Montague grammar. The semantics of verbs and times in Generative Semantics and in Montague's PTQ*. Dordrecht: Reidel.

Dowty, David (1982) Grammatical relations and Montague grammar. In Pauline Jacobson & Geoffrey Pullum (eds.) *The nature of syntactic representation*. Dordrecht: Reidel. 79–130.

Dowty, David (1985) On recent analyses of the semantics of control. *Linguistics and philosophy* 8. 291–331.

Dowty, David (1997) Non-constituent coordination, wrapping, and multimodal Categorial Grammars. In Maria Luisa Dalla Chiara, Kees Doets, Daniele Mundici & Johan Van Benthem (eds.) *Structures and norms in Science*. Dordrecht: Springer. 347–368.

Dowty, David (2003) The dual analysis of adjuncts/complements in Categorial Grammar. In Ewald Lang, Claudia Maienborn & Cathrine Fabricius-Hansen (eds.) *Modifying adjuncts*. Berlin: de Gruyter. 33–66.

Dowty, David (2007) Compositionality as an empirical problem. In Chris Baker & Pauline Jacobson (eds.) *Direct compositionality*. Oxford: OUP. 23–101.

Dowty, David, Robert Wall & Stanley Peters (1981) *Introduction to Montague semantics*. Dordrecht: Kluwer.

Eades, P. & C.F.X. de Mendonça (1995) Vertex splitting and tension-free layout. In F.J. Brandenburg (ed.) *Graph drawing*. Berlin: Springer. 202–211. https://doi.org/10.1007/BFb0021804.

Embick, David & Rolf Noyer (2007) Distributed Morphology and the syntax-morphology interface. In Gillian Ramchand & Charles Reiss (eds.) *The Oxford handbook of linguistic interfaces*. Oxford: OUP. 289–324.

Emonds, Joseph (1970) *Root and structure preserving transformations*. PhD dissertation, MIT.

Emonds, Joseph (1979) Appositive relatives have no properties. *Linguistic inquiry* 10(2). 211–243.

Emonds, Joseph (2007) *Discovering syntax: clause structures of English, German and Romance*. Berlin: de Gruyter.

Enderton, Herbert (1972) *A mathematical introduction to logic*. New York: Academic Press.

Engdahl, Elisabet (1986) *Constituent questions: with special reference to Swedish*. Dordrecht: Reidel.

Epstein, Samuel (1999) Un-principled syntax and the derivation of syntactic relations. In Samuel Epstein & Norbert Hornstein (eds.) *Working Minimalism*. Cambridge, Mass.: MIT Press. 317–345.

Epstein, Samuel (2015) On i(nternalist)-functional explanation in Minimalism. In Samuel Epstein, Hisatsugu Kitahara & T. Daniel Seely, *Explorations in maximizing syntactic minimization*. London: Routledge. 71–97.

Epstein, Samuel, Hisatsugu Kitahara & T. Daniel Seely (1998) *A derivational approach to syntactic relations*. Oxford: OUP.

Epstein, Samuel, Hisatsugu Kitahara & T. Daniel Seely (2012) Structure building that can't be. In Myriam Uribe-Etxebarria & Vidal Valmala (eds.) *Ways of structure building*. Oxford: OUP. 253–270

Epstein, Samuel, Hisatsugu Kitahara & T. Daniel Seely (2015) Simplest Merge generates set intersection: Implications for complementizer-trace explanation. In Samuel Epstein, Hisatsugu Kitahara & T. Daniel Seely, *Explorations in maximizing syntactic minimization*. London: Routledge. 175–194.

Epstein, Samuel & T. Daniel Seely (2002) Rule applications as cycles in a level free syntax. In Samuel Epstein & T. Daniel Seely (eds.) *Derivation and explanation in the Minimalist Program*. Oxford: Blackwell. 65–89.

Epstein, Samuel & T. Daniel Seely (2006) *Derivations in Minimalism*. Cambridge: CUP.

Erlewine, Michael Yoshitaka (2020) Anti-locality and subject extraction. *Glossa* 5(1). 84. https://doi.org/10.5334/gjgl.1079.

Erteschik-Shir, Nomi (1973) *On the nature of island constraints*. PhD dissertation, MIT.

Escandell Vidal, Mª Victoria & Manuel Leonetti (2000) Categorías conceptuales y semántica procedimental. In Marcos Martínez Hernández et al. (eds.) *Cien años de investigación semántica: de Michél Bréal a la actualidad. Tomo I*. Madrid: Ediciones Clásicas. 363–378.

Escandell Vidal, Mª Victoria & Manuel Leonetti (2006) Remarks on optimality-theoretic pragmatics. In Marta Carretero Lapeyre et al. (eds.) *A pleasure of life in words: a festschrisft for Angela Downing*. Madrid: Universidad Complutense. Vol. 1. 489–514.

Evers, Arnold (1975) *The transformational cycle in Dutch and German*. Indiana: Indiana University Linguistics Club.

Fabb, Nigel (1983) Three squibs on auxiliaries. In I. Haik & D. Massam (eds.) *MIT working papers in linguistics* 5. Cambridge, Mass.: MIT Press. 104–120.

Fabb, Nigel (1990) The difference between English restrictive and nonrestrictive relative clauses. *Journal of linguistics* 26(1). 57–77.

Falk, Yehuda (1984) The English auxiliary system: a Lexical-Functional analysis. *Language* 60(3). 483–509.

Falk, Yehuda (2000) Pivots and the theory of grammatical functions. In Miriam Butt & Tracy Holloway King (eds.) *Proceedings of the LFG00 conference*. Stanford: CSLI. http://web.stanford.edu/group/cslipublications/cslipublications/LFG/5/pdfs/lfg00falk.pdf. [Accessed on 20/06/2023]

Falk, Yehuda (2001) *Lexical-Functional Grammar: an introduction to parallel constraint-based syntax*. Stanford: CSLI.

Falk, Yehuda (2003) The English auxiliary system revisited. In Miriam Butt & Tracy Holloway King (eds.) *Proceedings of the LFG03 conference*. Stanford: CSLI. 184–204.

Falk, Yehuda (2009) Islands: a mixed analysis. In Miriam Butt & Tracy Holloway King (eds.) *Proceedings of the LFG09 conference*. Stanford: CSLI. http://web.stanford.edu/group/cslipublications/cslipublications/LFG/14/papers/lfg09falk.pdf. [Accessed on 15/02/2023]

Fang, Yu & Haitao Liu (2018) What factors are associated with dependency distances to ensure easy comprehension? A case study of *ba* sentences in Mandarin Chinese. *Language sciences* 67. 33–45.

Fanselow, Gisbert (2003) Münchhausen-style head movement and the analysis of verb-second. In A. Mahajan (ed.) *Head movement and syntactic theory*. UCLA & Universitat Potsdam Working Papers in Linguistics, Los Angeles & Potsdam. 40–76.

Ferrer-i-Cancho, Ramón (2014) Beyond description. Comment on "Approaching human language with complex networks" by Cong & Liu. *Physics of life reviews* 11(4). 621–623.

Fiengo, Robert (1977) On trace theory. *Linguistic inquiry* 8(1). 35–61.

Fillmore, Charles (1963) The position of embedding transformations in a grammar. *Word* 19(2). 208–231.

Findlay, Jamie (2016) The prepositional passive in Lexical Functional Grammar. In Doug Arnold Poland, Miriam Butt, Berthold Crysmann, Tracy Holloway King & Stefan Muller (eds.) *Proceedings of the Joint 2016 conference on Head-driven Phrase Structure Grammar and Lexical Functional Grammar*. Stanford: CSLI. 255–275.

von Fintel, Kai & Irene Heim (2011) Intensional semantics. Lecture notes, MIT. Available online at https://github.com/fintelkai/fintel-heim-intensional-notes/blob/master/fintel-heim-2011-intensional.pdf [Accessed on 19/12/2021].

Fox, Danny (2002) Antecedent-contained deletion and the copy theory of movement. *Linguistic inquiry* 33(1). 63–96.

Fox, Danny (2012) The semantics of questions: introductory remarks. Ms. MIT. Available online at: http://lingphil.mit.edu/papers/fox/firstclass.pdf [accessed on 20/10/2022].

Fox, Chris & Shalom Lappin (2005) *Foundations of intensional semantics*. Oxford: Blackwell.

Frank, Robert (1992) *Syntactic locality and Tree Adjoining Grammar: grammatical, acquisition and processing perspectives*. PhD dissertation, University of Pennsylvania.

Frank, Robert (2002) *Phrase structure composition and syntactic dependencies*. Cambridge, Mass.: MIT Press.

Frank, Robert (2006) Phase theory and Tree Adjoining Grammar. *Lingua* 116(2). 145–202.

Frank, Robert (2013) Tree Adjoining Grammar. In Marcel den Dikken (ed.) *The Cambridge handbook of Generative syntax*. Cambridge: CUP. 226–261.

Frank, Robert & Tim Hunter (2021) Variation in mild context-sensitivity: Derivational state and structural monotonicity. *Evolutionary linguistic theory* 3(2). 181–214.

Frank, Robert & Hadas Kotek (2022) Arguments for top-down derivations in syntax. *Proc Ling Soc Amer* 7(1). 5264. https://doi.org/10.3765/plsa.v7i1.5264.

Frank, Robert & Anthony Kroch (1995) Generalized transformations and the theory of grammar. *Studia linguistica* 49(2). 103–151.

Frampton, John & John Gutmann (1999) Cyclic computation: a computationally efficient Minimalist syntax. *Syntax* 2(1). 1–27.

Freidin, Robert (1992) *Foundations of Generative syntax*. Cambridge, Mass.: MIT Press.

Friedmann, Naama, Adriana Belletti & Luigi Rizzi (2009) Relativized relatives: types of intervention in the acquisition of A-bar dependencies. *Lingua* 119(1). 67–88.

Fukui, Naoki & Hiroki Narita (2014) Merge, labelling, and projection. In Andrew Carnie, Yosuke Sato & Daniel Siddiqi (eds.) *The Routledge handbook of syntax*. London: Routledge. 3–23.

García Fernández, Luis (2006) Perífrasis verbales en español. In Luis García Fernández (dir.) *Diccionario de perífrasis verbales*. Madrid: Gredos. 9–58.

García Fernández, Luis & Diego Gabriel Krivochen (2019a) Dependencias no locales y cadenas de verbos auxiliares. *Verba* 46. 207–244. http://dx.doi.org/10.15304/verba.46.4567.

García Fernández, Luis & Diego Gabriel Krivochen (2019b) *Las perífrasis verbales españolas en contraste*. Madrid: Arco Libros.

García Fernández, Luis & Diego Gabriel Krivochen (2020) Formas no finitas duplicadas en las cadenas de verbos auxiliares. *Revista internacional de lingüística iberoamericana* 35(1). 141–167.

García Fernández, Luis, Diego Gabriel Krivochen & Ana Bravo (2017) Aspectos de la semántica y sintaxis de las cadenas de verbos auxiliares en español. *Moenia* 23. 1–28.

García Fernández, Luis, Diego Gabriel Krivochen & Félix Martín Gómez (2020) Los elementos intermedios en las perífrasis verbales. *Lingüística española actual* XLII/2. 167–200.

Gazdar, Gerald (1981) Unbounded dependencies and coordinate structure. *Linguistic inquiry* 12(2). 155–184.

Gazdar, Gerald (1982) Phrase structure grammar. In Pauline Jacobson & Geoffrey Pullum (eds.) *The nature of syntactic representation*. Dordrecht: Reidel. 131–186.

Gazdar, Gerald, Ewan Klein, Geoffrey Pullum & Ivan Sag (1985) *Generalized Phrase Structure Grammar*. Cambridge, Mass.: Harvard University Press.

Gärtner, Hans-Martin (2002) *Generalized transformations and beyond. Reflections on Minimalist syntax*. Berlin: Akademie Verlag.

Gärtner, Hans-Martin (2014) Strange loops: Phrase-Linking Grammar meets Kaynean pronominalization. *Linguistische Berichte* 24. 1–13.

Gärtner, Hans-Martin (2021) Copies from "Standard Set Theory"? A note on the foundations of Minimalist syntax in reaction to Chomsky, Gallego and Ott (2019). *Journal of logic, language and information*. https://doi.org/10.1007/s10849--021--09342-x.

Geraci, Carlo (2020) Graft, remove or exfoliate? Towards a theory of structure reduction. Presented at GLOW 43. April 8th, 2020. https://osf.io/fph8a. [Accessed on 01/07/2023]

Giannakidou, Anastasia (2002) Licensing and sensitivity in polarity items: from downward entailment to nonveridicality. In Maria Andronis, Anne Pycha & Keiko Yoshimura (eds.). *CLS 38: Papers from the 38th annual meeting of the Chicago Linguistic Society, parasession on polarity and negation*. Available at http://home.uchicago.edu/~giannaki/pubs/cls.giannakidou.pdf.

Gibson, Jeanne & Eduardo Raposo (1986) Clause union, the Stratal Uniqueness Law and the Chômeur relation. *Natural language & linguistic theory* 4(3). 295–331.

Ginsburg, Jason (2016) Modelling of problems of projection: A non-countercyclic approach. *Glossa* 1(1). 1–46. http://dx.doi.org/10.5334/gjgl.22.

Gómez Torrego, Leonardo (1999). Los verbos auxiliares. Las perífrasis verbales de infinitive. In Ignacio Bosque & Violeta Demonte (dirs.) *Gramática descriptiva de la lengua española*, 3 vols. Madrid: Espasa, vol. 2. 3323–3389.

Goodall, Grant (1987) *Parallel structures in syntax: coordination, causatives, and restructuring*. Cambridge: CUP.

Gorn, Saul (1967) Handling the growth by definition of mechanical languages. *Proceedings of the April 18–20, 1967, spring joint computer conference*. New York: Association for Computing Machinery. 213–224.

Gould, Ronald (1988) *Graph theory*. California: The Benjamins/Cummings Publishing Company.

Graf, Thomas (2021) Minimalism and computational linguistics. To appear in Kleanthes Grohmann & Evelina Leivada (eds.) *Cambridge handbook of Minimalism*. https://ling.auf.net/lingbuzz/005855. [Accessed on 15/09/2022]

Graf, Thomas & Aniello De Santo (2019) Sensing tree automata as a model of syntactic dependencies. In *Proceedings of the 16th meeting on the mathematics of language*. Toronto: Association for Computational Linguistics. 12–26.

Greibach, Sheila (1965) A new normal-form theorem for context-free phrase structure grammars. *Journal of the ACM* 12(1). 42–52.

Grimshaw, Jane (2000) Locality and extended projection. In Peter Coopmans, Martin Everaert & Jane Grimshaw (eds.) *Lexical specification and insertion*. Amsterdam: John Benjamins. 115–133.

Groenendijk, Jeroen & Martin Stokhof (2011) Questions. In Johan Van Benthem & Alice ter Meulen (eds.) *Handbook of logic and language*, 2nd Edition. Amsterdam: Elsevier. 1059–1131.

Grohmann, Kleanthes (2003) *Prolific domains: on the anti-locality of movement dependencies*. Amsterdam: John Benjamins.

Gross, Jonathan & Jay Yellen (2014) Fundamentals of graph theory. In Jonathan Gross, Jay Yellen & Ping Zhang (eds.) *Handbook of graph theory* [2nd Edition]. London: Routledge. 2–20.

Gross, Jonathan, Jay Yellen & Mark Anderson (2018) *Graph theory and its applications* [3rd Edition]. CRC Press.

Gruber, Jeffrey (1965) *Studies in lexical relations*. PhD dissertation, MIT.

Guimarães, Maximiliano (2000) In defense of vacuous projections in Bare Phrase Structure. *University of Maryland working papers in linguistics* 9. 90–115.

Guimarães, Maximiliano (2004) *Derivation and representation of syntactic amalgams*. PhD dissertation, University of Maryland.

Hacquard, Valentine (2010) On the event relativity of modal auxiliaries. *Natural language semantics* 18. 79–114. https://doi.org/10.1007/s11050--010--9056--4.

Haegeman, Liliane (2009) Parenthetical adverbials: the radical orphanage approach. In Benjamin Shaer, Philippa Cook, Werner Frey & Maienborn Claudia (eds.) *Dislo-*

cated elements in discourse: syntactic, semantic and pragmatic perspectives. London: Routledge. 331–347.

Haider, Hubert (1996) Economy in syntax is projective economy. In Chris Wilder, Hans-Martin Gärtner & Manfred Bierwisch (eds.) *The Role of Economy Principles in Linguistic Theory*. Berlin: de Gruyter. 205–226.

Haider, Hubert (2018) On Minimalist theorizing and scientific ideology in grammar theory. Ms. https://ling.auf.net/lingbuzz/004967. [Accessed on 30/07/2019]

Haider, Hubert (2019) Grammatical rules are discrete, not weighted, and not vulnerable. In Ken Ramshøj Christensen, Henrik Jørgensen & Johanna L. Wood (eds.) *The sign of the V: papers in honour of Sten Vikner*. Aarhus University. 205–226.

Haider, Hubert (2020) A null theory of scrambling. *Zeitschrift für Sprachwissenschaft* 39(3). 375–405.

Hale, Kenneth & Samuel J. Keyser (1993) On argument structure and lexical expression of syntactic relations. In Kenneth Hale & Samuel J. Keyser (eds.) *The view from building 20: essays in linguistics in honor of Sylvain Bromberger*. Cambridge, Mass.: MIT Press. 53–109.

Hale, Kenneth & Samuel J. Keyser (2002) *Prolegomena to a theory of argument structure*. Cambridge, Mass.: MIT Press.

Hale, Kenneth & Samuel J. Keyser (2005) Aspect and the syntax of argument structure. In Nomi Erteschik-Shir & Tova Rapoport (eds.) *The syntax of aspect. Deriving thematic and aspectual interpretation*. Oxford: OUP. 11–41.

Hamblin, C.L. (1973) Questions in Montague English. *Foundations of language* 10. 41–53.

Han, Chung-Hye, David Potter & Dennis Storoshenko (2010) Non-local right node raising: an analysis using delayed tree-local MC-TAG. In *Proceedings of the 10th international workshop on Tree Adjoining Grammars and related formalisms* (TAG+10). Yale University. 9–16.

Han, Chung-Hye & Anoop Sarkar (2017) Coordination in TAG without the Conjoin operation. In *Proceedings of the 13th international workshop on Tree Adjoining Grammars and related formalisms* (TAG+13). 43–52.

Hankamer, Jorge (1979) *Deletion in coordinate structures*. New York: Garland.

Hankamer, Jorge & Ivan Sag (1976) Deep and surface anaphora. *Linguistic inquiry* 7(3). 391–426.

Haspelmath, Martin (2021) How to tear down the walls that separate linguists: continuing the quest for clarity about general linguistics. *Theoretical linguistics* 47(1–2). 137–154.

Harary, Frank, R.Z. Norman & Dorwin Cartwright (1965) *Structural models: an introduction to the theory of directed graphs*. New York: John Wiley and Sons.

Harley, Heidi (2003) Possession and the double object construction. *Linguistic variation yearbook* 2 (2002), 29–68.

Harley, Heidi (2011) A Minimalist approach to argument structure. In Cedric Boeckx (ed.) *The handbook of linguistic Minimalism*. Oxford: OUP. 427–448.

Harley, Heidi & Shigeru Miyagawa (2017) Syntax of ditransitives. In *Oxford research encyclopedia of linguistics*. https://doi.org/10.1093/acrefore/9780199384655.013.186 [Accessed on 07/02/2023].

Harris, Alice (2016) *Multiple exponence*. Oxford: OUP.

Harris, Zellig (1945) Discontinuous morphemes. *Language* 21(3). 121–127.

Harris, Zellig (1951) *Structural linguistics*. Chicago: University of Chicago Press.

Harris, Zellig (1957) Co-occurrence and transformation in linguistic structure. *Language* 33(3). 283–340.

Harwood, William (2014) Rise of the auxiliaries: a case for auxiliary raising vs. affix lowering. *The linguistic review* 31(2). 295–362.

Haug, Dag Trygve Truslew (2014) The anaphoric semantics of partial control. *Proceedings of SALT 24*. 213–233.

Hauser, Marc, Noam Chomsky & W. Tecumseh Fitch (2002) The faculty of language: What is it, who has it, and how did it evolve? *Science* 298: 1569–1579. https://doi.org/10.1126/science.298.5598.1569.

He, Chuansheng (2015) E-type interpretation without E-type pronoun: how Peirce's graphs capture the uniqueness implication of donkey pronouns in discourse anaphora. *Synthese* 192. 971–990.

Hefferon, Jim (2014) *Linear algebra*. Free access book. Available online at: http://joshua.smcvt.edu/linearalgebra [Accessed on 31/7/2022].

Hegarty, Michael (1993) Deriving clausal structure in Tree Adjoining Grammar. Ms. University of Pennsylvania.

Heim, Irene & Angelika Kratzer (1998) *Semantics in Generative Grammar*. Oxford: Blackwell.

Higginbotham, James (1983) Logical Form, binding, and nominals. *Linguistic inquiry* 14(3). 395–420.

Higginbotham, James (1985) On semantics. *Linguistic inquiry* 16(4). 547–593.

Higginbotham, James (2000) On events in linguistic semantics. In James Higginbotham, Fabio Pianesi & Achille Varzi (eds.) *Speaking of events*. Oxford: OUP. 49–79.

Higginbotham, James & Robert May (1981) Questions, quantifiers, and crossing. *The linguistic review* 1(1). 41–79.

Hockett, Charles (1954) Two models of grammatical description. *Word* 10. 210–234.

Hockett, Charles (1958) *A course in modern linguistics*. New York: Macmillan.

Hoffmann, Ludger (1998) Parenthesen. *Linguistische Berichte* 175. 299–328.

Hopcroft, John & Jeffrey Ullman (1969) *Formal languages and their relation to automata*. London: Addison-Wesley.

Horn, Laurence (1989) *A natural history of negation*. Chicago: University of Chicago Press.

Hornstein, Norbert (1994) An argument for Minimalism: the case of antecedent-contained deletion. *Linguistic inquiry* 25(3). 455–480.

Hornstein, Norbert (1999) Movement and control. *Linguistic inquiry* 30(1). 69–96.

Hornstein, Norbert (2001) *Move! A Minimalist theory of construal*. Oxford: Blackwell.

Hornstein, Norbert (2003) On control. In Randall Hendrick (ed.) *Minimalist syntax*. Oxford: Blackwell. 6–81.

Hornstein, Norbert & William Idsardi (2014) A program for the Minimalist Program. In Peter Kosta, Steven Franks, Teodora Radeva-Bork & Lilia Schürcks (eds.) *Minimalism and beyond: radicalizing the interfaces*. Amsterdam: John Benjamins. 9–36.

Huang, C.T. James (1982) *Logical relations in Chinese and the theory of grammar*. PhD dissertation, MIT.

Huck, Geoffrey (1984) *Discontinuity and word order in Categorial Grammar*. PhD dissertation, University of Chicago.

Huck, Geoffrey (1988) Phrasal verbs and postponement. In Richard Oehrle, Emmon Bach & Deirdre Wheeler (eds.) *Categorial Grammars and natural language structures*. Dordrecht: Reidel. 249–264.

Huddleston, Rodney & Geoffrey Pullum (2002) *The Cambridge grammar of the English language*. Cambridge: CUP.

Hudson, Richard (2007) *Language networks*. Oxford: OUP.

Hübler, Nataliia (2016) *Passive construction with intransitive verbs: typology and distribution*. MA dissertation, Kiel University.

Jackendoff, Ray (1977) *X′ syntax: a study of phrase structure*. Cambridge, Mass: MIT Press.

Jackendoff, Ray (1987) The status of thematic relations in linguistic theory. *Linguistic inquiry* 18(3). 369–411.

Jackendoff, Ray (2011) Alternative minimalist visions of language. In Robert Borsley & Kersti Börjars (eds.) *Non-transformational syntax*. Oxford: Wiley-Blackwell. 268–296.

Jacobson, Pauline (1977) *The syntax of crossing correference sentences*. PhD dissertation, University of California, Berkeley.

Jacobson, Pauline (1987) Phrase structure, grammatical relations, and discontinuous constituents. In Geoffrey Huck & Almerindo Ojeda (eds.), *Syntax and semantics 20: discontinuous constituency*. New York: Academic Press. 27–69.

Jacobson, Pauline (1990) Raising as function composition. *Linguistics and philosophy* 13. 423–475.

Jacobson, Pauline (1992) Raising without movement. In Richard Larson, Sabine Iatridou, Utpal Lahiri & James Higginbotham (eds.) *Control and grammar*. Dordrecht: Kluwer. 149–194.

Jacobson, Pauline (2000) Paycheck pronouns, Bach-Peters sentences, and variable-free semantics. *Natural language semantics* 8. 77–155.

Jacobson, Pauline (2012) Direct compositionality. In Wolfram Hinzen, Edouard Machery & Markus Werning (eds.) *The Oxford handbook of compositionality*. Oxford: OUP. 109–128.

Jaeger, Florian (2004) Binding in picture NPs revisited: evidence for a semantic principle of extended argumenthood. In Miriam Butt & Tracy Holloway King (eds.) *Proceedings of LFG04*. Stanford: CSLI. 268–288.

Jayaseelan, K.A. (2008) Bare Phrase Structure and specifier-less syntax. *Biolinguistics* 2(1). 87–106.

Jespersen, Otto (1985) [1937] *Analytic syntax*. Chicago: University of Chicago Press.

Jiménez Fernández, Angel (2015) Towards a typology of focus: subject position and microvariation at the discourse–syntax interface. *Ampersand* 2. 49–60.

Johnson, David & Paul Postal (1980) *Arc Pair Grammar*. Princeton, NJ: Princeton University Press.

Johnson, Kyle (2003) Towards an etiology of adjunct islands. *Nordlyd* 31(1). 187–215.

Johnson, Kyle (2009) Why movement? Ms. UMass. https://people.umass.edu/kbj/homepage/Content/islandspaper.pdf. [Accessed on 01/05/2022]

Johnson, Kyle (2014) Recoverability of deletion. In Kuniya Nasukawa & Henk van Riemsdijk (eds.) *Identity relations in the grammar*. Berlin: de Gruyter. 255–288.

Johnson, Kyle (2016) Toward a multidominant theory of movement. Lectures presented at ACTL, University College, June 2016. Available at https://people.umass.edu/kbj/homepage/Content/Multi_Movement.pdf. [Accessed on 03/05/2022]

Johnson, Kyle (2020) Rethinking linearization. In András Bárány, Theresa Biberauer, Jamie Douglas & Sten Vikner (eds.) *Syntactic architecture and its consequences II: between syntax and morphology*. Berlin: Language Science Press. 113–135.

Johnson, Mark (1988) *Attribute-value logic and the theory of grammar*. Stanford: CSLI.

Joshi, Aravind (1969) Properties of formal grammars with mixed types of rules and their linguistic relevance. Proceedings International Symposium on Computational Linguistics, Silnga Sfiby, Sweden. Available online at https://aclanthology.org/C69-4701.pdf. [Accessed on 21/06/2023]

Joshi, Aravind (1985) Tree adjoining grammars: How much context-sensitivity is required to provide reasonable structural descriptions? In David Dowty, Lauri Karttunen & Arnold Zwicky (eds.) *Natural language parsing*. Cambridge: CUP. 206–250.

Joshi, Aravind, S. Rao Kosaraju & H.M. Yamada (1972) String adjunct grammars II: equational representation, null symbols, and linguistic relevance. *Information and control* 21. 235–260.

Joshi, Aravind & Yves Schabes (1991) Tree-Adjoining Grammars and lexicalized grammars. *Technical Reports (CIS)*. Paper 445. http://repository.upenn.edu/cis_reports/445. [Accessed on 03/07/2023]

Joshi, Aravind & Yves Schabes (1997) Tree Adjoining Grammars. In Grzegorz Rozenberg & Arto Salomaa (eds.) *Handbook of formal grammars*. Berlin: Springer. 69–123.

Kahane, Sylvain & François Lareau (2016) Word ordering as a graph-rewriting process. In Annie Foret, Glyn Morrill, Reinhard Muskens, Rainer Osswald & Sylvain Pogodalla (eds.) *Formal grammar*. Berlin: Springer. 216–239. https://doi.org/10.1007/978--3--662--53042--9_13.

Kahane, Sylvain & Nicolas Mazziotta (2015) Syntactic polygraphs: a formalism extending both constituency and dependency. *Proceedings of the 14th meeting on the mathematics of language*. Association for Computational Linguistics. 152–164.

Kaplan, Ronald & Joan Bresnan (1982) Lexical Functional Grammar: a formal system for grammatical representation. In Joan Bresnan (ed.) *The mental representation of grammatical relations*. Cambridge, Mass.: MIT Press. 173–281.

Kaplan, Ronald & John T. Maxwell, III (1988) Constituent coordination in Lexical Functional Grammar. In *Proceedings of the international conference on Computational linguistics (COLING 88)*. International Committee on Computational Linguistics. 303–305.

Kaplan, Ronald & Annie Zaennen (1995) Long-distance dependencies, constituent structure, and functional uncertainty. In Mary Dalrymple, Ronald Kaplan, John Maxwell III & Annie Zaenen (eds.) *Formal issues in Lexical Functional Grammar*. Stanford: CSLI. 137–166.

Karttunen, Lauri (1971a) Definite descriptions with crossing reference. *Foundations of language* 7. 157–182.

Karttunen, Lauri (1971b) Some observations on factivity. *Papers in linguistics* 4(1). 55–69.

Karttunen, Lauri (1977) Syntax and semantics of questions. *Linguistics and philosophy* 1. 3–44.

Karttunen, Lauri & Martin Kay (1985) Structure sharing with binary trees. *Proceedings of the 23rd annual meeting of the Association for Computational Linguistics*. Chicago, Illinois: University of Chicago. 133–136.

Katz, Jerry & Paul Postal (1964) *An integrated theory of linguistic descriptions*. Cambridge, Mass.: MIT Press.

Kayne, Richard (1981a) On certain differences between French and English. *Linguistic inquiry* 12. 349–371.

Kayne, Richard (1981b) 'ECP' extensions. *Linguistic inquiry* 12(1). 93–134.

Kayne, Richard (1983) Connectedness. *Linguistic inquiry* 14(2). 223–249.

Kayne, Richard (1984) *Connectedness and binary branching*. Dordrecht: Foris.

Kayne, Richard (1994) *The antisymmetry of syntax*. Cambridge, Mass.: MIT Press.

Kayne, Richard (2002) Pronouns and their antecedents. In Samuel Epstein & T. Daniel Seely (eds.) *Derivation and explanation in the Minimalist Program*. Oxford: Blackwell. 133–166.

Kayne, Richard (2013) Why are there no directionality parameters? In Theresa Biberauer & Michelle Sheehan (eds.) *Theoretical approaches to disharmonic word order*. Oxford: OUP. 219–244.

Kayne, Richard (2018) The place of linear order in the language faculty. Talk delivered at the University of Venice, June 16th, 2018. Available online at: https://as.nyu.edu/content/dam/nyu-as/linguistics/documents/Kayne%200118%20Venice%20The%20Place%20of%20Linear%20Order%20in%20the%20Language%20Faculty.pdf [Accessed on 15/21/2020].

Kayne, Richard (2022) Antisymmetry and externalization. *Studies in Chinese linguistics* 43(1). 1–20.

Keenan, Edward (1997) The semantics of determiners. In Shalom Lappin (ed.) *The handbook of contemporary semantic theory* [1st Edition]. Oxford: Wiley-Blackwell. 41–63.

Keenan, Edward (2006) On the denotations of anaphors. *Research on language and computation* 5(1). 5–17.

Keenan, Edward & Bernard Comrie (1977) Noun phrase accessibility and Universal Grammar. *Linguistic inquiry* 8. 63–99.

Kempson, Ruth, Eleni Gregoromichelaki, Arash Eshghi & Julian Hough (2019) Ellipsis in Dynamic Syntax. In Jeroen van Craenenbroeck & Tanja Temmerman (eds.) *The handbook of ellipsis*. Oxford: OUP. 205–232.

Keshet, Ezra (2010) Split intensionality: a new scope theory of de re and de dicto. *Linguistics and philosophy* 33(4). 251–283.

Keshet, Ezra & Florian Schwarz (2019) De re / de dicto. In Jeanette Gundel & Barbara Abbott (eds.) *The Oxford handbook of reference*. Oxford: OUP. 168–202. https://doi.org/10.1093/oxfordhb/9780199687305.013.10.

Kibort, Anna (2008) On the syntax of ditransitive constructions. In Miriam Butt & Tracy Holloway King (eds.) *Proceedings of the LFG08 conference*. Stanford: CSLI.

King, Tracy Holloway (2016) Theoretical linguistics and grammar engineering as mutually constraining disciplines. In Doug Arnold, Miriam Butt, Berthold Crysmann, Tracy Holloway King & Stefan Müller (eds.) *Proceedings of the joint 2016 conference on Head-driven Phrase Structure Grammar and Lexical Functional Grammar*. Warsaw: Polish Academy of Sciences. 339–359.

Kiparsky, Paul & Carol Kiparsky (1970) Fact. In Manfred Bierwisch & K.E. Heidolph (eds.) *Progress in linguistics*. The Hague: Mouton. 143–173.

É. Kiss, Katalin (1999) The English cleft construction as a Focus Phrase. In Lunella Mereu (ed.) *Boundaries of morphology and syntax*. Amsterdam: John Benjamins. 217–229.

Kitahara, Hisatsugu (1997) *Elementary operations and optimal derivations*. Cambridge, Mass.: MIT Press.

Kluck, Mariles (2011) *Sentence amalgamation*. Utrecht: LOT.

Kluck, Mariles (2015) On representing anchored parentheses in syntax. In Andreas Trotzke & Josef Bayer (eds.) *Syntactic complexity across interfaces*. Berlin: de Gruyter. 107–136.

Kluck, Marlies, Dennis Ott & Mark de Vries (2014) Incomplete parenthesis: An overview. In Marlies Kluck, Dennis Ott & Mark de Vries (eds.) *Parenthesis and ellipsis*. Berlin: de Gruyter. 1–22.

Koizumi, Masatoshi (1995) *Phrase structure in minimalist syntax*. PhD dissertation, MIT.

Koopman, Hilda & Dominique Sportiche (2000) Variables and the bijection principle. In Hilda Koopman, *The syntax of specifiers and heads*. London: Routledge. 16–36.

Koopman, Hilda, Dominique Sportiche & Edward Stabler (2014) *An introduction to syntactic analysis and theory*. Oxford: Wiley-Blackwell.

Kornai, Andras (1985) Natural languages and the Chomsky hierarchy. In M. King (ed.) *Proceedings of the 2nd European conference of the Association for Computational Linguistics*. 1–7.

Koster, Jan (1987) *Domains and dynasties*. Dordrecht: Foris.

Kotek, Hadas (2020) *Composing questions*. Cambridge, Mass.: MIT Press.

Kotek, Hadas & Martin Hackl (2013) A new syntax for multiple wh-questions. Evidence from real time sentence processing. Ms. MIT. Available online at: https://lingbuzz.net/lingbuzz/001945/current.pdf [accessed 01/09/2022].

Koutsoudas, Andreas (1972) The strict order fallacy. *Language* 48(1). 88–96.

Koutsoudas, Andreas & Gerald Sanders (1979) On the universality of rules and rule ordering constraints. *Studia linguistica* 33(1). 57–78.

Koyama, Masanori, Michael Orrison & David Neel (2007) Irreducible graphs. *Journal of combinatorial mathematics and combinatorial computing* 62. 35–43.

Kracht, Marcus (2001) Syntax in chains. *Linguistics and philosophy* 24. 467–529.

Kracht, Marcus (2008) On the logic of LGB type structures. Part I: Multidominance structures. In Fritz Hamm & Stephan Kepser (eds.) *Logics for linguistic structures*. Berlin: Mouton. 105–142.

Kracht, Marcus (2013) Are logical languages compositional? *Studia logica* 101(6). 1319–1340.

Kratzer, Angelika (1991) Modality. In Arnim von Stechow & Dieter Wunderlich (eds.) *Semantics/Semantik: an international handbook of contemporary research*. Berlin: de Gruyter. 639–650.

Kremers, Joost (2009) Recursive linearization. *The linguistic review* 26. 135–166.

Krivine, Jean-Louis (1971) *Introduction to axiomatic set theory*. Dordrecht: Reidel.

Krivochen, Diego Gabriel (2013) Los verbos de ascenso como expresiones modales: el caso del español. *Anu.Filol.Est.Lingüíst.* 3. 33–56.

Krivochen, Diego Gabriel (2015a) On phrase structure building and labeling algorithms. *The linguistic review* 32(3). 515–572.

Krivochen, Diego Gabriel (2015b) Types vs. tokens: Displacement revisited. *Studia linguistica* 70(3). 250–296. https://doi.org/10.1111/stul.12044.

Krivochen, Diego Gabriel (2016a) Divide and ... conquer? On the limits of algorithmic approaches to syntactic structure. *Czech and Slovak linguistic review* 1 (2016). 15–38.

Krivochen, Diego Gabriel (2018) *Aspects of emergent cyclicity in language and computation.* PhD dissertation. University of Reading.

Krivochen, Diego Gabriel (2019a) On trans-derivational operations: Generative Semantics and Tree Adjoining Grammar. *Language sciences* 74. 47–76. https://doi.org/10.1016/j.langsci.2019.04.002.

Krivochen, Diego Gabriel (2019b) On intensive endophoric devices in English. *Studia anglica posnaniesia* 54 (2019). 81–112. https://doi.org/10.2478/stap-2019--0005.

Krivochen, Diego Gabriel (2020) On NEG lowering into quantifiers. *Acta linguistica hafniesia* 53 (1). 91–125.

Krivochen, Diego Gabriel (2021a) Mixed computation: grammar up and down the Chomsky hierarchy. *Evolutionary linguistic theory* 3(2). 216–245.

Krivochen, Diego Gabriel (2021b) I like this analysis, but I don't think every linguist will: syntactic NOT-transportation, VP ellipsis and VP pronominalisation. *Journal of the Spanish Association of Anglo-American studies* 43(2). 68–89.

Krivochen, Diego Gabriel (2022) Sobre la sintaxis de las oraciones de relativo: Un análisis de adjunción con estructura compartida. *Borealis* 11(3). 305–377. https://doi.org/10.7557/1.11.3.6661.

Krivochen, Diego Gabriel (2023a) The search for Minimal Search: a graph-theoretic view. *Biolinguistics* 17, Article e9793. https://doi.org/10.5964/bioling.9793.

Krivochen, Diego Gabriel (2023b) Towards a theory of syntactic workspaces: neighbourhoods and distances in a lexicalised grammar. The Linguistic Review 40(2). 311–360. *Linguistic review* (2023). Ms. https://doi.org/10.1515/tlr-2023-2004.

Krivochen, Diego Gabriel (forthcoming a) Different and proud of it: A TAG perspective on the coordination of unlike categories. Ms. Under review. https://ling.auf.net/lingbuzz/006490. [Accessed on 29/03/2022]

Krivochen, Diego Gabriel (forthcoming b) Constituents, arrays, and trees: two (more) models of grammatical description. Under review.

Krivochen, Diego Gabriel & Luis García Fernández (2019) On the position of subjects in Spanish periphrases: Subjecthood left and right. *Borealis* 8(1). 1–33. https://doi.org/10.7557/1.8.1.4687.

Krivochen, Diego Gabriel & Luis García Fernández (2022) On coordination and clitic climbing in Spanish auxiliary verb constructions. *Studies in Hispanic and Lusophone linguistics* 15(1). 111–139. https://doi.org/10.1515/shll-2022--2057.

Krivochen, Diego Gabriel, Luis García Fernández & Ana Bravo (in preparation) Auxiliary chains and clitic climbing in Spanish. Ms.

Krivochen, Diego Gabriel & James Douglas Saddy (2016) Structure mapping: uniformity vs. mixture in displacement. *Czech and Slovak linguistic review* 2 (2016). 16–45.

Krivochen, Diego Gabriel & Susan Schmerling (2016a) Two kinds of coordination and their theoretical implications: an aspect-based approach. Ms. Under review.

Krivochen, Diego Gabriel & Susan Schmerling (2016b) Mirage coordination in English. Ms. Under review. https://www.academia.edu/25110015/_Squib_Mirage_Coordinations_in_English. [Accessed on 18/09/2018]

Krivochen, Diego Gabriel & Susan Schmerling (2022) A Categorial Grammar for Spanish auxiliary chains. *Isogloss* 8(1). 1–49. https://doi.org/10.5565/rev/isogloss.126.

Krivochen, Diego Gabriel & Andrea Padovan (2021) Lexicalised locality: local domains and non-local dependencies in a lexicalised Tree Adjoining Grammar. *Philosophies* 6(3). 70. https://doi.org/10.3390/philosophies6030070.

Kroch, Anthony (1981) On the role of resumptive pronouns in amnestying island violations. In *Proceedings of the seventeenth regional meeting of the Chicago Linguistic Society*. Chicago Linguistic Society. 125–135.

Kroch, Anthony (2001) Asymmetries in long-distance extraction in a Tree-Adjoining Grammar. Ms. Available online at https://www.ling.upenn.edu/~kroch/online.html [accessed on 25/03/2016].

Kroch, Anthony & Aravind Joshi (1985) The linguistic relevance of a Tree Adjoining Grammar. Available online at http://babel.ling.upenn.edu/papers/faculty/tony_kroch/papers/relevance3.pdf [Accessed on 22/03/2016].

Kroch, Anthony & Aravind Joshi (1987) Analyzing extraposition in Tree Adjoining Grammar. In Geoffrey Huck & Almerindo Ojeda (eds.) *Syntax and semantics 20: discontinuous constituency*. New York: Academic Press. 107–151.

Kroch, Anthony & Beatrice Santorini (1991) The derived structure of the West Germanic verb raising construction. In Robert Freidin (ed.) *Principles and parameters in comparative grammar*. Cambridge, Mass.: MIT Press. 269–338.

Kural, Murat (2005) Tree traversal and word order. *Linguistic inquiry* 36(3). 367–387.

Kural, Murat & George Tsoulas (2004) Indices and the theory of grammar. Ms. https://www-users.york.ac.uk/~gt3/recent-mss/indices.pdf [accessed 04/12/2022].

Kuroda, Sige-Yuki (1968) English relativization and certain related problems. *Language* 44. 244–266.

Kuroda, Sige-Yuki (1976) A topological study of phrase structure languages. *Information and control* 30. 307–379.

Laca, Brenda (2004) Romance 'aspectual' periphrases: eventuality modification versus 'syntactic' aspect. In Jacquline Guéron & Jacqueline Lecarme (eds.) *The syntax of time*. Cambridge (Mass.): MIT Press. 425–440.

Ladusaw, William (1980) *Polarity sensitivity as inherent scope relations*. Bloomington, Indiana: University of Iowa, Indiana University Linguistics Club.

Lakoff, George (1965) *On the nature of syntactic irregularity*. PhD dissertation, Indiana University.

Lakoff, George (1976) Pronouns and reference. In James McCawley (ed.) *Notes from the linguistic underground*. New York: Academic Press. 275–336.

Lakoff, George & John Robert Ross (1967) [1976] Is Deep Structure necessary? Letter to Arnold Zwicky. In James McCawley (ed.) *Notes from the linguistic underground*. New York: Academic Press. 159–164.

Landau, Idan (2003) Movement out of control. *Linguistic inquiry* 34(3). 471–498.

Landau, Idan (2007) Movement-resistant aspects of control. In William Davies & Stanley Dubinsky (eds.) *New horizons in the analysis of control and raising*. Dordrecht: Springer. 293–325.

Landau, Idan (2010) The explicit syntax of implicit arguments. *Linguistic inquiry* 41(3). 357–388.

Landau, Idan (2013) *Control in Generative Grammar: a research companion*. Cambridge: CUP.

Langacker, Ronald (1969) On pronominalization and the chain of command. In David Reibel & Sanford Schane (eds.) *Modern studies in English*. Engelwood Cliffs, NJ: Prentice-Hall. 160–186.

Langendoen, D. Terence (1975) Finite-state parsing of Phrase-Structure Languages and the status of readjustment rules in grammar. *Linguistic inquiry* 6(4). 533–554.

Langendoen, D. Terence (2003) Merge. In Andrew Carnie, Mary Willie & Heidi Harley (eds.) *Formal approaches to function in grammar: in honor of Eloise Jelinek*. Amsterdam: John Benjamins. 307–318.

Langendoen, D. Terence & Paul Postal (1984) *The vastness of natural languages*. Oxford: Blackwell.

Larson, Richard (1987) 'Missing prepositions' and the analysis of English free relative clauses. *Linguistic inquiry* 18(2). 239–266.

Larson, Richard (1988) On the double object construction. *Linguistic inquiry* 19(3). 335–391.

Larson, Richard (1990) Double objects revisited: reply to Jackendoff. *Linguistic inquiry* 21(4). 589–632.

Larson, Richard (2014) *On shell structure*. Oxford: Blackwell.

Lasnik, Howard (1993) Antecedent contained deletion: a Minimalist perspective. Handout of a seminar at Harvard University. Available online at https://terpconnect.umd.edu/~lasnik/Handouts-Conf%20and%20colloq/Colloquia.Workshop/Lasnik92_Antecedent_contained_deletion.Harvard.pdf [Accessed on 22/10/2020].

Lasnik, Howard (2001) Subjects, objects, and the EPP. In William Davies & Stanley Dubinsky (eds.) *Objects and other subjects*. Dordrecht: Kluwer. 103–121.

Lasnik, Howard (2010) Quantifier lowering? Class handout, University of Maryland. Available online at https://terpconnect.umd.edu/~lasnik/LING819%202010/QL%20HO.pdf [Accessed on 14/10/2022].

Lasnik, Howard (2011) What kind of computing device is the human Language Faculty? In Anna-Maria Di Sciullo & Cedric Boeckx (eds.) *The Biolinguistic enterprise: new perspectives on the evolution and nature of the human Language Faculty*. Oxford: OUP. 354–365.

Lasnik, Howard & Kenshi Fukanoshi (2019) Ellipsis in transformational grammar. In Jeroen van Craenenbroeck & Tanja Temmerman (eds.) *The handbook of ellipsis*. Oxford: OUP. 46–74.

Lasnik, Howard & Joseph Kuppin (1977) A restrictive theory of transformational grammar. *Theoretical linguistics* 4. 173–196.

Lasnik, Howard & Mamoru Saito (1984) On the nature of proper government. *Linguistic inquiry* 15(2). 235–289.

Lasnik, Howard & Mamoru Saito (1991) On the subject of infinitives. *Proceedings of the Chicago Linguistic Society* (CLS) 27. 324–343.

Lasnik, Howard & Juan Uriagereka (1988) *A course in GB syntax*. Cambridge, Mass.: MIT Press.

Lasnik, Howard & Juan Uriagereka (2005) *A course in Minimalist syntax*. Oxford: Blackwell.

Lasnik, Howard & Juan Uriagereka (2012) Structure. In Ruth Kempson, Tim Fernando & Nicholas Asher (eds.) *Handbook of philosophy of science, volume 14: Philosophy of linguistics*. London: Elsevier. 33–61.

Lasnik, Howard & Juan Uriagereka (2022) *Structure*. Cambridge, Mass.: MIT Press.

Lawler, John (1972) A problem in participatory democracy. Indiana University Linguistics Club.

Lebeaux, David (1994) Where does Binding Theory apply? Ms. MIT.

Lebeaux, David (2009) *Where does Binding Theory apply?* Cambridge, Mass.: MIT Press.

Lechner, Winifred (1998) Two kinds of reconstruction. *Studia linguistica* 52(3): 276–310.

Lechner, Winifred (2013) Diagnosing XP movement. In Lisa Lai-Shen Cheng & Norbert Corver (eds.) *Diagnosing syntax*. Oxford: OUP. 235–248.

Lees, Robert (1976) What are transformations? In James McCawley (ed.) *Syntax and semantics 7: notes from the linguistic underground*. New York: Academic Press. 27–41.

Lees, Robert & Edward Klima (1963) Rules for English pronominalization. *Language* 39(1). 17–28.

Leonetti, Manuel & Victoria Escandell-Vidal (2015) La interfaz sintaxis-pragmática. In Ángel Gallego (ed.) *Perspectivas de sintaxis formal*. Madrid: Akal. 569–604.

Levin, Beth (1994) *English verb classes and alternations*. Cambridge, Mass.: MIT Press.

Levin, Beth (2014) Semantics and pragmatics of argument alternations. *Annual review of linguistics* 2015. 1–16.

Levine, Robert (1985) Right node (non-)raising. *Linguistic inquiry* 16(3). 492–497.

Lewis, David (1972) General Semantics. In Donald Davidson & Gilbert Harman (eds.) *Semantics of natural language*. Dordrecht: Reidel. 169–218.

Liang, Junying, Yuanyuan Fang, Qianxi Lv & Haitao Liu (2017) Dependency distance differences across interpreting types: implications for cognitive demand. *Frontiers in psychology*, 8. https://doi.org/10.3389/fpsyg.2017.02132.

Lindström, Per (1966) First-order predicate logic with generalized quantifiers. *Theoria* 32, 186–195.

Liu, Haitao, Chunshan Xu & Junying Liang (2017) Dependency distance: a new perspective on syntactic patterns in natural languages. *Physics of life reviews* 21. 171–193.

Lowe, John & Joseph Lovestrand (2020) Minimal phrase structure: a new formalized theory of phrase structure. *Journal of language modelling* 8(1). 1–52.

Lyons, John (1968) *Introduction to theoretical linguistics*. Cambridge: Academic Press.

Lyons, John (1977) *Semantics*. Cambridge: CUP.

Macauley, Matthew & Mortveit Henning (2009) Cycle equivalence of graph dynamical systems. *Nonlinearity* 22. 421–436.

MacFarlane, John (2017) Logical constants. In Edward Zalta (ed.) *The Stanford encyclopedia of philosophy*. http://plato.stanford.edu/archives/fall2015/entries/logical-constants/ [Accessed on 22/04/2019].

Maienborn, Claudia (2011) Event semantics. In Claudia Maienborn, Klaus von Heusinger & Paul Portner (eds.) *Semantics. An international handbook of natural language meaning*, Vol. 1. Berlin: de Gruyter. 802–829.

Manaster Ramer, Alexis & Walter Savitch (1997) Generative capacity matters. In *Proceedings from the fifth meeting on mathematics of language*. Available online at https://www.academia.edu/38564122/_1997_Generative_Capacity_Matters_with_Walter_Savitch_ [Accessed on 03/04/2022].

Manaster Ramer, Alexis & Wlodek Zadrozny (1990) Expressive power of grammatical formalisms. *COLING 1990 Volume 3: papers presented to the 13th international conference on computational linguistics*. 195–200.

Marantz, Alec (1984) *On the nature of grammatical relations*. Cambridge, Mass.: MIT Press.

Marcolli, Matilde (2014) Principles and Parameters: a coding theory perspective. arXiv:1407.7169

Marcolli, Matilde (2016) Syntactic parameters and a coding theory perspective on entropy and complexity of language families. *Entropy* 18. 110.

Martin, Roger, Román Orús & Juan Uriagereka (2019) Towards matrix syntax. *Catalan journal of linguistics*, Special issue. 27–44. https://doi.org/10.5565/rev/catjl.221.

Martín Gómez, Félix (2022) *Algunos aspectos de la gramática de* para *en español*. PhD dissertation, Universidad Complutense de Madrid.

Mateu Fontanals, Jaume (2002) *Argument structure: relational construal at the syntax-semantics interface*. PhD dissertation, Universitat Autònoma de Barcelona.

Mateu Fontanals, Jaume (2014) Argument structure. In Andrew Carnie, Dan Siddiqi & Yosuke Sato (eds.) *The Routledge handbook of syntax*. New York: Routledge. 24–41.

May, Robert (1985) *Logical Form: its structure and derivation*. Cambridge, Mass.: MIT Press.

Mayr, Clemens (2007) On the lack of subject-object asymmetries. *UPenn working papers in linguistics* 10(1). 1–14.

McCawley, James (1968) Concerning the base component of a transformational grammar. *Foundations of language* 4. 243–269.

McCawley, James (1970) Where do Noun Phrases come from? In Roderick Jacobs & Peter Rosenbaum (eds.) *Readings in English transformational grammar*. Waltham: Ginn & Co. 166–183.

McCawley, James (1973) *Grammar and meaning*. New York: Academic Press.

McCawley, James (1975) The category status of English modals. *Foundations of language* 12(4). 597–601.

McCawley, James (1981a) Language universals in linguistic argumentation. In James McCawley, *Thirty million theories of grammar*. Chicago: University of Chicago Press. 159–175.

McCawley, James (1981b) The nonexistence of syntactic categories. In James McCawley, *Thirty million theories of grammar*. Chicago: University of Chicago Press. 176–203.

McCawley, James (1981c) Introduction. In James McCawley, *Thirty million theories of grammar*. Chicago: University of Chicago Press. 1–9.

McCawley, James (1982) Parentheticals and Discontinuous Constituent Structure. *Linguistic inquiry* 13(1). 91–106.

McCawley, James (1987) Some additional evidence for discontinuity. In Geoffrey Huck & Almerindo Ojeda (eds.) *Syntax and semantics 20: discontinuous constituency*. New York: Academic Press. 185–202.

McCawley, James (1988) Review of *Knowledge of Language: its nature, origin, and use* by Noam Chomsky. *Language* 64(2). 355–365. https://doi.org/10.2307/415438.

McCawley, James (1998) *The syntactic phenomena of English*. [2nd Edition]. Chicago: University of Chicago Press.

McCloskey, James (1986) Right node raising and preposition stranding. *Linguistic inquiry* 17(1). 183–186.

McCloskey, James (2002) Resumption, successive cyclicity, and the locality of operations. In Samuel Epstein & T. Daniel Seely (eds.) *Derivation and explanation in the Minimalist Program*. Oxford: Blackwell. 184–226.

McCloskey, James (2006) Resumption. In Martin Everaert & Henk van Riemsdijk (eds.) *The Blackwell companion to syntax*. Oxford: Blackwell. 94–117.

McKinney-Bock, Katherine (2013) *Building phrase structure from items and contexts*. PhD dissertation, University of Southern California.

McKinney-Bock, Katherine & Jean-Roger Vergnaud (2014) Grafts and beyond: graph-theoretic syntax. In Katherine McKinney-Bock & María Luisa Zubizarreta (eds.) *Primitive elements of grammatical theory*. London: Routledge. 207–236.

Medeiros, David (2018) ULTRA: Universal Grammar as a Universal Parser. *Frontiers in psychology: language sciences* 9. https://doi.org/10.3389/fpsyg.2018.00155.

Medeiros, David (2021) Universal Supergrammar: *231 in neutral word order. Ms. University of Arizona. https://ling.auf.net/lingbuzz/006229.

Mel'čuk, Igor (1988) *Dependency syntax: theory and practice*. Albany: State University of New York Press.

Mel'čuk, Igor & Nikolai Pertsov (1987) *Surface syntax of English: A formal model with the Meaning-Text Framework*. Amsterdam: John Benjamins.

Melvod, Janis (1991) Factivity and definiteness. In Lisa L-S. Cheng & Hamida Demirdache (eds.) *More papers on Wh-movement*. MIT Working Papers in Linguistics 5. 97–117.

Merchant, Jason (2001) *The syntax of silence: sluicing, islands, and the theory of ellipsis*. Oxford: OUP.

Merchant, Jason (2008) Variable island repair under ellipsis. In Kyle Johnson (ed.), *Topics in ellipsis*. Cambridge: CUP. 132–153.

Messick, Troy & Gary Thoms (2016) Ellipsis, economy, and the (non)uniformity of traces. *Linguistic inquiry* 47(2). 306–332.

Moltmann, Friederike (1992) *Coordination and comparatives*. PhD dissertation, MIT.

Moltmann, Friederike (2017) A plural reference interpretation of three-dimensional syntactic trees. In Claire Halpert, Hadas Kotek & Coppe van Urk (eds.) *A Pesky Set. Papers for David Pesetsky*. Cambridge, Mass.: MIT Press. 103–109.

Montague, Richard (1970) Universal Grammar. *Theoria* 36, 373–398.

Montague, Richard (1973) The proper treatment of quantification in ordinary English. In Jaakko Hintikka, Julius Moravcsik & Patrick Suppes (eds.) *Approaches to natural language*. Dordrecht: Reidel. 221–242.

Montague, Richard (1974) English as a Formal Language. In Richmond Thomason (ed.) *Formal philosophy: selected papers of Richard Montague*. New Haven: Yale University Press. 188–221.

Morin, Yves-Charles & Michael O'Malley (1969) Multi-rooted vines in semantic representation. In Robert Binnick et al. (eds.) *Papers from the fifth regional meeting of the Chicago Linguistic Society*. University of Chicago. 178–185.

Moro, Andrea (2000) *Dynamic antisymmetry*. Cambridge, Mass.: MIT Press.

Moro, Andrea & Ian Roberts (2020) Unstable structures and Generalized Dynamic Antisymmetry. Ms. https://ling.auf.net/lingbuzz/005515.

Mortveit, Henning (2008) Graph dynamical systems: a mathematical framework for interaction-based systems, their analysis and simulations. Presented at *Discrete models in systems biology workshop SAMSI*, 5th December, 2008.

Munn, Alan (1993) *Topics in the syntax and semantics of coordinate structures*. PhD dissertation, University of Maryland.

Munn, Alan (2000) Three types of coordination asymmetries. In Kerstin Schwabe & Ning Zhang (eds.) *Ellipsis in conjunction*. Berlin: Max Niemeyer Verlag. 1–22.

Müller, Gereon (2011) *Constraints on displacement: a phase-based approach*. Amsterdam: John Benjamins.

Müller, Stefan (2000) The passive as a lexical rule. In Dan Flickinger & Andreas Kathol (eds.) *Proceedings of the 7th international HPSG conference*. Stanford: CLSI. 247–266.

Müller, Stefan (2013) Unifying everything: some remarks on simpler syntax, construction grammar, minimalism, and HPSG. *Language* 89(4). 920–950.

Müller, Stefan (2015) HPSG—a synopsis. In Tibor Kiss & Artemis Alexiadou (eds.) *Syntax—theory and analysis: an international handbook, Handbücher zur Sprach- und Kommunikationswissenschaft*. Berlin: de Gruyter. 937–973.

Müller, Stefan (2020) *Grammatical theory: from transformational grammar to constraint-based approaches*. Berlin: Language Science Press.

Mycock, Louise (2007) Constituent question formation and focus: a new typological perspective. *Transactions of the philological society* 105(2). 192–251.

Neeleman, Ad & Hand van de Koot (2002) The configurational matrix. *Linguistic inquiry* 33(4). 529–574.

Neeleman, Ad, Joy Philip, Misako Tanaka & Hans van de Koot (2020) Syntactic asymmetry and binary branching. Ms. UCL. https://www.researchgate.net/publication/345842867_Syntactic_Asymmetry_and_Binary_Branching. [Accessed on 02/03/2022]

Nivre, Joakim, Marie-Catherine de Marneffe, Filip Ginter, Jan Hajič, Christopher D. Manning, Sampo Pyysalo, Sebastian Schuster, Francis Tyers & Daniel Zeman (2020) Universal Dependencies v2: an evergrowing multilingual treebank collection. *Proceedings of the 12th language resources and evaluation conference*. 4034–4043.

Nykiel, Joanna & Jong-Bok Kim (2021) Ellipsis. In Stefan Müller, Anne Abeillé, Robert D. Borsley & Jean-Pierre Koenig (eds.) *Head-Driven Phrase Structure Grammar: the handbook*. Berlin: Language Science Press. 847–888.

Nunes, Jairo (2004) *Linearization of chains and sidewards movement*. Cambridge, Mass.: MIT Press.

Oehrle, Richard (2000) Context-sensitive node admissibility. *Grammars* 3, 275–293.

Ojeda, Almerindo (1987) Discontinuity, multidominance, and unbounded dependency in GPSG. In Geoffrey Huck & Almerindo Ojeda (eds.) *Syntax and semantics 20: discontinuous constituency*. New York: Academic Press. 257–282.

Ojeda, Almerindo (2005) Discontinuous constituents. In Keith Brown (ed.) *Encyclopedia of language and linguistics*. [2nd Edition]. Elsevier. Volume 3. 624–630.

Ordóñez, Francisco (2012) Clitics in Spanish. In José Ignacio Hualde, Antxon Olarrea &

Erin O'Rourke (eds.) *The handbook of Hispanic linguistics*. Oxford: Blackwell. 423–451.

Ore, Oysten (1990) *Graphs and their uses* [Edition revised and updated by Robin Wilson]. Providence, Rhode Island: the Mathematical Association of America.

Osborne, Timothy (2005) Beyond the constituent: a Dependency Grammar analysis of chains. *Folia linguistica XXXIX/3–4*. 251–297.

Osborne, Timothy (2006) Shared material and grammar: toward a Dependency Grammar theory of non-gapping coordination for English and German. *Zeitschrift für Sprachwissenschaft* 25. 39–93.

Osborne, Timothy (2008) Major constituents: and two dependency grammar constraints on sharing in coordination. *Linguistics* 46(6). 1109–1165.

Osborne, Timothy (2019) *A Dependency Grammar of English: an introduction and beyond*. Amsterdam: John Benjamins.

Osborne, Timothy & Ruochen Niu (2017) The Component Unit: introducing a novel unit of syntactic analysis. Paper presented at *Fourth international conference on Dependency Linguistics*, Università di Pisa, Italy. Available online: https://www.researchgate.net/publication/320010087_The_Component_Unit_Introducing_a_Novel_Unit_of_Syntactic_Analysis [Accessed on 23/08/2019].

Osborne, Timothy, Michael Putnam & Thomas Groß (2011) Bare phrase structure, label-less trees, and specifier-less syntax: is Minimalism becoming a dependency grammar? *The linguistic review* 28. 315–364.

Osborne, Timothy, Michael Putnam & Thomas Groß (2012) Catenae: introducing a novel unit of syntactic analysis. *Syntax* 15(4). 354–396.

Oshima, David (2007) On factive islands: pragmatic anomaly vs. pragmatic infelicity. In T. Washio, K. Satoh, H. Takeda & A. Inokuchi (eds) *New frontiers in artificial intelligence. JSAI 2006. Lecture notes in computer science*, vol. 4384. Berlin: Springer. https://doi.org/10.1007/978--3--540--69902--6_14.

Ott, Dennis (2016) Ellipsis in appositives. *Glossa* 1(1). 34. https://doi.org/10.5334/gjgl.37.

Ott, Dennis (2021) Phrase structure and its limits. To appear in Kleanthes Grohmann & Evelina Leivada (eds.) *Cambridge handbook of Minimalism*. Cambridge: CUP. https://ling.auf.net/lingbuzz/005891. [Accessed on 12/08/2022]

Padovan, Andrea (2016) Why a bed can be slept in but not under. Variation in passive V+P constructions. In Ermenegildo Bidese, Federica Cognola & Manuela Caterina Moroni (eds.) *Theoretical approach to linguistic variation*. Amsterdam: John Benjamins. 119–144.

Padovan, Andrea (2021a) A parallel between prepositional verbs in English and DOM constructions in Romance. Presented at *35th comparative Germanic syntax workshop*, University of Trento. Available online at https://www.researchgate.net/publication/352786471_A_parallel_between_prepositional_verbs_in_English_and_DOM_in_Romance [Accessed on 15/02/2022].

Padovan, Andrea (2021b) Local domains and non-local dependencies in a lexicalised Tree Adjoining Grammar. Talk presented at the Research Seminar in Linguistics, University of Frankfurt. 23/11/2021.

Panagiotidis, Phoevos (2001) *Pronouns, clitics and empty nouns*. Amsterdam: John Benjamins.

Panagiotidis, Phoevos (2021) Towards a (minimalist) theory of features. Ms. Available online at https://ling.auf.net/lingbuzz/005615 [Accessed on 02/10/2021].

Parsons, Terence (1990) *Events in the semantics of English. A study in subatomic semantics*. Cambridge, Mass.: MIT Press.

Partee, Barbara (1974) Some transformational extensions of Montague grammar. *Journal of philosophical logic* 2. 509–534.

Partee, Barbara (1975) Montague Grammar and Transformational Grammar. *Linguistic inquiry* 6(2). 203–300.

Partee, Barbara (1984) Compositionality. In Frank Landman & Frank Veltman (eds.) *Varieties of formal semantics*. Dordrecht: Foris. 281–312.

Pearson, Hazel (2016) The semantics of partial control. *Natural language and linguistic theory* 34. 691–738.

Perlmutter, David (1968) Deep and Surface Structure constraints in syntax. PhD dissertation, MIT.

Perlmutter, David (1978) Impersonal passives and the Unaccusative Hypothesis. *Proceedings of the annual meeting of the Berkeley Linguistics Society 38*. 157–189.

Perlmutter, David (1980) Relational grammar. In Edith Moravcsik & Jessica Wirth (eds.) *Syntax and semantics 13: current approaches to syntax*. New York: Academic Press. 195–229.

Perlmutter, David (1982) Syntactic representation, syntactic levels, and the notion of Subject. In Pauline Jacobson & Geoffrey Pullum (eds.) *The nature of syntactic representation*. Dordrecht: Reidel. 283–340.

Perlmutter, David (1983) Introduction. In David Perlmutter (ed.) *Studies in Relational Grammar 1*. Chicago: University of Chicago Press. ix–xv.

Perlmutter, David & Paul M. Postal (1983a) Toward a universal characterization of passivization. In David Perlmutter (ed.) *Studies in Relational Grammar 1*. Chicago: University of Chicago Press. 3–29.

Perlmutter, David & Paul M. Postal (1983b) Some proposed laws of basic clause structure. In David Perlmutter (ed.) *Studies in Relational Grammar 1*. Chicago: University of Chicago Press. 81–128.

Pesetsky, David (1995) *Zero syntax: experiencers and cascades*. Cambridge, Mass.: MIT Press.

Pesetsky, David & Esther Torrego (2001) T-to-C movement: causes and consequences. In Michael Kenstowicz (ed.) *Ken Hale: a life in language*. Cambridge, Mass.: MIT Press. 355–426.

Peters, Stanley & R.W. Ritchie (1973) On the generative power of transformational grammars. *Information sciences* 6. 49–83.

Peters, Stanley & R.W. Ritchie (1981) Phrase Linking Grammars. Ms. Stanford University and Palo Alto, California.

Peterson, Peter (2004) Non-restrictive relatives and other non-syntagmatic relations in a Lexical-Functional framework. In Miriam Butt & Tracy Holloway King (eds.) *Proceedings of LFG 2004*. Stanford: CSLI. 391–397.

Piattelli-Palmarini, Massimo & Heidi Harley (2004) Compositionality. Handout for the course *Topics in linguistics and philosophy*, University of Arizona. Available online at https://dingo.sbs.arizona.edu/~hharley/courses/596D/January14_MPPHandoutFIN.html [accessed 01/10/2022].

Pickering, Martin & Guy Barry (1993) Dependency categorial grammar and Coordination. *Linguistics* 31(5). 855–902.

Polinsky, Maria (2013) Raising and control. In Marcel den Dikken (ed.) *The Cambridge handbook of Generative syntax*. Cambridge: Cambridge University Press. 41–63.

Pollard, Carl (1997) The nature of constraint-based grammar. *Linguistic research* 15. 1–18. http://isli.khu.ac.kr/journal/content/data/15/1.pdf.

Pollard, Carl & Ivan Sag (1987) *Information-based syntax and semantics, Volume 1: fundamentals*. Stanford: CSLI.

Pollard, Carl & Ivan Sag (1994) *Head-Driven Phrase Structure Grammar*. Chicago: University of Chicago Press and Stanford: CSLI.

Pollock, Jean-Yves (1989) Verb movement, Universal Grammar, and the structure of IP. *Linguistic inquiry* 20(3). 365–424.

Post, Emil (1943) Formal reductions of the general combinatorial decision problem. *American journal of mathematics* 65(2). 197–215.

Post, Emil (1944) Recursively enumerable sets of positive integers and their decision problems. *Bulletin of the American Mathematical Society* 50. 284–316.

Postal, Paul (1964) *Constituent structure*. Bloomington, Indiana: University of Bloomington.

Postal, Paul (1969) On so-called 'pronouns' in English. In David Reidel & Sanford Schane (eds.) *Modern studies in English: readings in Transformational Grammar*. New Jersey: Prentice Hall. 201–224.

Postal, Paul (1971) *Crossover phenomena*. New York: Holt, Reinhart & Winston.

Postal, Paul (1972) On some rules that are not successive-cyclic. *Linguistic inquiry* 3(2). 211–222.

Postal, Paul (1974) *On raising*. Cambridge, Mass.: MIT Press.

Postal, Paul (1982) Some Arc Pair Grammar descriptions. In Pauline Jacobson & Geoffrey Pullum (eds.) *The nature of syntactic representation*. Dordrecht: Reidel. 341–426.

Postal, Paul (1986) *Studies of passive clauses*. New York: SUNY Press.

Postal, Paul (1998) *Three investigations on extraction*. Cambridge, Mass.: MIT Press.

Postal, Paul (2004a) A paradox in English syntax. In Paul Postal, *Skeptical linguistic sssays*. Oxford: OUP. 15–82.

Postal, Paul (2004b) A supposed account of strong crossover effects. In Paul Postal, *Skeptical linguistic essays*. Oxford: OUP. 205–232.

Postal, Paul (2004c) '(Virtually) conceptually necessary'. In *Skeptical linguistic essays*. Oxford: OUP. 323–336.

Postal, Paul (2010) *Edge-based clausal syntax*. Cambridge, Mass.: MIT Press.

Potts, Christopher (2002) The syntax and semantics of *as*-parentheticals. *Natural language and linguistic theory* 20. 623–689.

Potsdam, Eric & Jeffrey T. Runner (2001) Richard returns: copy raising and its implications. In Mary Andronis, Chris Ball, Heidi Elston & Sylvain Neuvel (eds.) *CLS 37: The main session*, Vol. 1. Chicago: Chicago Linguistic Society. 453–468.

Progovac, Ljiljana (1998) Structure for coordination. *Glot international* 3(7). 3–6.

Przepiórkowski, Adam (2022) Coordination of unlike grammatical cases (and unlike categories). *Language* 98(3). 592–634.

Pullum, Geoffrey (2007) The evolution of model-theoretic frameworks in linguistics. In James Rogers & Stephan Kepser (eds.) *Model-theoretic syntax at 10*. Dublin: Trinity College Dublin. 1–10.

Pullum, Geoffrey (2019) What grammars are, or ought to be. In Stefan Müller & Petya Osenova (eds.) *Proceedings of the 26th international conference on Head-Driven Phrase Structure Grammar*. Stanford: CSLI. 58–78.

Pullum, Geoffrey & Barbara Scholz (2001) On the distinction between model-theoretic and generative-enumerative syntactic frameworks. In Philippe de Groote, Glyn Morrill & Christian Retoré (eds.) *Logical aspects of computational linguistics: 4th international conference*. Berlin: Springer. 17–43.

Pullum, Geoffrey & Deirdre Wilson (1977) Autonomous syntax and the analysis of auxiliaries. *Language* 53(4). 741–788.

Putnam, Michael (2010) Exploring crash-proof grammars: an introduction. In Michael Putnam (ed.) *Exploring crash-proof grammars*. Amsterdam: John Benjamins. 1–12.

Putnam, Michael & Rui Chaves (2020) *Unbounded dependency constructions: theoretical and experimental perspectives*. Oxford: OUP.

Quirk, Randolph, Sidney Greenbaum, Geoffrey Leech & Jan Svartvik (1985) *A comprehensive grammar of the English language*. London: Longman.

Radford, Andrew (1997) *Syntax: a Minimalist introduction*. Cambridge: CUP.

RAE-ASALE (= Real Academia Española / Asociación de Academias de la Lengua Española) (2009) *Nueva gramática de la lengua española*. Madrid, Espasa.

Rambow, Owen (1993) *Mobile heads and strict lexicalization*. MA thesis, University of Pennsylvania.

Rambow, Owen & Aravind Joshi (1994) A formal look at Dependency Grammars and Phrase-Structure Grammars, with special consideration of word-order phenomena.

In Leo Wanner (ed.) *Current issues in Meaning-Text Theory*. London: Pinter. Available online at https://arxiv.org/pdf/cmp-lg/9410007.pdf%3Forigin%3Dpublication_detail [accessed 15/10/2022].

Ramchand, Gillian (2008) *Verb meaning and the lexicon*. Cambridge: CUP.

Ramchand, Gillian (2018) *Situations and syntactic structures*. Cambridge, Mass.: MIT Press.

Ramchand, Gillian & Peter Svenonius (2014) Deriving the 'functional hierarchy'. *Language sciences* 46, 152–174.

Reich, Peter (1969). The finiteness of natural languages. *Language* 45(4), 831–843.

Reinhart, Tanya (1976) *The syntactic domain of anaphora*. PhD dissertation, MIT.

Reinhart, Tanya (1983) *Anaphora and semantic interpretation*. Chicago: University of Chicago Press.

Reinhart, Tanya (1998) Wh-in situ in the framework of the Minimalist Program. *Natural language semantics* 6. 29–56.

Reinhart, Tanya (2006) *Interface strategies: optimal and costly computations*. Cambridge, Mass.: MIT Press.

Reinhart, Tanya & Eric Reuland (1993) Reflexivity. *Linguistic inquiry* 24(4). 657–720.

Reinhart, Tanya & Tal Siloni (2004) Against the unaccusative analysis of reflexives. In Artemis Alexiadou, Elena Anagnostopulou & Martin Everaert (eds.) *The unaccusativity puzzle*. Cambridge: CUP. 159–180.

Reuland, Eric (2005) Agreeing to bind. In Hans Broekhuis, Norbert Corver, Riny Huybregts, Ursula Kleinhenz & Jan Koster (eds.) *Organizing grammar. Studies in honor of Henk van Riemsdijk*. Berlin: de Gruyter. 505–513.

Reyle, Uwe & Christian Rohrer (1988) Introduction. In Uwe Reyle & Christian Rohrer (eds.) *Natural language parsing and linguistic theories*. Dordrecht: Reidel. 1–32.

Richards, Marc (2011) Deriving the edge: what's in a phase? *Syntax* 14(1). 74–95.

Richards, Norvin (1999) Feature cyclicity and ordering of multiple specifiers. In Samuel Epstein & Norbert Hornstein (eds.) *Working Minimalism*. Cambridge, Mass.: MIT Press. 127–158.

van Riemsdijk, Henk (2000) Free relatives inside out. Transparent free relatives as grafts. In Bozena Rozwadowska (ed.) *Proceedings of the 8th Annual Conference of the Polish Association for the Study of English*. 223–233.

van Riemsdijk, Henk (2006) Grafts follow from Merge. In Mara Frascarelli (ed.) *Phases of interpretation*. Berlin: Mouton. 17–44.

van Riemsdijk, Henk (2010) Grappling with Graft. In Jan-Wouter Zwart & Mark de Vries (eds.) *Structure Preserved: Festschrift for Jan Koster*. Amsterdam: John Benjamins. 289–298.

Ringen, Catherine (1972) On arguments for rule ordering. *Foundations of language* 8(2). 266–273.

Rivas, Alberto (1974) *A theory of clitics*. PhD dissertation, MIT.

Rizzi, Luigi (1990) *Relativized minimality*. Cambridge, Mass.: MIT Press.

Rizzi, Luigi (1996) Residual verb second and the *wh*-criterion. In Adriana Belletti & Luigi Rizzi (eds.) *Parameters and functional heads*. Oxford: OUP. 63–90.

Rizzi, Luigi (1997) The fine structure of the left periphery. In Liliane Haegeman (ed.) *Elements of grammar: a handbook of Generative syntax*. Dordrecht: Kluwer. 281–337.

Rizzi, Luigi (2004) Locality and left periphery. In Adriana Belletti (ed.) *Structures and beyond: the cartography of syntactic structures*. Oxford: OUP. 223–251.

Rizzi, Luigi (2013) Locality. *Lingua* 130. 169–186.

Rogers, Andy (1971) Three kinds of physical perception Vs. *CLS* 7. 206–222.

Rogers, Andy (1972) Another look at Flip perception Vs. *CLS* 8. 303–315.

Rogers, Andy (1974) A transderivational constraint on Richard? *Papers from the tenth regional meeting of the Chicago Linguistic Society*. 551–558.

Rogers, James (1994) *Studies in the logic of trees with applications to grammar formalisms*. PhD dissertation, University of Delaware.

Rogers, James (1997) 'Grammarless' phrase structure grammar. *Linguistics and philosophy* 20(6). 721–746.

Rogers, James (2003) Syntactic structures as multidimensional trees. *Research on language and computation* 1(4). 265–305.

Rooryck, Johan (1992) Negative and factive islands revisited. *Journal of linguistics* 28. 343–374.

Rosenbaum, Peter (1965) *The grammar of English predicate complement constructions*. PhD dissertation, MIT.

Ross, John Robert (1967) *Contraints on variables in syntax*. PhD dissertation, MIT.

Ross, John Robert (1968) Guess who? In Robert Binnick, Alice Davison, Georgia Green & Jerry Morgan (eds.) *Papers from the fifth regional meeting of the Chicago Linguistic Society*. Chicago: University of Chicago. 252–286.

Ross, John Robert (1969a) On the cyclic nature of English pronominalization. In David Reidel & Sanford Schane (eds.) *Modern studies in English: readings in Transformational Grammar*. New Jersey: Prentice Hall. 187–200.

Ross, John Robert (1969b) A proposed rule of tree pruning. In David Reidel & Sanford Schane (eds.) *Modern studies in English: readings in Transformational Grammar*. New Jersey: Prentice Hall. 288–299.

Ross, John Robert (1970a) Gapping and the order of constituents. In Manfred Bierwisch & Karl-Erich Heidolph (eds.) *Progress in linguistics*. The Hague: Mouton. 249–259.

Ross, John Robert (1970b) On declarative sentences. In Roderick Jacobs & Peter Rosenbaum (eds.) *Readings in English Transformational Grammar*. Waltham: Ginn & Co. 222–272.

Ross, John Robert (1973) Slifting. In Maurice Gross, Morris Halle & Marcel Schützenberger (eds.) *The formal analysis of natural languages*. The Hague: Mouton. 133–169.

Ross, John Robert (1984) Inner islands. In Claudia Brugman, Monica Macaulay et al. (eds.) *Proceedings of the 10th annual meeting of the Berkeley Linguistics Society*. Berkeley, 258–265.

Ross, Haj (2011) An automodular perspective on the frozenness of pseudoclefts, and *vice versa*. In Tista Bagchi, Katharine Beals & Etsuyo Yuasa (eds.) *Pragmatics and Autolexical Grammar: in honor of Jerry Sadock*. Amsterdam: John Benjamins. 243–260.

Ross, Haj (2012) A preliminary—but fattened—list of transformations. Ms. Available online at http://www-personal.umich.edu/~jlawler/haj/Preliminarybufattenedlistoftransformations.pdf [Accessed on 22/08/2016].

Runner, Jeffrey (2001) The double object construction at the interfaces. In J.S. Magnuson & K.M. Crosswhite (eds.) *University of Rochester working papers in the language sciences* 2(1). 23–51.

Runner, Jeffrey (2006) Lingering challenges to the raising-to-object and object control constructions. *Syntax* 9(2). 193–213.

Sabbagh, Joseph (2007) Ordering and linearizing rightward movement. *Natural language & linguistic theory* 25. 349–401.

Sabbagh, Joseph (2014) Right node raising. *Language and linguistics compass* 8(1). 24–35.

Saddy, Douglas (2018) Syntax and uncertainty. In Ángel Gallego & Roger Martin (eds.) *Language, syntax, and the natural sciences*. Cambridge: CUP. 316–332.

Saddy, Douglas & Juan Uriagereka (2004) Measuring language. *International journal of bifurcation and chaos* 14(2). 383–404.

Saddy, Douglas, Kelly Sloan & Diego Gabriel Krivochen (2019) Whoever that likes relatives … In Ken Ramshøj Christiansen, Henrik Jørgensen & Joanna L. Wood (eds.) *The sign of the V: papers in honour of Sten Vikner*. Aarhus University. 523–544.

Safir, Ken (2013) Syntax, binding, and patterns of anaphora. In Marcel den Dikken (ed.) *The Cambridge handbook of Generative syntax*. Cambridge: CUP. 515–576.

Safir, Ken (2014) One true anaphora. *Linguistic inquiry* 45(1). 91–124.

Sag, Ivan (2010) English filler-gap constructions. *Language* 86(3). 486–545.

Sag, Ivan, Gerald Gazdar, Tom Wasow & Steven Weisler (1985) Coordination and how to distinguish categories. *Natural language and linguistic theory* 3. 117–171.

Sag, Ivan, Ronald Kaplan, Lauri Karttunen, Martin Kay, Carl Pollard, Stuart Shieber & Annie Zaenen (1986) Unification and grammatical theory. *Proceedings of the west coast conference on formal linguistics*. Stanford Linguistics Association. 238–254.

Sag, Ivan, Tom Wasow & Emily Bender (2003) *Syntactic theory: a formal introduction*. Stanford: CSLI.

Saito, Mamoru (1989) Scrambling as semantically vacuous A'-movement. In Mark Baltin & Anthony Kroch (eds.) *Alternative conceptions of phrase structure*. Chicago: University of Chicago Press, 182–200.

Salzmann, Martin (2019) A new version of the matching analysis of relative clauses: combining deletion under recoverability with vehicle change. In Manfred Krifka & Mathias Schenner (eds.) *Reconstruction effects in relative clauses*. Berlin: de Gruyter. 187–223.

Sampson, Geoffrey (1975) The Single Mother Condition. *Journal of linguistics* 11(1). 1–11.

Sarkar, Anoop & Aravind Joshi (1997) Handling coordination in a Tree Adjoining Grammar. Technical report, University of Pennsylvania. Available online at https://www2.cs.sfu.ca/~anoop/papers/pdf/tag-coordination.pdf [Accessed on 03/02/2018].

Sauerland, Uli (1998) *The meaning of chains*. PhD dissertation, MIT.

Sauerland, Uli (2004) The interpretation of traces. *Natural language semantics* 12. 63–127.

Sauerland, Uli & Paul Elbourne (2002) Total reconstruction, PF movement, and derivational order. *Linguistic inquiry* 33(2). 283–319.

de Saussure, Ferdinand (1983) [1916] *Course in general linguistics*. La Salle, Illinois: Open Court.

Schachter, Paul (1973) Focus and relativization. *Language* 49(1). 19–46.

Schelfhout, Carla, Peter-Arno Coppen & Nelleke Oostidijk (2004) Transparent free relatives. In Sylvia Blaho, Luis Vincente & Mark De Vos (eds.) *Proceedings of ConSOLE XII*. 81–90.

Schmerling, Susan F. (1975) Asymmetric conjunction and rules of conversation. In Peter Cole & Jerry Morgan (eds.) *Syntax and semantics, vol. 3: speech acts*. New York: Academic Press. 211–231.

Schmerling, Susan F. (1976) Synonymy judgments as syntactic evidence. *Texas Linguistic Forum* 4. 118–131.

Schmerling, Susan F. (1979) A categorial analysis of Dyirbal ergativity. *Texas Linguistic Forum* 13. 96–112.

Schmerling, Susan F. (1982) How imperatives are special and how they aren't. In Robinson Schneider, Kevin Tuite & Robert Chametzky (eds.) *Papers from the parasession on nondeclaratives*. Chicago: Chicago Linguistic Society. 202–218.

Schmerling, Susan F. (1983a) Two theories of syntactic categories. *Linguistics and philosophy* 6. 393–421.

Schmerling, Susan F. (1983b) A new theory of English auxiliaries. In Frank Heny & Barry Richards (eds.) *Linguistic categories: auxiliaries and related puzzles*. Vol. 2. Dordrecht: Reidel. 1–53.

Schmerling, Susan F. (1988) On the definite article in German. November, 1988 revision of a paper presented at the *Symposium on determiners and A' binding*. Austin, TX. March 22–24, 1985.

Schmerling, Susan F. (2018a) *Sound and grammar: towards a neo-Sapirian theory of language*. Leiden: Brill.

Schmerling, Susan F. (2018b) Rhetorical meaning. *Linguistic frontiers* 1(1). https://doi.org/10.2478/lf-2018--0001.

Schmerling, Susan F. (2021) Eliminating bracketing paradoxes in phonologically driven syntax. Presented at the *Sensing syntax* seminar, University of British Columbia. https://ling.auf.net/lingbuzz/005865. [Accessed on 03/05/2021]

Schmerling, Susan F. & Diego Gabriel Krivochen (2017) On non-progressive 'being'. *Canadian journal of linguistics* 63(1). 112–119. https://doi.org/10.1017/cnj.2017.38.

Schuler, William, David Chiang, & Mark Dras (2000) Multi-component TAGs and notions of formal power. In *Proceedings of the 38th annual meeting of the Association for Computational Linguistics*. 448–455. https://doi.org/10.3115/1075218.1075275.

Schütze, Carson & Richard Stockwell (2019) Transparent free relatives with *who*: support for a unified analysis. In *Proceedings of the Linguistic Society of America* 4: 40. 1–6. https://doi.org/10.3765/plsa.v4i1.4548.

Scida, Emily (2004) *The inflected infinitive in Romance languages*. London: Routledge.

Scott, Gary-John (2002) Stacked adjectival modification and the structure of nominal phrases. In Guglielmo Cinque (ed.) *Functional structure in DP and IP*. Oxford: OUP. 91–120.

Seely, T. Daniel (2006) Merge, derivational c-command, and subcategorization in a label-free syntax. In Cedric Boeckx (ed.) *Minimalist essays*. Amsterdam: John Benjamins. 182–217.

Sells, Peter (2013) Lexical Functional Grammar. In Marcel den Dikken (ed.) *The Cambridge handbook of Generative syntax*. Cambridge: CUP. 162–201.

Shieber, Stuart (1985) Evidence against the context-freeness of natural language. *Linguistics and philosophy* 8(3). 333–343.

Shieber, Stuart (1986) *An introduction to unification-based approaches to grammar*. Brookline, Mass.: Microtome Publishing.

Shieber, Stuart (1988) Separating linguistic analyses from linguistic theories. In U. Reyle & C. Rohrer (eds.) *Natural language parsing and linguistic theories*. Dordrecht: Reidel. 33–68.

Shlonsky, Ur (1992) Resumptive pronouns as a last resort. *Linguistic inquiry* 23(3). 443–468.

Shu, Kevin, Sharjeel Aziz, Vy-Luan Huynh, David Warrick & Matilde Marcolli (2018) Syntactic phylogenetic trees. In Joseph Kouneiher (ed.) *Foundations of mathematics and physics one century after Hilbert*. New York: Springer. 417–441.

Siloussar, Natalia (2014) What is and what is not problematic about the T-model. In Peter Kosta, Steven Franks, Teodora Radeva-Bork, & Lilia Schürcks (eds.) *Minimalism and beyond: radicalizing the interfaces*. Amsterdam: John Benjamins. 350–362.

Simpson, Jane (1991) *Warlpiri morpho-syntax: a lexicalist approach*. Dordrecht: Kluwer.

Siva, Karthik, Jim Tao & Matilde Marcolli (2017) Syntactic parameters and spin glass models of language change. *Linguistic analysis* 41(3–4). 559–608.

Sloan, Kelly (1991) Quantifier-wh interaction. In Lisa L-S. Cheng & Hamidá Demirdache (eds.) *More papers on wh-movement*, MITWPL 15. 219–237.

Sloan, Kelly & Juan Uriagereka (1988) What does 'everyone' have scope over? Presented at GLOW Budapest.

Sloat, Clarence (1969) Proper nouns in English. *Language* 45(1). 26–30.

Smolensky, Paul, Matthew Goldrick & Donald Mathis (2014) Optimization and quantization in gradient symbol systems: A framework for integrating the continuous and the discrete in cognition. *Cognitive science* 38. 1102–1138.

Sowa, John (2017) Existential Graphs. MS 514 by Charles Sanders Peirce. Available online at http://www.jfsowa.com/peirce/ms514.htm [Accessed on 05/06/2021].

Sportiche, Dominique (1985) Remarks on crossover. *Linguistic inquiry* 16(3). 460–469.

Stabler, Edward (2011) Computational perspectives on minimalism. In Cedric Boeckx (ed.) *Oxford handbook of minimalism*. Oxford: OUP. 617–641.

Stabler, Edward (2013) Two models of minimalist, incremental syntactic analysis. *Topics in cognitive science* 5(3). 611–633.

Stanley, Richard (1967) Phonological redundancy rules. *Language*, 43. 393–436.

Starke, Michael (2004) On the inexistence of specifiers and the nature of heads. In Adriana Belletti (ed.) *Structures and beyond*. New York: OUP. 252–268.

Steedman, Mark (2019) Combinatory Categorial Grammar. In András Kertész, Edith Moravcsik & Csilla Rákosi (eds.) *Current approaches to syntax: a comparative handbook*. Berlin: de Gruyter. 389–420.

Steedman, Mark & Jason Baldridge (2011) Combinatory Categorial Grammar. In Robert Borsley & Kersti Börjars (eds.) *Non-transformational syntax*. Oxford: Blackwell. 181–224.

Stepanov, Arthur (2001) Late adjunction and Minimalist phrase structure. *Syntax* 4(2). 94–125.

Sternefeld, Wolfgang (1998a) Connectivity effects in pseudo-cleft sentences. In Artemis Alexiadou, Nanna Fuhrhop, Paul Law & Ursula Kleinhenz (eds.) *ZAS papers in linguistics* 10. 146–162.

Sternefeld, Wolfgang (1998b) The semantics of reconstruction and connectivity. *Arbeitspapiere des SFB 340*. Universität Stuttgart and Tübingen. 1–58.

Stockwell, Robert, Paul Schachter & Barbara Hall Partee (1973) *The major syntactic structures of English*. New York: Holt, Rinehart and Winston.

Stowell, Tim (1981) *Origins of phrase structure*. PhD dissertation, MIT.

Stroik, Tom (2009) *Locality in minimalist syntax*. Cambridge, Mass.: MIT Press.

Stroik, Tom & Michael Putnam (2013) *The structural design of language*. Oxford: OUP.

Stroik, Tom & Michael Putnam (2015) Is Simplest Merge too simple? Ms. University of Missouri-Kansas City & Penn State University.

Takahashi, Shoichi & Danny Fox (2005) MaxElide and the re-binding problem. In Efthymia Georgala & Jonathan Howell (eds.) *Semantics and linguistic theory* 15. 223–240.

Tesnière, Lucien (1959) *Éléments de syntaxe structural*. Paris: Klincksieck. [There is an

English translation: *Elements of structural syntax*. Translated by Timothy Osborne and Sylvain Kahane. Amsterdam: John Benjamins, 2015.]

Torrego, Esther (1984) On Inversion in Spanish and some of its effects. *Linguistic inquiry* 15(1). 103–129.

Torrego, Esther (2002) Arguments for a derivational approach to syntactic relations based on clitics. In Samuel Epstein & T. Daniel Seely (eds.) *Derivation and explanation in the Minimalist Program*. Malden, MA: Blackwell. 249–268.

Truswell, Robert (2014) Binding theory. In Andrew Carnie, Yosuke Sato & Daniel Siddiqi (eds.) *The Routledge handbook of syntax*. Vol. 1. London: Routledge. 214–238.

Truswell, Robert (2019) An adjunction theory of extraction from coordinate structures. Ms. Available online at http://robtruswell.com/assets/pdfs/ZAS_talk_2019.pdf [Accessed on 30/07/2020].

Turing, Alan (1936) On computable numbers, with an application to the Entscheidungsproblem, *Proc. London Math. Soc.* 42(2). 230–265.

Ura, Hiroyuki (2000) *Checking theory and grammatical functions in Universal Grammar*. Oxford: OUP.

Uriagereka, Juan (2002) Multiple Spell-Out. In Juan Uriagereka, *Derivations: exploring the dynamics of syntax*. London: Routledge. 45–65.

Uriagereka, Juan (2008) *Syntactic anchors: on semantic restructuring*. Cambridge: CUP.

Uriagereka, Juan (2011) A sketch of the grammar in non-classical conditions. Ms. UMD.

Uriagereka, Juan (2012) *Spell-Out and the Minimalist Program*. Oxford: OUP.

Van Steen, Maarten (2010) *Graph theory and complex networks: an introduction*. Available online at https://www.distributed-systems.net/index.php/books/gtcn/gtcn/ [Accessed on 03/02/2017].

Vijay-Shanker, K. & Aravind Joshi (1991) Unification-based Tree Adjoining Grammars. University of Pennsylvania Department of Computer and Information Science Technical Report No. MS–CIS–91–25. Available online at https://repository.upenn.edu/cgi/viewcontent.cgi?article=1799&context=cis_reports [Accessed on 18/07/2021].

Villata, Sandra & Julie Franck (2016) Semantic similarity effects on weak islands acceptability. *Research in Generative Grammar* 38. 269–285.

Vogel, Ralf & Markus Steinbach (1994) *Zum Konzept der Tiefenstruktur in der generativen Grammatik*. MA thesis, University of Frankfurt am Main.

de Vos, Mark (2005) *The syntax of verbal pseudo-coordination in English and Afrikaans*. PhD dissertation, Leiden University.

de Vos, Mark & Luis Vicente (2005) Coordination under right node raising. In John Alderete, Chung-hye Han & Alexei Kochetov (eds.) *Proceedings of the 24th west coast conference on formal linguistics*. Somerville, MA: Cascadilla Proceedings Project. 97–104.

de Vries, Mark (2007) Invisible constituents? Parentheses as B-Merged adverbial phrases. In Nicole Dehé & Yordanka Kavalova (eds.) *Parentheticals*. Amsterdam: John Benjamins. 203–234.

de Vries, Mark (2009a) On multidominance and linearization. *Biolinguistics* 3(4). 344–403.

de Vries, Mark (2009b) Specifying coordination: An investigation into the syntax of dislocation, extraposition and parenthesis. In Cynthia Dreyer (ed.) *Language and linguistics: emerging trends*. New York: Nova. 37–98.

Wall, Robert (1972) *Introduction to mathematical linguistics*. New York: Prentice Hall.

Wasow, Thomas (1979) *Anaphora in Generative Grammar*. Gent: E. Story-Scientia.

Watumull, Jeffrey (2012) A Turing program for linguistic theory. *Biolinguistics* 6(2). 222–245.

Wedekind, Jürgen & Ronald Kaplan (2020) Tractable Lexical-Functional Grammar. *Computational linguistics* 46(3). 515–569.

Wells, Rulon S. (1947) Immediate constituents. *Language* 23. 81–117.

Westcoat, Michael (2005) English non-syllabic auxiliary contractions: an analysis in LFG with lexical sharing. In Miriam Butt & Tracy Holloway King (eds.) *Proceedings of the LFG05 conference*. Stanford: CSLI. https://web.stanford.edu/group/cslipublications/cslipublications/LFG/10/pdfs/lfg05wescoat.pdf [Accessed 11/10/2022].

Westerhål, Dag (2015) Generalized quantifiers in natural language semantics. In Shalom Lappin & Chris Fox (eds.) *The handbook of contemporary semantic theory* [2nd Edition]. Oxford: Wiley-Blackwell. https://doi.org/10.1002/9781118882139.ch1.

Wexler, Kenneth & Peter Culicover (1980) *Formal principles of language acquisition*. Cambridge, Mass.: MIT Press.

Wiggins, Stephen (2003) *Introduction to applied nonlinear dynamical systems and chaos*. [2nd Edition] New York: Springer.

Wilder, Chris (1998) Transparent free relatives. *ZAS papers in linguistics* 10. 191–199.

Williams, Edwin (1978) Across-the-Board rule application. *Linguistic inquiry* 9(1). 31–43.

Williams, Edwin (1982) Another argument that the passive is transformational. *Linguistic inquiry*, 13(1). 160–163.

Williams, Edwin (2003) *Representation theory*. Cambridge, Mass.: MIT Press.

Wilson, Robin (1996) *Introduction to graph theory*. [4th edition]. London: Adison Wesley.

Wurmbrand, Susi (1999) Modal verbs must be raising verbs. In Sonya Bird, Andrew Carnie, Jason D. Haugen & Peter Norquest (eds.) *WCCFL 18 Proceedings*. Somerville, MA: Cascadilla Press. 599–612.

Wurmbrand, Susi (2011) Reverse Agree. Course notes from *Problems in Syntax* (Spring 2011), UConn. Available online at http://wurmbrand.uconn.edu/Papers/Agree-and-Merge.pdf [Accessed on 05/07/2020].

van Wyngaerd, Guido & Jan-Wouter Zwart (1999) Antecedent-contained deletion as deletion. *Linguistics in the Netherlands* 1999. 203–216.

XTAG group (2001) A lexicalized TAG for English. Technical report, University of Pennsylvania. Available online at https://repository.upenn.edu/cgi/viewcontent.cgi?article=1020&context=ircs_reports [Accessed on 06/07/2020].

Yasui, Myoko (2002) A graph-theoretic reanalysis of Bare Phrase Structure theory and its implications on parametric variation. Ms. Dokkyo University. Available online at https://www.academia.edu/1908645/A_graph_theoretic_reanalysis_of_bare_phrase_structure_theory_and_its_implications_on_parametric_variation [Accessed on 15/03/2013].

Zaenen, Annie (1983) On syntactic binding. *Linguistic inquiry* 14(3). 469–504.

Zaenen, Annie, Elisabet Engdahl & Joan Maling (1981) Resumptive pronouns can be syntactically bound. *Linguistic inquiry* 12(4). 679–682.

Zagona, Karen (2003) *The Syntax of Spanish*. Cambridge: CUP.

Zhang, Niina Ning (2010) *Coordination in syntax*. Cambridge: CUP.

Zwart, Jan-Wouter (2009) Prospects for top-down derivations. *Catalan journal of linguistics* 9. 161–187.

Zwicky, Arnold & Stephen Isard (1963) Some aspects of tree theory. Working Paper W-6674, The MITRE Corporation, Bedford, Mass. Available online at https://web.stanford.edu/~zwicky/some-aspects-of-tree-theory.pdf [Accessed on 29/03/2013].

Zyman, Erik (2023) On the definition of Merge. *Syntax*. Forthcoming. Available online at https://drive.google.com/open?id=1KWyZrOhrf-exnZpEqyYrmoUYdAqauZ6R&authuser=0 [Accessed on 04/02/2023].

General Index

www.ingramcontent.com/pod-product-compliance
Lightning Source LLC
LaVergne TN
LVHW012052011026
849162LV00018B/189